Birds of NEPAL

Revised Edition

HELM FIELD GUIDES

Birds of NEPAL

Revised Edition

Richard Grimmett, Carol Inskipp,
Tim Inskipp and Hem Sagar Baral

Illustrated by
Richard Allen, Adam Bowley, Clive Byers, Dan Cole, John Cox,
Gerald Driessens, Carl d'Silva, Martin Elliott, Kim Franklin, John Gale,
Alan Harris, Peter Hayman, Dave Nurney, Craig Robson, Christopher Schmidt,
Brian Small, Jan Wilczur, Tim Worfolk and Martin Woodcock

CHRISTOPHER HELM
LONDON

Christopher Helm
An imprint of Bloomsbury Publishing Plc

50 Bedford Square	1385 Broadway
London	New York
WC1B 3DP	NY 10018
UK	USA

www.bloomsbury.com

This edition first published in 2016

British Library Cataloguing-in-Publication Data
A catalogue record for this book is available from the British Library.

ISBN: PB: 978-1-4729-0571-0
ePDF: 978-1-4729-2568-8
ePub: 978-1-4729-0572-7

2 4 6 8 10 9 7 5 3 1

Designed by Julie Dando, Fluke Art

Printed and bound in India

Cover artwork
Front: Himalayan Monal (Alan Harris)
Back (top to bottom): Great Slaty Woodpecker (Carl D'Silva), Satyr Tragopan (Carl D'Silva), Spiny Babbler (Craig Robson), Ibisbill (Peter Hayman)

CONTENTS

ACKNOWLEDGEMENTS

The authors are grateful to the artists whose work illustrates this book, including Alan Harris who executed the cover, and Craig Robson, who drew the line drawings in the introduction. Numerous people generously provided Richard Grimmett assistance in the preparation of the original identification texts in our book *Birds of the Indian Subcontinent* (Christopher Helm, 1998) and these people are acknowledged in that work. Carol Inskipp is grateful to Tek Bahadur Gurung and Tika Giri from the Bird Education Society (BES) for providing information on the Chepang Hills trek and to Sagar Giri, from BES, for information on the society itself.

Chautara (resting point) with views of Annapurna Mountains. (*Hem Sagar Baral*)

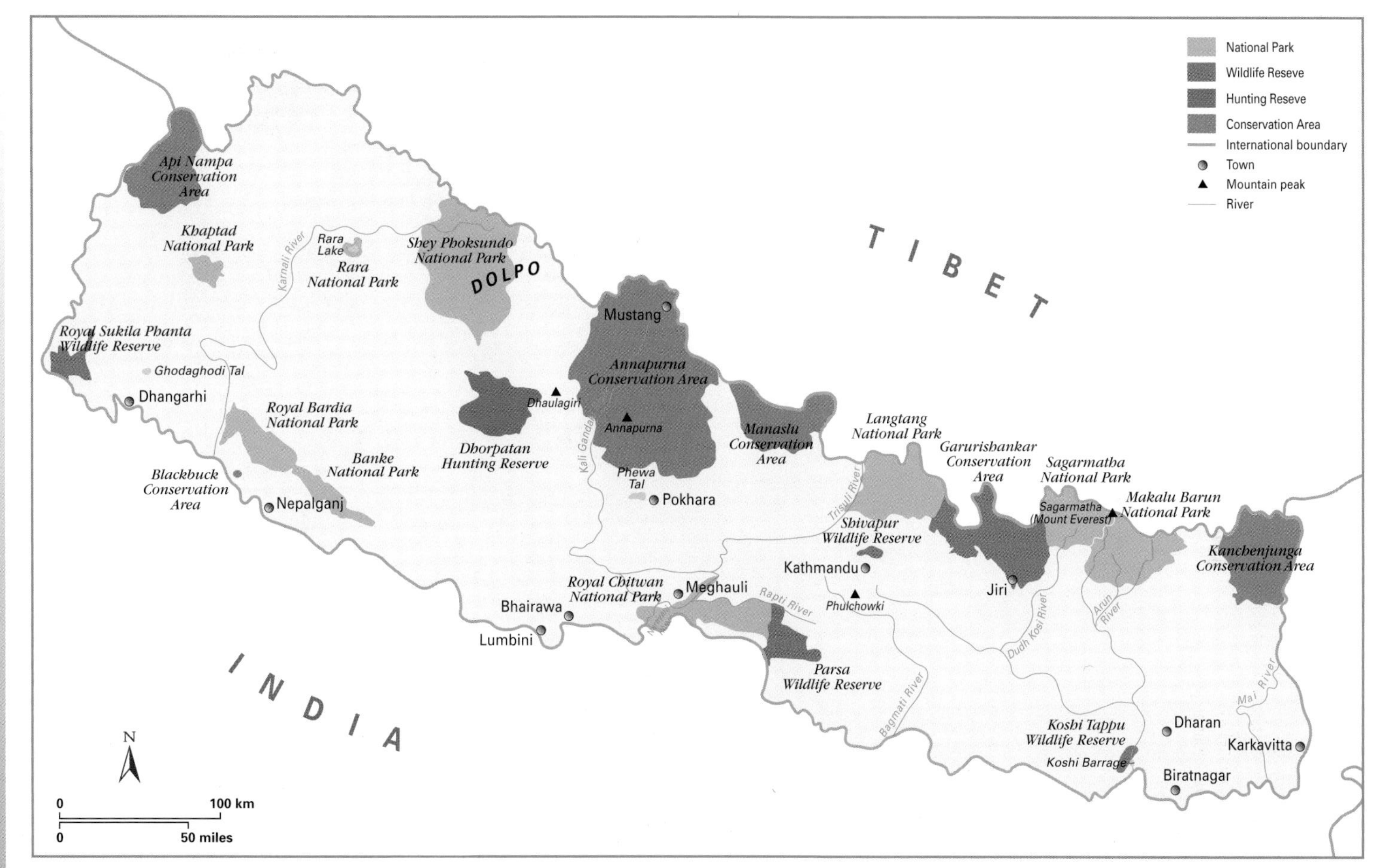
National Park
Wildlife Reseve
Hunting Reseve
Conservation Area
International boundary
Town
Mountain peak
River
TIBET
INDIA
Api Nampa Conservation Area
Khaptad National Park
Karnali River
Rara Lake
Rara National Park
Shey Phoksundo National Park
DOLPO
Royal Sukila Phanta Wildlife Reserve
Ghodaghodi Tal
Dhangarhi
Royal Bardia National Park
Banke National Park
Blackbuck Conservation Area
Nepalganj
Dhorpatan Hunting Reserve
Dhaulagiri
Mustang
Annapurna Conservation Area
Annapurna
Kali Gandaki
Phewa Tal
Pokhara
Manaslu Conservation Area
Langtang National Park
Trisuli River
Shivapur Wildlife Reserve
Kathmandu
Phulchowki
Garurishankar Conservation Area
Jiri
Sagarmatha National Park
Sagarmatha (Mount Everest)
Makalu Barun National Park
Kanchenjunga Conservation Area
Royal Chitwan National Park
Meghauli
Bhairawa
Lumbini
Rapti River
Parsa Wildlife Reserve
Bagmati River
Dudh Kosi River
Arun River
Mai River
Koshi Tappu Wildlife Reserve
Koshi Barrage
Dharan
Karkavitta
Biratnagar
N
0
100 km
0
50 miles

INTRODUCTION

Nepal is renowned for its high diversity of bird species, spectacular mountains and rich culture. Here you can enjoy birdwatching in one of the most beautiful places on Earth. This guide aims to help observers identify all of the bird species currently recorded in Nepal.

HOW TO USE THIS BOOK

Species included

All species that are known to have been reliably recorded in Nepal up to the end of June 2015 have been included. Descriptions of vagrants are given in Appendix 1.

Colour plates and species accounts

Species that occur regularly in Nepal are illustrated in colour and described in the species accounts. Vagrant and extirpated/extinct species are described in the appendices, with reference to distinguishing features from other more regularly recorded species where appropriate. The illustrations show distinctive sexual and racial variation whenever possible, as well as immature plumages.

The species accounts summarise the species' distribution, status, altitudinal range and habitats, and provide information on the most important identification characters, including voice and approximate body length.

A general description is given of the species' status as a resident, winter visitor, summer visitor, passage migrant or altitudinal migrant. Data on actual breeding records and non-breeding ranges are very few, so it has not been possible to present comprehensive details. In addition, space limitations have meant that the simple terms, 'breeds' and 'winters' are used to describe altitudinal ranges and habitats for many species. Note that many Himalayan species are recorded in summer over a wider altitudinal range than that within which they actually breed. Altitudes given are those for which the species have been recorded in Nepal, unless otherwise stated. Altitudes thought to be outside the normal range are given in brackets. The identification texts are based on Grimmett, R., Inskipp, C. & Inskipp, T. *Birds of the Indian Subcontinent*, Christopher Helm (2011). The vast majority of the illustrations have been taken from the same work and, wherever possible, the correct races for Nepal have been depicted. A small number of additional illustrations of races occurring in Nepal were executed for this book. Preparation of the text and plates for that work included extensive reference to museum specimens combined with considerable work in the field.

Distribution maps and species texts are on the page facing the relevant illustrations. The text comprises identification features (**ID**), including voice (**Voice**) for most species, as well as approximate body length of the species, including bill and tail, in centimetres. Length is expressed as a range when there is marked variation within the species (e.g. as a result of sexual dimorphism or subspecific differences). Habitat and habits (**HH**) that are useful for identification are also included where space allows. A taxonomic note (**TN**) or an alternative English name (**AN**) is given at the end where relevant.

Key to the maps

Taxonomy and nomenclature

Taxonomy and nomenclature largely follow Inskipp *et al.* (1996). However, many proposals for taxonomic change have been made since that date, particularly in Rasmussen & Anderton (2012). Some of these proposals for elevation to species status are adopted here, where they are supported by what are considered to be adequate published justifications in other sources:

Greater Flamingo *Phoenicopterus roseus* split from (Greater) American Flamingo *P. ruber*
Brown-cheeked Rail *Rallus indicus* split from Water Rail *R. aquaticus*
Eastern Grass Owl *Tyto longimembris* split from (African) Grass Owl *T. capensis*
Hume's Bush Warbler *Cettia brunnescens* split from Yellowish-bellied Bush Warbler *C. acanthizoides*
Hume's Whitethroat *Sylvia althaea* split from Lesser Whitethroat *S. curruca*
Hodgson's Treecreeper *Certhia hodgsoni* split from Eurasian Treecreeper *C. familiaris*
Great Myna *Acridotheres grandis* split from White-vented Myna *A. cinereus*
Black-throated Thrush *Turdus atrogularis* split from Dark-throated (Red-throated) Thrush *T. ruficollis*
Dusky Thrush *Turdus eunomus* split from Naumann's Thrush *T. naumanni*
Red-tailed Wheatear *Oenanthe chrysopygia* split from Kurdish Wheatear *O. xanthoprymna*
Taiga Flycatcher *Ficedula albicilla* split from Red-breasted Flycatcher *F. parva*

One proposed split has not been adopted because the justification available is considered to be inadequate:

Green Warbler *Phylloscopus nitidus* from Greenish Warbler *P. trochiloides*

A further nine do not merit consideration for elevation to species status because they were only provisionally split by Rasmussen & Anderton (2012):

Black-eared Kite *Milvus* [*migrans*] *lineatus*
Grey-headed Swamphen *Porphyrio* [*porphyrio*] *poliocephalus*
Eastern Black-tailed Godwit *Limosa* [*limosa*] *melanuroides*
Steppe Gull *Larus* [*heuglini*] *barabensis*
Fork-tailed Drongo Cuckoo *Surniculus* [*lugubris*] *dicruroides*
Eastern Jungle Crow *Corvus* [*macrorhynchos*] *levaillantii*
Indian Jungle Crow *Corvus* [*macrorhynchos*] *culminatus*
Indian Reed Warbler *Acrocephalus* [*stentoreus*] *brunnescens*
Siberian Chiffchaff *Phylloscopus* [*collybita*] *tristis*

Another 13 are proposals for splitting from polytypic species where the entire complex of subspecies needs to be reviewed before decisions are made:

Himalayan Buzzard *Buteo burmanicus* (= *refectus*) from Common Buzzard *B. buteo*
Ashy-headed Green Pigeon *Treron phayrei* from Pompadour Green Pigeon *T. pompadora*
Collared Scops Owl *Otus lettia* from (Collared) Indian Scops Owl *O. bakkamoena*
Himalayan Wood Owl *Strix nivicola* from Tawny Owl *S. aluco*
Scarlet Minivet *Pericrocotus speciosus* from (Scarlet) Orange Minivet *P. flammeus*
Spangled Drongo *Dicrurus bracteatus* from (Spangled) Hair-crested Drongo *D. hottentottus*
Cinereous Tit *Parus cinereus* from Great Tit *P. major*
Black-crested Bulbul *Pycnonotus flaviventris* from (Black-crested) Black-capped Bulbul *P. melanicterus*
Hill Prinia *Prinia superciliaris* from (Hill) Black-throated Prinia *P. atrogularis*
Chestnut-bellied Nuthatch *Sitta cinnamoventris* from (Chestnut-bellied) Indian Nuthatch *S. castanea*
Tibetan Blackbird *Turdus maximus* from Common Blackbird *T. merula*
Siberian Stonechat *Saxicola maurus* from Common Stonechat *S. torquatus*
Plain Flowerpecker *Dicaeum minullum* from (Plain) Nilgiri Flowerpecker *D. concolor*

The remaining 15 proposals fall into the category of requiring further research and compilation of data before their justification can be reassessed:

Bewick's Swan *Cygnus bewickii* from Tundra Swan *C. columbianus* (needs clarification of extent of interbreeding)
Eastern Marsh Harrier *Circus spilonotus* from (Eurasian) Western Marsh Harrier *C. aeruginosus* (needs clarification of extent of interbreeding)

Eastern Cattle Egret *Bubulcus coromandus* from (Western) Cattle Egret *B. ibis* (molecular differences more consistent with subspecific distinction)
Indian Thick-knee *Burhinus indicus* from Eurasian Thick-knee *B. oedicnemus* (needs molecular study)
Whistling Hawk Cuckoo *Hierococcyx nisicolor* from Hodgson's Hawk Cuckoo *H. fugax* (plumage and song differences relatively minor)
Indian Eagle owl *Bubo bengalensis* from Eurasian Eagle Owl *B. bubo* (Rasmussen & Anderton 2012 state that specific status 'is not fully established')
Grey Nightjar *Caprimulgus jotaka* from (Grey) Jungle Nightjar *C. indicus* (needs molecular study)
House Swift *Apus nipalensis* from (House) Little Swift *A. affinis* (the subspecies *nipalensis* has a square tail and is not part of the eastern complex with slightly forked tails)
Indian Golden Oriole *Oriolus kundoo* from Eurasian Golden Oriole *O. oriolus* (needs clarification of vocal differences)
Grey-throated Martin *Riparia chinensis* from Plain Martin *R. paludicola* (needs molecular study)
Swamp Prinia *Prinia cinerascens* from Rufous-vented Prinia *P. burnesii* (needs more information on morphometrics, and relationships of newly described subspecies *nepalicola*)
Eastern Orphean Warbler *Sylvia crassirostris* from (Western) Orphean Warbler *S. hortensis* (needs data on morphology and vocalisations in area of close approach/overlap)
Himalayan Shrike Babbler *Pteruthius ripleyi* and Blyth's Shrike Babbler *Pteruthius aeralatus* from White-browed Shrike Babbler *P. flaviscapis* (need clarification of distribution and vocalisations of these forms)
Himalayan Bluetail *Tarsiger rufilatus* from Red-flanked Bluetail *T. cyanurus* (needs molecular study)
Chestnut Munia *Lonchura atricapilla* from Black-headed (Tricoloured) Munia *L. malacca* (needs clarification of introgression)
Spotted Great Rosefinch *Carpodacus severtzovi* from Great Rosefinch *C. rubicilla* (needs molecular study)

In most cases, where the above potential splits represent additional forms to the parent species in the subcontinent, they are given separate accounts for clarity in the text. To indicate their provisional status the header text is in roman font (not bold), and the parent species name is included in parentheses after the generic name, e.g. Black-eared Kite *Milvus (migrans) lineatus.*

The sequence largely follows Dickinson (2003, *The Howard & Moore Complete Checklist of the Birds of the World*), although some species have been grouped out of this systematic order to enable useful comparisons to be made.

English names of birds represent another contentious issue, with almost as many different names for some species as there are books describing them. The names adopted here mainly follow Gill & Wright (2006 *Birds of the World: Recommended English Names*, and updates), the exceptions being those which, in the opinion of the authors, are more appropriate for the species concerned.

Plumage terminology

The figures opposite illustrate the main plumage tracts and bare-part features, and are based on Grant & Mullarney (1988–89). This terminology for bird topography has been used in the species accounts. Other terms used are defined in the Glossary. Juvenile plumage is the first plumage on fledging, and in many species it is looser and fluffier than subsequent plumages. In some families, juvenile plumage is retained only briefly after leaving the nest (e.g. pigeons), or hardly differs from adult plumage (e.g. many babblers), while in other groups it may be retained for the duration of long migrations or for many months (e.g. many waders). In some species (e.g. *Aquila* eagles), it may be several years before all of the juvenile feathers are finally moulted. The relevance of juvenile plumage to field identification therefore varies considerably. Some species reach adult plumage after their first post-juvenile moult (e.g. larks), whereas others go through a series of immature plumages. The term 'immature' has been employed more generally to denote plumages other than adult, and is used either where a more exact terminology has not been possible, or where more precision would give rise to unnecessary complexity. Terms such as 'first-winter' (resulting from a partial moult from juvenile plumage) or 'first-summer' (plumage acquired prior to the breeding season of the year after hatching) have, however, been used where it was felt that this would be useful.

Many species assume a more colourful breeding plumage, which is often more striking in the male compared to the female. This can be realised either through a partial (or in some species complete) body moult (e.g. waders) or results from the wearing-off of pale or dark feather fringes (e.g. redstarts and buntings).

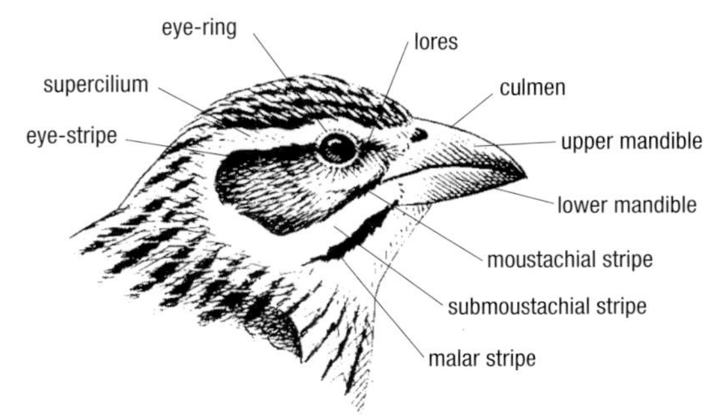
eye-ring
lores
supercilium
culmen
eye-stripe
upper mandible
lower mandible
moustachial stripe
submoustachial stripe
malar stripe

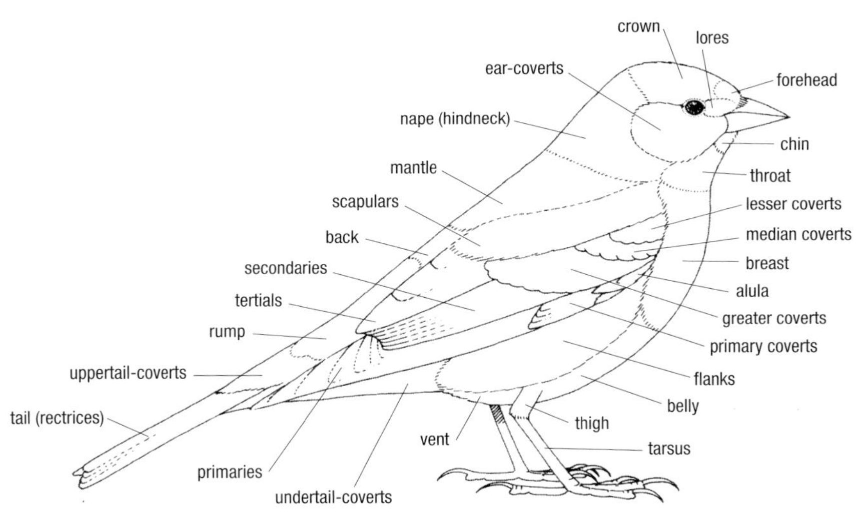
crown
lores
ear-coverts
forehead
nape (hindneck)
chin
mantle
throat
scapulars
lesser coverts
back
median coverts
breast
secondaries
alula
tertials
greater coverts
rump
primary coverts
uppertail-coverts
flanks
belly
tail (rectrices)
thigh
vent
tarsus
primaries
undertail-coverts

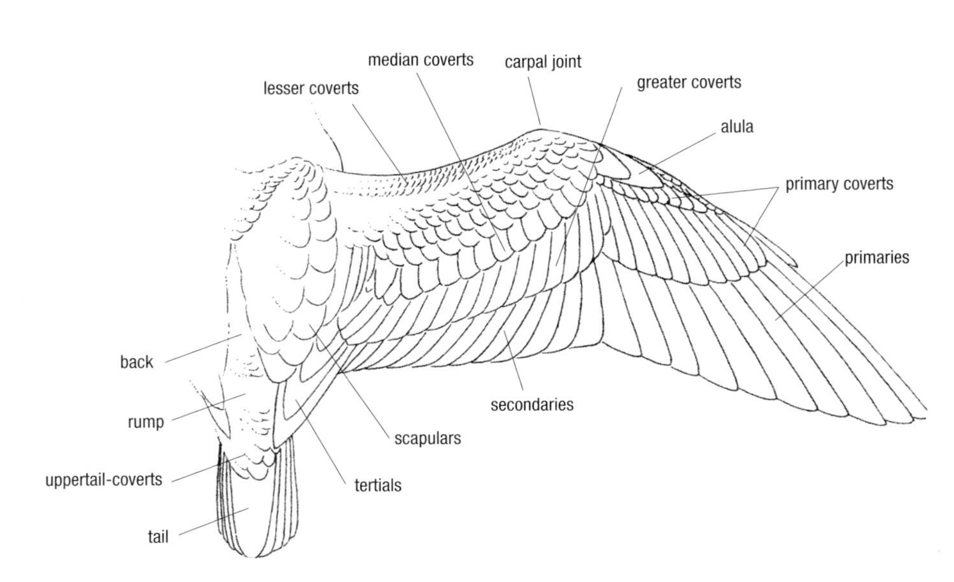
median coverts
carpal joint
lesser coverts
greater coverts
alula
primary coverts
primaries
back
secondaries
rump
scapulars
uppertail-coverts
tertials
tail

GLOSSARY

See also figures on p. 13, which cover bird topography.

Altitudinal migrant: a species which breeds at high altitudes (in mountains) and moves to lower levels and valleys in the non-breeding season.

Arboreal: tree-dwelling.

Axillaries: the feathers in the 'armpit' at the base of the underwing.

Cap: a well-defined patch of colour or bare skin on the top of the head.

Carpal: the bend of the wing, or carpal joint.

Carpal patch: a well-defined patch of colour on the underwing in the vicinity of the carpal joint.

Casque: an enlargement on the upper surface of the bill, in front of the head, as on hornbills.

Cere: a fleshy (often brightly coloured) structure at the base of the bill and enclosing the nostrils.

Culmen: the ridge of the upper mandible.

Eclipse plumage: a female-like plumage acquired by males of some species (e.g. ducks or some sunbirds) during or after breeding.

Edgings or edges: outer feather margins, which can frequently result in distinct paler or darker panels of colour on wings or tail.

Flight feathers: the primaries, secondaries and tail feathers (although not infrequently used to denote the primaries and secondaries alone).

Fringes: complete feather margins, which can frequently result in a scaly appearance to body feathers or wing-coverts.

Gape: the mouth and fleshy corner of the bill, which can extend back below the eye.

Graduated tail: a tail in which the longest feathers are the central pair and the shortest the outermost, with those in between being intermediate in length.

Gregarious: living in flocks or communities.

Gular pouch: a loose and pronounced area of skin extending from the throat (e.g. in hornbills).

Gular stripe: a usually very narrow (and often dark) stripe running down the centre of the throat.

Hackles: long and pointed neck feathers that can extend across the mantle and wing-coverts (e.g. on junglefowls).

Hand: the outer part of the wing, from the carpal joint to the tip of the wing.

Hepatic: used with reference to the rufous-brown morph of some (female) cuckoos.

Iris (plural **irides**): the coloured membrane that surrounds the pupil of the eye and which can be brightly coloured.

Lappet: a wattle, particularly one at the gape.

Leading edge: the front edge of the forewing.

Local: occurring or common within a small or restricted area.

Mandible: the lower or upper half of the bill.

Mask: a dark area of plumage surrounding the eye and often covering the ear-coverts.

Morph: a distinct plumage type that occurs alongside one or more other distinct plumage types of the same species.

Nomenclature: the scientific naming of species, subspecies, as well as genera, families and other categories in which species may be classified.

Nominate: the first-named race of a species, which has its scientific racial name the same as the specific name.

Nuchal: relating to the hindneck, used with reference to a patch or collar.

Ocelli: eye-like spots of iridescent colour; a distinctive feature in the plumage of peafowls.

Orbital ring: a narrow circular ring of feathering or bare skin surrounding the eye.

Plantation: a group of trees (usually exotic or non-native species) planted in close proximity to each other, used for timber or as a crop.

Primary projection: the extension of the primaries beyond the longest tertial on the closed wing; this can be of critical importance in identification (e.g. of larks or *Acrocephalus* warblers).

Race (subspecies): a geographical population whose members show constant differences (e.g. in plumage or size) from those of other populations of the same species.

Rectrices (singular **rectrix**): the tail feathers.

Remiges (singular **remex**): the primaries and secondaries.

Rictal bristles: bristles, often prominent, at the base of the bill.

Shaft-streak: a fine line of pale or dark colour in the plumage, produced by the feather shaft.

Speculum: the often glossy panel across the secondaries of, especially, dabbling ducks, often bordered by pale tips to these feathers and a greater coverts wing-bar.

Subspecies: see Race.

Subterminal band: a dark or pale band, usually broad, situated inside the outer part of a feather or feather tract (used especially in reference to the tail).

Taxonomy: the scientific classification of species, subspecies, genera, families and other categories.

Terai: this undulating alluvial, often marshy strip of land, 25–45 km wide, north of the Gangetic plain, extending from Uttarakhand through Nepal and northern West Bengal to Assam, naturally supports tall elephant grass interspersed with dense forest, but large areas have been drained and converted to cultivation.

Terminal band: a dark or pale band, usually broad, at the tip of a feather or feather tract (especially the tail), *cf.* Subterminal band.

Terrestrial: living or occurring mainly on the ground.

Trailing edge: the rear edge of the wing, often darker or paler than the rest of the wing, *cf.* Leading edge.

Vent: the area around the cloaca (anal opening), just behind the legs (should not be confused with the undertail-coverts).

Vermiculated: marked with narrow wavy lines, usually visible only at close range.

Wattle: a lobe of bare, often brightly coloured skin attached to the head (frequently at the bill base), as on mynas or wattled lapwings.

Wing-linings: the entire underwing-coverts.

Wing panel: a pale or dark band across the upperwing (often formed by pale edges to the remiges or coverts), broader and generally more diffuse than a wing-bar.

Wing-bar: generally, a narrow and well-defined dark or pale bar across the upperwing, often referring to a band formed by pale tips to the greater or median coverts (or both, as in 'double wing-bar').

Sagarmatha National Park. (*Philip Robinson*)

GEOGRAPHICAL SETTING

Most of Nepal lies in the central Himalayas and includes eight of the highest peaks in the world, each above 8,000 m. Nepal is surrounded by the world's two most populous countries, India and China, but is one of the smallest nations. The country covers an area of 147,181 km^2, little more than England and Wales combined, and averages about 870 km from east to west. Nepal exhibits more contrasts of landscape and culture than in most other countries many times its size. There is a narrow strip of lowlands, known as the *terai*, in the south that differs sharply from the rugged terrain of the rest of the country. Slightly higher up lies the dry *bhabar* zone that extends to about 300 m. Beyond the *bhabar* are the first Himalayan foothills known as the Siwalik or Churia hills, rising to 1,220 m. To the north lies the Mahabharat Lekh climbing to 2,740 m. Between these two ranges are the dun valleys or inner *terai*. Until relatively recently the *terai* and inner *terai* were malaria-ridden jungle and rich in wildlife, but today they hold almost all of Nepal's industry and most areas are highly cultivated. Beyond the Mahabharat Lekh lies a broad complex of midland hills and valleys including the Kathmandu Valley, the traditional heartland of Nepal. North of the midland regions lies the main Himalayan range including Sagarmatha (Mount Everest), at 8,848 m, which is no more than 160 km as the crow flies from the *terai* at 75 m. Some of the world's deepest river gorges, notably the Kali Gandaki, cut through the Himalayan range. In the north-west of the country, a drier trans-Himalayan range marks the boundary between Nepal and Tibet, and the southern edge of the Tibetan plateau. These peaks of 6,000 m to 7,000 m are less rugged, and wind-eroded landforms are dominant.

CLIMATE

Nepal has extremes of climate varying from tropical in the lowlands to arctic in the high peaks. At Meghauli near Chitwan, maximum temperatures reach 23°C to 33°C between October and March and 37°C between April and early June. In sharp contrast, maximum temperatures in Namche Bazaar in Khumbu (Everest region) vary from 6°C to 8°C from October to March and 11°C to 15°C between April and June. However, Nepal's capital, Kathmandu which lies at 1,330 m, has a mild climate. From October to March skies are clear and there is little rain. Maximum temperatures are 17°C to 25°C but may drop to near freezing at night. Maximum temperatures average 27°C in April and 30°C in June, and rain falls more frequently.

Nepal's climate is dominated by the monsoon of south Asia. About 90% of the rain falls between June and September. The region north of the Himalayan range, including Mustang, Manang and Dolpo districts, lies in the rainshadow and experiences very low precipitation. Western Nepal is generally drier than the east as the monsoon rains reach there later and last for a shorter period than in the east. However, there are pockets of high rainfall in the west caused by the topography, notably the area south of Annapurna, which is the wettest in the country. Rainfall also increases with altitude until the clouds have lost most of their moisture and then it decreases again. Slope aspect has a marked influence on climate, southern slopes being significantly warmer and sunnier than those facing north.

MAIN HABITATS AND BIRD SPECIES

The vegetation is classified following Dobremez, J. F. (1976) *Le Népal écologie et biogéographie*. Centre National de la Recherche Scientifique, Paris. Nepal's bird habitats can be roughly divided into forest, scrub, alpine habitats, wetlands, grasslands, agricultural land and that around human habitation.

Forests and scrub

Nepal has a very rich diversity of forest types. Forests and bushes hold the high proportion of 77% of Nepal's breeding birds.

Tropical forest lies between approximately 75 m and 1,000 m. Tropical forest is the richest in bird species. It includes Sal *Shorea robusta*, which is by far the most extensive forest, and tropical evergreen forest, which is restricted to narrow belts in damp shady areas in the centre and east, often within Sal forest. The globally threatened and restricted-range species Himalayan Wedge-billed Babbler has been recorded in the east. Yellow-vented Warbler is the only other restricted-range species that occurs in this forest type. It is probably mainly a winter visitor to Nepal, but has bred in dense, moist, broadleaved evergreen forest in the east. Malayan Night Heron and Jerdon's Baza are rare and local specialities.

Sal Forest, Bardia National Park. (*Hem Sagar Baral*)

Tropical forest, Chitwan National Park. (*Carol Inskipp*)

Subtropical forest lies between approximately 1,000 m and 2,000 m in the west and between 1,000 m and 1,700 m in the east.

Broadleaved subtropical forest is second in species-richness to tropical forest. *Schima–Castanopsis* is a moist broadleaved forest type that once covered much of subtropical central and eastern Nepal, but now only small patches remain in most places. Riverine forest grows along watercourses within *Schima–Castanopsis*. Alder *Alnus nipalensis* also grows along streams and often colonises abandoned cultivation and landslides. Three restricted-

Subtropical forest, Churia Hills. (*Carol Inskipp*)

Subtropical forest, Pokhara Valley. (*Carol Inskipp*)

range species breed: White-naped Yuhina and Yellow-vented Warbler, both of which are rare and local in the east, and Hoary-throated Barwing, although the last species mainly visits Nepal's subtropical forests in winter. Spot-bellied Eagle Owl, Blue-naped Pitta, Pied Thrush and Purple Cochoa are other breeding specialities.

Chir Pine *Pinus roxburghii* forest is widespread in the west, but in the centre and east is confined to drier situations. These forests are typically open with little or no understorey because of frequent fires. They are poor in bird species and hold no specialities. Breeding birds include Blue-capped Rock Thrush.

Lower temperate forest lies between approximately 2,000 m and 2,700 m in the west and between 1,700 m and 2,400 m in the east.

Moist broadleaved lower temperate forest grows in central and eastern Nepal and mainly comprises species of oaks *Quercus* with laurels Lauraceae. This habitat is important for the restricted-range species Broad-billed Warbler, which is very local and rare in the east, as well as the more widespread Hoary-throated Barwing, Nepal

Temperate forest, central Nepal. (*Hem Sagar Baral*)

Upper temperate forest, Langtang National Park. (*Hem Sagar Baral*)

Wren Babbler and White-naped Yuhina. Satyr Tragopan, Spot-bellied Eagle Owl, Yellow-rumped Honeyguide, Pied Thrush, Long-billed Thrush, Purple Cochoa, Slender-billed Scimitar Babbler, Fire-tailed Myzornis and Black-headed Shrike Babbler are other breeding specialities.

Dry broadleaved lower temperate forest occurs in the west and on drier slopes in the centre. Most remaining forest is degraded and supports a low number of bird species compared to moist forest types. Brown-fronted Woodpecker is characteristic of this dry forest.

Blue Pine *Pinus wallichiana* forest has a large altitudinal range, from 1,800 m to 4,000 m. It is widespread in the west and on dry slopes in the centre and east, but is species-poor in birds. Breeding birds include Yellow-billed Blue Magpie.

Upper temperate forest occurs between about 2,700 m and 3,100 m in the west and centre, and from 2,400 m to 2,800 m in the east. It is much less disturbed than forests lower down.

Broadleaved upper temperate forest Oak and mixed broadleaved forests grow in the centre and east. Rhododendrons are most widespread in high rainfall areas where they are dominant in many areas. They are also common in forests of fir *Abies*, hemlock *Tsuga* and birch *Betula*. Broadleaved upper temperate forest is relatively rich in birds. Hoary-throated Barwing is fairly common and widespread in mossy oak forests. Nepal Wren Babbler, Satyr Tragopan, Yellow-rumped Honeyguide, Long-billed Thrush, Slender-billed Scimitar Babbler and Fire-tailed Myzornis all occur in the breeding season.

Upper temperate coniferous forest Blue Pine is the most common and widespread conifer in the upper temperate zone. In moister areas it is mixed with other conifers, such as spruce *Picea smithiana* and fir *Abies pindrow*, which sometimes also form pure stands. In drier regions it is associated with juniper *Juniperus indica*. Characteristic species include White-throated Tit, White-cheeked Nuthatch and Kashmir Nuthatch.

Subalpine forest lies between 3,000 m and 4,200 m in the west, and 3,000 m to 3,800 m in the east. This includes some of the least disturbed forest, especially that of fir *Abies spectabilis*. Fir usually forms a continuous

Broadleaved temperate forest, Makalu Barun National Park. (*Carol Inskipp*)

Himalayan fir forest, central Nepal. (*Hem Sagar Baral*)

Alpine habitat, Langtang National Park. (*Hem Sagar Baral*)

belt between 3,000 m and 3,500 m on the southern slopes of the main ranges in central Nepal. Birch *Betula utilis* forest occurs between 3,300 m and the treeline, and both fir and birch usually possess a rhododendron understorey. In very wet sites *Rhododendron* spp. forest often replaces other forest types and forms shrubberies at higher altitudes. In drier areas juniper *Juniperus* occurs both as a tree and a small shrub. Specialities include Satyr Tragopan, Gould's Shortwing, Rufous-breasted Bush Robin, Slender-billed Scimitar Babbler and Fire-tailed Myzornis.

Bamboo *Arundinaria* spp. and *Bambusa* spp. flourishes in very high rainfall areas and in a few places, such as the Modi Khola Valley, south of Annapurna, it forms pure dense stands. Bamboo is an important component of forests for many birds. These include Broad-billed Warbler and Golden-breasted Fulvetta in temperate forests, and Fulvous Parrotbill and Great Parrotbill in subalpine forest.

Alpine habitats lie between the treeline (3,800 m in the east and 4,200 m in the west) and the region of permanent snow. Alpine scrub, including rhododendron and juniper, grows as high as 4,870 m in places. Only a small number of bird species breed, notably the globally threatened Wood Snipe, which occurs in alpine meadows with scattered bushes and a few streams. Other breeding species include Tibetan Snowcock, Robin Accentor and Grandala, which occur as high as 5,500 m.

Tibetan steppe zone

North and north-west of Dhaulagiri and Annapurna and in northern Humla district the country is almost treeless with a climate and flora of Tibetan character. The dominant vegetation is of shrubs, especially *Caragana*, grasses and alpine flora. Characteristic species include White-browed Tit Warbler and Red-fronted Serin, and in summer Desert Wheatear and Blue Rock Thrush.

Wetlands

Koshi Tappu Wildlife Reserve and Koshi Barrage area in the south-east *terai* is listed as a Ramsar Site of international importance for migrating wildfowl, waders, gulls and terns. Koshi comprises a large expanse of open

Rara Lake National Park. (*Carol Inskipp*)

water, marshes, grassland and scrub. In recent years, numbers of these species have sharply declined but several globally threatened wetland species still occur: Baer's Pochard, Pallas's Fish Eagle, Greater Spotted Eagle, Eastern Imperial Eagle, Woolly-necked Stork, Lesser Adjutant, and also Swamp Francolin, which inhabits tall grasses and swamps.

Lakes are scattered at all altitudes throughout the country, including Rara Lake (3,050 m) in the north-west, Phewa Tal (915 m) near Pokhara, and Jagdishpur Reservoir, Gaidahawa Tal and Ghodaghodi Tal in the western *terai*. These act as staging posts for small numbers of migrating wetland birds; those in the lowlands and foothills also support resident and wintering species.

Rivers and streams are generally fast-flowing and support a good variety of breeding species, including kingfishers, forktails, dippers, wagtails, redstarts and Blue Whistling Thrush.

There are numerous small ponds and marshes, often in cultivation and around habitation, throughout Nepal. The globally threatened Sarus Crane occurs in open cultivation in well-watered country with marshes and pools in the western *terai*.

Grasslands

Only small grassland areas remain and these are almost all within protected areas in the lowlands. These are important for several globally threatened species that breed or probably breed: Swamp Francolin, Bengal Florican, Jerdon's Babbler, Slender-billed Babbler, Grey-crowned Prinia, Bristled Grassbird and Finn's Weaver (very local). Hodgson's Bushchat is a regular but localised winter visitor.

Agricultural land

Only a small number of species inhabit agricultural land compared to the natural habitats of forests, wetlands and grasslands that it replaced. Nearly all of them are widespread and common, either in Nepal or in the plains of India south of the border. One exception is the globally threatened Sarus Crane. The globally threatened Lesser Adjutant also regularly forages in agricultural fields. Typical breeding species of agricultural land in the tropical

Lowland grassland is an important bird habitat. (*Carol Inskipp*)

and subtropical zones are Spotted Dove and Paddyfield Pipit. Oriental Turtle Dove and Grey Bushchat are common in the temperate zone.

Urban areas

A small number of species are especially associated with human habitation, including White-rumped Vulture, Spotted Owlet, House Crow, Asian Pied Starling, Common Myna and *Passer* sparrows – House Sparrow, Eurasian Tree Sparrow and Russet Sparrow. Three aerial species commonly nest under the eaves of houses: House Swift, Red-rumped Swallow and Barn Swallow. The last species even nests in some streets of Kathmandu.

Farmland bird and farmer, Lumbini, central lowland Nepal. (*Hem Sagar Baral*)

IMPORTANCE FOR BIRDS

Nepal's avifauna is highly diverse considering the size of the country. A total of 884 species has been recorded, including about 600 that probably breed or have bred.

Vagrants

A total of 75 vagrant species has been recorded to date. These are listed and described, together with habitat and voice details, in Appendix 1.

Extinct species

The following species were all recorded in the 19th century by Hodgson, but are now extirpated in Nepal: Jungle Bush Quail, Rufous-necked Hornbill, White-bellied Heron, Silver-breasted Broadbill, Green Cochoa, Red-faced Liocichla and Black-breasted Parrotbill. Hodgson also collected Pink-headed Duck in Nepal, but this species may be extinct worldwide. These species are also described in Appendix 1.

Additional species

An additional 15 species collected in the 19th century by Brian Hodgson were listed for Nepal by Gray (1863), but may have originated from forests in India close to the Nepalese border (Inskipp & Inskipp 1991). These species are listed in Appendix 2.

Endemic species

Spiny Babbler is the only endemic species. It has been found from east to west Nepal and may occur west into India, but there are no definite records. Nepal Wren Babbler, which was only recently described to science (Martens & Eck 1991), was previously thought to be endemic, but has also since been recorded in northern India.

Reasons for species-richness

Nepal's species-richness can be partially explained by the wide range of altitudes in the country, from 75 m above sea level in the *terai* up to the summit of the Himalayas, the world's highest mountains. Other factors are Nepal's highly varied topography and climate, and associated diverse vegetation. Also important is the country's geographical position in a region of overlap between two biogeographical provinces: the Indo-Malayan (South and Southeast Asia) and Palearctic. As a result, species typical of both realms occur.

Restricted-range species

There are three areas of especially high biological diversity in Nepal that are of global importance. These are known as the Western Himalayas, Central Himalayas and Eastern Himalayas Endemic Bird Areas (EBAs) and are four of eight such areas identified in the Indian Subcontinent. BirdLife International identified EBAs throughout the world. First, BirdLife analysed the distribution patterns of birds with restricted ranges, i.e. landbird species which have, throughout historical times (i.e. post-1800), had a total global breeding range of < 50,000 km^2 (about the size of Sri Lanka) (Stattersfield *et al.* 1998). BirdLife's analysis showed that restricted-range species tend to occur in places that are often islands or isolated patches of a particular habitat.

The Western Himalayas EBA is important for its west Himalayan temperate forests (broadleaf, mixed coniferous/broadleaf and coniferous). This EBA extends along the mountain chain from western Nepal (west of the Kali Gandaki Valley) through Uttarakhand, Himachal Pradesh, Jammu and Kashmir in north-west India and northern Pakistan, and then south-west along the mountains in the border region between Pakistan and Afghanistan. Six of the 11 restricted-range species in this EBA occur in Nepal: Cheer Pheasant, Kashmir Flycatcher, Tytler's Leaf Warbler, White-throated Tit, Kashmir Nuthatch, and Spectacled Finch.

The Central Himalayas EBA lies entirely within Nepal; its key habitats are moist temperate forest and dense secondary forest and scrub. It supports three restricted-range species: Nepal Wren Babbler, Hoary-throated Barwing (also found in the eastern Himalayas) and Spiny Babbler.

The Eastern Himalayas EBA is important because of its wet forests. It encompasses southern and eastern Bhutan and extends into north-eastern India, eastern Nepal, south-eastern Tibet, the Chittagong Hills in Bangladesh, the Chin Hills in western Myanmar and northeastern Myanmar to south-western China. Six of the

22 restricted-range species in this EBA occur in Nepal: Rufous-throated Wren Babbler, Himalayan Wedge-billed Babbler, Hoary-throated Barwing, White-naped Yuhina, Yellow-vented Warbler and Broad-billed Warbler. All are forest species.

The Assam Plains EBA comprises the plains and foothills of the Brahmaputra watershed in the north-east of the Indian subcontinent and includes extreme eastern Nepal. The original vegetation of the plains was seasonally inundated floodplain forest and grassland, with an adjacent strip of *terai* at the base of the foothills; this land was often marshy and supported tall elephant grass and forest. Most of the plains and foothills, however, have been converted to agricultural land. Black-breasted Parrotbill is one of the three restricted-range species in this EBA that was recorded in Nepal, but is now extirpated.

Globally threatened species

Thirty-nine species recorded in Nepal have been identified as globally threatened (BirdLife International 2015). These are: Swamp Francolin, Cheer Pheasant, Bengal Florican, Great Slaty Woodpecker, Sarus Crane, Wood Snipe, Lesser Adjutant, Woolly-necked Stork, White-rumped Vulture, Slender-billed Vulture, Red-headed Vulture, Egyptian Vulture, Indian Spotted Eagle, Grey-crowned Prinia, Bristled Grassbird, Jerdon's Babbler, Slender-billed Babbler and Finn's Weaver, which currently breed or probably breed. Baer's Pochard, Common Pochard, Eastern Imperial Eagle, Pallas's Fish Eagle, Greater Spotted Eagle, Steppe Eagle, Saker Falcon and Hodgson's Bushchat are regular winter visitors and passage migrants. Yellow-breasted Bunting is now an uncommon winter visitor. Kashmir Flycatcher is a rare passage migrant; Lesser Florican and Black-bellied Tern are very rare and local visitors. Greater Adjutant has not been recorded since 1995 and Indian Skimmer has not been seen since 2006. Black-necked Crane was previously thought to be a vagrant, but in 2014 was found possibly nesting in northern Humla (Acharya & Ghimirey in press). Long-tailed Duck and Indian Vulture are vagrants. Rufous-necked Hornbill, White-bellied Heron and Black-breasted Parrotbill have been extirpated in Nepal, and Pink-headed Duck may now be globally extinct.

Important Bird and Biodiversity Areas

A total of 36 Important Bird and Biodiversity Areas (IBAs) has been identified in Nepal. IBAs are key sites of international significance for bird and biodiversity conservation, which have been identified using standardised global criteria. Three IBAs are the most important for globally threatened species: Sukla Phanta and Kosi Tappu Wildlife Reserves, and Chitwan National Park and Buffer Zone, which are protected areas in the lowlands. Annapurna Conservation Area and Makalu Barun National Park and Buffer Zone are the two IBAs that are the most important for restricted-range species and biome-restricted species, and lie in the mountains. Half of IBAs are wholly within protected areas, one is partially protected, but the large number of 16 are unprotected (BCN and DNPWC in prep.).

Makalu Barun National Park. (*Carol Inskipp*)

MIGRATION

Many Himalayan residents are altitudinal migrants, the level to which they descend in winter frequently depending on weather conditions. For instance, Red-billed Chough occurs as high as 7,950 m, and usually remains above 2,440 m in winter, but has been noted as low as 1,450 m in cold weather. Some residents are sedentary throughout the year, while others undertake irregular movements, either locally or more widely in the region, depending on food supply, for example Speckled Wood Pigeon.

About 62 species are summer visitors or partial migrants including species of cuckoos, swifts, bee-eaters, *Phylloscopus* warblers, flycatchers and drongos. Many species winter further south in the subcontinent, including Common Hoopoe, Blue-tailed Bee-eater, Asian Brown Flycatcher and Asian Paradise-flycatcher. The winter quarters of some of these summer visitors is poorly known, for example Himalayan Cuckoo, Fork-tailed Swift and Dark-sided Flycatcher. Other species move south-eastwards, perhaps as far as Malaysia and Indonesia, for example Crow-billed Drongo, whereas Lesser Cuckoo and Common Swift winter in Africa.

Nepal attracts about 150 winter visitors, originating mainly from northern and Central Asia, some of which are also passage migrants. These include ducks, waders, birds of prey, gulls, terns, thrushes, bush warblers, *Acrocephalus*, *Locustella* and *Phylloscopus* warblers, pipits, wagtails, finches and buntings.

Some birds breeding in the Palearctic, mainly non-passerines, migrate directly across the Himalayas to winter in the subcontinent. Birds have been seen flying over the highest parts of these ranges; for example a flock of Bar-headed Geese was recorded flying as high as 9,375 m above Sagarmatha. Other birds follow the main valleys, such as those of the Kali Gandaki, Dudh Kosi and Arun. Spot-winged Starling undertakes east–west movements along the Himalayas. Large numbers of birds of prey, especially *Aquila* eagles, have also been found to use the Himalayas as an east–west flyway when migrating from their breeding grounds. The raptors leave the Tibetan Plateau and northern Asia and head south to uncertain wintering grounds (for some species possibly as far as sub-Saharan Africa) following the Himalayan range of Nepal and India. Each year, between 2012 and 2014, 10,000 to 14,000 individuals were counted, comprising 37 migrating raptor species. The majority were Steppe Eagles, and the globally threatened Pallas's Fish Eagle, Eastern Imperial Eagle and Greater Spotted Eagle; Egyptian Vulture, White-rumped Vulture, Saker Falcon, and Near Threatened Himalayan Vulture and Cinereous Vulture were also seen (Subedi 2015).

BIRDWATCHING AREAS

The following areas are some of the most interesting localities and treks for birdwatching in Nepal, but there are many others; for example, all of Nepal's protected areas have very good bird habitat.

Key: r = resident, w = winter visitor, s = summer visitor, pm = passage migrant

Chitwan National Park

Location: South-central Nepal.
Habitat: Grasslands, sal and riverine forests on low ground, sal forest on hills and sal/Chir Pine on ridges, rivers and small lakes.
Best time to visit: October–May, especially late April.
Birds: Oriental Pied and Great Hornbills, Red-headed Trogon, Blue-eared Kingfisher (rare), White-rumped Spinetail, Spot-bellied and Dusky Eagle Owls (rare), Bengal Florican, Grey-headed Fish Eagle, Greater Spotted Eagle (w), Collared Falconet, Darter, Lesser Adjutant, Woolly-necked Stork, Hooded Pitta (s), Indian Pitta (s), White-throated Bulbul, Black-backed Forktail, Golden-headed Cisticola, Pale-footed Bush Warbler, Chestnut-crowned Bush Warbler (w), Grey-crowned Prinia, Dusky and Smoky Warblers (w), Bristled Grassbird (s), Rufous-rumped Grassbird, Rufous-necked Laughingthrush, Jerdon's Babbler (rare), and Chestnut-capped, Yellow-eyed and Slender-billed Babblers (best location in Indian Subcontinent for this species).

Phulchowki Mountain (2,760 m)

This locality is one of the best birdwatching sites in Nepal. At the time of writing (June 2015), security is an issue, although the situation is likely to improve. Please check with a Nepalese travel agent before visiting.
Location: Edge of Kathmandu Valley, east-central Nepal.

Marshy reeds, Chitwan National Park .(*Hem Sagar Baral*)

Habitat: Subtropical broadleaved and moist broadleaved temperate forests.
Best time to visit: October–early May.
Birds: Good variety of subtropical and temperate forest species, including Striated Bulbul, Orange-bellied Leafbird, Long-billed Thrush (w, rare), Black-faced Warbler, Striated and Rufous-chinned Laughingthrushes, Grey-throated Babbler, Himalayan Cutia, Hoary-throated Barwing, Black-throated Parrotbill (rare), Nepal Fulvetta, Yellow-bellied Flowerpecker (w), Golden-naped Finch (w?).

Shivapuri Nagarjun National Park (2,730 m)

Location: Edge of Kathmandu Valley, east-central Nepal.
Habitat: Dense scrub and subtropical broadleaved forest on lower slopes, broadleaved temperate forest higher up.
Best time to visit: October–early May.
Birds: Good variety of subtropical and temperate forest species including Long-tailed Broadbill (s?), Spotted Forktail, Black-faced Warbler, Pale Blue Flycatcher (s), Spiny Babbler, Grey-throated Babbler, Nepal Fulvetta, Grey-sided Laughingthrush, Scarlet Finch, Golden-naped Finch and Brown Bullfinch.

Koshi Tappu Wildlife Reserve

Location: South-east lowlands.
Habitat: Marshes, impounded river, wet grasslands, scrub and riverine forest.
Best time to visit: Late November–late May.
Birds: Passage migrant and wintering wildfowl, including Bar-headed Goose, Baer's Pochard (rare), Falcated Duck (rare), Pacific Golden Plover and Oriental Pratincole (rare); also a good variety of birds of prey, especially in winter and on passage including Pallas's Fish Eagle, White-tailed Eagle, White-rumped Vulture, Pied Harrier, Greater Spotted Eagle and Red-necked Falcon. Regular winter visitors and passage migrants include Baillon's Crake and Spotted Bush Warbler, Clamorous Reed, Dusky and Smoky Warblers, Siberian Rubythroat, while Watercock and Bristled Grassbird are summer visitors. Residents include Swamp Francolin, Black, Cinnamon and Yellow Bitterns, Indian Courser, Lesser Adjutant, White-tailed Stonechat, Graceful and Rufous-vented Prinia (the only location for this endemic subspecies) and Striated Grassbird.

Rani Tal, Sukla Phanta Wildlife Reserve. (*Hem Sagar Baral*)

Sukla Phanta Wildlife Reserve

Location: Far western lowlands.
Habitats: Grasslands, sal forest, marshes and small lakes.
Best time to visit: October–May, especially from December onwards.
Birds: Most of Nepal's grassland specialities and a good variety of lowland sal forest species. Swamp Francolin, Bengal Florican, Dusky Eagle Owl (rare), Black Bittern, Lesser Adjutant, Woolly-necked Stork, White-naped and Great Slaty Woodpeckers, Rufous-rumped Grassbird, Bristled Grassbird (s), Yellow-bellied Prinia, Spotted Bush Warbler (w), Dusky and Smoky Warblers (w), Jerdon's Bushchat (s), Hodgson's Bushchat (w and pm – best locality in Indian Subcontinent for this species), Tawny-bellied, Chestnut-capped, Yellow-eyed and Jerdon's Babblers, Finn's and Black-breasted Weavers.

Bardia National Park

Location: Far western lowlands.
Habitats: Sal forest, deciduous hill forest and Chir pine; grasslands, river and riverine habitats.
Best time to visit: October–April, especially from December onwards.
Birds: Good variety of forest, grassland and river species, including White-naped and Great Slaty Woodpeckers, Oriental Pied Hornbill, Darter, Lesser Adjutant, Woolly-necked Stork, Indian Pitta, Tickell's Blue Flycatcher, Chestnut-capped and Yellow-eyed Babblers, Silver-eared Mesia.

Lumbini and Jagdishpur

Location: Western lowlands.
Habitats: Farmlands and wetlands in and around Lumbini; reservoir at Jagdishpur.
Best time to visit: October–April, especially from December onwards.
Birds: Good variety of waterbirds, birds of prey and farmland birds: Sarus Crane, Indian Spotted Eagle, vultures, Common and Large Grey Babblers, Brown Rock Chat, Short-eared Owl (w, pm), Yellow-crowned Woodpecker, Rufous-tailed and Crested Larks, Falcated Duck (w, rare), Cotton Pygmy Goose, Pheasant-tailed Jacana, Grey-headed Lapwing (w).

Pokhara Valley

Location: West-central Nepal.
Habitats: Subtropical broadleaved forest, lake, rivers and farmlands.
Best time to visit: October–May.
Birds: Kalij Pheasant, nearly all *Aquila* eagles (pm), vultures (eight species) and other birds of prey, Chestnut-headed and Grey-bellied Tesias (w), Long-billed Thrush (w), Blue-naped Pitta (rare, s), Long-tailed Broadbill, Spiny Babbler, Spotted, Little and Slaty-backed Forktails, Green Magpie, Upland Pipit, Yellow-breasted Bunting (w), waterbirds (pm).

The hills above Kande (= Khare) and Thoolakharka are excellent for raptor migration, mainly between late October and mid November (see Subedi 2015).

Chepang Hill–Siraichuli trek [information kindly provided by Bird Education Society (BES), Sauraha]

Location: Mahabharat Hills central Nepal, just north of Chitwan National Park.
Habitats: Subtropical broadleaved and Chir Pine forests.
Best time for trek: October–early June.
Recommended time for trek: Four days.
Route: Sauraha – Saktikhor (one-hour drive) – Upardang Gadhi (1,275 m) overnight – Chisapanitar – Siraichuli 1,945 m (overnight) – Saktikhor (overnight) – Sauraha. Home stay is available on this trek.
Birds: More than 300 species have been recorded on this trek by BES including Long-tailed Broadbill, Red-headed Trogon, Mountain Hawk Eagle, Himalayan, Cinereous and Red-headed Vultures, Grey-chinned Minivet, Blue-capped Rock Thrush (s), Pale Blue Flycatcher, Spiny Babbler, Striated and Blue-winged Laughingthrushes, Himalayan Cutia, Silver-eared Mesia, Black-chinned Yuhina, White-browed Shrike Babbler, Streaked and Little Spiderhunters, and Black-throated and Crimson Sunbirds.

Trekking in Langtang National Park

Following the 2015 earthquakes, avalanches destroyed Langtang village and other settlements in the upper Langtang Valley, and large parts of the trails in the park. Many teahouses were also damaged. At the time of

Langtang National Park. (*Hem Sagar Baral*)

writing (March 2016), some trekking trails and lodges have re-opened. Please check with the Langtang National Park authority or your Nepal trekking agent before starting these treks.
Location: East-central Nepal
Habitats: Lower temperate and upper temperate broadleaved forests; subalpine forests of fir, rhododendron, birch and juniper; alpine scrub of rhododendron and juniper, and alpine grasslands.
Best time for treks: Late April–early June.
Recommended time: 3–4 weeks.

Gosainkund

Route: Sundarijal, Kathmandu Valley (1,265 m) – Pati Bhanjyang (1,770 m) – Kutumsang (2,470 m) – Thare Pati (3,505 m) – Ghopte Cave (3,565 m) – Laurebina Pass (4,600 m) – Gosainkund lakes (4,380 m) – Sing Gomba (= Chandanbari) (3,255 m) – Dhunche (1,965 m).
Birds: Snow Partridge (Gosainkund lakes), Blood Pheasant, Himalayan Monal, Spotted Nutcracker, Long-billed Thrush, Gould's Shortwing (near Ghopte), Rufous-breasted Bush Robin, Grandala, Smoky Warbler (above Ghopte), Spotted Laughingthrush, Fire-tailed Myzornis (near Ghopte), Spot-winged Rosefinch, Red-fronted Rosefinch (Laurebina Pass).

Langtang Valley

Route: Dhunche (1,965 m) – Syabru (2,120 m) – Kyanjin, upper Langtang Valley (3,750 m).
Birds: Blood Pheasant, Satyr Tragopan (Syabru, rare), Himalayan Monal, Brown-capped Woodpecker, Yellow-rumped Honeyguide, Wood Snipe, Ibisbill (Kyanjin), Pied Thrush (s), Nepal Wren Babbler (Dhunche – below Lama Hotel), Mrs Gould's Sunbird, Yellow-bellied Flowerpecker, Spot-winged Rosefinch, Scarlet Finch.

Trekking in Annapurna Conservation Area

Location: West-central Nepal.

Thak Khola trek

Route: Pokhara (915 m) – Birethanti (1,065 m) (mainly by bus to Naya Pul) – Tirkhedhungha (1,525 m) – Ghorepani (2,775 m) – side trips to Poon Hill and along trail to Tatopani (1,190 m) – Ghasa (2,000 m) – Jomsom (2,715 m) –

Annapurna Conservation Area. (*Kim Robinson*)

Kagbeni (2,805 m) – Muktinath (3,800 m). Nowadays there is a drivable road to Jomsom. This trek can easily be linked to the Annapurna Sanctuary trek by using the trail from Ghorepani to Ghandruk.
Habitats: Broadleaved subtropical forest, Chir pine lower down and blue pine forest higher up, broadleaved lower and upper temperate forests, coniferous upper temperate, subalpine forests of fir, birch and *Caragana* scrub in the Tibetan steppe zone.
Best time for trek: December–March; also October (for migrants)
Recommended time: 17 days.
Birds: White-rumped, Himalayan, Red-headed, Cinereous and Egyptian Vultures, Bearded Vulture, Greater Spotted Eagle (w, Pokhara–Ghorepani), Crested Kingfisher, Little, Slaty-backed and Spotted Forktails (especially between Birethanti and Tirkhedhungha), Koklass Pheasant (Ghorepani and Ghasa), White-browed and Rufous-breasted Bush Robins, Hume's Bush Warbler, Great, Brown and Black-throated Parrotbills (Ghorepani, Ghorepani–Ghandruk), Spotted Laughingthrush, Spot-winged Rosefinch (Ghorepani and Ghasa), Variegated Laughingthrush (Ghasa), Crimson-browed Finch (Ghorepani), White-throated and Güldenstädt's Redstarts, White-browed Tit Warbler, Brown, Robin, Altai and Alpine Accentors, Red-fronted Serin, Streaked and Great Rosefinches (Tukche–Muktinath), Solitary Snipe (Muktinath and nearby Jharkhot), and Tibetan and Himalayan Snowcocks (above Muktinath on trail to Thorong La).

Annapurna Sanctuary trek

Route: Pokhara – Phedi (bus), Phedi – Pothana (2,035 m) – Landruk (1,715 m) – Chhomrong (2,050 m) – Khuldi (2,550 m) – Bagar (3,300 m) – Annapurna Base Camp (4,130 m). On the return, there is an option via Ghandruk to Naya Pul, from where there are buses to Pokhara.
Best time for trek: March, April, October, early November.
Recommended time: 14 days.
Habitats: Moist broadleaved lower temperate forest; upper temperate forest of oak and rhododendron, bamboo jungles, subalpine birch and rhododendron; alpine scrub and grasslands.
Birds: Cinereous Vulture (w), Yellow-rumped Honeyguide, Rufous-breasted Bush Robin, Grandala, Slender-billed Scimitar Babbler, Golden Babbler, Black-headed Shrike Babbler (scarce), Golden-breasted Fulvetta, Great and Fulvous Parrotbills, Yellow-bellied Flowerpecker, Spot-winged, Streaked and Red-fronted Rosefinches.

Makalu Barun National Park

Location: Eastern Nepal
Best time for trek: April and May.
Habitats: Moist broadleaved subtropical, lower temperate and upper temperate forests, subalpine forests of rhododendron, fir and birch, rhododendron shrubberies and alpine meadows.

Arun Valley

Route: Kathmandu – Tumlingtar (550 m; flight) – Khandbari (900 m) – Bhotebas (1,900 m) – Num (1,700 m) – Seduwa (1,500 m) – Navagaon (2,200 m) – Tashigaon (2,300 m) – Shunin Oral (3,300 m) – Kongma (3,800 m) – return by same route.
Recommended time for trek: 19 days.

Barun Valley extension

Route: Kongma – Makalu Base Camp – Kongma.
Recommended time for trek: Six days.
Birds: Snow Partridge, Blood Pheasant, Himalayan Monal, Yellow-rumped Honeyguide, Solitary Snipe, Wood Snipe (Barun Valley), Gould's Shortwing (Barun Valley), White-browed and Rufous-breasted Bush Robins, Grandala, Purple Cochoa, Sultan Tit and Yellow-vented Warbler (lower elevation), Broad-billed Warbler, Rufous-chinned, Spotted, Grey-sided and Blue-winged Laughingthrushes, Slender-billed Scimitar Babbler, Rufous-throated and Nepal Wren Babblers, Fire-tailed Myzornis, White-naped Yuhina, Great, Fulvous and Black-throated Parrotbills, Yellow-bellied Flowerpecker, Maroon-backed Accentor, Crimson-browed and Golden-naped Finches.

WHAT BIRDWATCHERS CAN DO

By recording their observations, birdwatchers can play a valuable role by increasing knowledge of the country's birds and helping to conserve them. Studies of globally threatened species would be especially useful. Much is still to be learned concerning the breeding behaviour of many of Nepal's birds. Please post any information that corrects or adds to that presented here on www.ebird.org.

CHANGES IN BIRD DISTRIBUTION AND STATUS

Since the early 1990s, there have been significant changes in the known distribution and status of Nepal's birds. This partly reflects much increased survey effort, especially since the first edition of this field guide in 2000. Surveys have been undertaken by Bird Conservation Nepal, the Bird Education Society, Department of National Parks and Wildlife Conservation, as well as individuals. This has greatly increased our knowledge of areas that have not been previously surveyed, e.g. Api Nampa, Kanchenjunga, Manaslu and Gaurishankar Conservation Areas and the Limi Valley, or areas that have previously been subject to little recording effort such as Dang district. Major efforts have been made to survey new protected areas, Important Bird and Biodiversity Areas (IBAs) and potential IBAs. These surveys have filled in previous gaps in distribution such as the mid-west, far north-west and far north-east. Some areas are now under surveillance very regularly or almost continually such as Chitwan National Park and Buffer Zone, and the Koshi area. Seventeen species have been added to the Nepal list since 2000.

This survey work has revealed the presence of four new probable residents for Nepal, which were probably overlooked previously. These include the Near Threatened Himalayan Wedge-billed Babbler, which was first recorded near Dharan in 1996 (Karki & Choudhary 1997). Long-billed Wren Babbler was first recorded in Makalu Barun National Park in 1995 (Cox & Sherpa 1998) and confirmed in 2006 (Cox 2010). Mottled Wood Owl was discovered in Bardia National Park in June 2015 (Ram Shahi & Benj Smelt). The most remarkable find was a small population of a new subspecies of the Near Threatened Rufous-vented Prinia *Prinia burnesii nepalicola* from grassland on islands in the Koshi River (Baral *et al.* 2007, 2008).

Field surveys, as well as a comprehensive review of all Nepal's bird records since 1990, for the Nepal national bird Red Data Book (research to be completed by the end of 2015) have revealed that some species have not been seen for a considerable period of time or have become very rare. These include a number of wetland species that have declined mainly as a result of over-fishing and habitat loss, such as Indian Skimmer and Caspian Tern (last known records 2006), Gull-billed Tern (last known record 2007), and River Tern and Black-bellied Tern (now very rare and local).

Some specialities of moist broadleaved evergreen or semi-evergreen forests have not been recorded for some years as a result of habitat loss and degradation, especially in the east. These include Mountain Imperial Pigeon (last known record 1996), Blue-eared Barbet (last known record 2006), Yellow-cheeked Tit (last known record 1997), White-hooded Babbler (last known record 1989), Long-tailed Sibia (last known record 2006) and Rufous-backed Sibia (last known record 2007). Two lowland bamboo specialists have not been found since the early 1980s, almost certainly because of habitat loss: Rufous-faced Warbler (last known record 1982) and Pale-headed Woodpecker (last known record 1981). The only known record of the lowland grassland specialist, Eastern Grass Owl, since the 1990s is an injured bird from Bardia National Park in 2011.

Some species have become absent from the far east because of habitat loss, although they have still been recorded (although rarely and very locally) further west, notably Little Spiderhunter and Silver-eared Mesia.

A number of species are now being regularly recorded at higher altitudes than previously, presumably as a result of climate change, e.g. Greater Coucal and Pale Blue Flycatcher.

Vulture populations have decreased dramatically in Nepal since the 1990s, as they have in the rest of the Indian Subcontinent because of poisoning by diclofenac, a livestock drug. This has resulted in White-rumped Vulture, Slender-billed Vulture, Red-headed Vulture and Egyptian Vulture becoming listed as globally threatened, and Bearded Vulture, Himalayan Vulture and Cinereous Vulture being listed as Near Threatened. Widespread, concerted and determined conservation efforts within Nepal and elsewhere in the subcontinent are bringing about a recovery in some species at least.

The 2010 assessment of the state of Nepal's birds (BCN and DNPWC 2011) concluded that 16 additional species were nationally threatened compared to the 2004 assessment of 133 threatened species (Baral & Inskipp

2004). There was considered to be a much a larger number of Critically Endangered species (61) in 2010 than in 2004 (40 species), as well as significantly higher numbers of Endangered species (38) compared to the 2004 total of 32 species. The number of species categorised as Vulnerable was higher in 2004 (61) compared to 50 in 2010.

More time was available for the 2010 study, however, and more records were received from a larger number of contributors, revealing a truer picture of the state of Nepal's birds. The more detailed analysis undertaken in 2010 revealed that some species were much more threatened in 2004 than now. Even so, overall, the situation for birds had worsened.

Since 2004, more research work has been carried out to identify threats and the extent of their impact on Nepal's birds: for example, the impacts of diclofenac (a non-steroidal anti-inflammatory drug used to treat livestock) on White-rumped Vulture and Slender-billed Vulture; finding new vulture nesting sites, including in Arghakhanchi District, where five vulture species breed (Bhusal 2010); research work on owls, which has revealed the threats from trade; and a desk study which revealed the strong impacts of agricultural changes (Inskipp & Baral 2011).

It is clear that some threats are increasing, for example loss and degradation of forests, pressure on grasslands, the spread and intensification of agriculture and, most especially, the wide range of threats facing wetlands. However, positive responses are increasing too, for instance the raising of conservation awareness, spread of community forestry, and many projects benefiting local livelihoods.

White-capped Redstart. (*Sagar Giri*)

BIRD CONSERVATION

This magnificent picture of thine is very charming indeed, O Lord
Of the forest, when these birds are happily enjoying themselves upon trust.
Oh! And their sweet musical notes are heard distinctly as they sing
Beautifully during the season of fragrant flower (the Spring).

Mahabharat 12.150.14–15

RELIGIOUS ATTITUDES AND TRADITIONAL PROTECTION

Attitudes to wildlife that stem from the teachings of Hinduism and Buddhism have undoubtedly helped to conserve Nepal's rich natural diversity. Nepal's birds are generally tame, at least outside heavily disturbed areas.

Wildlife, forests and water are viewed primarily as important resources, and so people have traditionally created many ways to preserve, manage and use them. In every aspect of communal interest, such as forestry and pasture, local people established unwritten laws that have become a way of life. Special permission was required to cut down trees in community forests, for example. Defaulters were fined and the revenue collected was used in community welfare. Forests protected for religious reasons, for example in Pashupatinath, Swayambhunath, Dakshinkali and Chapagaon, are important wildlife habitats. These traditional conservation methods were effective but, unfortunately, many have broken down in more recent times, giving rise to increasing deforestation and environmental degradation.

CURRENT THREATS

Nepal's rich biodiversity is seriously at risk. The principal reason is an expanding and poor population that remains reliant on natural resources and hence continues to put severe pressure on the environment. In 2010, the alarming number of 149 bird species (17%) of Nepal's birds was considered nationally threatened. Eight species are believed to have been recently extirpated from Nepal (see Appendix 1). Habitat loss was the major threat to 86% of the birds at risk nationally (BCN and DNPWC 2011).

A 2010 study of the impacts of agriculture on Nepal's birds concluded that the spread of agriculture and changes in agricultural practices are the major root causes of loss and damage to natural habitats – grasslands, wetlands and forests and their bird species (Inskipp & Baral 2011). Urbanisation and the construction of an increasing road network are other significant threats to natural habitats. Threats extend into protected areas, albeit to a lesser extent. Almost all protected areas and buffer zones face pressure from livestock grazing, fodder collection and fuelwood gathering by local villagers whose basic needs have not always been met. There is a general lack of awareness amongst poorer people of the importance and conservation of wildlife. Park staff often lack sufficient resources to implement regulations and improve their own management skills.

Threats to forest birds

At one time Nepal was extensively forested, but the latest assessment gave a forest area of only 25% and other wooded area (mainly shrubs) of 13% (FAO 2010). Forest losses have been so widespread and extensive in the lower and middle hills that some forests have become very fragmented. As a result many bird species no longer lack the continuum of habitats that they require to move altitudinally with the seasons, and their distributional range is becoming restricted.

Major causes of forest losses are: the spread of settlements and cultivation; unregulated construction, including roads and illegal building of schools, temples and hospitals etc., as well as government-planned conversion of forests for the development of projects such as road construction, electric transmission lines and reservoirs.

One quarter of Nepal's forest area is heavily degraded (World Bank 2008), which has led to loss of biodiversity, increased landslides, and soil erosion. The main causes of forest degradation are: unsustainable overharvesting of biological resources to meet persistently high demands for fuel; felling for construction timber, and collection of fodder and other forest products. Illicit felling of commercially valuable trees, the trans-boundary timber trade, uncontrolled forest fires and overgrazing by domestic livestock are also major problems.

The vast majority of Nepalese people depend on forests for their essential requirements of fuel, animal fodder and other basic materials; fuelwood, for example, provides 78% of Nepal's domestic energy (Bhusal 2010). There are initiatives to provide alternative energy sources to wood, for example small-scale solar power and hydropower schemes. However, the issue of firewood can be rather difficult to solve, even if alternative energy sources are available, as people living in rural areas/villages mostly prefer to use wood as fuel.

With 55% of nationally threatened birds depending on forests, loss and deterioration of the latter are major threats to the country's birds. Many of the threatened species require plenty of undergrowth, moist conditions or trees covered in epiphytes. Hornbills and Great Slaty Woodpecker depend on mature trees (BCN and DNPWC 2011).

Nepal's large annual precipitation and dense river networks provide high potential for hydroelectricity. However, construction of hydroelectric dams brings substantial threats to wildlife and people. They can inundate important habitats, act as barriers to wildlife migration, lead to associated developments, displace people into new sensitive habitats, and can also alter local microclimates.

Threats to wetland birds

Wetland birds face a wide range of threats in Nepal. These threats are increasing and include: drainage and encroachment for agriculture, settlement and infrastructure development; diversion and abstraction of water for irrigation; unsustainable exploitation of wetland resources, including overfishing and destructive fishing; widespread extraction of gravel from streams and riverbeds; water pollution from households, industrial discharges and agricultural run-off; invasion of alien species into wetland ecosystems; illegal hunting and trapping of birds and other wildlife; siltation, and channelling and damming of rivers (MoFSC 2014).

Many observers have noted a decline in wetland birds in recent years. The Annual Waterbird Counts are now providing data to illustrate trends on a national level; for example there has been a sharp fall in bird numbers at the internationally important wetland at Koshi, which was first noted around 1990 (Choudhary 2003). This trend was confirmed by subsequent Annual Waterbird Counts at Chitwan and in the Pokhara Valley.

Threats to grassland birds

Cultivation has significantly reduced the once extensive subtropical grasslands that border Nepal's lowland rivers and formed the northern extension of the huge Gangetic plain. Remaining areas are fragmented and almost entirely lie within protected areas. Within protected areas, Nepal's grassland specialist birds, for instance the globally threatened Bengal Florican, Swamp Francolin and Hodgson's Bushchat, are suffering from inappropriate grassland management, including untimely and intensive annual cutting, burning and ploughing, which alter species composition and are aimed at the conservation of mammals, not birds (Baral 2001).

Controlling illegal grazing and cutting activities is a difficult task for Nepal's park managers. Livestock grazing is by far the greatest threat to lowland grasslands in protected areas. As tall grass is so useful, and because of the lack of tall grassland outside reserves, illegal cutting continues around all protected areas. Nowadays fires seem essential to maintain lowland grassland ecosystems in protected areas and are a very useful management tool to maintain biodiversity. However, fires carried out during the breeding season can be extremely damaging to nests and eggs and, if burning is too comprehensive, no shelter remains for grassland wildlife (Baral 2001).

Overexploitation

Hunting and trapping

Hunting is contributing to the decline of some species, including the globally threatened Lesser Adjutant, Sarus Crane, Spot-billed Pelican and Cheer Pheasant. In some areas, hunting is on the increase as traditional values wane. The slaughter of many larks and buntings for sale as snacks (locally known as *bagedi*) in village inns has been observed in the *terai*. Illegal bird trading goes on near Koshi Barrage year-round. Buyers from Bihar in India purchase birds for food, including ducks, jacanas, doves, moorhens, egrets, munias, larks and francolins. Once widespread throughout lowland Nepal, Great Hornbill is now confined to Chitwan in central Nepal and Bardia in the west. The oil from this bird's casque and the beak itself are much valued in traditional medicine (Fleming *et al.* 1984). Hunters and villagers routinely shoot such large and conspicuous birds both for medicine and for food.

Some globally threatened species, such as Lesser Adjutant and Sarus Crane, live close to human settlements making them easy targets. The establishment of protected areas does not necessarily ensure their conservation. What is needed is improved conservation awareness amongst local people. The increased illegal bird trade in the Kathmandu Valley is a threat to some bird species (Thapa & Thakuri 2009).

Overfishing

Overfishing, which has led to a marked decline in prey, is a serious threat to many fish-eating bird species, for example Black-bellied Tern, gulls, Indian Skimmer and fish-eating raptors and owls such as Grey-headed Fish Eagle, Lesser Fish Eagle and Tawny Fish Owl (BCN and DNPWC 2011).

Overexploitation of forest products

A number of mountain IBAs are suffering from very large numbers of people illegally gathering non-timber forest products, including yarsagumba *Cordyceps sinensis*, a fungus that parasitises moth larvae and is highly valued as a herbal remedy in Chinese medicine. Every year, these huge influxes of people are causing high levels of wildlife disturbance and poaching mammals and birds, especially pheasants. Forests are becoming reduced and thinned as a result of their collecting fuelwood. Wildlife disturbance is especially high because these forest products are gathered during the breeding season for most birds and some mammals.

Climate change

Climate change is a major threat to biodiversity and is having a significant impact on the Himalayan environment. However, the impacts of climate change on Nepal's biodiversity are poorly understood. The ranges of many species are very likely to move upwards from their current locations. Species with specialist habitat requirements will be especially vulnerable as they have nowhere to go when the climate warms. High-mountain ecosystems are likely to be worst affected by climate change (MoFSC 2014).

Some of Nepal's threatened birds are largely confined to protected areas, notably grassland species. As the climate changes, habitats in these protected areas may eventually become no longer suitable for these species. However, as natural habitats outside protected areas have been converted to agriculture or developed, grassland species will have no suitable habitat to colonise (BCN and DNPWC 2011).

Invasive alien species

Introduced species of fauna and flora are a common threat to native birds and other wildlife worldwide. In Nepal, many pools and small lakes have become overgrown by an exotic plant species, Water Hyacinth *Eichhornia crassipes*. The species produces free-floating mats, causing a sharp decline in the number of open-water dwelling species, especially cormorants, grebes and many ducks. Water Hyacinth can also lead to low dissolved oxygen levels and thereby reduce insect and fish populations on which many bird species feed (Dahal 2007).

More recently, another alien plant, *Mikania micrantha*, has been having devastating effects in some areas, notably in Chitwan National Park and Koshi Tappu Wildlife Reserve. It is a climber that can cover trees and shrubs, as well as the entire forest floor, making it impossible for terrestrial species, such as thrushes and pipits to feed (Baral 2002).

Chemical poisoning

In Europe, use of some pesticides has been demonstrated to cause widespread declines of numerous bird species, many of which were previously common, including birds of prey and finches.

Persistent chemical pesticides have been banned for use in agriculture and public health since April 2001 in Nepal. However, there is serious concern concerning the illegal import of pesticides. This was illustrated by an investigation of threats to Sarus Crane in Lumbini, where Paudel (2009 a,b) carried out a survey of local markets and found that a wide range of pesticides was available and a disturbingly large range of insecticides was being used in the area. There is widespread documentation of farmers' lack of awareness of pesticides, including impacts on the environment and the ongoing need for farmers' education and development of a safety culture in pesticide use (Palikhe 2005).

Poisoning by diclofenac, a drug used to treat livestock ailments, was identified as the cause of drastic vulture declines in the Indian Subcontinent, including Nepal (Oaks *et al.* 2004). In a bid to stop the illegal use of diclofenac for veterinary use after its ban by the Nepal government in 2006, Diclofenac Free Zones have been created in many of Nepal's districts. In 2007 Bird Conservation Nepal (BCN) set up the first community-managed Vulture Safe Feeding Site at Pithauli and a number of such sites have now been established across Nepal. Pharmaceutical firms are encouraged to promote a safe alternative called meloxicam (Swan *et al.* 2006). BCN has undertaken a widespread awareness programme on vulture conservation. The use of diclofenac has since declined by 90% across parts of Nepal, although its complete elimination from the scavenger food chain has yet to be achieved (Gilbert *et al.* 2007).

NATURAL DISASTERS

Natural disasters such as the 2008 monsoon flooding of the Koshi River have had major impacts on wildlife and the environment. More recently, the large earthquakes of April and May 2015, and numerous accompanying aftershocks, have seriously damaged or destroyed infrastructure, villages, towns and parts of Kathmandu, as well as causing major environmental damage to between one third to one half of the country. Further encroachment of forests and degradation of habitats are likely, as large numbers of hard-pressed people will be forced to find new settlements and farmland.

CONSERVATION MEASURES

Nepal's Protected Area system

Nepal's Protected Area system covers 23.33% of the country and almost 4% is covered by buffer zones. The protected area system includes ten national parks: Chitwan, Bardia, Sagarmatha, Shey Phoksundo, Langtang, Makalu Barun, Rara, Khaptad, Shivapuri Nagarjun and Banke; as well as three wildlife reserves: Sukla Phanta, Parsa and Koshi Tappu; six Conservation Areas: Annapurna, Api Nampa, Manaslu, Kanchenjunga, Blackbuck and Gaurishankar; and one hunting reserve: Dhorpatan. Nepal is the second most important country in Asia for the percentage of its surface area that is protected, illustrating Nepal's enlightened attitude to both biodiversity and landscape conservation.

The Department of National Parks and Wildlife Conservation (DNPWC) is the body authorised to maintain and conserve protected areas in Nepal. In addition, Annapurna, Gaurishankar and Manaslu Conservation Areas are managed by the National Trust for Nature Conservation, a not-for-profit organisation, working in Nepal for nature conservation. The Nepal army guards most of the national parks and wildlife reserves. Involvement of local people in the management of protected areas has been further promoted by the establishment of buffer zones around most of these sites.

In 2006 the government handed over management responsibility of Kanchenjunga Conservation Area to local communities. Local people are also involved in the management of Annapurna, Manaslu and Gaurishankar Conservation Areas through conservation committees. These initiatives have put Nepal at the forefront in linking communities to benefits from protected areas (MoFSC 2014).

Government policies and legislation

There are many government policies and laws that support conservation efforts in Nepal. These include the National Conservation Strategy for Nepal, which was endorsed as policy in 1988. Policy resolutions cover the basic requirements of the people, as well as the need to safeguard natural and aesthetic values, and to maintain the country's cultural heritage.

The Convention on Biological Diversity (CBD) was ratified by Nepal in November 1993. In 2002, Nepal developed a comprehensive Nepal Biodiversity Strategy to fulfil its obligations to the CBD. The latest strategy covers the period 2014–20.

Community forests

In 1978, the Panchayat Forests and Panchayat Protected Forests regulations were introduced, enabling the Forest Department to return control and ownership of forests to local communities. By June 2013, more than half of

Kanchenjunga Conservation Area. (*Carol Inskipp*)

Nepal's forest area was directly managed as community forests throughout nearly all of the country. Communities are encouraged to protect forest resources and to plant trees on unproductive land, and are being trained in biodiversity monitoring. This approach has proven to be an effective means of conserving forests and biodiversity in some areas, especially where pressures on forests are high.

Wetland management

The Ramsar Convention on Wetlands of International Importance especially as Waterfowl Habitat was ratified by Nepal in December 1987. The Ramsar Convention is an intergovernmental treaty that provides the framework for national action and international cooperation for the conservation and wise use of wetlands and their resources.

In 2003 the Nepal government formulated a National Wetland Policy. This was replaced by the National Wetland Policy 2012. This policy's objectives are to conserve biodiversity and protect the environment via conservation of wetlands, involving local people in the management of wetlands, and conservation, rehabilitation and effective management of wetland areas; supporting the wellbeing of wetland dependent communities; and enhancing the knowledge and capacity of stakeholders, along with maintaining good governance in management of wetland areas.

By 2015 the Nepal government had designated nine Ramsar Sites: Koshi Tappu Wildlife Reserve and Koshi Barrage, Jagdishpur Reservoir, Ghodaghodi Lake complex, Rara Lake, Gokyo and associated lakes, Gosaikunda and associated lakes, Phoksundo, Beeshazar and associated lakes, and Mai Pokhari. However, there are an additional six wetlands within IBAs that could also qualify.

The National Lake Conservation Development Committee was formed in 2006 with the objectives of conserving Nepal's lakes, resolving conflicts, making policy recommendations, and taking responsibility for national and international coordination on issues relating to lakes.

Ghodaghodi Lake, a Ramsar Site and IBA, far west Nepal. (*Hem Sagar Baral*)

Grassland protection

At the time of writing, there are no management guidelines or regulations for bird conservation in lowland grasslands. However, grassland management guidelines will be a major output of a Zoological Society of London/DNPWC/NTNC project in Sukla Phanta Wildlife Reserve, which started in June 2015 and was funded by the Darwin Initiative, UK. Currently the DNPWC allows local people to harvest grasses for a limited period each year within lowland protected areas (Chitwan, Bardia, Sukla Phanta and Koshi Tappu).

Species protection

At the time of writing only nine bird species are protected by law. The protected animal list dates back more than 40 years and is therefore very out of date. There is an urgent need to revise and expand this list.

Improving livelihoods

In recent years there have been many valuable projects aiming to improve the livelihoods of local communities and thereby reduce their dependency on resources obtained from protected areas. Examples include projects in Koshi Tappu Wildlife Reserve, Jagdishpur Reservoir, Ghodaghodi Lake and Bardia National Park, carried out by Bird Conservation Nepal, and numerous projects initiated by the National Trust for Nature Conservation and the World Wide Fund for Nature Nepal Program.

Raising awareness

Government of Nepal ministries, including the Department of National Parks and Wildlife Conservation, have been implementing targeted awareness-raising programmes using TV and radio, regular publications, training, visits, and study tours. The Department of Forests is using different media (including television and radio) to raise awareness of forest fires, uncontrolled grazing, and afforestation.

Many NGOs and individuals are working to change local people's attitudes towards biodiversity by working with them to recognise the importance of conserving biodiversity for their own livelihoods and wellbeing. Bird Conservation Nepal, Bird Education Society, Himalayan Nature and also Raju Acharya's widespread campaign to raise awareness of the plight of owls in Nepal are important examples of these efforts. Methods used include television and radio programmes, public awareness campaigns on special days such as World Environment Day (5 June), National Conservation Day (23 September), World Wetlands Day (2 February), World Migratory Bird Day (10–11 May) and International Vulture Awareness Day (first Saturday of September), street theatre, exhibitions, information boards, and distribution of brochures and newsletters.

There is growing concern amongst the numerous young Nepalese working in the field of conservation, resulting in many critical management and awareness issues being addressed. The phenomenal growth in the number of people interested in bird conservation, especially since the first edition of this book in 2000, is most encouraging. Nepal may have lost a good deal of its native biota, and will doubtless lose more but, nonetheless, it has a good chance of saving much of what remains.

Spiny Babbler (*Jyotendra Jyu Thakuri*)

NATIONAL ORGANISATIONS

Department of National Parks and Wildlife Conservation (DNPWC)

PO Box 860, Kathmandu. www.dnpwc.gov.np

The DNPWC is the Government authority responsible for wildlife conservation. It manages all of Nepal's protected areas except for some Conservation Areas. The overall goal of the DNPWC is to conserve wildlife and outstanding landscapes of ecological importance for the wellbeing of the people.

The primary objective of the Department is to conserve the country's major representative ecosystems, unique natural and cultural heritage, and protect valuable and endangered wildlife species.

The specific objectives of the Department are to:

- conserve rare and endangered wildlife, including floral and faunal diversity, by maintaining representative ecosystems;
- conserve and manage outstanding landscapes of ecological importance;
- support the livelihood of local people through buffer zone and conservation area management programmes; and
- promote ecotourism consistent with biodiversity conservation.

Bird Conservation Nepal (BCN)

PO Box 12465, Kathmandu. www.birdlifenepal.org

Established in 1982, BCN is the leading organisation in Nepal focused on the conservation of birds, their habitats and sites. It seeks to promote interest in birds amongst the general public, encourage research on birds and identify major threats to birds' continued survival. As a result, BCN is the foremost scientific authority providing accurate information on birds and their habitats throughout Nepal. BCN provides scientific data and expertise on birds for the government of Nepal through the DNPWC, and works closely for bird and biodiversity conservation throughout the country.

BCN is a membership-based organisation with a Founder President, patrons, life members, friends of BCN and several supporters. The membership provides strength to the society and is drawn from people of all walks of life from students, professionals and conservationists. Members act collectively to set the organisation's strategic agenda.

BCN is committed to showing the value of birds and their special relationship with people. As such, BCN strongly advocates the need for peoples' participation as future stewards to attain long-term conservation goals.

As the Nepalese Partner of BirdLife International, a network of more than 120 organisations around the world, BCN also works on a worldwide agenda to conserve the world's birds and their habitats.

Regular Saturday birdwatching trips are organised to various bird habitats in the Kathmandu Valley.

Bird Education Society (BES)

BES, Sauraha. www.besnepal.org

BES is a grassroots organisation concerned with the conservation of birds and wildlife in Nepal. BES undertakes awareness programmes in villages and educational activities in schools, as well as bird surveys and research programmes across Nepal. BES was established in 1994 with the objective to create awareness of bird conservation among schoolchildren and far-flung village communities through bird surveys, studies, publications and public meetings. BES looks forward to building a stronger, vibrant society through partnership with all conservation organisations that safeguard birds and wildlife in Nepal.

BES's mission is to:

- generate conservation awareness by organising activities related to the conservation of birds and their habitat in both schools and communities;
- conduct bird surveys and publish reports based on the information gathered;
- discourage illegal activities such as hunting, trapping and poisoning that threaten bird populations; and
- emphasise and support activities related to environmental protection and wildlife conservation.

Friends of Bird (FoB)

PO Box 13248, Kathmandu. www.friendsofbird.net.np

FoB is a non-profit and informal forum for bird lovers and enthusiasts through which members can discuss and share their knowledge, information and experiences about birdwatching. Saturday field outings are held in the Kathmandu Valley every fortnight to complement those held by BCN.

Himalayan Nature (HN)

PO Box 10918, Kathmandu. www.himalayannature.org

Founded in 2000, Himalayan Nature is a conservation research institute, initiating scientific research on Himalayan floral and faunal diversity and the broader environment. It is a non-profit charity actively working on emerging issues in the conservation of natural resources, and livelihood improvement of people in the Himalayan region.

Himalayan Nature (HN) is a science-based organisation and takes an independent view of biodiversity conservation issues. HN runs Koshi Bird Observatory, north of Koshi Tappu Wildlife Reserve, where aims include providing field-based training to local youths for bird and biodiversity monitoring and conservation. HN also runs the Nepal Raptor Migration Project along the Thoolakharka raptor migration corridor to the south of Annapurna Himalaya as a long-term project.

HN's main objectives are to:

- prioritise and implement biodiversity conservation programmes that provide tangible benefits to local people and improve their living conditions;
- conduct scientific and participatory research on flora, fauna and ecosystem dynamics;
- promote meaningful participation and awareness of local people in biodiversity research and conservation;
- foster dialogue, networking and partnership among local, national and international stakeholders to deal with biodiversity issues at a landscape level; and
- promote eco-friendly activities in the region, e.g. eco-tourism.

National Trust for Nature Conservation (NTNC)

PO Box 3712, Khumaltar, Lalitpur, Kathmandu. www.ntnc.org.np

The NTNC was established in 1982 as an autonomous and not-for-profit organisation, mandated to work in the field of nature conservation in Nepal. The trust's mission statement is to promote, conserve and manage nature in all its diversity, balancing human needs with the environment on a sustainable basis for posterity – ensuring maximum community participation with due cognisance of the linkages between economics, environment and ethics through a process in which people are both the principal actors and beneficiaries. Currently, the trust's projects are divided into three geographical areas – the lowlands, the mid-hills (Kathmandu Valley) and high mountains. The trust's activities in the lowlands are based in and around Chitwan National Park, Bardia National Park and Sukla Phanta Wildlife Reserve, located in the central, western and far-western regions of Nepal, through the Biodiversity Conservation Center in Chitwan, the Bardia Conservation Program in Bardia and the Sukla Phanta Conservation Program in Kanchanpur. The Annapurna Conservation Area Project, Manaslu Conservation Area Project and Gaurishankar Conservation Area Project are three protected areas managed by the trust in the mountain region. The Central Zoo is the trust's only project in the Kathmandu Valley. The trust has also established an Energy and Climate Change Unit to address the emerging issues of climate change through mitigation and adaptation, and renewable energy technologies. The trust works closely with many international organisations.

Friends of Nature (FON Nepal)

PO Box 23491, Sundhara, Kathmandu. www.fonnepal.org
Email: info@fonnepal.org, naturesfren@yahoo.com

FON Nepal is a youth-led, non-profitmaking, non-political organisation working in the field of environment and biodiversity conservation. It conducts research on wildlife and other components of biodiversity in unexplored areas to assess conservation problems, while also developing innovative solutions and implementing these.
It promotes 'green school', 'capacity building' and 'environmental education' programmes for university graduates and schools students. FON Nepal holds an annual 'Nepal Owl Festival' (first week of February) to raise conservation awareness of the plight of owls and their benefits to communities and the environment.

Pokhara Bird Society

PO Box 163, Lakeside-6, Pokhara, Kaski district. Email: pokharabirdsociety@gmail.com
Pokhara Bird Society is a group of avid birders, photographers and naturalists based in Pokhara and nearby. It aims to create awareness amongst local stakeholders about the conservation of birds and their habitats. The society also monitors the current status of birds in the region.

Koshi Bird Society (KBS)

Email: koshibirdsociety@gmail.com
KBS is the local birdwatching group at Koshi. It focuses on the conservation of birds, their habitats and sites.

Bardia Nature Conservation Club (BNCC)

Thakurdwara 6, Bardia district. Email: bncc_bardianatureinfo@yahoo.com
BNCC aims to increase awareness of conservation and biodiversity in and around Bardia National Park through social work and by surveying the flora and fauna.

Bird Conservation Network

Ghodaghodi Municipality-1, Sukhad, Kailali district. Email: drchaudhary11@gmail.com
The Bird Conservation Network is based at Ghodaghodi Tal, Kailali district. Local community members of this organisation lead bird conservation and awareness on Ghodaghodi and associated lakes in the Kailali District.

Biodiversity Conservation Society Nepal (BIOCOS NEPAL)

PO Box 12834, Dhobighat, Ringroad, Lalitpur. E-mail: info@biocosnepal.org.np www.biocosnepal.org.np/
BIOCOS NEPAL aims to provide more effort in biological resources conservation. Major emphasis is given to participatory conservation and to species management so that conservation can be effective.

INTERNATIONAL ORGANISATIONS

BirdLife International

The David Attenborough Building, Pembroke Street, Cambridge, CB2 3QZ. www.birdlife.org
BirdLife International (formerly the International Council for Bird Preservation) is now the world's leading authority on the status of the world's birds, their habitats and the urgent problems facing them.

Wetlands International – South Asia Office

A-25, Second Floor, Defence Colony, Delhi 110024, India. www.south-asia.wetlands.org
Wetlands International promotes the protection and sustainable utilisation of wetlands and wetland resources. Its mission is to sustain and restore wetlands, their resources and biodiversity.

Oriental Bird Club (OBC)

The Lodge, Sandy, Beds. SG19 2DL, UK. www.orientalbirdclub.org
OBC's aims are to:

- encourage an interest in wild birds of the Oriental region and their conservation;
- promote the work of regional bird and nature societies; and
- collate and publish information on Oriental birds

World Pheasant Association, South Central Asia (WPA)

www.pheasant.org.uk/southcentralasia
WPA's objectives are to:

- promote the conservation of galliform species that are rare or in danger of extinction;
- advance the education of the public in the knowledge of such species; and
- conduct studies and research on captive and wild birds, and to publish the results of all such research.

In Nepal, WPA has a special interest in Pipar and Santel in the Annapurna Conservation Area.

ZSL Nepal Programme (ZSL)

PO Box 5867, Uttar Dhoka Marg, Lazimpat, Kathmandu. www.zsl.org

The Zoological Society of London has a long-standing relationship with Nepal, having provided technical advisers, supported capacity building and implemented joint conservation projects in the field for well over two decades. The ZSL Nepal Programme was launched in 2014 and is working for the conservation of key threatened species and ecosystems of Nepal; strengthening capacity of national partners and communities; conservation of Evolutionary Distinct and Globally Endangered (EDGE) species, and providing policy support to the government of Nepal.

WWF Nepal Program

PO Box 7660, Baluwatar, Kathmandu. Email: info@wwfnepal.org www.wwfnepal.org

The World Wide Fund for Nature (WWF) began its support to the government of Nepal in 1967 with the protection of species and their habitat. In the 1980s, integrated conservation development projects paved the way for people-centred conservation. Since the late 1990s and 2000, the historic landscape-level approach for conservation was adopted that spread across ecoregions or ecoregion complexes in adjoining countries. WWF Nepal strives to attain greater results in the field of biodiversity conservation against the backdrop of climate change in the coming period. It puts community benefit and community participation at the centre of conservation interventions so as to ensure programme sustainability and demonstrate conservation impact.

International Center for Integrated Mountain Development (ICIMOD)

Khumaltar, Lalitpur, PO Box 3226, Kathmandu. www.icimod.org

ICIMOD is the international centre devoted to integrated mountain development. Its mission is to enable sustainable and resilient mountain development for improved and equitable livelihoods through knowledge and regional cooperation.

IUCN – The International Union for Conservation of Nature

Kupondole, Lalitpur, PO Box 3923, Kathmandu. www.iucn.org/nepal

IUCN has been assisting conservation efforts in Nepal since the late 1960s. With strong support from civil society, government and donors, IUCN has been able to contribute greatly in linking conservation with better livelihoods, mobilising local communities and generating tangible results to promote biodiversity conservation, environmental justice and sustainable livelihoods in Nepal.

Fire-tailed Myzornis. (*Gunjan Arora*)

REFERENCES

Acharya, R. & Ghimirey, Y. (in press) Limi Valley: a new birding hotspot for Nepal. *BirdingASIA*.

Baral, H. S. (2001) *Community structure and habitat associations of lowland grassland birds in Nepal.* Ph.D. thesis. University of Amsterdam, the Netherlands.

Baral, H. S. (2002) Invasive weed threatens protected area. *Danphe* 11(3): 10–11.

Baral, H. S., Basnet, S., Chaudhary, B., Chaudhary, H., Giri, T. & GC, S. (2007) A new subspecies of Rufous-vented Prinia *Prinia burnesii* (Aves: Cisticolidae) from Nepal. *Danphe* 16(4): 1–10. http://www.birdlifenepal.org/publication.php

Baral, H. S., Basnet, S., Chaudhary, B., Chaudhary, H., Giri, T. & GC, S. (2008) A substitute name for *Prinia burnesii nipalensis. Danphe* (17(1): 1. http://www.birdlifenepal.org/publication.php

Baral, H. S. & Inskipp, C. (2004) *The state of Nepal's birds 2004.* Department of National Parks and Wildlife Conservation, Bird Conservation Nepal & IUCN Nepal, Kathmandu.

Bhusal, K. (2010) Fuelwood as a source of energy in Nepal. http://forestrycomponents.blogspot.co.uk/2010/03/fuel-wood-as-source-of-energy-in-nepal.html

Bird Conservation Nepal and Department of National Parks and Wildlife Conservation (2011) *State of Nepal's birds 2010.* Kathmandu.

Bird Conservation Nepal and Department of National Parks and Wildlife Conservation (in prep.) *Important Bird and Biodiversity Areas: key sites for conservation.* Kathmandu.

BirdLife international (2015) www.birdlife.org

Choudhary, H. (2003) One-day bird survey at Koshi Tappu Wildlife Reserve. *Danphe* 12(1/2): 6.

Cox, J. H. (2010) Confirmation of Long-billed Wren Babbler *Rimator malacoptilus* in Nepal. *Forktail* 26: 134–136.

Cox, J. H. & Sherpa, C. (1998) Long-billed Wren Babbler *Rimator malacoptilus*: a new species for Nepal. *Ibisbill* 1: 118–120.

Dahal, B. R. (2007) Effects of Water Hyacinth *Eichhornia crassipes* on aquatic birds at Koshi Tappu Wildlife Reserve, south-east Nepal. *Danphe* 16(1): 64–65. http://www.birdlifenepal.org/publication.php

Dobremez, J. F. (1976) *Le Népal écologie et biogéographie.* Centre National de la Recherche, Paris.

Fleming, R. L. Fleming, R. L. & Bangdel, L. S. (1984) *Birds of Nepal.* Third edn. Avalok, Kathmandu.

Food and Agriculture Organization (2010) *Global forest resources assessment.* Forest Resources Assessment Program, Rome. http://www.fao.org/docrep/013/i1757e/i1757e.pdf

Gilbert, M., Watson, R. T., Ahmed, S., Asim, M. & Johnson, J. A. (2007) Vulture restaurants and their role in reducing diclofenac exposure in Asian vultures. *Bird Conserv. Intern.* 17: 63–77.

Grant, P. & Mullarney, K. (1988–89) The new approach to bird identification *Birding World* 1: 266–267, 350–354, 387–391, 422–425, 2: 15–17, 65–68, 97–99, 132–134, 180–184, 204–206.

Gray, J. R. (1863) *Catalogue of the specimens and drawings of mammals, birds, reptiles and fishes of Nepal and Tibet presented by B. H. Hodgson Esq. to the British Museum.* Second edn. London.

Grimmett, R., Inskipp, C. & Inskipp, T. (1998) *Birds of the Indian Subcontinent.* Christopher Helm, London.

Inskipp, C. & Baral, H. S. (2011) Potential impacts of agriculture on Nepal bird. *Our Nature* (2010) 8: 270–312. http://www.nepjol.info/index.php/ON

Inskipp, C. & Inskipp, T. (1991) *A guide to the birds of Nepal.* Second edn. Christopher Helm, London.

Inskipp, T., Duckworth, W. & Lindsey, N. (1996) *An annotated checklist of the birds of the Oriental region.* Oriental Bird Club, Sandy.

Karki, R. & Choudhary, B. (1997) Wedge-billed Wren Babbler *Sphenocichla humei*: a new species for Nepal. *Danphe* 6(3): 5.

Martens, J. & Eck, S. (1991) *Pnoepyga immaculata* n. sp. eine neue bodenbewohnende Timalie aus dem Nepal – Himalaya. *J. Orn.* 132: 179–198.

Ministry of Forests and Soil Conservation (2014) *Nepal national biodiversity strategy and action plan 2014–2020.* Kathmandu. https://www.cbd.int/doc/world/np/np-nbsap-v2-en.pdf

Oaks, J. L., Gilbert, M., Virani, M. Z., Watson, R. T., Meteyer, C. U., Rideout, B. A., Shivaprasad, H. L., Ahmed, S., Chaudhary, M. J. I., Arshad, M., Mahamood, S., Ali, A. & Khan, A. A. (2004) Diclofenac residues as the cause of vulture population decline in Pakistan. *Nature* doi: 10.1038/nature02317.

Palikhe, B. R. (2005) Pesticide management in Nepal. In view of Code of Conduct. Paper presented at the Regional Workshop on International Code of Conduct on the Distribution and Use of Pesticides: Implementation, Monitoring and Observance. Bangkok, Thailand, 26–28 July 2005.

Paudel, S. (2009a) Study on threats to Sarus Crane *Grus antigone antigone* in farmlands in Lumbini, an Important Bird Area of Nepal – AEC/OBC Award 2007. *BirdingASIA* 12: 9–10.

Paudel, S. (2009b) Study on threats to Sarus Crane (*Grus antigone antigone*) in farmlands in Lumbini, an Important Bird Area of Nepal. Final report to Oriental Bird Club, Sandy.

Rasmussen, P. C. & Anderton, J. C. (2012) *Birds of South Asia. The Ripley Guide.* Smithsonian Institution, Washington DC, Michigan State University, Ann Arbor, and Lynx Edicions, Barcelona.

Stattersfield, A. J., Crosby M. J., Long, A. J. & Wege, D. C. (1998) *Endemic Bird Areas of the world: priorities for biodiversity conservation.* BirdLife International, Cambridge.

Subedi, T (2015) East to west migration of Steppe Eagle *Aquila nipalensis* and other raptors in Nepal (abundance, timing and class determinations). Report to National Birds of Prey Trust, UK. https://www.dropbox.com/s/twqn0d7pmhln50j/Final%20report_Raptor%20Migration%20Study_Thoolakharka%20Nepal2014%20%20.pdf?dl=0

Swan, G., Naidoo, V., Cuthbert, R., Green, R. E., Pain, D. J., Swarup, D., Prakash, V., Taggart, M., Bekker, L., Das, D., Diekmann, J., Diekmann, M., Killian, E., Meharg, A., Patra, R. C., Saini, M. & Wolter, K. (2006) Removing the threat of diclofenac to critically endangered Asian vultures. *PLoS Biol.* 43: 395–402.

Thapa, I. & Thakuri, J. J. (2009) Study of wild bird trade issues in Nepal. Report by BCN to the World Parrot Trust and WWF Nepal, Kathmandu.

World Bank (2008) *Nepal – Country Environmental Analysis: Strengthening Institutions and Management Systems for Enhanced Environmental Governance.* World Bank, Washington DC.

Greater Flameback (*Sagar Giri*)

FAMILY SUMMARIES

PARTRIDGES, PHEASANTS AND ALLIES Phasianidae

These heavy-bodied birds feed and nest on the ground, but many species roost in trees at night. They are good runners, often preferring to escape on foot rather than taking to the air. Their flight is powerful and fast, but except in the case of the migratory quail, it cannot be sustained for long periods. Typically, they forage by scratching the ground with strong feet to expose food hidden among dead leaves or in the soil. They mainly eat seeds, fruit, buds, roots and leaves, complemented by invertebrates.

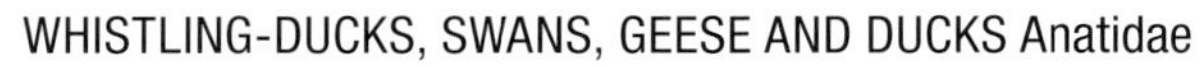

WHISTLING-DUCKS, SWANS, GEESE AND DUCKS Anatidae

Aquatic and highly gregarious, typically migrating, feeding, roosting and resting together, often in mixed flocks. Most species are chiefly vegetarian when adult, feeding on seeds, algae, plants and roots, often supplemented by aquatic invertebrates. Their main foraging methods are diving, surface-feeding or dabbling, and grazing. They also upend, wade, filter and sieve water and debris for food and probe with the bill. They have a direct flight with sustained fast wingbeats, and characteristically they often fly in V-formation.

GREBES Podicipedidae

Aquatic birds adapted for diving from the surface and swimming under water to catch fish and aquatic invertebrates. Their strong legs are placed near the rear of their almost tailless body, and the feet are lobed. In flight grebes have an elongated appearance, with the neck extended, and feet hanging lower than the humped back. They usually feed singly, but may form loose congregations in the non-breeding season.

STORKS Ciconiidae

Large or very large birds with long bills, necks and legs, long and broad wings and short tails. In flight, the legs are extended and the neck is outstretched. They have a powerful, slow-flapping flight and frequently soar for long periods, often at great heights. They capture fish, frogs, snakes, lizards, large insects, crustaceans and molluscs while walking slowly in marshes, at the edge of lakes and rivers, and in grassland.

FLAMINGOS Phoenicopteridae

Large wading birds with long necks, very long legs, webbed feet and pink plumage. The bill is highly specialised for filter-feeding. Flamingos often occur in huge numbers and are found mainly on salt lakes and lagoons.

IBISES AND SPOONBILLS Threskiornithidae

Large birds with long necks and legs, partly webbed feet and long broad wings. Ibises have long, decurved bills and forage by probing in shallow water, mud and grass. Spoonbills have long spatulate bills, and catch floating prey in shallow water; there is only one spoonbill species in Nepal.

HERONS AND BITTERNS Ardeidae

Medium-sized to large birds with long legs for wading. The diurnal herons have slender bodies and long heads and necks; the night herons are more squat, with shorter necks and legs. They fly with leisurely flaps, with the legs outstretched and projecting beyond the tail, and nearly always with neck and head drawn back. They frequent marshes and the shores of lakes and rivers. Typically, herons feed by standing motionless at the water's edge waiting for prey to swim within reach, or by slow stalking in shallow water or on land.

Bitterns usually skulk in reedbeds, although occasionally one may forage in the open, and they can clamber about reed stems with agility. Normally they are solitary and crepuscular, and are most often seen flying low over reedbeds with slow wing-beats, soon dropping into cover again. When in danger bitterns freeze, pointing the head and neck upward and compressing their feathers so that the whole body appears elongated. The bitterns are characterised by their booming territorial calls. Herons and bitterns feed on a wide variety of aquatic prey.

PELICANS Pelecanidae

Large aquatic, gregarious fish-eating birds. The wings are long and broad, and the tail is short and rounded. They have characteristic long, straight, flattened bills, hooked at the tip, and with a large expandable pouch suspended beneath the lower mandible. Many pelicans often fish cooperatively by swimming forward in a semicircular formation, driving the fish into shallow water; each bird then scoops up fish from the water into its pouch before swallowing the food. Pelicans fly either in V-formation or in lines, and often soar for considerable periods in thermals. They are powerful fliers, proceeding by steady flaps and with the head drawn back between the shoulders. When swimming the closed wings are typically held above the back.

DARTERS Anhingidae

Large aquatic birds adapted for hunting fish underwater. Darters have long slender necks and heads, long wings and very long tails. Only one species in the family occurs in the region.

CORMORANTS Phalacrocoracidae

Medium-sized to large aquatic birds. They are long-necked, with hook-tipped bills of moderate length and long, stiff tails. Cormorants swim with the body low in the water, with the neck straight and the head and bill pointing a little upwards. They eat mainly fish, which are caught by underwater pursuit. In flight, the neck is extended and the head is held slightly above the horizontal. Typically they often perch for long periods in upright posture with spread wings and tail on trees, posts or rocks.

FALCONS Falconidae

Small to medium-sized birds of prey which resemble the Accipitridae in having hooked bills, sharp curved talons, and remarkable powers of sight and flight. Like other raptors they are mainly diurnal, although a few are crepuscular. Some falcons kill flying birds in a surprise attack, often by stooping at great speed (e.g. Peregrine); others hover and then swoop on prey on the ground (e.g. Common Kestrel) and several species hawk insects in flight (e.g. Eurasian Hobby).

OSPREY, HAWKS, EAGLES, HARRIERS AND VULTURES etc Accipitridae

A large and varied family of raptors, ranging from the Besra to the huge Griffon Vulture. In most species, the vultures being an exception, the female is larger than the male, and is often duller and brownish. The Accipitridae feed on mammals, birds, reptiles, amphibians, fish, crabs, molluscs and insects – dead or alive. All have hooked, sharp-tipped bills and very acute sight, and all except the vultures have powerful feet with long curved claws. They frequent all habitat types, ranging from dense forests, deserts and mountains to fresh waters.

BUSTARDS Otididae

Medium-sized to large terrestrial birds that inhabit grasslands, semi desert, and desert. They have fairly long legs, stout bodies, long necks, and crests and neck plumes, which are

exhibited in display. The wings are broad and long, and in flight the neck is outstretched. Their flight is powerful and can be very fast. When feeding, bustards have a steady, deliberate gait. They are more or less omnivorous, and feed opportunistically on large insects, such as grasshoppers and locusts, young birds, shoots, leaves, seeds and fruits. Males perform elaborate and spectacular displays in the breeding season.

RAILS, CRAKES, GALLINULES AND COOTS Rallidae

Small to medium-sized birds, with moderate to long legs for wading and short rounded wings. With the exception of the Common Moorhen and Eurasian Coot, which spend much time swimming in the open, rails are mainly terrestrial. Many occur in marshes. They fly reluctantly and feebly, with legs dangling, for a short distance and then drop into cover again. Most are heard more often than seen, and are generally voluble at dusk and at night. Their calls consist of strident or raucous repeated notes. They eat insects, crustaceans, amphibians, fish and vegetable matter.

BUTTONQUAILS Turnicidae

Small, plump terrestrial birds. They are found in a wide variety of habitats having a dry, often sandy substrate and low ground cover under which they can readily run or walk. Buttonquails are very secretive and fly with great reluctance, with weak whirring beats low over the ground, dropping quickly into cover. They feed on grass and weed seeds, grain, greenery and small insects, picking food from the ground surface, or scratching with the feet.

CRANES Gruidae

Stately long-necked, long-legged birds with tapering bodies, and long inner secondaries which hang over the tail. The flight is powerful, with the head and neck extended forwards and legs and feet stretched out behind. Flocks of cranes often fly in V-formation; they sometimes soar at considerable heights. Most cranes are gregarious outside the breeding season, and flocks are often very noisy. Cranes have a characteristic resonant and far-reaching musical trumpet-like call. A wide variety of plant and animal food is taken. The bill is used to probe and dig for plant roots and to graze and glean vegetable material above the ground. Both sexes have a spectacular and beautiful dance that takes place throughout the year.

THICK-KNEES Burhinidae

Medium-sized to large waders, mainly crepuscular or nocturnal, and with cryptically patterned plumage. They eat invertebrates and small animals.

OYSTERCATCHERS Haematopodidae

Oystercatchers are waders that usually inhabit the seashore and are only vagrants inland. They have all-black or black and white plumage. Their bills are long, stout, orange-red and adapted for opening shells of bivalve molluscs. Eurasian Oystercatcher is the only family member recorded in the region.

PLOVERS AND LAPWINGS Charadriidae

Plovers and lapwings are small to medium-sized waders with rounded heads, short necks and short bills. Typically, they forage by running in short spurts, pausing and standing erect, then stooping to pick up invertebrate prey. Their flight is swift and direct.

IBISBILL Ibidorhynchidae

Ibisbill is a wader with a distinctive red decurved bill used for probing for small aquatic animals among stones in the shallows of mountain rivers. It is the only species in its family.

STILTS AND AVOCETS Recurvirostridae

Stilts and avocets are waders that have characteristic long bills, and longer legs in proportion to the body than any other birds except flamingos. They inhabit marshes, lakes and pools. Pied Avocet is the only avocet species in the region.

PAINTED-SNIPES Rostratulidae

Waders that frequent marshes and superficially resemble snipes, but have spectacular plumages. Greater Painted-snipe is the only species in the family recorded in the region.

JACANAS Jacanidae

Jacanas characteristically have very long toes, which enable them to walk over floating vegetation. They inhabit freshwater lakes, ponds and marshes.

SNIPES, CURLEWS AND SANDPIPERS Scolopacidae

Woodcocks and snipes are small to medium-sized waders with very long bills, fairly long legs and cryptically patterned plumages. They feed mainly by probing with their bills in soft substrates and also by picking from the surface. Their diet consists mostly of small invertebrates. If approached, they usually crouch at first on the ground and 'freeze', preferring to rely on their protective plumage pattern to escape detection. They generally inhabit marshy ground. Godwits and curlews are wading birds with quite long to very long legs and a long bill. They feed on small aquatic invertebrates. Sandpipers and stints are small to medium-sized, rather plump waders with medium to longish bill, and medium-long legs.

PRATINCOLES AND COURSERS Glareolidae

Coursers and pratincoles have arched and pointed bills, wide gapes and long, pointed wings. Coursers are long-legged and resemble plovers; they feed on the ground. Most pratincoles are short-legged; they catch most of their prey in the air, although they also feed on the ground. All pratincoles live near water, whereas coursers frequent dry grassland and dry stony areas.

GULLS, TERNS AND SKIMMERS Laridae

Gulls are medium-sized to large birds with relatively long, narrow wings, usually a stout bill, moderately long legs and webbed feet. Immatures are brownish and cryptically patterned. In flight, gulls are graceful and soar easily in updraughts. All species swim buoyantly and well. They are highly adaptable, and most species are opportunistic feeders with a varied diet, including invertebrates. Most species are gregarious.

Terns are small to medium-sized aerial birds with gull-like bodies, but are generally more delicately built. The wings are long and pointed, typically narrower than those of the gulls, and the flight is buoyant and graceful. Terns are highly vocal and most species are gregarious. Two groups of terns occur in Nepal: the *Sterna* terns and the *Chlidonias* or marsh terns. The *Sterna* terns generally have deeply forked tails. They mainly eat small fish and crabs caught by hovering and then plunge-diving from the air, often submerging completely, but they also pick prey from the surface. Marsh terns lack a prominent tail-fork and, compared with *Sterna* terns, are smaller, more compact and short-tailed, and have a more erratic and rather stiff-winged flight. Typically, marsh terns hawk insects or swoop down to pick small prey from the water surface. Noddies are pelagic terns, with distinctive wedge-shaped and slightly forked tails.

Skimmers are distinguished by their long, strong scissor-like bills with elongated lower mandibles. They feed by skimming the water surface with the bill open and lower mandible partly immersed to snap up fish. Indian Skimmer is the only species recorded in the region.

SANDGROUSE Pteroclidae

Cryptically patterned terrestrial birds resembling the pigeons in size and shape. The wings of sandgrouse are long and pointed. Most sandgrouse are wary and, when disturbed, rise with a clatter of wings, flying off rapidly and directly with fast and regular wing-beats. They walk and run well, foraging mainly for small hard seeds picked up from the ground and sometimes also eating green leaves, shoots, fruits and berries, small bulbs and insects. They need to drink every day, and will sometimes travel over long distances to waterholes. Most sandgrouse have regular drinking times which are characteristic of each species, and they often visit traditional watering places, sometimes gathering in quite large numbers. Most species are gregarious except during the breeding season.

PIGEONS AND DOVES Columbidae

Pigeons and doves have stout compact bodies, rather short necks, and small heads and bills. Their flight is swift and direct, with fast wing-beats. Most species are gregarious outside the breeding season. Seeds, fruits, buds and leaves form their main diet, but many species also eat small invertebrates. They have soft plaintive cooing or booming voices that are often monotonously repeated.

PARROTS AND PARAKEETS Psittacidae

Parrots have short necks and short, stout hooked bills with the upper mandible strongly curved and overlapping the lower mandible. Most parrots are noisy and highly gregarious. They associate in family parties and small flocks and gather in large numbers at concentrations of food, such as paddyfields. Their diet is almost entirely vegetarian: fruit, seeds, buds, nectar and pollen. The flight of *Psittacula* parakeets is swift, powerful and direct. The hanging parrots are much smaller with short tails lacking streamers. They habitually sleep upside-down.

CUCKOOS, MALKOHAS AND COUCALS Cuculidae

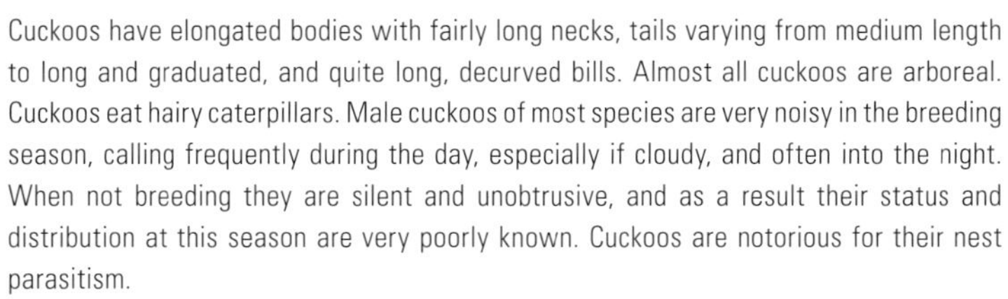

Cuckoos have elongated bodies with fairly long necks, tails varying from medium length to long and graduated, and quite long, decurved bills. Almost all cuckoos are arboreal. Cuckoos eat hairy caterpillars. Male cuckoos of most species are very noisy in the breeding season, calling frequently during the day, especially if cloudy, and often into the night. When not breeding they are silent and unobtrusive, and as a result their status and distribution at this season are very poorly known. Cuckoos are notorious for their nest parasitism.

Malkohas are larger than cuckoos, plumper-bodied with stouter bills and very long graduated tails. They are usually seen singly or in pairs in the middle storey. Malkohas raise their own young.

Coucals are large, skulking birds with long graduated tails and weak flight. They are terrestrial, frequenting dense undergrowth, bamboo, tall grassland or scrub jungle. Coucals eat small animals and invertebrates.

BARN OWLS Tytonidae; TYPICAL OWLS Strigidae

Owls have large and rounded heads, big forward-facing eyes surrounded by a broad facial disc, and short tails. Most are nocturnal and cryptically coloured and patterned, making them inconspicuous when resting during the day. When hunting, owls either quarter the ground or scan and listen for prey from a perch. Their diet consists of small animals and invertebrates. Owls are usually located by their distinctive and often weird calls, which are diagnostic of the species and advertise their presence and territories.

NIGHTJARS Caprimulgidae

Small to medium-sized birds with long, pointed wings, and gaping mouths with long bristles that help to catch insects in flight. Nightjars are crepuscular and nocturnal in habit, with soft, owl-like, cryptically patterned plumage. By day they perch on the ground or lengthwise on a branch, and are difficult to detect. They eat flying insects that are caught on the wing. Typically, they fly erratically to and fro over and among vegetation, occasionally wheeling, gliding and hovering to pick insects from foliage. Most easily located by the calls.

SWIFTS Apodidae; TREESWIFTS Hemiprocnidae

Swifts have long pointed wings, compact bodies, short bills with a wide gape and very short legs. They spend most of the day swooping and wheeling in the sky with great agility and grace. Typical swift flight is a series of rapid shallow wing-beats interspersed with short glides. They feed entirely in the air, drink and bathe while swooping low over water, and regularly pass the night in the air. Swifts eat mainly tiny insects, caught by flying back and forth among aerial concentrations of these with their large mouths open; they also pursue individual insects.

HOOPOES Upupidae

Hoopoes have a distinctive appearance, with long decurved bills, short legs and rounded wings. They are insectivorous and forage by pecking and probing the ground. Flight is undulating, slow and butterfly-like. Common Hoopoe is the only species in the family recorded in the region.

TROGONS Trogonidae

Brightly coloured, short-necked, medium-sized birds with long tails, short rounded wings and rather short, broad bills. They usually keep singly or in widely separated pairs. Characteristically, they perch almost motionless in upright posture for long periods in the middle or lower storey of dense forests. Trogons are mainly insectivorous and also eat leaves and berries. They capture flying insects on the wing when moving from one vantage point to another, twisting with the agility of a flycatcher.

ROLLERS Coraciidae

Stoutly built, medium-sized birds with large heads and short necks, which mainly eat large insects. Typically, they keep singly or in widely spaced pairs. Flight is buoyant, with rather rapid deliberate wing-beats.

KINGFISHERS Alcedinidae

Small to medium-sized birds, with large heads, long strong bills and short legs. Most kingfishers spend long periods perched singly or in well-separated pairs, watching intently before plunging swiftly downwards to seize prey with bill; they usually return to the same perch. They eat mainly fish, tadpoles and invertebrates; larger species also eat frogs, snakes, crabs, lizards and rodents. Their flight is direct and strong, with rapid wing-beats and often close to the surface.

BEE-EATERS Meropidae

Brightly coloured birds with long decurved bills, pointed wings and very short legs. They catch large flying insects on the wing, by making short, swift sallies like a flycatcher from an exposed perch such as a treetop, branch, post or telegraph wire; insects are pursued in a lively chase with a swift and agile flight. Some species also hawk insects in flight like swallows. Most species are sociable. Their flight is graceful and undulating, a few rapid wing-beats followed by a glide.

HORNBILLS Bucerotidae

Medium-sized to large birds with massive bills with variable-sized casque. Mainly arboreal, feeding chiefly on wild figs *Ficus*, berries and drupes, supplemented by small animals and insects. Flight is powerful and slow, and for most species consists of a few wing-beats followed by a sailing glide with the wing-tips upturned. In all but the smaller species, the wing-beats make a distinctive loud puffing sound audible for some distance. Hornbills often fly one after another in follow-my-leader fashion. Usually found in pairs or small parties, sometimes in flocks of up to 30 or more where food is abundant.

ASIAN BARBETS Ramphastidae

Arboreal, and usually found in the treetops. Despite their bright coloration, they can be very difficult to see, especially when silent, their plumage blending remarkably well with tree foliage. They often sit motionless for long periods. Barbets call persistently and monotonously in the breeding season, sometimes throughout the day; in the non-breeding season they are usually silent. They are chiefly frugivorous, many species favouring figs *Ficus*. Their flight is strong and direct, with deep woodpecker-like undulations.

HONEYGUIDES Indicatoridae

Small, inconspicuous birds that inhabit forest or forest edge. A peculiarity of the family is that they also eat wax, usually from bee combs. They spend long periods perched upright and motionless and feed by clinging to bee combs, often upside-down, and by aerial sallies. Yellow-rumped Honeyguide is the only family member in the region.

WRYNECK, PICULETS AND WOODPECKERS Picidae

Chiefly arboreal, and usually seen clinging to, and climbing up, vertical trunks and lateral branches. Typically, they work up trunks and along branches in jerky spurts, directly or in spirals. Some species feed regularly on the ground, searching mainly for termites and ants. Most species have powerful bills, for boring into wood to extract insects and for excavating nest holes. Woodpeckers feed chiefly on ants, termites, and grubs and pupae of wood-boring beetles. Most woodpeckers also hammer rapidly against tree trunks with their bill, producing a loud rattle, known as 'drumming', which is used to advertise and in defence of their territories. Their flight is strong and direct, with marked undulations. Many species can be located by their characteristic loud calls.

BROADBILLS Eurylaimidae

Small to medium-sized plump birds with rounded wings and short legs, most species having a distinctively broad bill. Typically they inhabit the middle storey of forest and feed mainly on invertebrates gleaned from leaves and branches. Broadbills are active when foraging, but are often unobtrusive and lethargic at other times.

PITTAS Pittidae

Brilliantly coloured, terrestrial forest passerines. They are of medium size, stocky and long-legged, with short square tails, stout bills and an erect carriage. Most of their time is spent foraging for invertebrates on the forest floor, flicking leaves and other vegetation, and probing with their strong bill into leaf litter and damp earth. Pittas usually progress on the ground by long hopping bounds. Typically, they are skulking and are often most easily located by their high-pitched whistling calls or songs. They sing in trees or bushes.

WOODSHRIKES AND ALLIES Tephrodornithidae

Woodshrikes are medium-sized, arboreal, insectivorous passerines. The bill is stout and hooked, the wings are rounded, and the tail is short.

Flycatcher-shrikes are small pied birds with arboreal flycatching habits and an upright stance when perched. Bar-winged Flycatcher-shrike is the only species of this genus recorded in the region.

WOODSWALLOWS Artamidae

Woodswallows are plump birds with long, pointed wings, short tail and legs, and wide gapes. They feed on insects, usually captured in flight, and spend prolonged periods on the wing. They perch close together on a bare branch or wire, and often waggle the tail from side to side.

IORAS Aegithinidae

Ioras are small, lively group of passerines that feed in trees, mainly on insects and especially on caterpillars.

CUCKOOSHRIKES, MINIVETS AND ALLIES Campephagidae

Cuckooshrikes are arboreal, insectivorous birds that usually keep high in the trees. They are of medium size, with long pointed wings, moderately long rounded tails, and an upright carriage when perched.

Minivets are small to medium-sized, brightly coloured passerines with moderately long tails and an upright stance when perched. They are arboreal, and feed on insects by flitting about in the foliage to glean prey from leaves, buds and bark, sometimes hovering in front of a sprig or making short aerial sallies. They usually keep in pairs in the breeding season, and in small parties when not breeding. When feeding and in flight, they continually utter contact calls.

SHRIKES Laniidae

Medium-sized, predatory passerines with strong, stout bills, hooked at the tip of the upper mandible, strong legs and feet, large heads, and long tails with graduated tips. Shrikes search for prey from a vantage point, such as the top of a bush or small tree or post. They swoop down to catch invertebrates or small animals from the ground or in flight. Over long distances their flight is typically undulating. Their calls are harsh, but most have quite musical songs and are good mimics. Shrikes typically inhabit open country with scattered bushes or light scrub.

DRONGOS Dicruridae

Medium-sized passerines with characteristic black and often glossy plumage, long, often deeply forked tails, and a very upright stance when perched. They are mainly arboreal and insectivorous, catching larger-winged insects by aerial sallies from a perch. Usually found singly or in pairs. Their direct flight is swift, strong and undulating. Drongos are rather noisy, and have a varied repertoire of harsh calls and pleasant whistles; some species are good mimics.

ORIOLES Oriolidae

Medium-sized arboreal passerines that usually keep hidden in the leafy canopy. Orioles have beautiful, fluty, whistling songs and harsh grating calls. They are usually seen singly, in pairs or in family parties. Their flight is powerful and undulating, with fast wing-beats. They feed mainly on insects and fruit.

FANTAILS Rhipiduridae

Small, confiding, arboreal birds, perpetually on the move in search of insects. Characteristically, they erect and spread their tails like fans, and droop the wings, while pirouetting and turning from side to side with jerky, restless movements. When foraging, they flit from branch to branch, making frequent aerial sallies after winged insects. They call continually. Fantails are usually found singly or in pairs, and often join mixed hunting parties with other insectivorous birds.

MONARCHS Monarchidae

Most species are small to medium-sized, with long, pointed wings and a medium-length to long tail. They feed mainly on insects. Black-naped Monarch and Asian Paradise-flycatcher are the only two species in the family recorded in the region.

CROWS, MAGPIES AND JAYS Corvidae

These are all robust perching birds which differ considerably from each other in appearance, but which have a number of features in common: a fairly long straight bill, very strong feet and legs, and a tuft of nasal bristles extending over the base of the upper mandible. The sexes are alike or almost alike in plumage. They are strong fliers. Most are gregarious, especially when feeding and roosting. Typically, they are noisy birds, uttering loud and discordant squawks, croaks or screeches. Crows are highly inquisitive and adaptable.

TITS Paridae; LONG-TAILED TITS Aegithalidae

With the exception of Groundpecker, tits and long-tailed tits are small, active, highly acrobatic passerines with short bills and strong feet. Their flight over long distances is undulating. They are mainly insectivorous, although many species also depend on seeds, particularly from trees in winter, and some also eat fruit. They probe bark crevices, search branches and leaves, and frequently hang upside-down from twigs. Tits are chiefly arboreal, but also descend to the ground to feed, hopping about and flicking aside leaves and other debris. They are very gregarious; in the non-breeding season most species join roving flocks of other insectivorous birds. Groundpecker is a thrush-sized terrestrial bird unlike other tits, with strong decurved bill, long legs and an upright stance.

MARTINS AND SWALLOWS Hirundinidae

Gregarious, rather small passerines with a distinctive slender, streamlined body, long, pointed wings and small bills. The long-tailed species are often called swallows, and the shorter-tailed species termed martins. All hawk day-flying insects in swift, agile, sustained flight, sometimes high in the air. Many species have a deeply forked tail, which affords better manoeuvrability. Hirundines catch most of their food while flying in the open. They perch readily on exposed branches and wires.

LARKS Alaudidae

Terrestrial cryptically coloured passerines, generally small-sized, which usually walk and run on the ground and often have a very elongated hindclaw. Their flight is strong and undulating. Larks take a wide variety of food, including insects, molluscs, arthropods, seeds, flowers, buds and leaves. Many species have a melodious song, which is often delivered in a distinctive, steeply climbing or circling aerial display, but also from a conspicuous low perch. They live in a wide range of open habitats, including grassland and cultivation.

WAXWINGS Bombycillidae

Waxwings have soft plumage, crested heads, short broad-based bills and short, strong legs and feet. Outside the breeding season, they are found wherever fruits are available. Only Bohemian Waxwing has been recorded in the region.

BULBULS Pycnonotidae

Medium-sized passerines with soft, fluffy plumage, rather short and rounded wings, medium-long to long tails, slender bills and short, weak legs. Bulbuls feed on berries and other fruits, often supplemented by insects, and sometimes also nectar and buds of trees and shrubs. Many species are noisy, especially when feeding. Typically, bulbuls have a variety of cheerful, loud, chattering, babbling and whistling calls. Most species are gregarious in the non-breeding season.

CISTICOLAS, PRINIAS AND ALLIES Cisticolidae

Cisticolas are a group of tiny, short-tailed, insectivorous passerines. The tail is longer in winter than in summer. They are often found in grassy habitats, and many have aerial displays. Two species occur in Nepal.

Prinias have long, graduated tails that is longer in winter than in summer. Most inhabit grassland, marsh vegetation or scrub. They forage by gleaning insects and spiders from vegetation, and some species also feed on the ground. When perched, the tail is often held cocked and slightly fanned. Their flight is weak and jerky.

Tailorbirds have long, decurved bills, short wings and graduated tails, the latter held characteristically cocked.

Grassbirds are brownish warblers with longish tails. They inhabit damp tall grassland. The males perform song flights in the breeding season.

WARBLERS Sylviidae

Bush warblers are medium-sized warblers with rounded wings and tail that inhabit marshes, grassland and forest undergrowth. They are usually found singly. Bush warblers call frequently, and are usually heard more often than seen. *Cettia* species have surprisingly loud voices, and some can be identified by their distinctive melodious songs. Bush warblers seek insects and spiders by actively flitting and hopping about in vegetation close to the ground. They are reluctant to fly, and usually cover only short distances at low level before dropping into dense cover again. When excited, they flick their wings and tail.

Locustella warblers are very skulking, medium-sized warblers with rounded tails, usually found singly. Characteristically, they keep low down or on the ground among dense vegetation, walking furtively and scurrying off when startled. They fly at low level, flitting between plants, or rather jerkily over longer distances, ending in a sudden dive into cover.

Acrocephalus warblers are medium-sized to large warblers with prominent bills and rounded tails. They usually occur singly. Many species are skulking, typically keeping low down in dense vegetation. Most frequent marshy habitats, and are able to clamber about readily in reeds and other vertical stems of marsh plants. Their songs are harsh and often monotonous.

Iduna warblers are medium-sized warblers with large bills, square-ended tails and a distinctive domed head shape with a rather sloping forehead and peaked crown. Their songs are harsh and varied. They clamber about vegetation with a rather clumsy action.

Leaf warblers are rather small, slim and short-billed warblers. Useful identification features are voice, strength of supercilium, colour of underparts, rump, bill and legs, and presence or absence of wing-bars, of coronal bands or of white on the tail. The coloration of upperparts and underparts and the presence or prominence of wing-bars are affected by wear. Leaf warblers are fast moving and restless, hopping and creeping about actively and often flicking the wings. They mostly glean small insects and spiders from foliage, twigs and branches, often first disturbing prey by hovering and fluttering; they also make short fly-catching sallies.

Sylvia warblers are small to medium-sized passerines with fine bills. Typically, they inhabit bushes and scrub and feed chiefly by gleaning insects from foliage and twigs; they sometimes also consume berries in autumn and winter.

BABBLERS Timaliidae

A large and diverse group of small to medium-sized passerines. They have soft, loose plumage, short or fairly short wings, and strong feet and legs. The sexes are alike in most species. Members of this tribe associate in flocks outside the breeding season, and some species do so throughout the year. Babbler flocks are frequently a component of mixed-species feeding parties. Most babblers have a wide range of chatters, rattles and whistles; some have a melodious song. Many are terrestrial or inhabit bushes or grass close to the ground, while other species are arboreal. Babblers are chiefly insectivorous, and augment their diet with fruits, seeds and nectar. Arboreal species collect food from leaves, moss, lichen and bark; terrestrial species forage by probing, digging, and tossing aside dead foliage.

Laughingthrushes are medium-sized, long-tailed babblers that are gregarious even in the breeding season. At the first sign of danger, they characteristically break into a concert of loud hissing, chattering and squealing. They often feed on the ground, moving along with long springy hops, rummaging among leaf litter, flicking leaves aside and into the air, and digging for food with their strong bills. Their flight is short and clumsy, the birds flying from tree to tree in follow-my-leader fashion.

WHITE-EYES Zosteropidae

Small or very small insectivorous passerines with slightly decurved and pointed bills, brush-tipped tongues, and a white ring around each eye. White-eyes frequent forest, forest edge, and bushes in gardens.

KINGLETS Regulidae

Tiny passerines with bright crown feathers, represented by only one species in the subcontinent, Goldcrest. Typically inhabits the canopy of coniferous forest, frequently hovering to catch insects. Often in mixed feeding parties.

WRENS Troglodytidae

Small, plump, insectivorous passerines with only species in the subcontinent, Winter Wren. Has rather short, blunt wings, strong legs and the tail characteristically held erect.

DIPPERS Cinclidae

Rotund birds with short wings and tails, dippers are adapted for feeding on invertebrates in or under running water. They fly low over the water surface on rapidly whirring wings.

NUTHATCHES AND WALLCREEPER Sittidae

Nuthatches and Wallcreeper are small, energetic, compact passerines with short tails, large strong feet and long bills. The Wallcreeper is adept at clambering over rock faces. Nuthatches are also agile tree climbers. They can move with ease upwards, downwards, sideways and upside-down over trunks or branches progressing by a series of jerky hops. Unlike woodpeckers and treecreepers, they usually begin near the top of a tree and work down the main trunk or larger branches, often head-first, and do not use the tail as a prop. Their flight is direct over short distances, and undulating over longer ones. Nuthatches capture insects, spiders, seeds and nuts. They are often found singly or in pairs; outside the breeding season, they often join foraging flocks of other insectivorous birds.

TREECREEPERS Certhiidae

Small, quiet, arboreal passerines with slender, decurved bills and stiff tails that they use as a prop when climbing, like that of the woodpeckers. Treecreepers forage by creeping up vertical trunks and along the underside of branches, spiralling upwards in a series of jerks in search of insects and spiders; on reaching the top of a tree, they fly to the base of the next one. Their flight is undulating and weak, and is usually only over short distances. Treecreepers are non-gregarious, but outside the nesting season they usually join mixed hunting parties of other insectivorous birds. They inhabit broadleaved and coniferous forest, woodland, groves, and gardens with trees. Thin high-pitched contact calls are used continually.

STARLINGS AND MYNAS Sturnidae

Robust, medium-sized passerines with strong legs and bills, moderately long wings and square tails. The flight is direct; strong and fast in the more pointed-winged species (*Sturnus*), and rather slower with more deliberate flapping in the more rounded-winged ones. Most species walk with an upright stance in a characteristic, purposeful jaunty fashion, broken by occasional short runs and hops. Their calls are often loud, harsh and grating, and the song of many species is a variety of whistles; mimicry is common. Most are highly gregarious at times. Some starlings are mainly arboreal and feed on fruits and insects; others are chiefly ground-feeders, and are omnivorous. Many are closely associated with human cultivation and habitation.

THRUSHES, COCHOAS AND SHORTWINGS Turdidae

Thrushes are medium-sized passerines with rather long, strong legs, slender bills and fairly long wings. On the ground they progress by hopping. All are insectivorous, and many eat fruit as well. Some species are chiefly terrestrial and others arboreal. Most thrushes have loud and varied songs, which are used to proclaim and defend their territories when breeding. Many species gather in flocks outside the breeding season.

Cochoas are fairly large, robust, colourful, thrush-like birds with fairly broad bills. Shy, unobtrusive, arboreal and frugivorous.

Shortwings are small chat-like thrushes with short rounded wings, almost square tails and strong legs. They are mainly terrestrial, and inhabit low bushes, undergrowth or thickets. Shortwings are chiefly insectivorous and found singly or in pairs.

CHATS AND OLD WORLD FLYCATCHERS Muscicapidae

Chats are a diverse group of small/medium-sized passerines that includes the chats, blue robins, magpie robins, redstarts, forktails, wheatears and rock thrushes. Most are terrestrial or partly terrestrial, some are arboreal, and some are closely associated with water. Their main diet is insects, and they also consume fruits, especially berries. They forage mainly by hopping about on the ground in search of prey, or by perching on a low vantage point and then dropping to the ground onto insects or making short sallies to catch them in the air. Found singly or in pairs.

Flycatchers are small insectivorous birds with small, flattened bills, and bristles at the gape that help in the capture of flying insects. They normally have a very upright stance when perched. Many species frequently flick the tail and hold the wings slightly drooped. Generally, flycatchers frequent trees and bushes. Some species regularly perch on a vantage point, from which they catch insects in mid-air in short aerial sallies or by dropping to the ground, often returning to the same perch. Other species capture insects while flitting among branches or by picking them from foliage. Flycatchers are usually found singly or in pairs; a few join mixed hunting parties of other insectivorous birds.

FAIRY BLUEBIRDS Irenidae

Medium-sized passerines with fairly long, slender bills, the upper mandible decurved at the tip; arboreal, typically frequenting thick foliage in the canopy. They search leaves for insects and also feed on berries and nectar. Their flight is swift, usually over a short distance. Represented by only one species in the subcontinent, Asian Fairy Bluebird.

LEAFBIRDS Chloropseidae

Medium-sized green and yellow birds with slender downcurved bills. They feed on nectar on flowering trees, fruit and invertebrates.

FLOWERPECKERS Dicaeidae

Flowerpeckers are very small passerines with short bills and tails, and with tongues adapted for nectar-feeding. They usually frequent the tree canopy and feed mainly on soft fruits, berries and nectar; also on small insects and spiders. Many species are especially fond of mistletoe *Loranthus* berries. Flowerpeckers are very active, continually flying about restlessly, and twisting and turning in different attitudes when perched, while calling frequently with high-pitched notes. Normally they live singly or in pairs; some species form small parties in the non-breeding season.

SUNBIRDS AND SPIDERHUNTERS Nectariniidae

Sunbirds have bills and tongues adapted to feed on nectar; they also eat small insects and spiders. The bill is long, thin and curved for probing the corollas of flowers. The tongue is very long, tubular and extensible far beyond the bill, and is used to draw out nectar. Sunbirds feed mainly at the blossoms of flowering trees and shrubs. They flit and dart actively from flower to flower, clambering over the blossoms, often hovering momentarily in front of them, and clinging acrobatically to twigs. Sunbirds usually keep singly or in pairs, although several may congregate in flowering trees, and some species join mixed foraging flocks. They have sharp, metallic calls and high-pitched trilling and twittering songs.

Spiderhunters are small, robust arboreal forest birds with very long decurved bills. Very active with fast dashing flight. Usually found singly or in pairs. They feed on nectar and small invertebrates.

SPARROWS, PETRONIAS AND SNOWFINCHES Passeridae

Small passerines with thick, conical bills. This family includes *Passer*, the true sparrows, some of which are closely associated with human habitation and *Petronia*, the rock sparrows, which inhabit dry rocky country or light scrub. Most species feed on seeds, taken on or near the ground. The *Passer* sparrows are rather noisy, using a variety of harsh, chirping notes; the others have more varied songs and rather harsh calls. Snowfinches have distinctive white patches in their plumage and inhabit barren alpine areas. They are almost entirely terrestrial in feeding habits. Outside the breeding season they gather in large flocks.

WEAVERS Ploceidae

Small, rather plump, finch-like passerines with large, conical bills. Adults feed chiefly on seeds and grain, supplemented by invertebrates; the young are often fed on invertebrates. Weavers inhabit grassland, marshes, cultivation and very open woodland. They are highly gregarious, roosting and nesting communally, and are noted for their elaborate roofed nests.

AVADAVATS AND MUNIAS Estrildidae

Small, slim passerines with short, stout, conical bills. They feed chiefly on small seeds, which they pick up from the ground or gather by clinging to stems and pulling the seeds directly from seed heads. Their gait is a hop or occasionally a walk. Outside the breeding season all species are gregarious. Flight is fast and undulating.

ACCENTORS Prunellidae

Small, compact birds resembling *Passer* sparrows in appearance, but with more slender and pointed bills. Accentors forage quietly and unobtrusively on the ground, moving by hopping or in shuffling walk; some species also run. In summer accentors are chiefly insectivorous, and in winter they feed mainly on seeds. Their flight is usually low over the ground and sustained only over short distances.

PIPITS AND WAGTAILS Motacillidae

Small, slender, terrestrial birds with long legs, relatively long toes and thin, pointed bills. Some wagtails exhibit wide geographical plumage variation. All walk with a deliberate gait and run rapidly. The flight is undulating and strong. Most wagtails wag the tail up and down, and so do some pipits. They feed mainly by picking insects from the ground as they walk along, or by making short rapid runs to capture insects they have flushed; they also catch prey in mid-air. Song flights are characteristic of many pipits. Both wagtails and pipits call in flight, and this is often a useful identification feature. They are usually found singly or in pairs in the breeding season and in scattered flocks in autumn and winter.

FINCHES Fringillidae

Small to medium-sized passerines with strong, conical bills used for eating seeds. They forage on the ground; some species also feed on seedheads of tall herbs, and blossoms or berries of bushes and trees. Finches are highly gregarious outside the breeding season. Their flight is fast and undulating.

BUNTINGS Emberizidae

Small to medium-sized, terrestrial passerines with strong, conical bills designed for shelling seeds, usually of grasses; adults also eat insects in summer. They forage by hopping or creeping on the ground. Their flight is undulating. Buntings are usually gregarious outside the breeding season, feeding and roosting in flocks. Buntings occur in a wide variety of open habitats.

PLATE 1: SNOWCOCKS AND PARTRIDGES

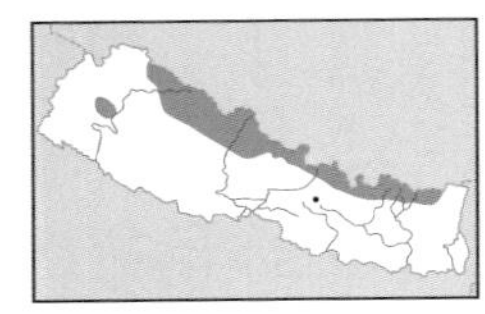

Snow Partridge *Lerwa lerwa* 38 cm

Locally fairly common resident; breeds 4,000–5,000 m, non-breeding season 3,050–4,880 m. **ID** Head, neck, upper breast and upperparts finely vermiculated dark brown and white, with a chestnut wash. Underparts heavily streaked chestnut. Bill, legs and feet are red. In flight, shows narrow white trailing edge to wings, blackish primaries and finely barred tail. Smaller size, and finely barred upperparts, help distinguish from the two snowcock species that occur in similar habitat. **Voice** Loud, harsh and frequently repeated whistle. **HH** Usually in pairs in spring and in parties of up to 30 in non-breeding season. Very tame where not hunted. When disturbed, plunges downhill with much wing-clattering. High-altitude steep rocky and grassy slopes with dwarf scrub.

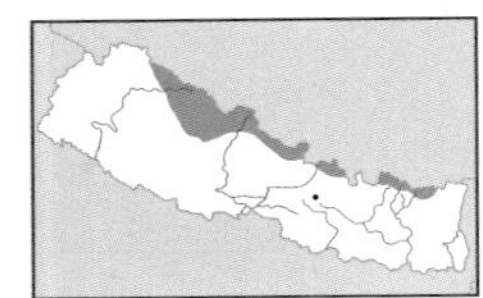

Tibetan Snowcock *Tetraogallus tibetanus* 51 cm

Locally fairly common resident; breeds 4,500–5,550 m; down to 3,650 m in non-breeding season. **ID** From Himalayan Snowcock by prominent white patch on ear-coverts offset against grey of head and neck, double band of grey across upper breast (absent, or just a single band, in some birds), white underparts with broad black flanks stripes, and whitish fringes to coverts and scapulars. In flight, wing pattern is very different, Tibetan showing only a small amount of white in primaries but extensive white in secondaries. Also has chestnut coloration on rump and uppertail-coverts. **Voice** Similar to Himalayan's: a subdued chuckling becoming louder and reaching a climax, and a whistle and call reminiscent of Eurasian Curlew. **HH** Habits similar to Himalayan's. Rocky alpine pastures, stony ridges, grassy slopes and steep hillsides amongst tumbled boulders.

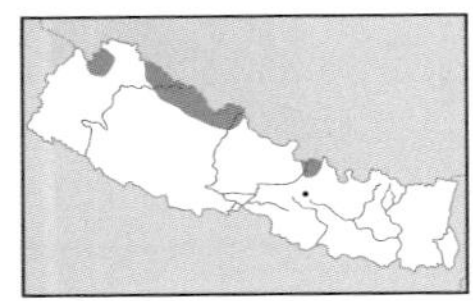

Himalayan Snowcock *Tetraogallus himalayensis* 54–72 cm

Resident of uncertain status. Mainly occurs in west and west-central Nepal and in Langtang National Park; 4,250–5,900 m. **ID** Distinguished from smaller Tibetan by broad dark chestnut stripes on sides of neck that join to form band across upper breast, greyish-white breast (variably barred with black) contrasting strongly with dark grey underparts; strong contrast between pale grey crown/mantle and dark grey back. In flight, Himalayan shows extensive white in primaries but little or no white in secondaries, and greyish rump and uppertail-coverts. **Voice** A far-carrying inflected whistle ending in two shorter, rising whistles, *cour-lee-whi-whi*, repeated at intervals, and reminiscent of Eurasian Curlew; also a *chok, chok, chok* which often accelerates into a rapid chatter. **HH** In the west its altitudinal range overlaps with Tibetan Snowcock, but is usually slightly higher. Keeps in pairs in breeding season and at other times in flocks of up to 30 birds. Escapes by running uphill or, if pressed, by flying extremely swiftly downhill, covering a long distance before settling again. High-altitude rocky slopes and alpine meadows, and on open mountains among dwarf junipers.

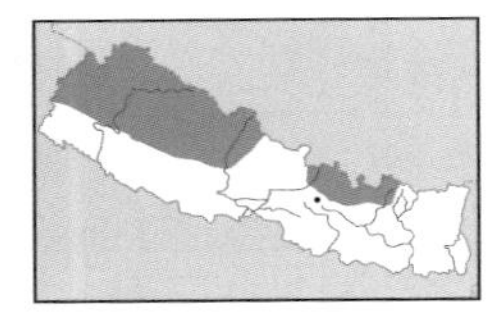

Chukar Partridge *Alectoris chukar* 38 cm

Locally common resident now found from far west east to Manang, west-central Nepal; 1,300–3,960 m. **ID** A stocky, medium-sized partridge. Has black stripe through eye that extends to form black gorget, encircling creamy-white throat; broad chestnut and black rib-like bars on flanks, and bright red bill and legs. Displays rufous corners to tail in flight. **Voice** Utters a rapidly repeated *chuck, chuck-aa*; when flushed, an anxious 'rolled together' *chuck, chuck, chuck*. **HH** Keeps in flocks of up to 30 birds outside breeding season. If flushed the birds disperse, flying very fast and strongly, downhill. Moves rapidly and with agility over rough ground. Open, arid rocky hills, barren hillsides with scattered scrub, grassy slopes, dry terraced cultivation and stony ravines, often near water.

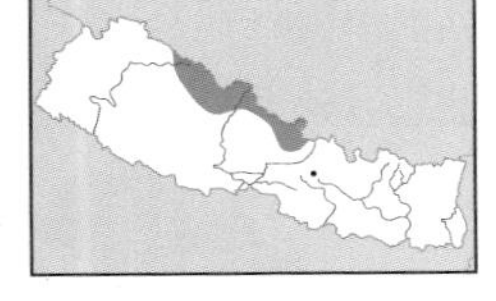

Tibetan Partridge *Perdix hodgsoniae* 31 cm

Uncommon resident; in non-breeding season 3,700–4,100 m, breeds up to 5,000 m. **ID** Unmistakable, with bold, black ear-coverts patch on whitish face, rufous collar not quite complete across breast, bold black barring on underparts forming variable black patch on belly, and rufous barring on flanks. Shows rufous in primaries and tail in flight. **Voice** A rattling and repeated *scherrrrreck-scherrrrreck*; when flushed a shrill *chee, chee, chee, chee*. **HH** Usually in pairs in the breeding season and in large flocks at other times. When disturbed birds run fast uphill, calling loudly. If pressed, they scatter in different directions and dive downhill, re-establishing contact by uttering buzzing calls to each other. High-altitude semi-desert and rocky slopes with scattered dwarf scrub.

Snow Partridge

Tibetan Snowcock

Himalayan Snowcock

Chukar Partridge

Tibetan Partridge

PLATE 2: FRANCOLINS AND PARTRIDGES

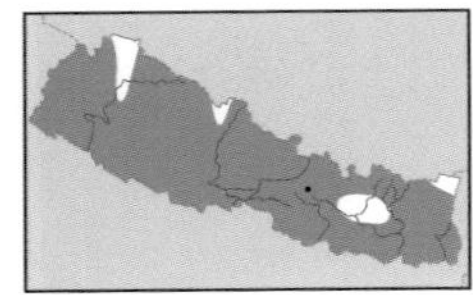

Black Francolin *Francolinus francolinus* 34 cm

Locally fairly common resident up to 2,050 m, subject to altitudinal movements. **ID** Male has black face with white ear-coverts patch, rufous collar, black upper mantle spotted with white, and black underparts with white spotting on flanks. Female has rufous patch on hindneck, buffish supercilium and cheeks divided by dark stripe behind eye, and blackish barring to white underparts. Shows blackish tail in flight. **Voice** A loud, penetrating, repeated, harsh *kar-kar, kee, ke-kee.* **HH** Found singly, in pairs or in groups of up to five birds. Rests and roosts in dense ground cover. If much disturbed, escapes by running away swiftly, but otherwise flushes easily, flying off strongly and at great speed. Active in early mornings and late afternoons. Calls at any time of day during breeding season, often from a tree or stump. Requires good ground cover and water close by. Cultivation, tall grass and scrub, especially near rivers.

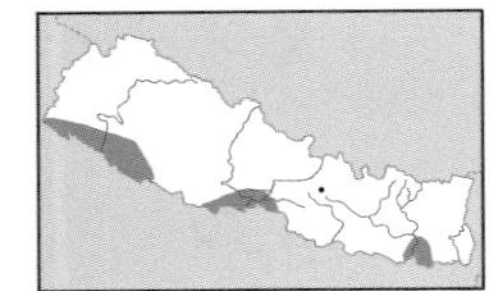

Grey Francolin *Francolinus pondicerianus* 33 cm

Local resident mainly occurring in western lowlands, with a few recent records from far east; 90–200 m. **ID** Rather plain buffish face, and buffish-white throat with fine necklace of dark spotting. Upperparts finely barred with buff, chestnut and brown. Underparts buffish and finely barred with dark brown. Shows rufous tail in flight. **Voice** A rapidly repeated *khateeja-khateeja-khateeja*; also softer, more whistling *kila-kila-kila*, and a high, whirring *khirr-khirr.* **HH** Typically in pairs or in groups of up to eight birds which roost together in small thorny trees or shrubs. Digs and scratches in the ground with bill and feet. Very fast on its legs and usually escapes by running; seldom flies. When pressed rises with a loud whirr of wings, scatters in different directions and alights again after only 50–100 m. Male usually calls from the ground. Dry grassy and scrubby areas, often near cultivation; also riverine scrub.

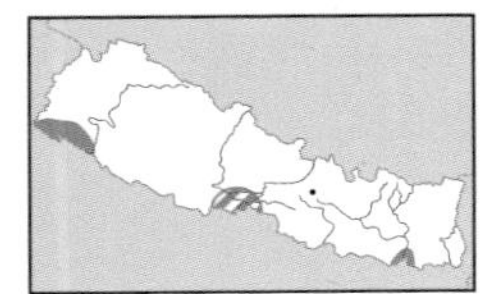

Swamp Francolin *Francolinus gularis* 37 cm

Very local resident; 75–250 m. **ID** Rufous-orange throat, buff supercilium and cheek stripe (separated by dark eye-stripe), finely barred upperparts, and bold white streaking on underparts. Sexes similar. Shows rufous primaries and tail in flight. **Voice** A loud *kew-care* when alarmed, occasional *qua, qua, qua* ascending in tone, and a harsh *chukeroo, chukeroo, chukeroo* preceded by several chuckles and croaks. May sound similar to Grey Francolin but louder. **HH** Found in pairs or in groups of up to six birds. In marshes often wades through shallow water or mud, and climbs on reeds in deep water. Reluctant to fly, but if flushed, it rises clumsily and noisily with loud chuckling and whirring of wings. Roosts in thorny trees and on broken reeds in swamps. Tall wet grassland and swamps. Globally threatened.

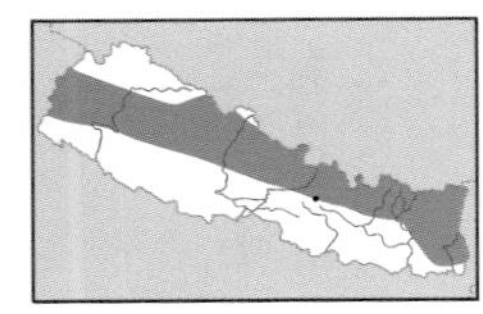

Hill Partridge *Arborophila torqueola* 28 cm

Locally fairly common resident; 1,830–3,550 m. **ID** Male from Rufous-throated Partridge by rufous crown and ear-coverts, black eye-patch and eye-stripe, white neck sides streaked with black, and white collar. Female lacks rufous crown and white collar of male. From Rufous-throated by buff supercilium, black barring on mantle, and lacks black border between rufous-orange foreneck and grey breast. Legs and feet are dark (red in Rufous-throated). **Voice** A mournful drawn-out whistle, followed by double whistles. **HH** Usually in pairs, or, outside breeding season, in flocks of 5–10 birds. Often secretive and keeps to dense cover. Digs for food among leaves and humus on forest floor, birds keeping contact by calling as they move through the forest. Roosts in groups in trees. If disturbed, usually moves away on foot; if flushed, flies strongly, zigzagging around trees, and usually settles or takes refuge on a leafy branch. Once flushed, birds are often very reluctant to fly again. Ravines and slopes in damp, dense forests of oak and other broadleaved, evergreen trees.

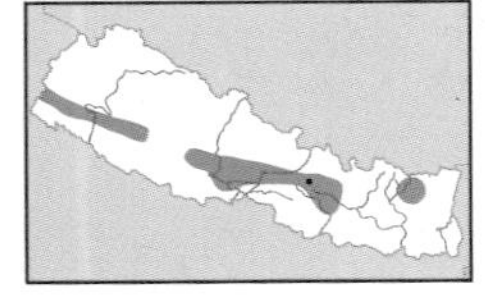

Rufous-throated Partridge *Arborophila rufogularis* 27 cm

Rare and local resident; 1,450–1,830 m (250–2,050 m). **ID** Superficially resembles female Hill and best told by greyish-white supercilium, diffuse white moustachial stripe, rufous hindneck, unbarred mantle, black border between rufous-orange foreneck and grey breast, and reddish legs and feet. Sexes are more similar than in Hill. **Voice** A mournful double whistle, *wheea-whu*, repeated constantly and on ascending scale. **HH** Habits similar to those of Hill Partridge. Dense understorey of broadleaved, evergreen forests.

♀
♂
Black Francolin
Grey Francolin
Swamp Francolin
♂
♀
Hill Partridge
♀
♂
Rufous-throated Partridge

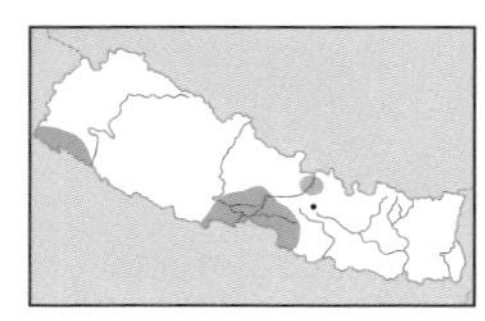

Common Quail *Coturnix coturnix* 20 cm

Rare winter visitor, mainly to lowlands, and passage migrant; 75–915 m (–2,900 m). **ID** Male has black 'anchor' mark on throat and buff gorget, although head pattern is variable and black anchor lacking in some. Some males have rufous face and throat, with or without black anchor. Female has less striking head pattern and lacks black anchor. See account for Rain Quail in Appendix 1. **Voice** Song a far-carrying *whit, whit-tit* repeated in quick succession. When flushed a trilling, slightly rising *whreeee* followed by short quiet *chak* notes. **HH** Usually singly or in pairs. Very secretive and keeps out of sight. If flushed rises with rapid, whirring wingbeats and brief glides on downturned wings, flying low in an arc before plunging into cover again after only 100–200 m. Standing crops, field edges and grassland.

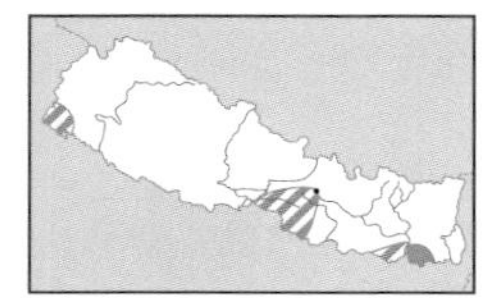

King Quail *Coturnix chinensis* 14 cm

Very rare; 75–1,350 m. **ID** Small size. Male has black-and-white patterned throat, slaty-blue flanks and chestnut belly. Female similar to Common Quail but noticeably smaller, with rufous-buff forehead and supercilium, barred breast and flanks, and more uniform upperparts. **Voice** Typical call a high-pitched series of 2–3 descending piping notes, *ti-yu* or *quee-kee-kew*; repeated *tir* notes may be uttered when flushed. **HH** Usually flushes singly, dropping into grass after a short flight and very reluctant to fly again. Habits similar to Common's. Favours wetter habitats than other quails: edges of marshes, tall grassland and fields. **AN** Blue-breasted Quail.

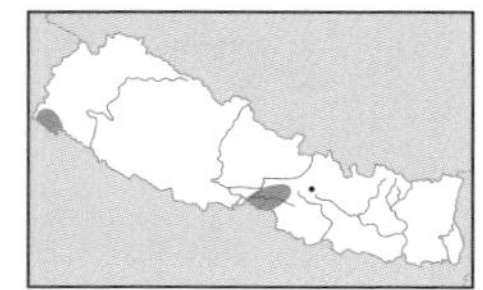

Small Buttonquail *Turnix sylvaticus* 13 cm

Rare and very local, presumably resident, Chitwan National Park and buffer zone, Sukla Phanta Wildlife Reserve; 150–250 m. **ID** Very small with pointed tail. Bill greyish, and legs pinkish to greyish. More heavily marked above than Yellow-legged Buttonquail; buff edges to scapulars and tertials form prominent lines. Has rufous hindneck and mantle fringed with buff. Buff-fringed, dark-centred coverts result in spotted appearance, but less prominently so than in Yellow-legged. Underparts similar to many Yellow-legged, with orange-buff lower throat and breast, and bold black spotting on sides of breast (becoming chestnut spotting on flanks). Female has brighter and more extensive rufous on neck than male. **Voice** Female has a booming call that lasts one second and is repeated every 1–2 seconds for half a minute or more. **HH** Habits similar to Barred. Tall grassland. **AN** Common Buttonquail.

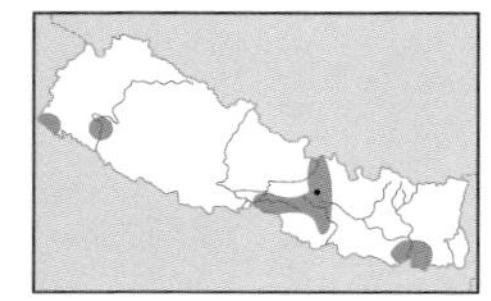

Yellow-legged Buttonquail *Turnix tanki* 15–16 cm

Rare and local resident; 75–250 m (–915 m). Yellow legs and bill (with variable dark culmen and tip). Bold black spotting to buff coverts and upper flanks. Upperparts more uniform than Small, varying from grey and finely dotted, to being more heavily marked with black and diffusely marked with rufous (and buff edges to some feathers, although not as prominent on Small). Some (breeding?) females distinctive, with unmarked rufous nape and upper mantle, rufous-orange throat, sides of neck and breast, and blackish crown. Other females less striking with buff crown-stripe, indistinct rufous collar, and orange-buff breast. Rufous collar lacking in male. **Voice** Low-pitched hoot, repeated with increasing strength to resemble a human-like moan. **HH** Habits similar to Barred. Usually runs when frightened. If closely approached, it crouches then buzzes explosively into the air, completes a short arc, drops to the ground and runs again. Scrub and grassland.

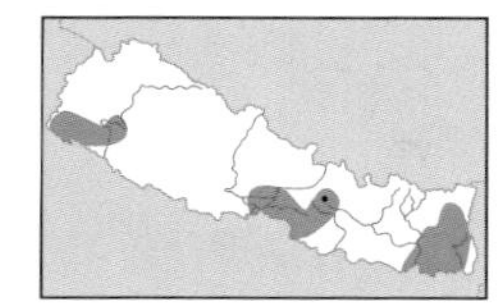

Barred Buttonquail *Turnix suscitator* 15 cm

Locally frequent resident; 75–300 m (–2,050 m). **ID** Grey legs, and bold black barring on sides of neck, breast and wing-coverts. Orange-buff flanks and belly clearly demarcated from barred breast. Most show buff crown-stripe, and speckled supercilium is often apparent. Female usually has black throat and centre of breast. Male usually has greyish- or buffish-white throat. **Voice** Advertising call of female is a *drr-r-r-r-r-r* similar to the sound of a distant motorcycle, lasting for 15 seconds or more, and usually prefaced by 3–4 long, deep *groo* notes, as well as a far-carrying booming call, *hoon-hoon-hoon-hoon*. **HH** Usually found singly or occasionally in pairs. Very secretive and prefers to escape by walking away quickly. Flies with great reluctance, on weak, whirring beats low over the ground, dropping quickly into cover. Typically can be found in the same place day after day. Feeds by picking food from the ground, or scratching with feet. Fond of dustbathing and sunbathing. Scrub, grassland and field edges.

♀
♂
Common Quail
♀
♂
King Quail
ad
Small Buttonquail
♂
♀
Yellow-legged Buttonquail
♂
♀
Barred Buttonquail

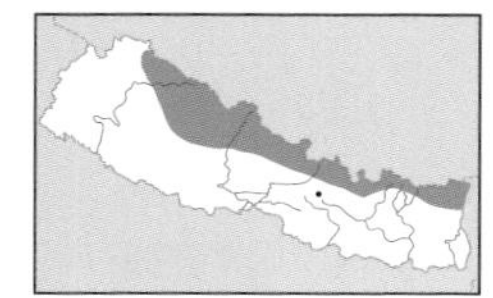

Blood Pheasant *Ithaginis cruentus* 38 cm

Locally fairly common resident mainly found in protected areas; subject to altitudinal movements, 3,200–4,400 m. **ID** Crested head, and red orbital skin and legs/feet. Male has blood-red throat, black mask, grey upperparts streaked with white, and greenish underparts; plumage splashed with red, particularly on breast and uppertail- and undertail-coverts. Female has grey crest and nape, rufous-orange face, dark brown upperparts, and rufous-brown underparts. **Voice** A repeated *chuck*, and a loud, grating *kzeeuk-cheeu-cheeu-chee*. **HH** Keeps in coveys of 5–30 or more. Tame and confiding. Rarely flies except to go to roost. Forages actively throughout day, scratching the ground like a domestic fowl and turning over leaves and grasses with its feet; can dig down through snow to reach vegetation beneath. Roosts in trees, thickets or on the ground. Dense bamboo clumps, open forests or scrub of rhododendron and birch or juniper, often near water.

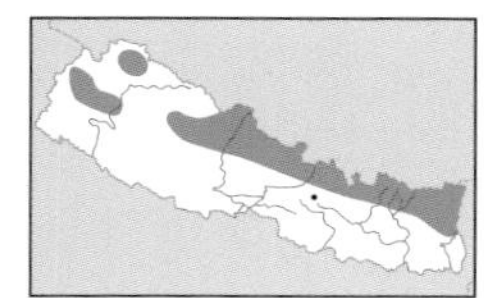

Satyr Tragopan *Tragopan satyra* ♂ 67–72 cm, ♀ 57.5 cm

Uncommon resident mainly occurring in protected areas; 2,590–3,800 m in breeding season, winters down to 2,100 m. **ID** Male has red neck and red underparts with black-bordered white spots, and olive-brown upperparts profusely spotted with white. Bare facial skin and throat are blue surrounded by black. Female is rufous-brown with white streaking and spotting. Immature male is more like female, but has black on head, and red on neck, upper mantle and breast. **Voice** A repeated deep, wailing drawn-out call *wah, waah! oo-ah! oo-aaaaa!* rising in volume and becoming more protracted until it becomes almost a shriek; also a *wah, wah* uttered at any time of day. When alarmed or flushed, gives a more anxious *wak wak*. **HH** Usually found singly or in pairs. Normally extremely wary and skulking, but sometimes searches for food in forest glades. Forages mainly in early morning and from late afternoon to dusk. Roosts in trees. Territorial in breeding season. When displaying, males repeatedly expand and contract their horns and colourfully patterned lappets. Moist broadleaved and rhododendron forest with dense undergrowth including bamboo.

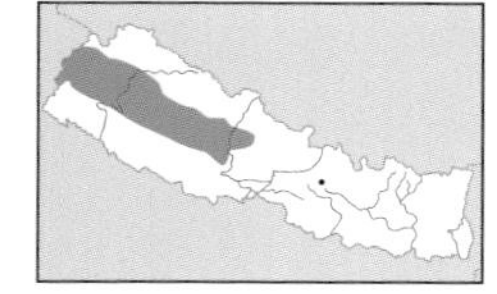

Koklass Pheasant *Pucrasia macrolopha* ♂ 58–64 cm, ♀ 52.5–56 cm

Local resident in the west; breeds 2,680–3,200 m (–3,500 m), winters down to 2,135 m. **ID** Male has bottle-green head and ear-tufts, white patch on neck, chestnut on breast, and streaked appearance to body. Female has white throat, short buff ear-tufts, and streaked body. Both sexes have wedge-shaped tail. Race in Nepal is mainly *nipalensis*; birds in the far west may be *macrolopha*. **Voice** A far-carrying and loud *kok, kark, kuk... kukuk* uttered mainly at dawn and repeated at intervals. When flushed, a harsh *kwak-kwak-kwak*. **HH** Usually occurs singly or in pairs, but in winter several may congregate. Very secretive and wary. When disturbed, runs away quickly through undergrowth, or bursts upwards giving a noisy alarm before hurtling downslope, twisting between the trees at great speed. Feeds mainly in early mornings and late afternoons in grassy glades. Roosts in trees, from where males crow at dawn almost throughout the year. Highly territorial during breeding season. Conifer, oak and rhododendron forest with dense undergrowth of bushes and ringal bamboos.

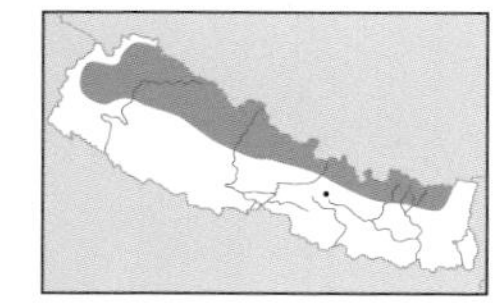

Himalayan Monal *Lophophorus impejanus* ♂ 70 cm, ♀ 63.5 cm

National bird of Nepal, known as Danphe in the country. Locally common resident mainly found in protected areas; subject to altitudinal movements, 3,300–4,570 m (–2,560 m). **ID** Male is iridescent green, copper and purple, with small white patch on back, cinnamon-brown tail, and spatulate-tipped crest. Female has pale streaking on underparts, prominent white throat, short crest, and bright blue orbital skin. In flight, shows whitish 'horseshoe' on uppertail-coverts, and narrow white tip to tail. **Voice** A series of upward-inflected whistles, *kur-leiu* or *kleeh-vick*, alternated with a higher-pitched *kleeh*; reminiscent of snowcocks and Eurasian Curlew. In alarm *kleeh-wick-kleeh-wick*, alternating with *kwick-kwick*. **HH** Loosely gregarious even in breeding season, often three or four together. Feeds mainly by digging with its strong bill, and can dig deep in snow if necessary. Although cautious, it is less shy than most Himalayan pheasants and can often be seen in summer foraging in alpine pastures. Steep grassy slopes and open rocky slopes above treeline in summer, descends to lower altitudes in rhododendron forest during winter; also at edges of forest and pastures, and in forest clearings.

♀
♂
Blood Pheasant
♀
♂
Satyr Tragopan
♂
macrolopha
♀
nipalensis
♂
Koklass Pheasant
♀
♂
Himalayan Monal

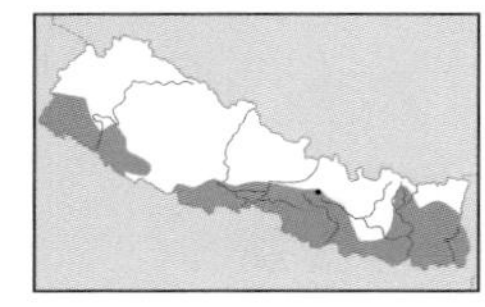

Red Junglefowl *Gallus gallus* ♂ 65–75 cm, ♀ 42–46 cm

Locally fairly common resident, mainly found in protected areas; 75–1,270 m. **ID** Male has rufous-orange hackles, blackish-brown underparts, rufous wing panel, white tail base, and long greenish-black, sickle-shaped tail. There is an eclipse plumage, after summer moult, when the hackles are replaced by short, dark brown feathers, and the central tail feathers are lacking. Female has 'shawl' of elongated (edged golden-buff, black-centred) feathers, rufous head, and naked reddish face. Immature male much duller than adult male; hackles less developed (with black centres); lacks elongated central tail feathers. **Voice** Male's loud *cock-a-doodle-doo* is very similar to a crowing domestic cockerel; both sexes make cackling and clucking notes. **HH** Keeps in small groups. Often wary and very secretive. If flushed, rises, cackling, with a clatter of wings. In early mornings and late afternoons, comes into open to forage. Roosts in trees and bamboo clumps. Inhabits forest undergrowth and forest edges.

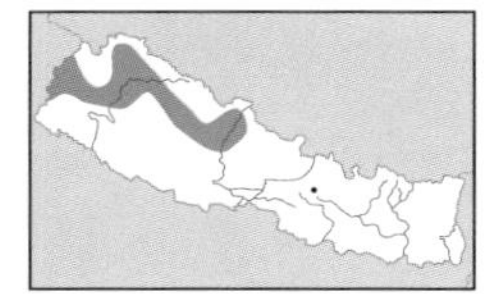

Cheer Pheasant *Catreus wallichii* ♂ 90–118 cm, ♀ 61–76 cm

Local, scarce resident in the west; 1,445–3,050 m. **ID** Long, broadly barred tail, pronounced crest, and red facial skin. Male is more cleanly and strongly marked than female, with pronounced barring on mantle, unmarked neck, rufous rump, and broader barring on tail. Female browner above, more heavily barred on breast, and has grey-brown rump. **Voice** Far-carrying, loud *chir-a-pir, chir-a-pir, chir, chir chirwa, chirwa*. Also, dusk and pre-dawn, high piercing whistles, *chewewoo*, interspersed with short *chut* calls and short staccato notes. **HH** Usually keeps in coveys of about five outside breeding season. Extremely wary and skulking. When disturbed prefers to run off rapidly or crouch in thick undergrowth rather than fly. If flushed, rises noisily and dives downhill. Birds often roost together in trees. Precipitous terrain with scrub, tall grasses and stunted trees, particularly where interspersed by rocky crags. Globally threatened.

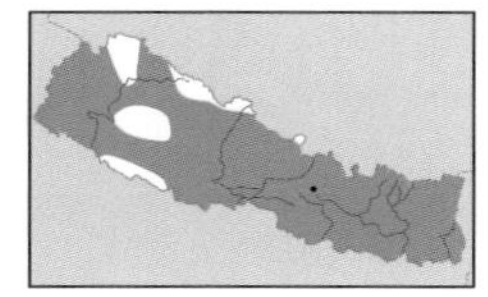

Kalij Pheasant *Lophura leucomelanos* ♂ 65–73 cm, ♀ 50–60 cm

Locally fairly common and widespread resident; subject to altitudinal movements, 245–3,700 m. Both sexes have red facial skin and downcurved tail. Three intergrading races occur: *L. l. leucomelanos* which is endemic to Nepal (male has blue-black crest and white barring on rump); *L. l. hamiltonii* of W Nepal (male has white or grey-brown crest, broad white barring on rump, and heavily scaled upperparts), and *L. l. melanota* of E Nepal (male has blue-black crest, and blue-black rump that lacks pale scaling). Female is reddish-brown, with greyish-buff fringes producing scaly appearance. **Voice** A loud, whistling chuckle or *chirrup*. When flushed gives guinea-pig-like squeaks and chuckles, sharply repeated *koorchi, koorchi, koorchi* or a whistling *psee-psee-psee-psee*. **HH** Found in pairs or family parties. Spends much time digging and scratching for food. Emerges into open to forage in early mornings and late afternoons. Roosts in trees. All forest types with dense undergrowth and thickly overgrown steep gullies; usually not far from water.

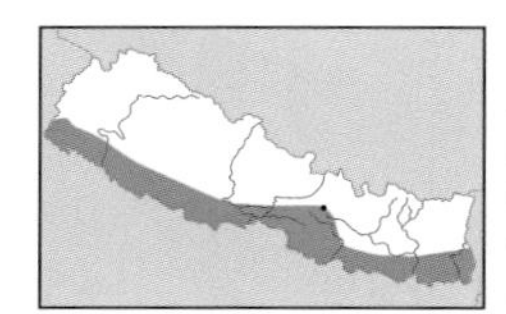

Indian Peafowl *Pavo cristatus* ♂ 180–230 cm, ♀ 90–100 cm

Locally common resident, mainly in protected areas, 120–305 m. **ID** Male has blue neck and breast, and spectacular glossy green train of elongated uppertail-coverts feathers with numerous ocelli. Female lacks train; has whitish face and throat, bronze-green neck, brown upperparts and white belly. Primaries of female are brown (chestnut in male). First-year male lacks train and is similar to female, but head and neck are usually blue, and primaries are chestnut with dark brown mottling. Second-year male more closely resembles adult male but has a short train, which lacks ocelli and is barred green and brown. Length of train increases until fifth or sixth year. **Voice** Trumpeting, far-carrying and mournful *kee-ow, kee-ow, kee-ow*. Also a series of short, gasping screams, *ka-an... ka-an... ka-an*, repeated 6–8 times, and *kok-kok* and *cain-kok* when alarmed. **HH** Gregarious, keeping in small flocks of usually one cock and 3–5 hens when breeding and often in separate parties of adult males and of females with immatures in non-breeding season. Roosts in tall trees. Emerges from dense thickets in early mornings and afternoons to feed. Quite shy and secretive where hunted. In the wild inhabits dense riverine vegetation and undergrowth in sal forest, often near streams; where semi-feral found in villages and cultivation.

Red Junglefowl
♂
♀
Cheer Pheasant
♂
♀
♂
♀
hamilttonii
♂
Kalij Pheasant
♂
♀
Indian Peafowl
♂
leucomelanos
Kalij Pheasant
♂
melanota

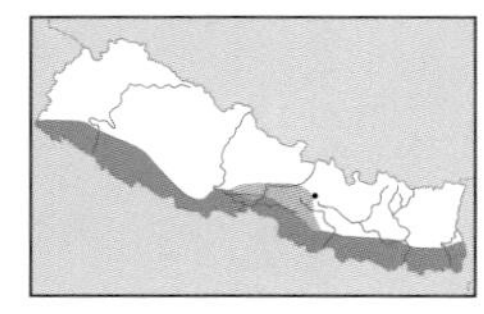

Lesser Whistling-duck *Dendrocygna javanica* 42 cm

Local breeding resident; 75–915 m. **ID** Has rather weak, deep-flapping flight, appears very dark on upperwing and underwing. From larger Fulvous (see Appendix 1) by greyish-buff head and neck, dark brown crown, lack of well-defined dark line down hindneck, bright chestnut patch on forewing, and chestnut uppertail-coverts. **Voice** Incessant twittering call in flight; at rest, a clear whistled *whi-whee*, also a subdued quacking. **HH** Flooded grassland and paddyfields, freshwater marshes, and shallow pools and lakes with plentiful fringe cover, emergent vegetation and partially submerged trees.

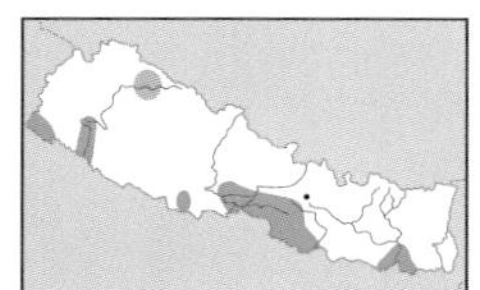

Greylag Goose *Anser anser* 75–90 cm

Very uncommon passage migrant and rare winter visitor; 75–275 m (–3,050 m on passage). **ID** Large, grey goose, with stout pink bill and pink legs and feet. Juvenile similar to adult, but has less prominent pale fringes to upperparts, flanks and belly. Shows pale grey forewing in flight. See Appendix 1 to compare Bean and White-fronted Geese, which are vagrants to Nepal. **Voice** Utters loud cackling and honking, deeper than in other 'grey' geese with repeated deep *aahng-ahng-ung*. **HH** Forages mainly at night and early morning; spends day swimming on large lakes or rivers, or loafing on spits or open fields. Wet grassland, crops, lakes and large rivers.

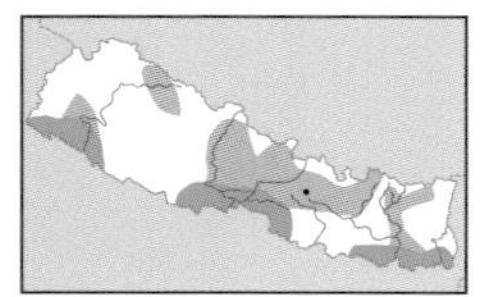

Bar-headed Goose *Anser indicus* 71–76 cm

Fairly common passage migrant and winter visitor; 75–275 m (–9,375 m on passage). **ID** Yellowish legs and black-tipped yellow bill. Adult has white head with black banding over crown, and white line down grey neck. Juvenile has white face and dark grey crown and hindneck. Plumage paler steel-grey, with more uniform pale grey forewing compared to Greylag. **Voice** Honking flight call, but notes more nasal and more slowly uttered compared to Greylag. **HH** Feeds mainly at night in cultivation or grassland on riverbanks; roosts by day on sandbanks of large rivers. Winters by large open lakes and rivers, especially on sandbanks and sandy islets.

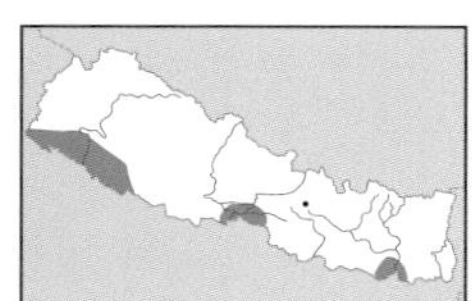

Knob-billed Duck *Sarkidiornis melanotos* 56–76 cm

Very rare, resident in the west; 75–275 m. **ID** Whitish head, speckled black, and whitish underparts with incomplete narrow breast-band. Upperwing and underwing blackish. Male has blackish upperparts glossed bronze, blue and green, with fleshy 'comb' at base of bill and yellowish-buff wash to sides of head and neck in summer; comb much reduced in winter. Female much smaller with duller upperparts and no comb. Juvenile has pale supercilium contrasting with dark crown and eye-stripe, buff scaling on upperparts, and rufous-buff underparts with dark scaling on sides of breast. **Voice** Generally silent, but utters low croak when flushed. **HH** Grazes in marshes and wet grassland, also wades and dabbles in shallows. Pools with plentiful aquatic vegetation in well-wooded areas. **AN** Comb Duck.

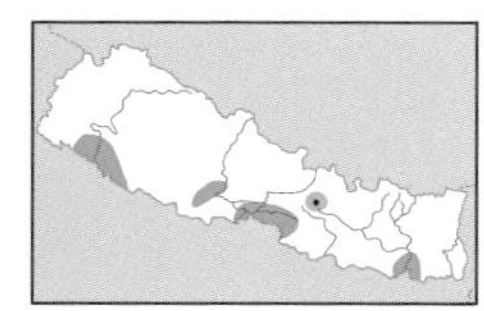

Common Shelduck *Tadorna tadorna* 58–67 cm

Rare winter visitor and spring passage migrant; 75–275 m (–915 m). **ID** Adult has greenish-black head and neck, and largely white body with chestnut breast-band and black scapular stripe. White upperwing- and underwing-coverts contrast with black remiges in flight. Female slightly smaller than male, has narrower chestnut breast-band and lacks knob on bill. Adult eclipse duller and greyer, with less distinct breast-band. Juvenile lacks breast-band and has sooty-brown crown, hindneck and upperparts, and white forehead, cheeks, foreneck and underparts. Flight pattern similar to adult (though less contrasting), but shows white trailing edge to secondaries. **Voice** Low whistling call (male); rapid *gag-ag-ag-ag-ag* (female). **HH** Dabbles on mud, also wades or upends in shallows. Open freshwater lakes and large rivers.

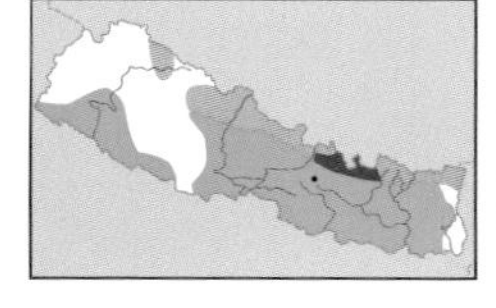

Ruddy Shelduck *Tadorna ferruginea* 61–67 cm

Common winter visitor throughout lowlands 75–305 m; breeds on Himalayan lakes, also a passage migrant up to 5,385 m. **ID** Rusty-orange, with buff to orange head; white upperwing- and underwing-coverts contrast with black remiges in flight. Breeding male has black neck-collar, which is less distinct or absent in non-breeding plumage. Female very similar to male, but lacks neck-collar and often has diffuse whitish face patch. Juvenile as female, but has browner and duller upperparts and underparts, and greyish tone to head. **Voice** Honking *aakh* and trumpeted *pok-pok-pok-pok* when taking off. **HH** Grazes on banks of rivers and lakes, also wades, dabbles and upends in shallows. Breeds around high-altitude lakes and swamps, winters by large open lakes and rivers, especially those with sandbanks and sandy islets.

ad
Lesser Whistling-duck
ad
Greylag Goose
ad
juv
Bar-headed Goose
Knob-billed Duck
♂
♀
juv
♂
Common Shelduck
♂
♀
♂
Ruddy Shelduck

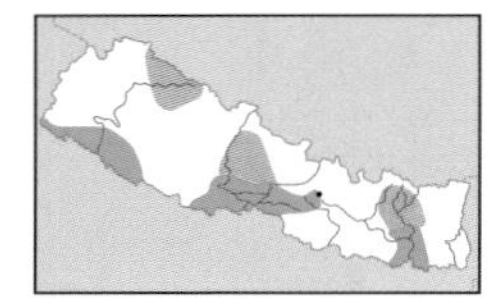

Gadwall *Anas strepera* 39–43 cm

Locally fairly common winter visitor and passage migrant; 75–3,050 m (–4,570 m on passage). **ID** White patch on inner secondaries in all plumages (although can be indistinct in female); lacks metallic speculum shown by Mallard. Male is mainly grey, with white belly and black rear end; bill dark grey. Female similar to female Mallard; orange sides to dark bill, clear-cut white belly and white inner secondaries are best features. Eclipse male similar to female, but has more uniform grey upperparts and pale grey (rather than blackish-grey) tertials, and upperwing pattern of breeding male. **Voice** Generally silent outside breeding season. **HH** Normally found in small parties. Feeds mainly by dipping head into shallow water; sometimes also by upending. Usually shy and wary, keeps close to emergent vegetation. Freshwater marshes and lakes with extensive aquatic and emergent vegetation.

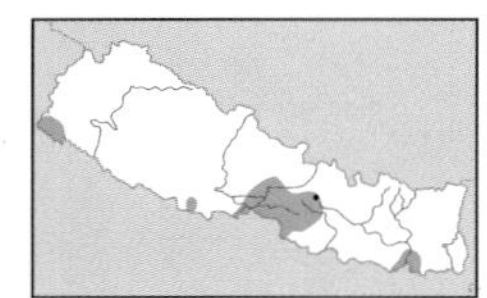

Falcated Duck *Anas falcata* 48–54 cm

Very rare winter visitor and passage migrant; 75–915 m. **ID** Male has bottle-green head with maned hindneck, elongated black-and-grey tertials, and black-bordered yellow patch at sides of vent; pale grey forewing in flight. Female has rather plain greyish head (with maned appearance), a dark bill, and greyish-white fringes to exposed tertials; greyish forewing and white greater coverts bar in flight, but does not show striking white belly. Eclipse male similar to female, but has dark crown, hindneck and upperparts, and pale grey forewing. **Voice** Distinctive, loud, piercing whistle in flight. **HH** Usually found singly or in small flocks. Feeds mainly by dabbling and upending; usually keeps close to emergent vegetation. Often shy and wary. Lakes, reservoirs and large rivers.

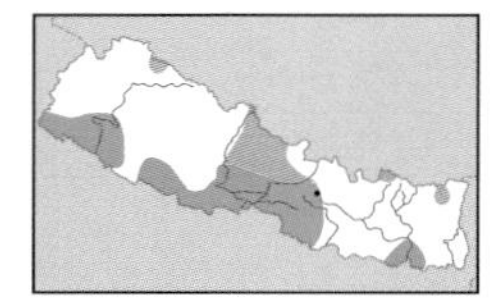

Eurasian Wigeon *Anas penelope* 45–51 cm

Winter visitor and passage migrant; 75–915 m (–4,570 m on passage). **ID** Male has yellow forehead and forecrown, chestnut head, and pinkish breast; white forewing in flight. Female has rather uniform brownish head, breast and flanks. In all plumages, shows white belly and rather pointed tail in flight. Eclipse male similar to female, but is more rufous on head and breast, and has white forewing. **Voice** Male has distinctive whistled *wheeooo* call. **HH** Highly gregarious. Feeds chiefly by grazing on waterside grasslands and in wet paddyfields in compact parties; grazes more than other ducks. Also feeds by dabbling and upending in shallow water. Other habitats are open lakes, reservoirs, rivers, pools and marshes.

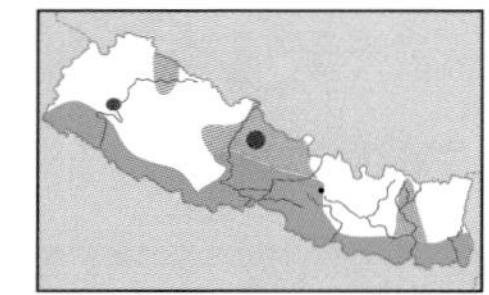

Mallard *Anas platyrhynchos* 50–65 cm

Mainly a winter visitor and passage migrant, 75–3,050 m; also a rare breeding species at 2,620 m. In all plumages has white-bordered purplish speculum. Male has yellow bill, dark green head and purplish-chestnut breast, mainly grey body, and black rear end. Female is pale brown, boldly patterned with dark brown. Bill variable, patterned mainly in dull orange and dark brown. Eclipse male similar to female, but has (less heavily marked) rusty-brown breast, blackish (glossed green) crown and eye-stripe, and uniform olive-yellow bill. **Voice** Male has soft rasping *kreep* and female a distinctive laughing *quack-quack-quack-quack*. **HH** Sociable duck; gathers in flocks of up to 40–50 birds. Where hunted, rests by day on large open waterbodies and feeds at night. Often flights at dusk to flooded paddyfields and marshes, where it feeds by dabbling, head-dipping, grazing or upending. Marshes and reed-fringed lakes.

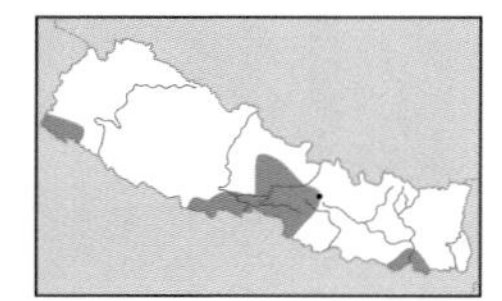

Indian Spot-billed Duck *Anas poecilorhyncha* 58–63 cm

Resident and winter visitor; 75–915 m (–3,290 m); rare breeder at Ghodaghodi Lake. **ID** From other *Anas* species by yellow-tipped black bill, greyish-white head and neck, with black crown and eye-stripe, blackish spotting on breast, white scalloping on flanks, and largely white tertials. In flight, wings appear dark except white on tertials and white underwing-coverts. Male has prominent red loral spot and is more strongly marked than female and juvenile (red loral spot less conspicuous on female and lacking on juvenile). See Appendix 1 for Eastern Spot-billed Duck. **Voice** As Mallard's, see above. **HH** Habits very similar to Mallard's, but is less quick taking off from water. Feeds by dabbling, head-dipping, upending, and walking amongst marsh vegetation. Marshes, lakes and pools with extensive emergent vegetation.

Gadwall
Falcated Duck
Eurasian Wigeon
Mallard
Indian Spot-billed Duck
♂
♀

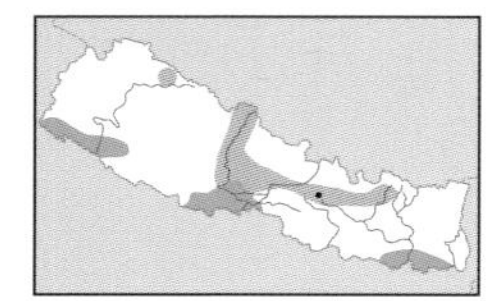

Northern Shoveler *Anas clypeata* 44–52 cm

Mainly a passage migrant; also a winter visitor; 75–1,350 m (–4,570 m on passage). **ID** Long spatulate bill and bluish forewing. Male has dark green head, white breast, chestnut flanks and blue forewing. Female recalls female Mallard, but has greyish-blue forewing and lacks white trailing edge. Eclipse male recalls female, but is more rufous-brown, especially on flanks and belly, and has upperwing pattern of breeding male. In sub-eclipse resembles breeding male, but has black scaling on breast and flanks, and whitish facial crescent between bill and eye. **Voice** Usually silent; female has short descending series of quacks. **HH** Sociable; usually keeps in pairs or small parties. Often feeds by sweeping bill from side to side while swimming, sifting water for minute organisms and aquatic seeds. Also upends or fully immerses its head and neck while swimming; sometimes wades or dives. All types of fresh waters.

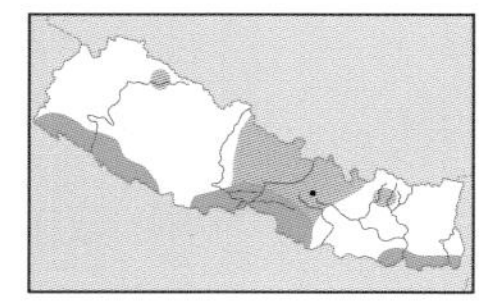

Northern Pintail *Anas acuta* 51–56 cm

Passage migrant and winter visitor; 75–915 m (–4,650 m). **ID** Long neck and pointed tail. Male has chocolate-brown head, with white stripe down sides of neck. Female has comparatively uniform buffish head, slender grey bill, and (as male) shows white trailing edge to secondaries and greyish underwing in flight. Eclipse male resembles female, but has grey tertials, and bill pattern and upperwing pattern as breeding male. **Voice** Male utters mellow *prop-proop* recalling male Common Teal; female gives descending series of weak quacks, recalling Mallard, and a low croak when flushed. **HH** Highly gregarious. Feeds mainly by upending, dabbling and head-dipping in shallow water; also grazes on land. Forages at night and early morning and evening in marshes and flooded paddyfields; roosts by day on open waters with aquatic vegetation.

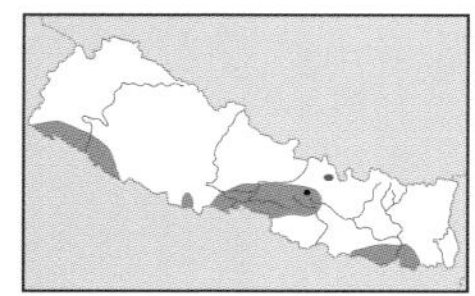

Cotton Pygmy-goose *Nettapus coromandelianus* 30–37 cm

Local breeding resident and summer visitor; 75–915 m. Small size. **ID** Male has broad white band across wing, and female has white trailing edge to wing. Male has white head and neck, black cap, greenish-black upperparts, and black breast-band. Eclipse male, female and juvenile are duller and have dark stripe through eye. **Voice** Male utters sharp staccato cackle *car-car-carawak* or *quack-quacky duck* at rest and in flight. Female makes weak *quack*. **HH** Generally in pairs in breeding season and in small flocks at other times. Forages by dabbling and grazing among floating vegetation; picks food from surface and dips head and neck underwater. Usually escapes predators by flying off strongly and rapidly. Perches readily in trees. Reed-edged pools partly covered with vegetation and flooded paddyfields.

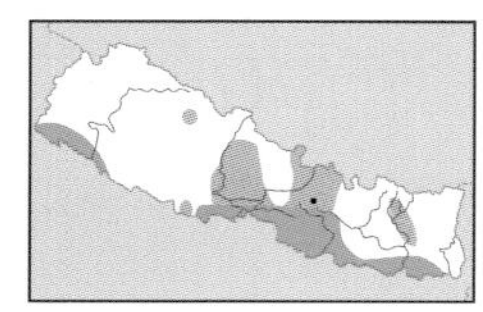

Garganey *Anas querquedula* 37–41 cm

Mainly a passage migrant and uncommon winter visitor; 75–915 m (–4,570 m on passage). **ID** Male has white stripe behind eye, and brown breast contrasting with grey flanks; blue-grey forewing in flight. Female has more patterned head than female Common Teal, with pale supercilium, whitish loral spot, pale line below dark eye-stripe, dark cheek-bar, and whiter throat; in flight shows prominent white belly, pale grey forewing and broad white trailing edge to wing. Eclipse male similar to female, but has upperwing pattern of breeding male. **Voice** Male has dry crackling call in alarm; female has Common Teal-like quack. **HH** Gregarious. Feeds by dabbling, head-dipping in shallow water and picking from surface; sometimes upends. Usually shy and prefers to keep to emergent vegetation. All types of wetlands with plenty of vegetation.

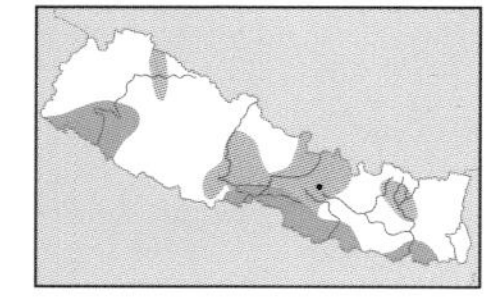

Common Teal *Anas crecca* 34–38 cm

Common winter visitor and passage migrant; 75–1,350 m (–4,300 m on passage). **ID** Male has chestnut head with green band behind eye, white stripe along scapulars, and yellowish patch on undertail-coverts. Female has rather uniform head, lacking pale loral spot and dark cheek-bar of female Garganey, with less prominent supercilium; further, bill often shows orange at base, and has prominent white streak at sides of undertail-coverts. Eclipse male and juvenile much as female. In flight, both sexes have broad white band on greater coverts, and green speculum with narrow white trailing edge; forewing brown. **Voice** Male has distinctive, soft, throaty whistle, *preep-preep*. Female usually silent, but utters a sharp *quack* when flushed. **HH** Found in small and large flocks. Feeds by dabbling, head-dipping and upending, also by grazing on marshes and foraging in fields by night. All kinds of shallow wetlands.

♂
♀
♂
Northern Shoveler
♀
♂
♂
Northern Pintail
♂
♀
Cotton
Pygmy-goose
♂ eclipse
♂
♀
♂
♀
♀
♂
Garganey
♂
♀
♂
♀
Common Teal

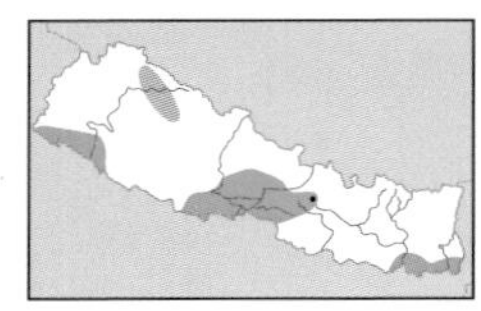

Red-crested Pochard *Netta rufina* 53–57 cm

Frequent winter visitor and passage migrant; 75–3,050 m. **ID** Large, with square-shaped head. Shape at rest and in flight more like dabbling duck. Male has red bill, rusty-orange head, and white flanks which contrast with black breast and ventral region. Female has pale cheeks contrasting with brown cap, and brown bill with pink towards tip. Both sexes have largely white flight feathers on upperwing, and whitish underwing. Eclipse male very similar to female, but has reddish iris and bill. **Voice** Silent away from breeding grounds. **HH** Feeds chiefly by diving; occasionally by upending and head-dipping. Large lakes with deep open water and plentiful submerged and fringing vegetation; occasionally rivers.

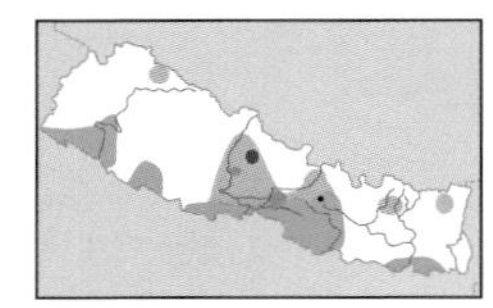

Common Pochard *Aythya ferina* 42–49 cm

Passage migrant and fairly common winter visitor; 75–915 m (–4,570 m on passage). **ID** Large, with domed head. Pale grey flight feathers and grey forewing produce different upperwing pattern from other *Aythya* ducks. Male has chestnut head, black breast, and grey upperparts and flanks. Female has brownish head and breast contrasting with paler brownish-grey upperparts and flanks; usually shows indistinct pale patch on lores, and pale throat and streak behind eye. Eye of female dark and bill has grey central band. Does not show white undertail-coverts of Ferruginous Duck. Eclipse male and immature male recall breeding male, but are duller with browner breast. **Voice** Silent away from breeding grounds. **HH** Highly gregarious, often in flocks of several hundreds. Feeds chiefly by diving in open water, sometimes also by upending or dabbling. Where disturbed, mainly a nocturnal feeder. Lakes, reservoirs with large areas of open water and large rivers. Globally threatened.

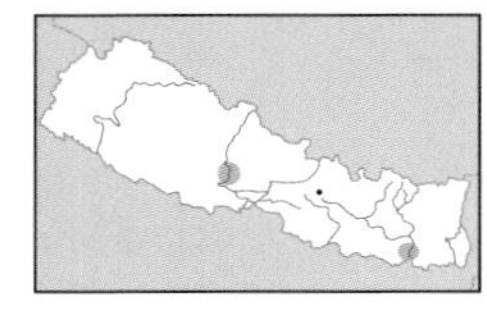

Baer's Pochard *Aythya baeri* 41–46 cm

Very rare and very local passage migrant; 100 m. **ID** Greenish cast to dark head and neck, which contrast with chestnut-brown breast. White patch on fore flanks visible above water and white undertail-coverts. Male has white iris. Female and immature male have duller head and breast than adult male. Female has dark iris and pale and diffuse chestnut-brown loral spot. **Voice** Silent away from breeding areas. **HH** Little known in wild. A shy duck, usually found singly, in pairs or small parties. Feeds mainly by diving. Lakes and large rivers. Globally threatened.

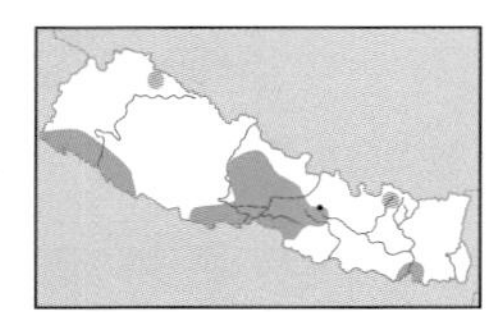

Ferruginous Duck *Aythya nyroca* 38–42 cm

Locally distributed; mainly a passage migrant, also a winter visitor; 75–915 m (–4,900 m on passage). Smallest *Aythya* duck, with dome-shaped head. Breeding male unmistakable, with rich chestnut head, neck and breast, and white iris. Female is chestnut-brown on head, neck, breast and flanks, with dark iris. Eclipse male resembles female, but is brighter on head and breast, and has white iris. In flight, shows extensive white wing-bar extending further onto outer primaries than in other *Aythya* species, and striking white belly (less pronounced in female). **Voice** Silent away from breeding grounds. **HH** Shy and feeds mainly at night. Secretive and not easily flushed, preferring to hide amongst aquatic vegetation. Freshwater pools and reservoirs with extensive submerged vegetation. **AN** Ferruginous Pochard.

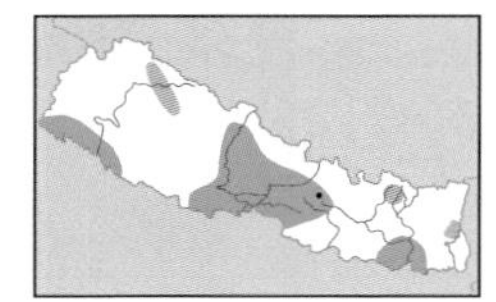

Tufted Duck *Aythya fuligula* 40–47 cm

Frequent winter visitor and passage migrant, 75–915 m; generally uncommon up to 4,900 m. **ID** Breeding male is glossy black, with prominent crest and white flanks. Eclipse/immature male duller, with greyish flanks, and less pronounced crest. Female is dusky-brown, with paler flanks; some females may show scaup-like white face patch, but they usually also show tufted nape and squarer head. Female has yellow iris; dark in female Common and Baer's Pochards and Ferruginous Duck. See Appendix 1 for Greater Scaup. **Voice** Silent away from breeding grounds. **HH** Gregarious; sometimes in flocks of several hundreds. Feeds during day mainly by diving; also upends, dips head or picks items from surface. Lakes, reservoirs and large rivers with large open areas and deep enough to permit diving.

♀
♂
♀
Red-crested Pochard
♀
♂
♂ immature
♀
Common Pochard
♂
♀
Baer's Pochard
♂
♀

Ferruginous Duck
♀ with scaup-like head
♀

♂
♀
♂ immature
♂
Tufted Duck

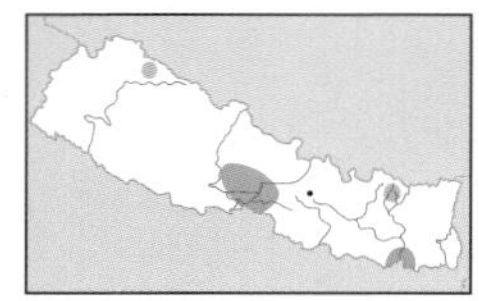

Common Goldeneye *Bucephala clangula* 42–50 cm

Rare winter visitor and passage migrant; 75–3,050 m. **ID** Stocky, with bulbous head. Male has dark green head, with large white patch on lores, and black-and-white patterned upperparts. Female has brown head, indistinct whitish collar, and grey body, with white wing patch usually visible at rest. Has 'golden' eye (as does male) and pink band at tip of bill. Immature male resembles female, but has pale loral spot and some white in scapulars. Eclipse male resembles female, but wing pattern as breeding male. In flight, both sexes show distinctive white pattern on wing. **Voice** Silent except when displaying. **HH** Swims with body flattened, and partially spreads wings when diving. Swift in flight, the wings producing a distinctive whistling sound. Feeds mainly by diving in daytime, group members often submerging simultaneously, occasionally dabbles and upends. Open-water areas in lakes and large rivers.

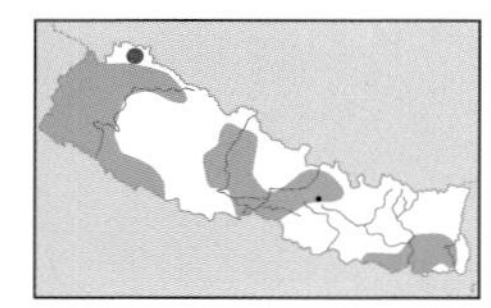

Goosander *Mergus merganser* 58–72 cm

Fairly common and widespread winter visitor, 75–3,050 m; also recorded in Humla, far north-west in summer, 4,800 m. **ID** Male has dark green head and whitish breast and flanks (with variable pink wash). Shows extensive white patch on wing-coverts and secondaries in flight. Female, and eclipse/immature male, have chestnut head and upper neck with shaggy crest, which contrasts with white throat and greyish neck, and white secondaries in flight. Eclipse male has upperwing pattern like breeding male. **Voice** Silent except when displaying. **HH** Sociable; usually in small parties. Forages in daytime, often fishing cooperatively. Feeds mainly by diving; usually after scanning with head submerged. An expert swimmer and diver. When feeding, swims with body low in water, and when resting floats high and buoyantly. Flight usually follows river course. Lakes and fast- and slow-moving rivers. **AN** Common Merganser.

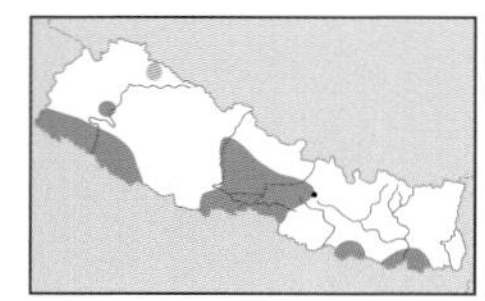

Little Grebe *Tachybaptus ruficollis* 25–29 cm

Fairly common resident and winter visitor; 75–3,050 m. **ID** Small size, often with puffed-up rear end. Shows whitish secondaries in flight. In breeding plumage, has rufous cheeks and neck-sides and yellow patch at base of bill. In non-breeding plumage, has buff cheeks, foreneck and flanks. Juvenile similar to non-breeding but has brown stripes on cheeks. **Voice** Drawn-out whinnying trill in breeding season, and a sharp *wit wit* in alarm. **HH** Often seen singly or in pairs among aquatic vegetation when breeding; in non-breeding season more frequently seen on open water and sometimes in small loose groups. Swims buoyantly. Often detected by its bubbling call. Lakes, ponds, reservoirs, ditches and slow-moving rivers.

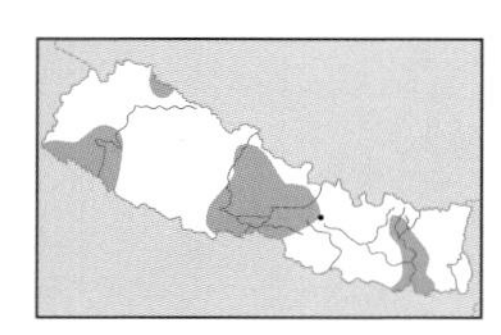

Great Crested Grebe *Podiceps cristatus* 46–51 cm

Locally frequent winter visitor; 75–3,050 m (–4,800 m). **ID** Large and slender-necked, with pinkish bill. Black crown does not extend to eye, and has white cheeks and foreneck in non-breeding plumage. Rufous-orange ear-tufts and white cheeks and foreneck in breeding plumage. Juvenile similar to non-breeding, but has brown striping on cheeks. **Voice** Silent in non-breeding season. **HH** An expert diver. Usually feeds singly. Rises clumsily from water, pattering over surface, and flying with rapid wingbeats. Swims with body low in water and neck held erect. In flight has an elongated appearance, with neck extended. Favours large areas of deep open water.

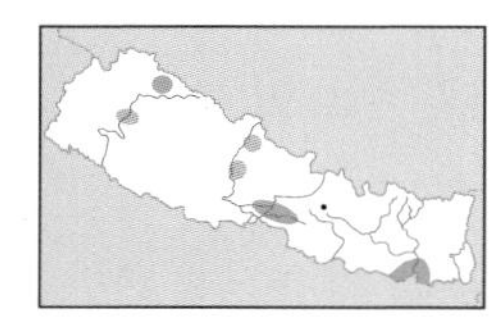

Black-necked Grebe *Podiceps nigricollis* 28–34 cm

Local and scarce winter visitor; 75–3,050 m. Steep forehead, with crown typically peaking at front or centre. In non-breeding plumage, head pattern more contrasting than Little; black of crown extends below eye, ear-coverts dusky-grey, and striking white throat curves up behind ear-coverts. Has yellow ear-tufts, black neck and breast, and rufous flanks in breeding plumage. Juvenile as non-breeding, but may show buff wash to cheeks and foreneck, and more closely resembles Little. Typically swims with curved neck. **Voice** Silent in non-breeding season. **HH** Habits similar to those of Little Grebe, but prefers reedbeds in shallows to open water. Reed-fringed lakes with emergent vegetation. **AN** Eared Grebe.

Common Goldeneye
♂
♀
♂
♀
Goosander
♂
♀
♂
♀
non-br
br
Little Grebe
non-br
br
non-br
Great Crested Grebe
non-br
br
Black-necked Grebe

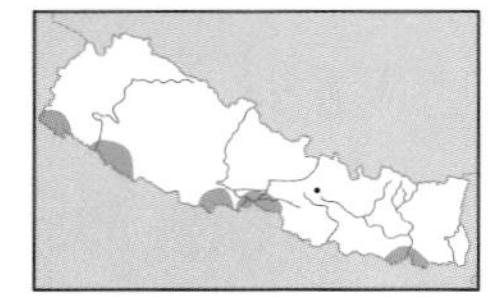

Painted Stork *Mycteria leucocephala* 93–100 cm

Very rare non-breeding visitor; 75–250 m. **ID** Adult has downcurved yellow bill, bare orange head (redder in breeding season), and pinkish legs; white barring on mainly black upperwing-coverts, pinkish tertials, and black barring on breast. In flight, underwing appears mainly dark, with whitish barring on coverts. Juvenile dirty greyish-white, with grey-brown (feathered) head and neck, and brown lesser coverts; bill and legs duller than adult's. Has distinctive appearance in flight, with extended drooping neck and downcurved bill, long wings with rather deep flapping beats, and long trailing legs. **Voice** Largely silent away from nesting colonies. **HH** Like other storks has a powerfully slow-flapping flight and frequently soars for long periods, often at great heights. Found singly and in small parties up to about eight birds, often on large marshes. Forages by wading slowly in shallow water with bill open and partly submerged, feeling for prey. Often stirs water with its foot and occasionally flicks a wing to drive prey between its mandibles. Roosts gregariously in trees if available, otherwise on open sandbanks or on mud.

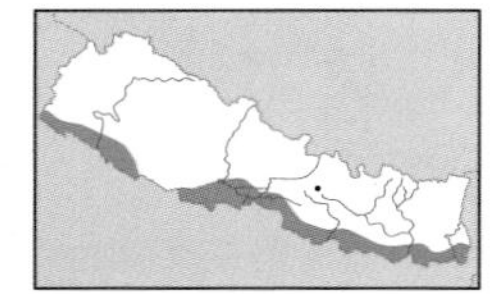

Asian Openbill *Anastomus oscitans* 68 cm

Widespread resident in the lowlands; 75–250 m (–1,400 m). **ID** Stout, dull-coloured 'open bill'. Largely white (breeding) or greyish-white (non-breeding), with black flight feathers and tail; legs usually dull pink, brighter in breeding condition. Juvenile has brownish-grey head, neck and breast, and brownish mantle and scapulars slightly paler than blackish flight feathers. Forages singly or in small to medium-sized flocks. **Voice** Largely silent away from the nest. **HH** Usually seeks food by submerging its head and open bill into shallow water and probing bottom mud; the bill is quickly closed on any prey. Feeds mainly on molluscs. The gap in the bill is an adaption to extract the soft body and viscera of molluscs from the shell. Moves locally depending on water conditions. Like other storks, it regularly soars on thermals on sunny days, circling for hours high overhead. Forms large colonies, sometimes with other waterbirds. Nests in trees, often over water. Inhabits marshes, shallow lakes and reservoirs; also flooded paddyfields; rarely on riverbanks.

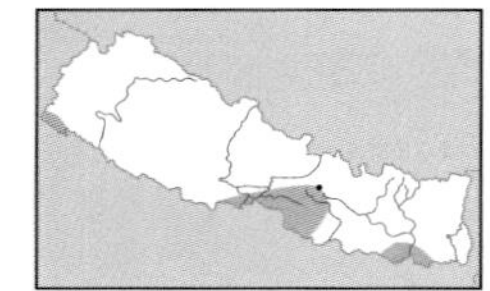

White Stork *Ciconia ciconia* 100–125 cm

Very rare passage migrant; 75–305 m (–915 m). **ID** Mainly white, with black flight feathers and striking red bill and legs. Generally has cleaner black-and-white appearance than Asian Openbill; note tail is white (black in Asian Openbill). Juvenile similar to adult but has brown greater coverts and duller brownish-red bill and legs. **Voice** Largely silent away from the nest. **HH** Mainly found singly in Nepal. Habits similar to those of other storks. Flight a few flaps followed by a glide and appears leisurely, but is fast and strong. Soars on thermals and glides in circles aloft for many hours at a stretch. Often roosts at night in bare treetops. Usually shy and difficult to approach. Stalks deliberately on dry or moist ground in search of prey. Grassland and damp ploughed or fallow fields.

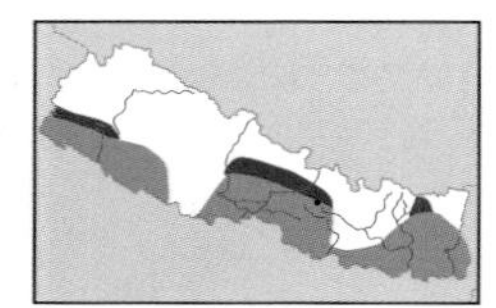

Woolly-necked Stork *Ciconia episcopus* 75–92 cm

Fairly common and widespread resident; 75–915 m (–1,800 m). **ID** Stocky, largely blackish stork with 'woolly' white neck, black 'skullcap', and white vent and undertail-coverts. Adult has black of body and wings glossed greenish-blue, purple and copper. Bill is black, with variable amounts of red, and legs and feet are dull red. Juvenile similarly patterned, but has duller brown body and wings, and feathered forehead. In flight, upperwing and underwing entirely dark. **Voice** Largely silent away from the nest. **HH** Usually found singly or in pairs, occasionally in small parties. Habits similar to those of other storks. Has a slow-flapping flight and characteristic stork habit of soaring and circling high up during heat of day. Hunts on dry or marshy ground and wet grasslands; rarely wades. Builds solitary nest on a horizontal branch or near the top of a tree. Inhabits flooded grassland, marshes, irrigated fields and riverine areas; usually near open wooded country.

Painted Stork
imm
ad
imm
ad
br
non-br
br
Asian Openbill
ad
imm
ad
White Stork
ad
Woolly-necked Stork

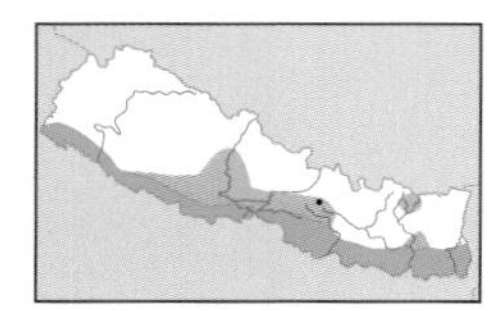

Black Stork *Ciconia nigra* 90–100 cm

Widespread winter visitor and passage migrant; 75–1,000 m (–2,925 m). **ID** Adult mainly glossy black, with white lower breast and belly, and red bill and legs; in flight, white underparts and axillaries contrast strongly with black neck and underwing. Juvenile has brown head, neck and upperparts flecked with white; bill and legs greyish-green. **Voice** Largely silent away from the nest. **HH** In pairs or small parties. Often very shy and wary. Forages by walking with measured strides in shallow water. Flight a few flaps followed by a glide and is fast and strong. Soars on thermals, with legs and neck extended, for hours at time. Marshes, rivers and cut-over paddyfields.

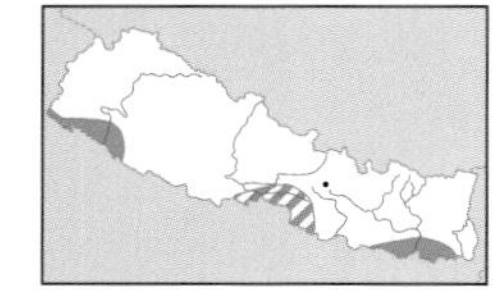

Black-necked Stork *Ephippiorhynchus asiaticus* 129–150 cm

Very rare and very local resident in lowlands; 75–300 m. **ID** Large, black-and-white stork with long red legs and huge black bill. In flight, wings white except broad black band across coverts, and tail black. Male has brown iris; yellow in female. Juvenile has fawn-brown head, neck and mantle, mainly brown wing-coverts, and mainly blackish-brown flight feathers; legs dark. **Voice** Largely silent away from the nest. **HH** Forages singly, in well-separated pairs in sight of each other, or in family parties after the breeding season. Usually very wary. When foraging, walks sedately in marshes or wades slowly while probing aquatic vegetation and in shallow water with its bill open at the tip. Frequently soars for long periods, often at great heights, the legs extended and the neck outstretched like other storks. Builds solitary nest in a large tree. Inhabits marshes and large rivers.

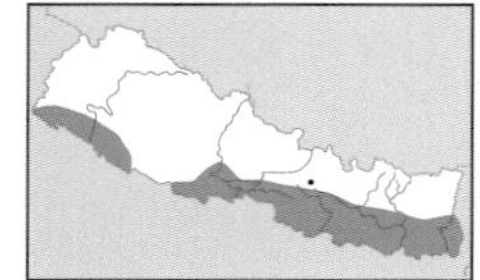

Lesser Adjutant *Leptoptilos javanicus* 110–120 cm

Local resident in lowlands, now mainly in the east; 75–250 m (–1,450 m). **ID** Flies with neck retracted, as Greater Adjutant, giving rise to different profile compared with other storks. Smaller than Greater, with slimmer bill that has straighter ridge to culmen. From adult breeding Greater Adjutant by smaller size, glossy black mantle and wings (lacking paler panel across greater coverts – although this is much less distinct in non-breeding and immature Greater), and white undertail-coverts; neck ruff is largely black (appearing as black patch on sides of breast in flight). Further, has pale frontal plate, denser hair-like feathering on back of head (forming small crest) and down hindneck, and lacks neck pouch. Adult breeding has red tinge to face and neck, copper spots at tips of median coverts, and narrow white fringes to scapulars and inner greater coverts. Juvenile similar to adult, but upperparts are dull black, and head and neck duller and more densely feathered. **Voice** Largely silent away from the nest. **HH** Usually found singly. Forages by walking slowly on dry ground or in shallow water, and grabs prey with its bill. Semi-colonial or colonial when nesting, usually in small numbers. Builds nest in large tree. Frequents flooded fields, mainly paddyfields, marshes and pools. Globally threatened.

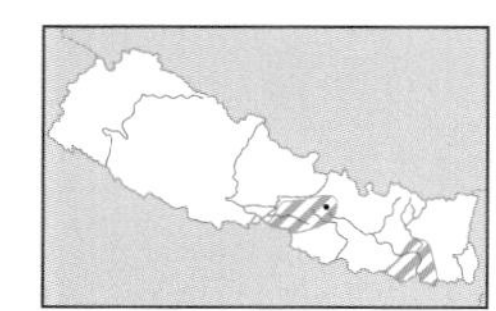

Greater Adjutant *Leptoptilos dubius* 120–150 cm

Former rare and erratic non-breeding visitor to the centre and east; no known records since 1995; 75–245 m (–1,500 m). **ID** Larger than Lesser Adjutant, with stouter, conical bill with convex ridge to culmen. Adult breeding from adult Lesser Adjutant by larger size, bluish-grey (rather than glossy black) mantle, prominent silvery-grey panel across greater coverts and tertials, mainly white neck ruff (lacking or with less pronounced black patch on sides of breast in flight), and grey undertail-coverts. Further, has blackish face and forehead (with appearance of dried blood), more sparsely feathered head and neck (lacking small crest), and larger neck pouch (visible only when inflated). Adult non-breeding has darker grey mantle and wing-coverts (which barely contrast with rest of wing). Immature similar to adult non-breeding, but upperparts including wings are browner, has brownish (rather than whitish) iris, and head and neck are more densely feathered. **Voice** Largely silent away from the nest. **HH** Usually seen singly. Habits similar to those of Lesser Adjutant, but less shy and, unlike that species, feeds partly on carrion. Also hunts small live animals in typical stork fashion, by walking slowly in marshes and shallow waters. Marshes and open fields. Globally threatened.

imm
ad
ad
Black Stork
♀
♂
ad
imm
imm
Black-necked Stork
ad
ad
br
br
imm
non-br
br
Lesser Adjutant
Greater Adjutant

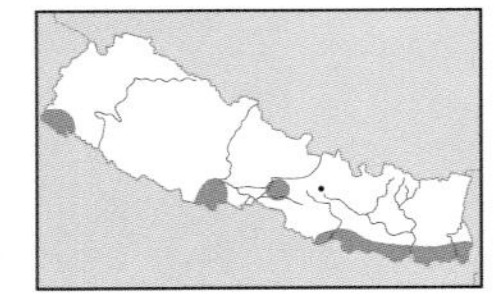

Black-headed Ibis *Threskiornis melanocephalus* 75 cm

Resident, mainly recorded in south-east lowlands; may be nomadic depending on water and feeding conditions; 75–200 m. **ID** Stocky, mainly white ibis with stout downcurved black bill. Adult breeding has naked black head, white lower-neck plumes, variable yellow wash to mantle and breast, and grey on scapulars and elongated tertials. In flight, shows stripe of bare red skin on underside of white forewing and on flanks. Adult non-breeding has all-white body and lacks neck plumes. Immature has grey feathering on head and neck, and black-tipped wings. **Voice** Usually silent away from breeding colonies. **HH** Seen mainly in parties. Forages in shallow water and marshes, often in well-scattered flocks. Walks about or often wades belly-deep while rapidly probing water and mud. Readily perches and roosts in trees. Inhabits flooded fields, marshes, rivers and pools.

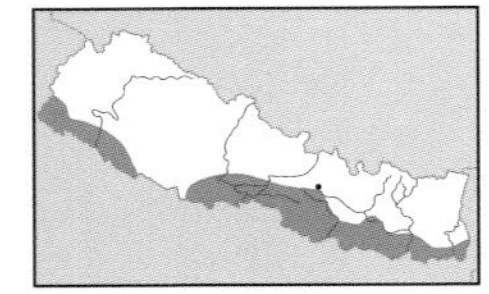

Red-naped Ibis *Pseudibis papillosa* 68 cm

Fairly common and widespread resident; 75–275 m (–915 m). **ID** Stocky, dark ibis with relatively stout downcurved bill. Has white shoulder patch and reddish legs. Appears bulky and broad-winged in flight, with only the feet extending beyond tail. Adult has naked black head with red nape, and is dark brown with green-and-purple gloss. Immature dark brown, including feathered head. See Appendix 1 for comparison with Glossy Ibis. **Voice** Mainly silent away from breeding colonies. **HH** Habits similar to those of Black-headed Ibis, but prefers to feed in drier habitats and is less gregarious. Usually found singly or in small parties of up to ten. Often nests singly, but sometimes in small, unmixed colonies. Builds a large platform nest in a large tree; old nests of kites, storks and vultures are often used. Frequents riverbanks and open fields; sometimes in dry areas. **AN** Black Ibis.

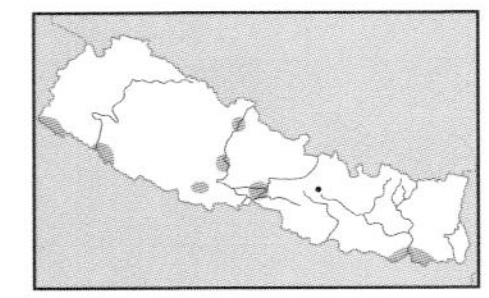

Eurasian Spoonbill *Platalea leucorodia* 80–90 cm

Very rare and very local winter visitor and passage migrant; almost all recent records from Koshi Barrage and Koshi Tappu Wildlife Reserve in lowlands. **ID** White, with spatulate-tipped bill. In flight, neck is outstretched, and flapping is rather stiff and interspersed by gliding. Adult has black bill with yellow tip; crest and yellow breast patch when breeding. Juvenile has pink bill; in flight, shows black tips to primaries. **Voice** Usually silent away from breeding colonies. **HH** Usually in small parties or flocks. Spends much of day resting on one leg or sleeping with bill tucked under a wing. Forages mainly in mornings and evenings and at night. Wades actively in shallow water, making rhythmic side-to-side sweeps of its bill and sifting floating and swimming prey. Marshes, large rivers and large lakes.

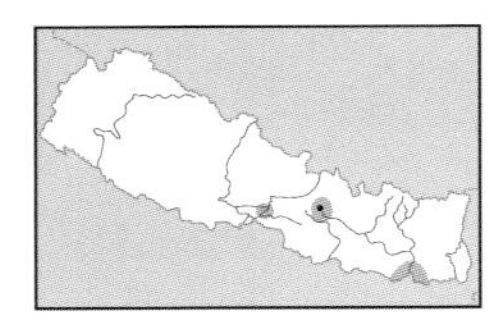

Spot-billed Pelican *Pelecanus philippensis* 140 cm

Rare and irregular non-breeding visitor with almost all known records from Koshi in the far east; 75–150 m. **ID** Much smaller than Great White Pelican (see Appendix 1), with dingier appearance, rather uniform pinkish bill and pouch (except in breeding condition), and black spotting on upper mandible (except juveniles). Pale circumorbital skin looks cut off from bill (as if wearing goggles). Tufted crest/hindneck usually apparent even in young birds. Underwing pattern quite different from Great White, with little contrast between wing-coverts and flight feathers, and paler greater coverts producing distinct central panel. Adult breeding has cinnamon-pink rump, underwing-coverts and undertail-coverts; head and neck appear greyish; purplish skin in front of eye, and pouch is pink to dull purple and blotched black. Adult non-breeding dirtier greyish-white, with paler pouch and facial skin. Immature has variable grey-brown markings on upperparts. Juvenile has brownish head and neck, brown mantle and upperwing-coverts (fringed pale buff), and brown flight feathers; spotting on bill initially lacking (and still indistinct at 12 months). **Voice** Usually silent away from breeding colonies. **HH** Recently found alone or in small flocks of up to four birds. Fishes alone or cooperatively by swimming forward in a semi-circular formation, driving fish into shallow waters. Each bird then scoops up fish from water into its pouch, before swallowing them. A powerful flier, proceeding by steady flaps with the head drawn back between the shoulders. Large expanses of water on the Koshi River.

Black-headed Ibis
imm
ad
ad
Red-naped Ibis
br
juv
Eurasian Spoonbill
imm
non-br
br
juv
Spot-billed Pelican

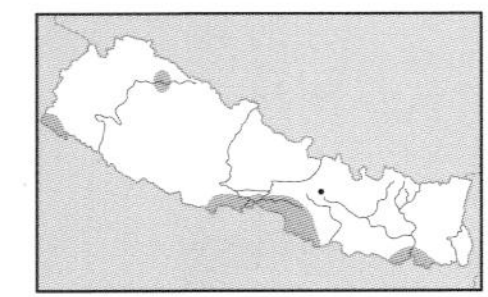

Great Bittern *Botaurus stellaris* 70–80 cm

Very rare passage migrant, mainly in the lowlands, only recorded recently from Koshi area in the far east; 75–250 m (–3,050 m). **ID** Stocky, broad-winged, with stout-looking neck and head. Golden-brown and cryptically patterned, with boldly streaked neck and breast, and black crown and moustachial stripe. In flight, wing-coverts appear paler than brown (barred) flight feathers. **Voice** Typically silent away from breeding areas. **HH** Habits are those of typical bittern. Normally crepuscular. Usually remains hidden in reedbeds and is most often seen flying low over reed tops, soon dropping into cover again. Hunts alone, by walking stealthily through vegetation, often with intervals of standing motionless. Can clamber on reed stems with agility. Freezes when in danger, pointing bill and neck upwards and compressing its feathers so that the whole body appears elongated. Dense tall wet beds of *Phragmites* reeds or *Typha* bulrushes in lakes and marshes. **AN** Eurasian Bittern.

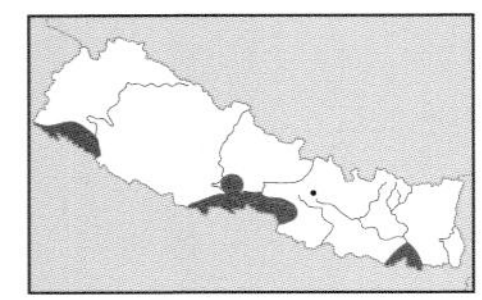

Yellow Bittern *Ixobrychus sinensis* 38 cm

Local and uncommon, mainly a summer visitor, rare in winter; 75–915 m. **ID** Yellowish-buff wing-coverts contrast with dark brown flight feathers. Male has pinkish-brown mantle/scapulars, and face and sides of neck are vinaceous. Female similar to male, but has rufous streaking on black crown, variable rufous-orange streaking on foreneck and breast, and buff streaking on rufous-brown mantle and scapulars. Juvenile appears buff with bold dark streaking to upperparts including wing-coverts; foreneck and breast heavily streaked. **Voice** Territorial call a low-pitched *ou-ou*. **HH** Habits are those of a typical bittern. Solitary. Most active at dusk; usually spends day concealed in thick waterside vegetation, but may be seen during the daytime in cloudy weather. Forages by creeping through dense vegetation or by standing and waiting at edges of cover. If disturbed, often freezes with head and bill pointing vertically skywards. Nest is usually built in a dense reedbed. Inhabits reedbeds and marshes.

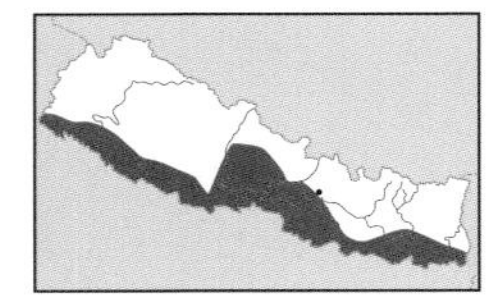

Cinnamon Bittern *Ixobrychus cinnamomeus* 38 cm

Frequent, mainly a summer visitor, but recorded in all months; chiefly 75–250 m; has bred 75–1,370 m. **ID** Uniform-looking cinnamon-rufous flight feathers and tail in all plumages. Male has cinnamon-rufous crown, hindneck and mantle/scapulars. Female has browner crown and mantle, and brown streaking on foreneck and breast. Juvenile has buff mottling on dark brown upperparts, and is heavily streaked dark brown on underparts. **Voice** Territorial call a loud *kok-kok*. **HH** Habits very similar to those of Yellow; often found in same locality and same habitat as that species. Nest is built on bent-over reeds about 1 m above water or mud. Inhabits reedbeds in lakes and marshes.

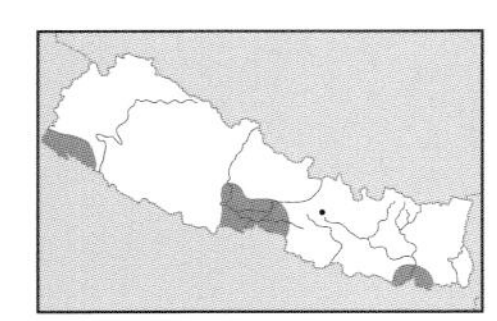

Black Bittern *Dupetor flavicollis* 58 cm

Local rare resident recorded from far west to the far east; 75–250 m. **ID** Male has blackish upperparts, with yellowish malar and sides of neck, and dark streaking on underparts. Female similar but browner upperparts and chestnut-streaked underparts. Juvenile similar but has distinct fringes to upperparts. **Voice** Territorial call a loud booming. **HH** Habits are those of a typical bittern. Chiefly nocturnal and crepuscular; skulks in dense swamps during day. Most often seen flying at dawn and dusk and in cloudy weather. Nests in reeds or dense thicket in a marsh. Inhabits forest pools, marshes and reed-fringed lakes.

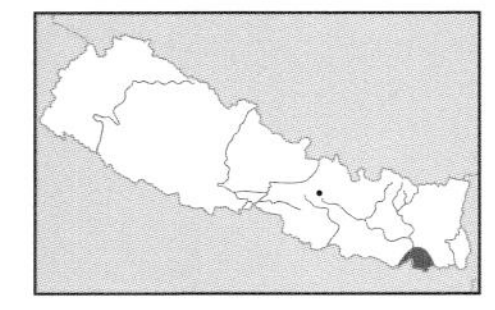

Malayan Night Heron *Gorsachius melanolophus* 51 cm

Very rare and very local summer visitor to the east; 150 m. **ID** Stocky, with stout bill and short neck. Adult has black crown and crest, rufous sides to head and neck, and rufous-brown upperparts. Juvenile is greyish in coloration, and finely vermiculated white, black and rufous-buff, with bold white spotting on crown and crest. **Voice** Usually silent. A sequence of 10–11 deep *oo* notes about 1.5 seconds apart recorded in Thailand. **HH** Shy and mainly nocturnal. Skulks in damp places in undergrowth in dense forest during the day. If flushed, flies off silently into a nearby thickly foliaged tree. Builds nest in a small tree overhanging a stream in thick forest. Inhabits wet areas in dense broadleaved forest.

ad
Great Bittern
♂
juv
♂
Yellow Bittern
♂
♀
juv
Cinnamon Bittern
Black Bittern
♂
juv
♂
ad
imm
Malayan Night Heron

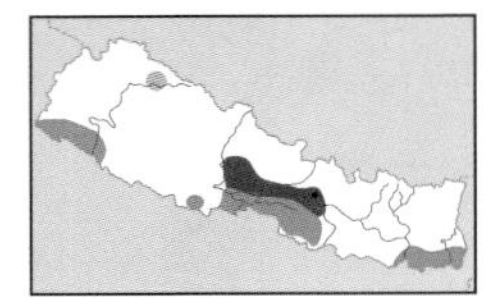

Black-crowned Night Heron *Nycticorax nycticorax* 58–65 cm

Locally common summer visitor and resident; 75–1,370 m. **ID** Stocky, with thick neck. Adult has black crown and mantle contrasting with grey wings and whitish underparts. Juvenile brownish and is boldly streaked and spotted. Immature similar to adult but has browner mantle/scapulars, and variable streaking on underparts. **Voice** A distinctive, deep and rather abrupt *wouck* in flight. **HH** Nocturnal and crepuscular except when feeding young. Usually spends day perched hunched in a densely foliaged tree or reedbed. Most frequently seen at dusk, flying singly or in small groups from its daytime roost. Like other herons flies with leisurely flaps, with the legs projecting beyond the tail and nearly always with head and neck drawn back. Mainly forages alone, sometimes in a loose group. Ponds, tanks and lakes.

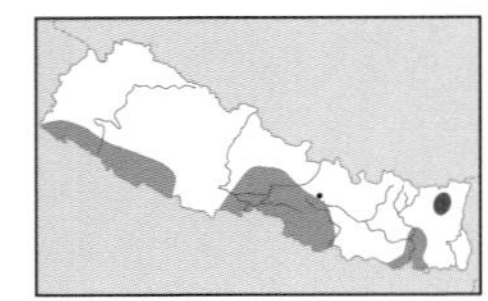

Striated Heron *Butorides striata* 40–48 cm

Frequent resident and summer visitor; 75–915 m. **ID** Small, stocky and short-legged heron. Adult has black crown and crest, dark greenish upperparts and greyish underparts. Juvenile has buff streaking and spotting on upperparts and dark-streaked underparts. Immature similar to juvenile with uniform brown crown and mantle. **Voice** Usually silent, but can give a *k-yow k-yow* or *k-yek k-yek*. **HH** Normally frequents same area day after day. Hunts alone and in typical heron fashion. Often crepuscular although sometimes active during day, especially in overcast weather. In daytime mainly keeps to thick vegetation on banks of rivers and pools, and often seen perching on branches overhanging water. Pools, lakes, streams and rivers with dense shrubby vegetation on banks. **AN** Little Heron.

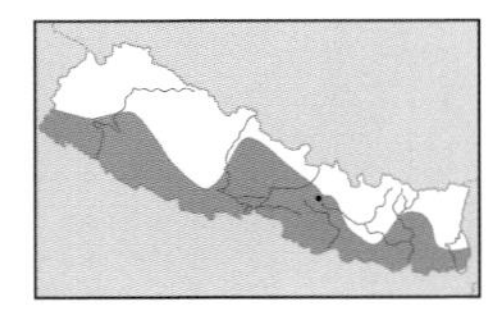

Indian Pond Heron *Ardeola grayii* 42–45 cm

Common and widespread resident; 75–1,525 m. **ID** Whitish wings contrast with dark mantle/scapulars. Adult breeding has yellowish-buff head and neck, and maroon-brown mantle/scapulars. Head, neck and breast streaked/spotted in non-breeding and immature plumages. **Voice** A high, harsh squawk when flushed. **HH** Usually solitary when hunting, but will gather in large numbers at drying-out pools to feed on stranded fish. Like other herons, typically hunts by standing motionless at water's edge, waiting for prey to swim within reach, or by slow stalking in shallow water or on land. Roosts communally. Tame and inconspicuous when perched, but flies up with a startling flash of white wings. Marshes, flooded paddyfields, lakes, village tanks, streams and ditches.

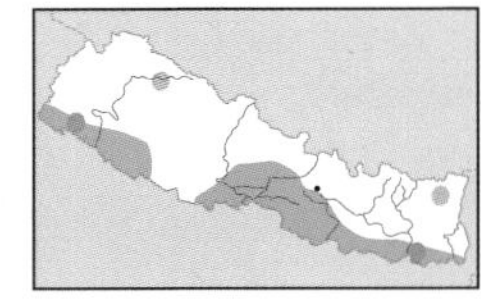

Grey Heron *Ardea cinerea* 90–98 cm

Rare resident and fairly common winter visitor; widespread below 915 m, but recorded up to 3,050 m. **ID** A large, mainly grey heron, lacking any brown or rufous in its plumage. In flight, black flight feathers contrast with grey upperwing- and underwing-coverts, and shows prominent white leading edge to wing when head-on. Adult has yellow bill, whitish head and neck with black head plumes, and black patches on belly. In breeding season, whitish scapular plumes and bill and legs become orange or reddish. Immature duller than adult, with grey crown, reduced black 'crest', greyer neck, less pronounced black patches on sides of belly, and duller bill and legs. Juvenile has dark grey cap with slight crest, dirty grey neck and breast, lacks black patches on belly sides, lacks plumes, and has dark legs. See Appendix 1 for comparison with White-bellied Heron. **Voice** Often calls in flight, a loud *frarnk*. **HH** A typical diurnal heron. Usually forages alone; occasionally gathers in loose parties at good feeding areas. Roosts communally in winter. Prefers to hunt in open unlike Purple Heron. Lakes, large rivers and marshes.

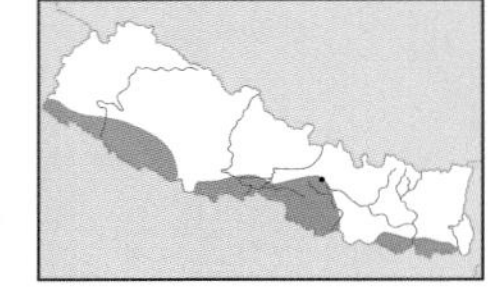

Purple Heron *Ardea purpurea* 78–90 cm

Widespread, locally fairly common, mainly a resident and monsoon visitor; 75–300m (–1,370 m). **ID** Rakish, with long, thin neck. In flight, compared to Grey Heron, bulge of recoiled neck is very pronounced, protruding feet large, underwing-coverts purplish (adult) or buff (juvenile) and lacks white leading edge to wing. Adult has chestnut head and neck with black stripes, grey mantle and upperwing-coverts, and dark chestnut belly and underwing-coverts. Juvenile has black crown, buffish neck, and brownish mantle and upperwing-coverts with rufous-buff fringes. **Voice** Flight call similar to Grey's, but a higher-pitched and quieter *frarnk*. **HH** Active mainly in early mornings and evenings; sometimes also feeds by day. Shyer than Grey, normally feeding amongst dense aquatic vegetation. Most often seen in flight. Hunts alone, usually by standing motionless and waiting; less often by slow stalking in shallow water. Dense reedbeds in lakes and marshes.

ad
Black-crowned Night Heron
juv
ad
Striated Heron
juv
ad
ad
Grey Heron
imm
non-br
br
Indian Pond Heron
ad
Purple Heron
juv

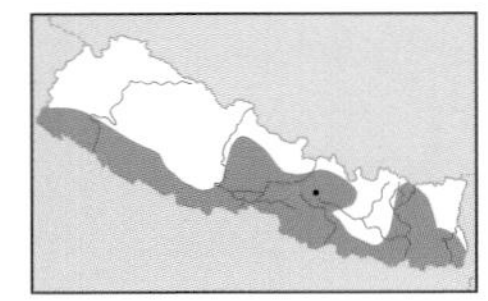

Cattle Egret *Bubulcus ibis* 48–53 cm

Common and widespread resident; 75–1,525 m. **ID** Small and stocky with short yellow bill and short dark legs. Has orange-buff on head, neck and mantle in breeding plumage; base of bill and legs become reddish in breeding condition. All white in non-breeding plumage. **Voice** Usually silent away from breeding colony. **HH** Gregarious when feeding and roosting. Typically seen in flocks around domestic stock and with Wild Buffalo, feeding on insects disturbed by the animals; often rides on animals' backs, picking parasitic insects and flies from their hides. Also forages in flooded fields. Unlike other egrets, feeds mainly on insects, also tadpoles and lizards. Breeds colonially in large trees, not necessarily close to water; sometimes with other herons and egrets. Inhabits damp grassland, paddyfields, lakes, pools and marshes. **AN** Eastern Cattle Egret, if eastern form *coromandus* is split.

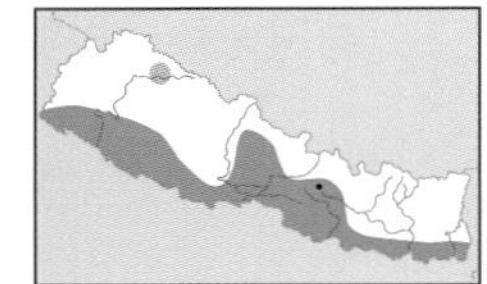

Great Egret *Casmerodius albus* 90–102 cm

Locally fairly common and widespread resident; 75–915 m (–3,050 m). Compared to Intermediate, is larger and longer-billed, and looks thinner-necked with more angular and pronounced kink to neck. Black line of gape extends behind eye. Bill black, lores blue and tibia reddish in breeding plumage when has prominent plumes on mantle. In non-breeding plumage, bill yellow and lores pale green. **Voice** Normally silent, but occasionally utters low *kraak*; gives various deep guttural calls and softer notes during display. **HH** A typical diurnal heron. Like other herons, feeds by standing motionless at water's edge, waiting for prey to swim within reach, or by slow stalking in shallow water or on land. Prey normally grasped and killed by battering, less often speared. Generally less sociable than other egrets, and often solitary when hunting, but will feed communally at concentrated food sources. Roosts communally. Breeds colonially with other herons and cormorants. Builds nest in solitary tree or in a grove, either standing in water or on dry land. Inhabits rivers, lakes, marshes, pools and wet fields.

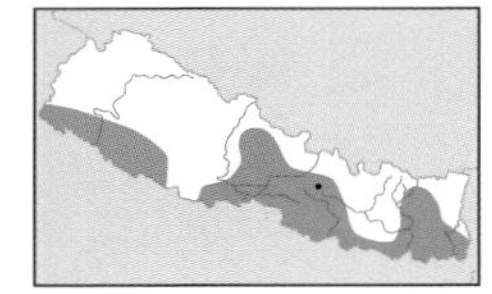

Intermediate Egret *Mesophoyx intermedia* 65–72 cm

Mainly resident and widespread, locally fairly common or locally common; 75–915 m (–4,695 m). **ID** Smaller than Great, with shorter bill and neck. Black gape-line does not extend beyond eye. Bill black and lores yellow-green during courtship, with pronounced plumes on breast and mantle. Has black-tipped yellow bill and yellow lores outside breeding season. **Voice** Normally silent; gives distinctive buzzing calls during display. **HH** A typical diurnal heron. Flies with leisurely flaps, with legs projecting beyond the tail and nearly always with head and neck drawn back like other herons. Usually in small flocks, which separate when foraging. Hunts chiefly by slow stalking. Roosts communally. Breeds colonially with other herons and cormorants. Builds nest in solitary tree or in a grove, either standing in water or on dry land. Inhabits marshes, flooded grassland, well-vegetated pools; also shores of lakes, reservoirs and slow-moving rivers.

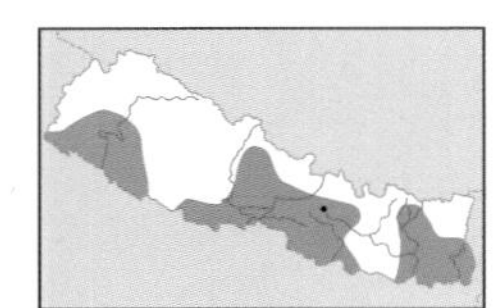

Little Egret *Egretta garzetta* 55–65 cm

Common and widespread resident; 75–1,525 m. **ID** Slim and graceful. Has black bill, black legs with yellow feet, and greyish or yellowish lores. In breeding plumage has two elongated nape plumes, and mantle plumes, while lores and feet become reddish during courtship. Bill in non-breeding and immature plumages can be paler and pinkish or greyish at base, or dull yellowish on some. **Voice** Normally silent, except throaty squawk when disturbed and various guttural calls at colonies. **HH** A typical diurnal heron. Often in flocks when foraging and more sociable than the two larger egrets; also found alone. Roosts communally. Breeds colonially with other herons and cormorants. Builds nest in solitary tree or grove, either standing in water or on dry land. Inhabits lakes, rivers, pools, marshes and flooded paddyfields.

br
non-br
br
coromandus
Cattle Egret
br
non-br
br
Great Egret
non-br
non-br
br
Intermediate Egret
non-br
br
non-br
Little Egret

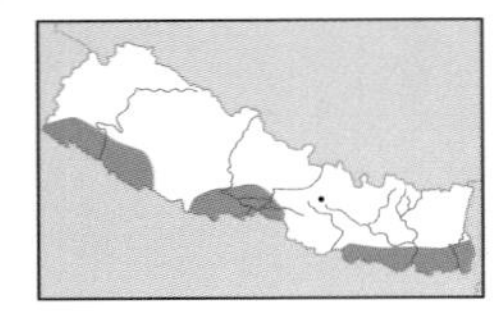

Little Cormorant *Phalacrocorax niger* 51 cm

Resident, winter visitor and passage migrant; locally common; 75–300 m. **ID** Much smaller than Great Cormorant. Has shortish bill, rectangular-shaped head (with steep forehead), short neck and long-looking tail. Lacks yellow gular pouch. Adult breeding all black, with white plumes on sides of head. Bill, eyes, facial skin and pouch black. Non-breeding browner (and lacks white head plumes), with whitish chin, and paler bill and pouch. Immature has whitish chin and throat, and foreneck and breast a shade paler than upperparts, with some pale fringes. See Appendix 1 for comparison with Indian Cormorant. Like other cormorants swims with body low in water, neck straight and head and bill pointing slightly upwards. In flight neck extended and head held slightly above the horizontal. **Voice** Usually silent except in vicinity of nest. **HH** A typical cormorant. Like other cormorants, a good swimmer, aided by its strong legs set far back on body and by its large webbed feet, which are used for propulsion. On smaller waters occurs singly or in small groups; on large waters often gathers in flocks. Breeds in mixed colonies with other waterbirds in trees. Builds a typical cormorant nest comprising a platform of sticks lined with weeds. Inhabits rivers, lakes, reservoirs and pools.

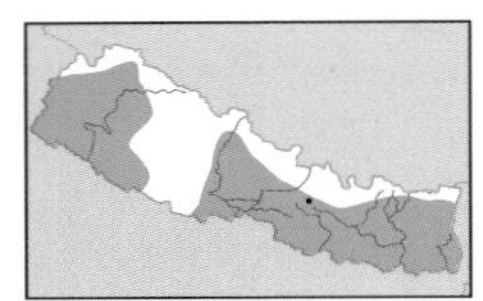

Great Cormorant *Phalacrocorax carbo* 80–100 cm

Fairly common and widespread winter visitor; 75–1,370 m (–3,960 m). **ID** Much larger and bulkier than Little. Has thick neck, large and angular head, and stout bill. Adult breeding glossy black, with dark gular skin, red spot at base of bill, white cheeks and throat, extensive white plumes covering much of head, and white thigh patch. Non-breeding lacks white head plumes and thigh patch. Base of bill and gular skin yellow. Immature similar but browner with underparts dark or extensively whitish or pale buff. See Appendix 1 for comparison with Indian Cormorant. Flies and swims like a typical cormorant, see Little. **Voice** Usually silent except in vicinity of nest. **HH** A typical cormorant. Like other cormorants, often perches for long periods in an upright posture, with spread wings and tail. Usually found alone or in small groups, although sometimes joins fishing flocks of other cormorants. Often roosts communally in winter. Lakes and large rivers.

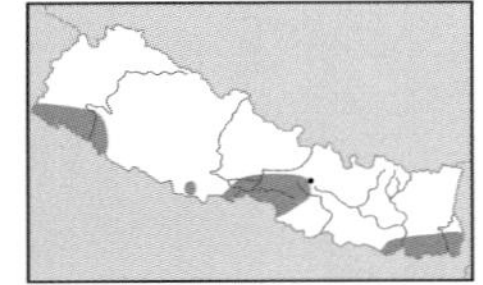

Darter *Anhinga melanogaster* 85–97 cm

Local breeding resident and visitor; 75–300 m (–1,370 m). **ID** Long, slim head and neck, dagger-like bill, and long tail. Adult breeding has dark brown crown and hindneck, white stripe down side of neck, blackish breast and underparts, lanceolate white scapular streaks, and white streaking on wing-coverts. Duller in non-breeding plumage. Immature browner with indistinct neck stripe and buff fringes to coverts form pale panel on upperwing. **Voice** Usually silent except in vicinity of nest, when utters a variety of harsh rattling and grunting calls. **HH** Seen singly, in scattered pairs and sometimes in larger groups. Spends much time drying its spread wings and tail while on a favoured perch. Often swims with head and neck above water and body below. Unlike cormorants, does not leap upwards before diving, but slowly submerges producing hardly a ripple. Pursues fish underwater, and catches them by spearing. Often nests colonially with other large waterbirds. Nest an untidy platform of twigs lined with weeds, built on trees standing in or near water. Inhabits lakes, ponds, slow-moving rivers and marshes. **AN** Oriental Darter if African and Australasian forms are split.

non-br
imm
br
Little Cormorant
non-br
br
imm
br
Great Cormorant
♂ br
♂ br
imm
Darter

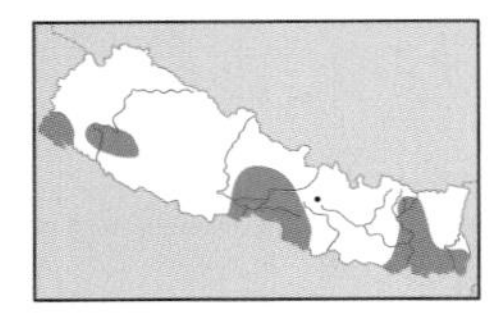

Collared Falconet *Microhierax caerulescens* 18 cm

Local and uncommon or frequent resident; 75–1,050 m. **ID** Very small, with broad wings and long, square-ended tail. Flies with rapid beats interspersed by long glides. Rather shrike-like when perched. Adult has white collar, black crown and eye-stripe, and rufous-orange underparts. Juvenile has rufous-orange on forehead and supercilium, white throat and yellowish bill. **Voice** High *kli-kli-kli* or *killi-killi-killi*. **HH** Perches on dead branches and makes short, swift darting sorties to seize prey. Edges and clearings of broadleaved forest.

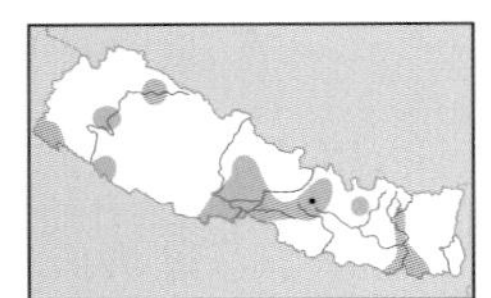

Lesser Kestrel *Falco naumanni* 29–32 cm

Mainly October/November passage migrant, also irregular and uncommon winter visitor and rare spring passage migrant; 75–2,745 m (–3,700 m). **ID** Slightly smaller and slimmer than Common Kestrel. Claws whitish (black in Common). Male has uniform blue-grey head (without dark moustachial stripe), unmarked rufous upperparts, blue-grey greater coverts, and almost plain orange-buff underparts. In flight, underwing whiter with more clearly pronounced dark wingtips; tail often looks more wedge-shaped. First-year male more like Common, but has unmarked rufous mantle and scapulars. Female and juvenile have less distinct moustachial stripe than Common, and lack any suggestion of dark eye-stripe; underwing less heavily marked with more pronounced dark wingtips. **Voice** Less piercing and more slurred than Common's. **HH** Hunting manner rather like Common's, but usually in small or large flocks, and is more agile and graceful in flight. Roosts communally. Open country.

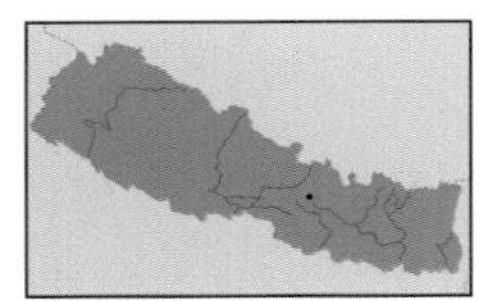

Common Kestrel *Falco tinnunculus* 32–35 cm

Common resident, winter visitor and passage migrant; 75–5,200 m. **ID** Long, rather broad tail; wingtips more rounded than on most falcons. Male has greyish head with diffuse dark moustachial stripe, rufous upperparts heavily marked with black, and grey tail with black subterminal band. Female and juvenile have rufous crown and nape streaked black, diffuse and narrow dark moustachial stripe, rufous upperparts heavily marked with black, and dark barring on rufous tail. **Voice** High-pitched, shrill *kee-kee-kee*. **HH** Usually found alone or in pairs. Hovers with rapidly beating wings and fanned tail, while scanning ground for prey. Open country.

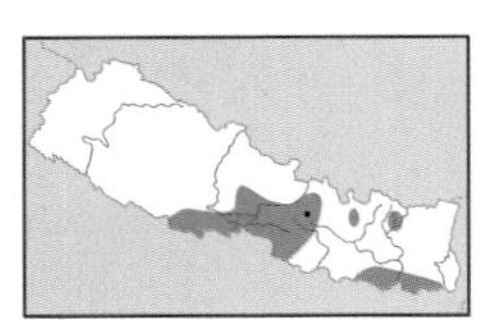

Red-necked Falcon *Falco chicquera* 31–36 cm

Resident, uncommon at Koshi, rare elsewhere; 75–1,400 m. **ID** Adult has rufous crown, nape and narrow moustachial stripe, pale blue-grey upperparts with fine dark barring, white underparts finely barred black, and grey tail with broad black subterminal band. In flight, blackish primaries contrast with rest of upperwing. Sexes alike, but female larger. Juvenile similar but darker, with fine dark shaft-streaking on crown, fine rufous fringes to upperparts, and fine rufous-brown barring on underparts. **Voice** Shrill *ki-ki-ki-ki-ki*, rasping *yak, yak, yak* and screaming *tiririri, tiriririeee*. **HH** Flight usually fast and dashing. Often hunts cooperatively in pairs, one bird pursuing prey and the other cutting off its escape. Open country with trees.

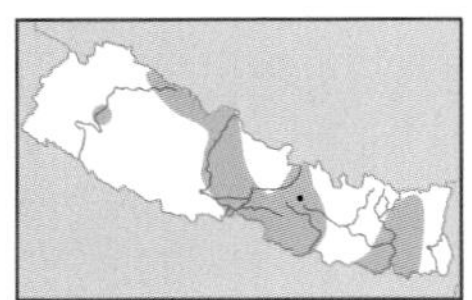

Amur Falcon *Falco amurensis* 28–31 cm

Uncommon passage migrant, mainly October/November, but recorded in all months; 75–2,900 m (–4,420 m). **ID** In all plumages, has red to pale orange cere, eye-ring, legs and feet. Shape similar to Eurasian Hobby but has slightly more rounded wingtips and longer tail. Male dark grey, with rufous thighs and undertail-coverts, and white underwing-coverts. First-year male shows mix of adult male and juvenile characters. Female has dark grey upperparts, short moustachial stripe, whitish underparts with some dark barring and spotting, and orange-buff thighs and undertail-coverts; uppertail barred; underwing white with strong dark barring and dark trailing edge. Juvenile similar to female but has rufous-buff fringes to upperparts, rufous-buff streaking on crown, and boldly streaked underparts. **Voice** Shrill, screaming *kew-kew-kew* when settling to roost; may continue throughout night. **HH** Highly gregarious and crepuscular. Roosts communally, often with Lesser Kestrels. Hunts by hawking insects and by hovering. Open country.

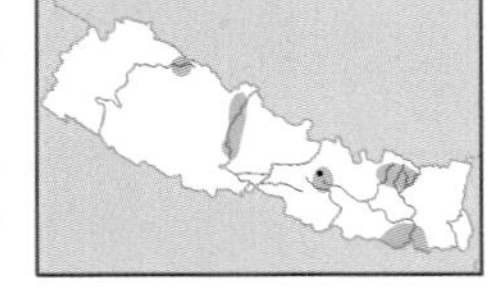

Merlin *Falco columbarius* 25–30 cm

Rare passage migrant; 75–4,000 m. **ID** Small and compact, with short, pointed wings. Fine supercilium and weak moustachial. Male has blue-grey upperparts, black subterminal tail-band, diffuse patch of rufous-orange on nape, and rufous-orange streaking on underparts. Female and juvenile have brown upperparts with variable buffish markings, heavily streaked underparts, and strongly barred uppertail. **Voice** Usually silent away from breeding grounds. **HH** Chiefly hunts in low flight with fast wingbeats and short glides. Open country.

ad
ad
Collared Falconet
Lesser Kestrel
♂
♀
♂
♀
♂
♀
♀
♂
♀
Common Kestrel
♂
ad
ad
Red-necked Falcon
♂
Amur Falcon
♀
♂
♀
Merlin

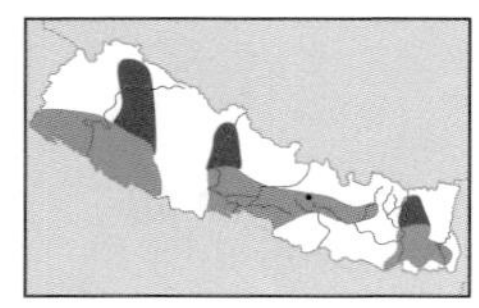

Eurasian Hobby *Falco subbuteo* 30–36 cm

Uncommon; resident, partial migrant, passage migrant and winter visitor; 75–3,050 m (–4,200 m). **ID** Slim, with long pointed wings and mid-length tail. Hunting flight swift and powerful, with stiff beats interspersed by short glides, acrobatic in pursuit of prey. Adult has broad black moustachial stripe, cream underparts with bold blackish streaking, and rufous thighs and undertail-coverts. Juvenile has dark brown upperparts with buffish fringes, pale buffish underparts which are more heavily streaked, and lacks rufous thighs and undertail-coverts. **Voice** Calls *kew-kew-kew* or *ki ki-ki* at nest site. **HH** Markedly crepuscular. Often perches on isolated trees. Well-wooded areas; also open country and cultivation in winter.

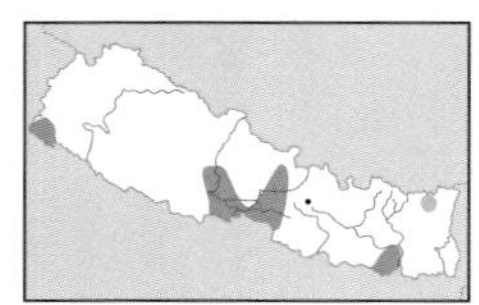

Oriental Hobby *Falco severus* 27–30 cm

Very rare resident; 75–1,900 m. **ID** Similar to Eurasian in structure, flight action and appearance, although slightly stockier, with shorter tail. Slimmer wings and body than Peregrine. Adult has complete blackish hood, bluish-black upperparts and sides of breast (suggesting half-collar), and unmarked rufous underparts and underwing-coverts. Straight cut to black cheeks and absence of any barring on underparts help to distinguish from *peregrinator* subspecies of Peregrine. Juvenile has browner upperparts, and heavily streaked rufous-buff underparts. **Voice** Rapid *ki-ki-ki-ki*. **HH** Habits very similar to Eurasian Hobby's. Forested hills.

Laggar Falcon *Falco jugger* 43–46 cm

Very rare, status uncertain; 75–1,400 m, mainly in lowlands. **ID** Large falcon; smaller, slimmer-winged and less powerful than Saker Falcon. Adult has rufous crown, dark eye stripe, narrow but long and prominent dark moustachial stripe, brownish-grey to dark brown upperparts (can be greyer than illustrated), and rather uniform uppertail. Underparts and underwing-coverts vary, can be largely white or heavily streaked, but lower flanks and thighs usually wholly dark brown; typically has dark panel across underwing-coverts. Juvenile similar to adult, but crown duller, moustachial broader, and underparts very heavily streaked (almost entirely dark on belly, flanks and underwing-coverts), and has greyish bare parts; differs from juvenile Peregrine in paler crown, finer moustachial stripe, more heavily marked underparts, and unbarred uppertail. **Voice** Shrill *whi-ee-ee* in breeding season. **HH** Usually seen perched on a regularly used vantage point. Also circles high overhead. Hunts mainly by flying rapidly and low, seizing prey on the ground. Open dry country and cultivation.

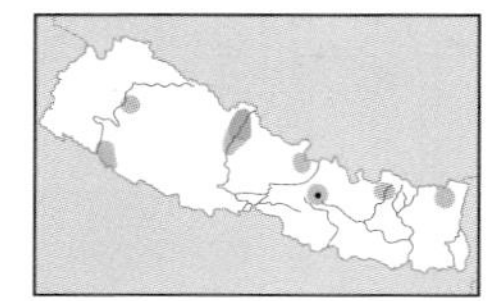

Saker Falcon *Falco cherrug* 50–58 cm

Rare winter visitor and passage migrant; 1,525–3,795 m (–75 m). **ID** Large falcon with long wings and long tail. Wingbeats slow in level flight, with lazier action than Peregrine. At rest, tail extends noticeably beyond closed wings (wings fall just short of tail tip on Laggar and are equal to tail on Peregrine). *F. c. milvipes*, which is the race recorded in Nepal, has broad orange-buff barring on upperparts and is rather different from Laggar. Additional differences include paler crown, less distinct moustachial, and less heavily-marked underparts (with flanks and thighs usually clearly streaked and not appearing wholly brown, although some overlap exists). **Voice** Normally silent away from breeding grounds. **HH** When hunting flies fast and low and strikes prey on ground; also stoops on aerial prey like Peregrine. Semi-desert and open, dry scrubby areas mainly in mountains. Globally threatened.

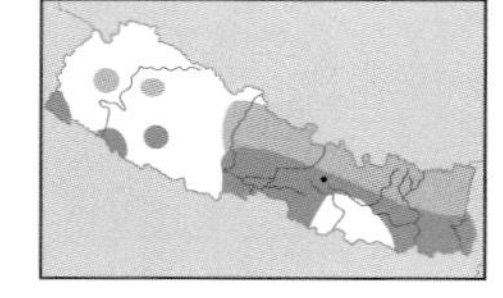

Peregrine Falcon *Falco peregrinus* 38–51 cm

Locally frequent, generally uncommon resident, winter visitor and passage migrant; summers 1,500–3,000 m (–4,200 m), winters down to 75 m; up to 4,785 m on passage. **ID** Heavy looking falcon with broad-based and pointed wings and short, broad-based tail. Flight strong, with stiff, shallow beats and occasional short glides. *F. p. calidus*, a winter visitor, has slate-grey upperparts, broad and clean-cut black moustachial stripe, and whitish underparts with narrow blackish barring; juvenile *calidus* (not illustrated) has browner upperparts, heavily streaked underparts, broad moustachial stripe, and barred uppertail. May show pale supercilium. Resident *F. p. peregrinator* has dark grey upperparts with more extensive black hood (and less pronounced moustachial stripe), and rufous underparts with dark barring on belly and thighs; juvenile *peregrinator* has darker brownish-black upperparts than adult, and paler underparts with heavy streaking. **Voice** That of *peregrinator* is unrecorded, except *chir-r-r-r* close to nest. Generally silent away from breeding areas. **HH** Pursues flying prey rapidly, finally rises above it and stoops with terrific force, wings almost closed. Breeds in open rugged hills and mountains, also lakes and large rivers in winter.

ad
Oriental Hobby
ad
imm
Eurasian Hobby
juv
ad
juv
ad
Laggar Falcon
juv
ad
ad
Saker Falcon
ad
calidus
Peregrine Falcon
ad
ad
juv
peregrinator

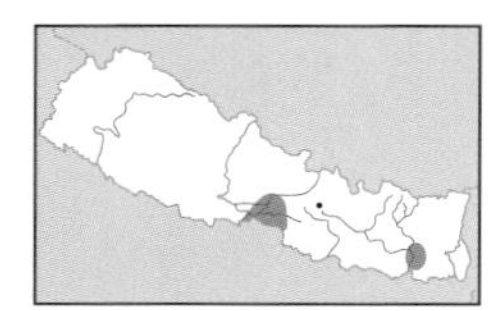

Jerdon's Baza ***Aviceda jerdoni*** 46 cm

Very rare and very local, has bred; 150–250 m. **ID** Long and erect white-tipped crest. Broad wings (pinched in at base) and fairly long tail. Greyish (male) or pale rufous (female) head, indistinct gular stripe, rufous-barred underparts and underwing-coverts, and bold barring on primary tips. At rest, closed wings extend well down tail. Juvenile has dark-streaked head and breast, and narrower dark barring on tail. **Voice** A *kip-kip-kip* or *kikiya, kikiya* uttered in display flight; also a plaintive mewing *pee-ow*. **HH** Crepuscular and elusive, keeping mostly within cover. Rather sluggish. Spends long periods perched in a tree on lookout for prey. Broadleaved evergreen forest.

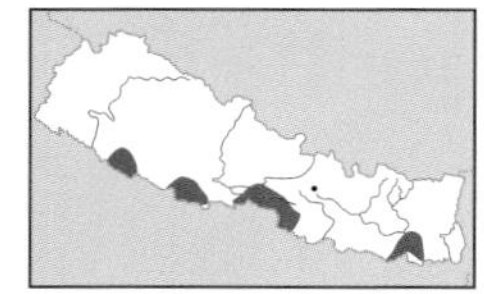

Black Baza ***Aviceda leuphotes*** 33 cm

Local and very uncommon summer visitor; 75–450 m (–1,280 m). **ID** Largely black, with long crest, white breast-band, rufous barring on underparts, and greyish underside to primaries contrasting with black underwing-coverts. Wings broad and rounded and tail medium length. Flight corvid-like, interspersed by short glides on flat wings. Male has more extensive patch of white on upperwing (extending onto secondaries) compared with female. **Voice** Loud, shrill, high-pitched *tcheeoua*, often repeated. **HH** Perches upright high in canopy of a tall forest tree. Makes short flights to capture prey. Broadleaved evergreen forest, often near glades or broad streams.

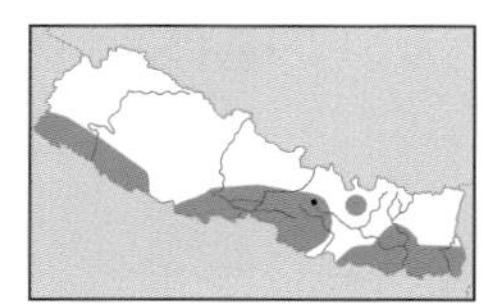

Black-winged Kite ***Elanus caeruleus*** 31–35 cm

Fairly common resident; 75–300 m (–1,550 m). **ID** Small size. Grey and white with black 'shoulders' and black eye-patch. Wings pointed and tail rather short. Flight buoyant, with much hovering. Juvenile has brownish-grey upperparts with pale fringes, and less distinct shoulder patch. **Voice** Weak whistled notes. **HH** Usually found alone or in widely spaced pairs, and in same area day after day. Spends much time perched on prominent vantage points. Hunts by quartering open ground, hovering at intervals with wings held high over back, beating rather slowly, and feet often trailing. Open country: cultivation, grassland and open scrubland. **AN** Black-shouldered Kite.

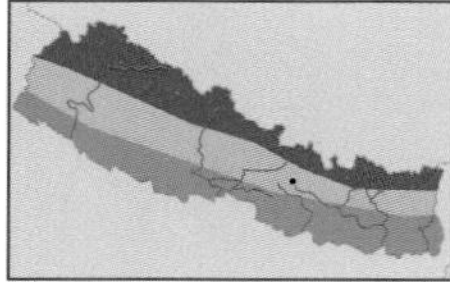

including Black-eared Kite

Black Kite ***Milvus migrans*** 58–66 cm

Common and widespread resident and passage migrant; 75–2,300 m. **ID** Shallow tail-fork. Much manoeuvring of arched wings and twisting of tail in flight. Dark rufous-brown, with variable whitish crescent on primary bases of underwing, and a pale band across median coverts on upperwing. Juvenile has broad whitish or buffish streaking on head and underparts. **Voice** Shrill, almost musical whistle *ewe-wir-r-r-r-r*. **HH** Gregarious throughout year, birds often soaring together and roosting communally, sometimes in large numbers. Closely associated with habitation. Feeds mainly on refuse and offal, but is omnivorous. A bold scavenger and can swoop and turn with amazing dexterity when snatching food. Mainly occurs around cities, towns and villages, also mountains.

Black-eared Kite ***Milvus (migrans) lineatus*** 61–66 cm

Common resident; 75–2,135 m (winter) up to 5,150 m (summer). **ID** Larger than Black Kite, with broader wings and generally more prominent whitish patch at base of primaries on underwing (but this is variable in Black). Shows more pronounced dark mask, with paler crown and throat. Belly and vent also paler. Adult has dark iris (yellow in adult Black). Juvenile more heavily streaked than juvenile Black. **Voice** Squealing whistles. **HH** Habits similar to Black Kite's, but more a montane bird, less closely associated with people and less of a municipal scavenger. Open country, mountains, around water; also around habitation.

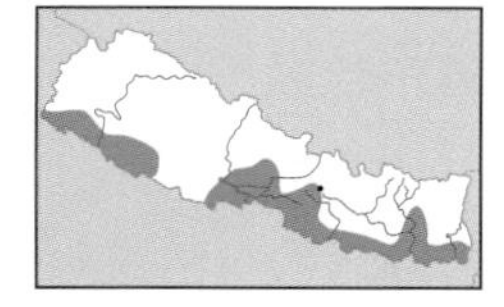

Brahminy Kite ***Haliastur indus*** 48 cm

Very rare, possibly resident; 75–915 m (–1,370 m). **ID** Small size and kite-like flight. Wings usually angled at carpals. Tail rounded. Adult mainly chestnut, with white head, neck and breast. Juvenile mainly brown, with pale streaking on head, mantle and breast, large pale patch at base of primaries on underwing, and pale brown and unmarked undertail. **Voice** A nasal, drawn-out, slightly undulating *kyerrh* or a squeal. **HH** Frequently perches on a tall tree overlooking water or flies slowly above ground looking for prey. Found in the vicinity of water: wetlands and flooded paddyfields.

♀
juv
Jerdon's Baza
ad
Black Baza
ad
juv
ad
Black-winged
Kite
ad
juv
Black Kite
ad
Black-eared Kite
ad
juv
juv
ad
juv
ad
ad
juv
Brahminy Kite

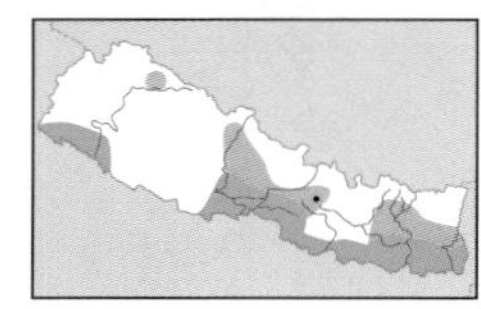

Osprey *Pandion haliaetus* 55–58 cm

Generally frequent and locally fairly common winter visitor, passage migrant and non-breeding resident; 75–915 m (–3,965 m). **ID** Long wings, typically angled at carpals, and short tail. Head often held downwards. Whitish head with black stripe through eye, uniform dark upperparts with pale barring on tail, white underparts and underwing-coverts, and black carpal patches. Juvenile similar to adult but has buff tips to feathers of upperparts. **Voice** A shrill cheeping whistle, but mostly silent away from nest. **HH** Usually solitary. Often perches on stakes, dead trees or prominent rocks standing in or near water. Feeds entirely on fish, captured in a powerful shallow dive with feet first to grasp prey. Lakes and large rivers.

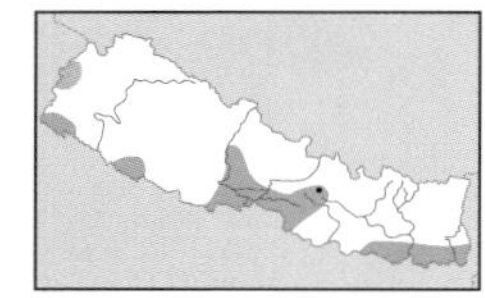

Pallas's Fish Eagle *Haliaeetus leucoryphus* 76–84 cm

Very rare winter visitor and passage migrant; 75–305 m (–2,745 m). **ID** Soars and glides on flat wings. Long, broad wings and (rather small) protruding head and neck. Adult has pale head and neck, dark brown upperwing and underwing, and mainly white tail with broad black terminal band. Juvenile less bulky, looks slimmer-winged, longer-tailed and smaller-billed than juvenile White-tailed; has dark mask, pale band on underwing-coverts, pale patch on underside of inner primaries, all-dark tail (lacking pale inner webs of White-tailed), and pale crescent on uppertail-coverts. Older immatures have more uniform underwing and whitish tail with mottled dark band. **Voice** Commonest call a hoarse bark, *kvo kvok kvok*. **HH** Rather sluggish, perching for long periods on a tree, post or sandbank near water. Feeds mainly on fish snatched near surface, also waterbirds hunted by gliding low over water, and snakes, frogs and carrion. Lakes and large rivers. Globally threatened.

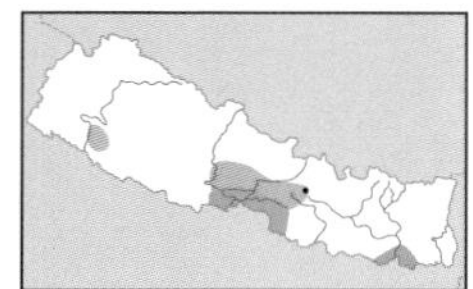

White-tailed Eagle *Haliaeetus albicilla* 70–90 cm

Very rare winter visitor and passage migrant; 75–1,370 m. **ID** Huge, with broad parallel-edged wings, short wedge-shaped tail, protruding head and neck, and heavy bill. Soars and glides with wings level. Adult has yellow bill, pale head, and white tail. Juvenile mainly blackish-brown with whitish centres to tail feathers, pale patch on axillaries, and variable pale band on underwing-coverts; bill becomes yellow with age. **Voice** Rather vocal; usual call a quick series of metallic yapping notes. **HH** Spends much of time perched on stump or ground near water. Captures fish, its main prey, by flying low over water and seizing them near surface; also occasionally hunts ducks and small mammals by flying low along shores. Lakes and large rivers.

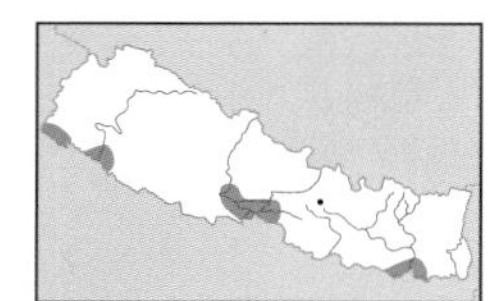

Grey-headed Fish Eagle *Icthyophaga ichthyaetus* 69–74 cm

Rare and local breeding resident; 75–250 m (–915 m). **ID** Larger than Lesser, with longer tail. Wingtips fall short of tail tip at rest. Adult from Lesser by largely white tail with broad black subterminal band. Also darker and browner upperparts and deeper rufous-brown breast. Juvenile has boldly streaked head and underparts, diffuse brown tail barring, and pale underwing with dark-barred flight feathers and pronounced dark trailing edge. **Voice** A squawk *kwok*, *kuwok* or *kuwonk*, or harsh shrieks. In display flight, gives an eerie series of far-carrying, dreamy *tiu-weeeu* notes. **HH** Found singly or in pairs. Usually seen perched low on trees or rocks overlooking water, and has regular perches. Typically flies only short distances, and rarely soars. Hunts mainly from a perch, occasionally in flight. Feeds almost entirely on fish seized near surface. Nest a huge platform of sticks built atop a large forest tree. Inhabits slow-moving waters and lakes in wooded country.

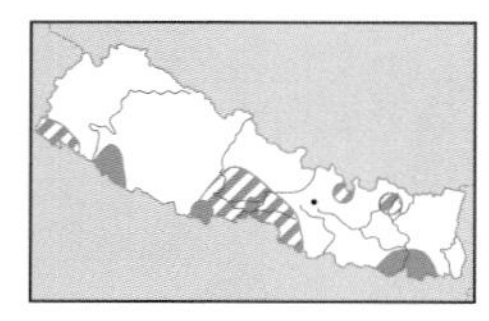

Lesser Fish Eagle *Icthyophaga humilis* 64 cm

Very rare, probably resident; 100–250 m (–915 m). **ID** Small, with broad wings and short tail, and rather small and protruding head and neck. Soars with wings held slightly raised and curved forwards. Smaller than Grey-headed Fish Eagle, with shorter tail. Wingtips almost reach tail tip at rest (not shown in plate). Adult differs from Grey-headed in having dark tail (appears uniform brown above; from below shows greyish base with slightly darker subterminal band). Juvenile browner than adult, with pale underwing and paler base to tail. Lacks prominent streaking on head and underparts of juvenile Grey-headed Fish Eagle, has less pronounced barring on underside of flight feathers and less pronounced dark tail-band. **Voice** Gives a characteristic penetrating plaintive wail, *pheeow-pheeoow-pheeow* and a shorter *pheeo-pheeo* intermittently repeated during breeding season. **HH** Habits very similar to Grey-headed. Favours swift-flowing forested upper reaches of rivers; also fast-flowing shallow rivers and streams with deep pools.

ad
Osprey
ad
juv
Pallas's Fish Eagle
ad
ad
juv
ad
juv
White-tailed Eagle
ad
juv
ad
juv
Grey-headed Fish Eagle
ad
juv
ad
juv
Lesser Fish Eagle

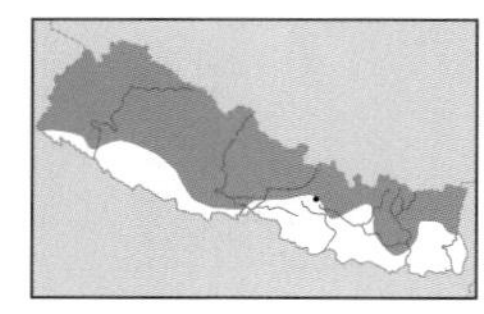

Bearded Vulture *Gypaetus barbatus* 100–115 cm

Widespread resident in Himalayas; 250–5,600 m (–7,500 m). **ID** Huge size, long and narrow pointed wings, and large wedge-shaped tail. Adult has pale head with black mask and beard, greyish-black upperparts, wings and tail, and cream or rufous-orange underparts contrasting with black underwing-coverts. Juvenile has blackish head and neck, and grey-brown underparts. **Voice** Generally silent, but thin screams and whistles in display. **HH** Spends long periods soaring majestically and gracefully over mountainsides. Highly manoeuvrable in flight; rarely flaps wings, and often glides close to ground following curve of mountain slopes. Has unique habit of splitting bones by dropping them onto rock slabs. Eats bone fragments; also carrion, and scavenges around mountain villages. Nest an enormous platform of sticks on a cliff ledge. Inhabits mountains and Trans-Himalayan Tibetan steppe desert. **AN** Lammergeier.

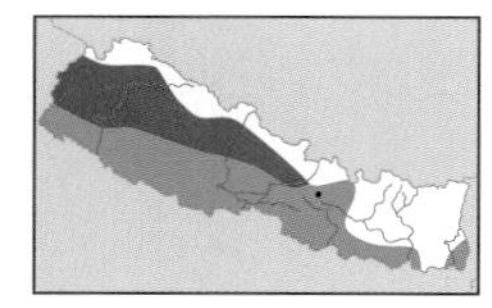

Egyptian Vulture *Neophron percnopterus* 60–70 cm

Resident, widespread and locally fairly common in west and west-central Nepal, but very rare in east; 75–2,900 m (–3,810 m). **ID** Small vulture with long, pointed wings, small pointed head, and wedge-shaped tail. Adult mainly dirty white, with bare yellowish face and black flight feathers. Juvenile blackish-brown with bare grey face. With maturity, tail, body and wing-coverts become whiter and face yellower. **Voice** Practically silent, but can give mewing sounds, hisses and low grunts. **HH** An opportunistic scavenger; less dependent on large carcasses than other vultures. Spends day soaring and gliding in search of food, or perched around habitation and rubbish dumps. Gathers at carcasses and rubbish dumps with other larger vultures; stays on outside of throng. Nest a platform of sticks on the ledge of a cliff or ruin. Inhabits towns, villages and city outskirts, especially around dumps and slaughterhouses. Globally threatened.

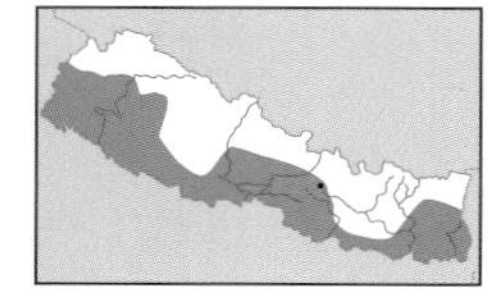

White-rumped Vulture *Gyps bengalensis* 75–85 cm

Resident; rare in the centre and east, generally uncommon in west, though after sharp declines in late 20th and early 21st centuries, populations are recovering at some sites in response to conservation measures; 75–1,800 m (–3,100 m). **ID** Smallest of *Gyps* vultures. Adult mainly blackish, with white neck-ruff, white rump and back, and white underwing-coverts. Juvenile dark brown with streaked underparts and upperwing-coverts, dark rump and back, whitish head and neck, and all-dark bill. In flight, underparts and underwing-coverts of juvenile darker than Slender-billed. Juvenile similar in colour to juvenile Himalayan, but smaller and less heavily built, with narrower wings and shorter tail; underparts less heavily streaked, and lacks prominent streaking on mantle and scapulars. **Voice** Croaks, grunts, hisses and squeals at colonies, roosts and carcasses. **HH** Gregarious all year. Groups spend much time hunched on treetops or buildings, often with other vulture species. Often soars at great height in search of food. Feeds almost entirely on carrion, mainly by scavenging at rubbish dumps and slaughterhouses or by seeking dead animals. Roosts communally at traditional sites in groves. Nests colonially at equally traditional sites. Builds platform nest of sticks and twigs, often with green leaves, in top of large tree. Found around habitation, cultivation and open country. Globally threatened.

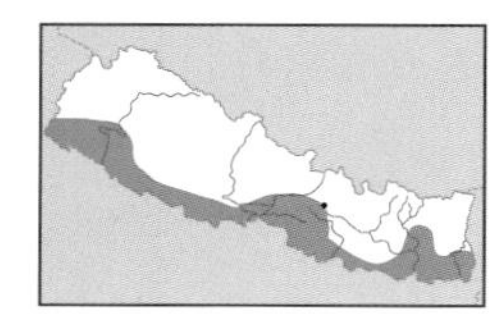

Slender-billed Vulture *Gyps tenuirostris* 93–100 cm

Local resident; now extremely rare in the east and uncommon in centre and west after sharp decline in late 20th and early 21st centuries; 150–1,525 m. **ID** Bill, head, neck and body more slender than in Griffon. Adult has dark bill and cere with pale culmen, lacks any down on black head and neck, and has dirty white ruff that is rather small and ragged (by comparison, Griffon has extensive white down on head and neck, mainly yellowish bill, and extensive white ruff). Upperparts are colder grey-brown, and has white thighs. In flight, from below, prominent white thighs, trailing edge to wing appears rounded and pinched-in at body, outer primaries appear noticeably longer than inner primaries, undertail-coverts appear dark (pale in Griffon) and in flight feet reach tip of tail (falling short in Griffon). Juvenile similar to adult; mainly dark bill, some white down on head and neck, and pale streaking on underparts. See Appendix 1 for differences from Indian Vulture. **Voice** Hissing and cackling sounds. **HH** Habits like White-rumped and often associates with that species. Nest a platform of twigs, sometimes with green leaves, on a cliff ledge or in large leafy tree. Inhabits cultivation, open country and around habitation, especially villages. Globally threatened.

ad
imm
imm
Bearded Vulture
ad
juv
ad
imm
ad
imm
Egyptian Vulture
juv
ad
ad
ad
juv
White-rumped Vulture
juv
ad
juv
ad
Slender-billed Vulture
ad
juv

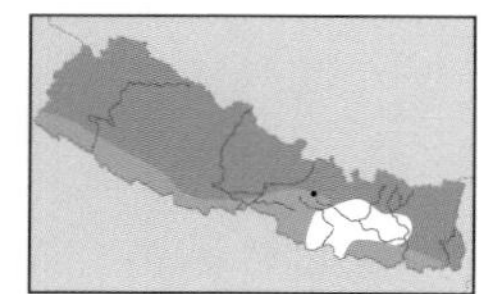

Himalayan Vulture *Gyps himalayensis* 115–125 cm

Widespread resident, subject to seasonal altitudinal movements, and partial migrant; 75–6,100 m. **ID** Larger than Griffon Vulture, with broader body and slightly longer tail. Wing-coverts and body pale buffish, contrasting strongly with dark flight feathers and tail, and ruff is buffish. Underparts lack pronounced streaking. Legs and feet pinkish with dark claws, and has yellowish bill and pale blue cere and facial skin (blackish in Griffon). Juvenile has brown-feathered ruff, with bill and cere initially black, dark brown body and upperwing-coverts boldly and prominently streaked buff (wing-coverts almost concolorous with flight feathers), and back and rump also dark brown. Streaked upperparts and underparts and pronounced white banding across underwing-coverts are best distinctions of juvenile from Cinereous Vulture; very similar in plumage to juvenile White-rumped, but much larger and more heavily built, with broader wings and longer tail, underparts more heavily streaked, and streaked mantle and scapulars. **Voice** A variety of grunts and hisses. **HH** Roosts gregariously on crags, and when currents are suitable takes to air and soars over mountains at great heights and often over long distances. Aggressive when feeding, and dominates all other vultures except Cinereous. Nests alone or in small colonies, nest a large platform of sticks and rubbish on a cliff ledge. Breeds in mountains and Trans-Himalayan Tibetan steppe desert, especially along routes well used by pack animals; winters down to plains. **AN** Himalayan Griffon.

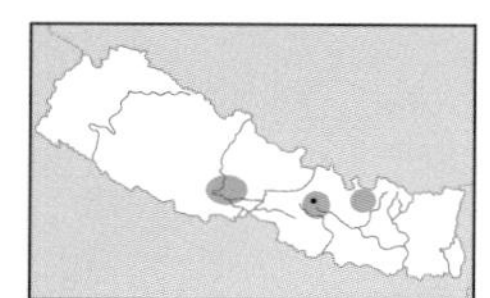

Griffon Vulture *Gyps fulvus* 95–105 cm

Probably mainly a passage migrant, but a rare winter visitor in small numbers. **ID** Larger than Slender-billed, with stockier head and neck, and stouter bill. Key features of adult are yellowish bill with blackish cere, whitish head and neck, fluffy white ruff, rufescent-buff upperparts, rufous-brown underparts and thighs with prominent pale streaking, and dark grey legs and feet. Rufous-brown underwing-coverts usually show prominent whitish banding, especially on median coverts. Immature richer rufous-brown on upperparts and upperwing-coverts (with prominent pale streaking) than adult; has rufous-brown feathered neck-ruff, more whitish down over grey head and neck, blackish bill, and dark iris (pale yellowish-brown in adult). **Voice** Normally rather silent but can emit varied grunting, whistling, hissing and sobbing sounds. **HH** Gathers at carcasses, often with other vultures. Regularly makes long journeys in search of food. Semi-desert, dry open plains and hills. **AN** Eurasian Griffon.

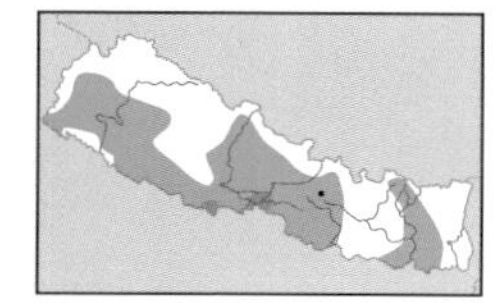

Cinereous Vulture *Aegypius monachus* 100–110 cm

Winter visitor and passage migrant; now very uncommon in the centre and west, and rare and very local in east, following sharp declines in late 20th and early 21st centuries. **ID** Very large vulture with broad, parallel-edged wings. Soars on flat wings (*Gyps* vultures soar with wings held in shallow V). At distance appears typically uniformly dark, except pale areas on head and bill. Adult blackish-brown with paler brown ruff; may show paler band on greater underwing-coverts, but underwing darker and more uniform than in *Gyps* species. Juvenile blacker and more uniform than adult. **Voice** Usually silent. **HH** Normally solitary by day. Usually roosts communally on ground, often on an open escarpment close to a steep slope, in readiness for suitable thermals. Dominates all other vultures at carcass. Open country. **AN** Eurasian Black Vulture.

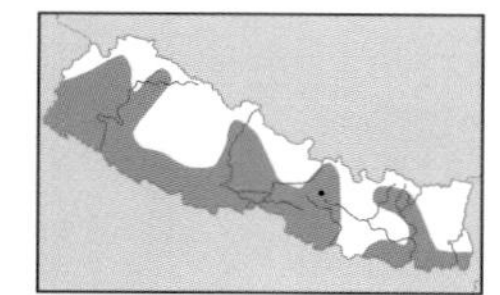

Red-headed Vulture *Sarcogyps calvus* 85 cm

Resident, still widespread in mid-west to far west, and locally frequent there and in west-central Nepal, but virtually absent east of Kathmandu, following sharp declines in late 20th and early 21st centuries; 75–2,050 m (–3,050 m). **ID** Comparatively slim pointed wings. Adult mainly black with bare reddish head and cere, white patches at base of neck and upper thighs, and reddish legs and feet; in flight, greyish-white bases to secondaries form broad panel (particularly on underwing). Juvenile browner with white down on head; pinkish head and feet, white patch on upper thighs, and whitish undertail-coverts are best features. **Voice** Mostly silent, but squeaks, hisses and grunts like other vultures. **HH** Usually alone or in pairs. Frequently feeds on carcasses of small animals overlooked by other large vultures; also feeds timidly with other vultures at carcasses of larger animals. Open country near habitation, and well-wooded hills. Globally threatened.

ad
juv
juv
ad
Himalayan Vulture
juv
ad
juv
ad
Griffon Vulture
ad
ad
juv
juv
juv
ad
ad
Cinereous Vulture
imm
ad
ad
ad
imm
imm
Red-headed Vulture

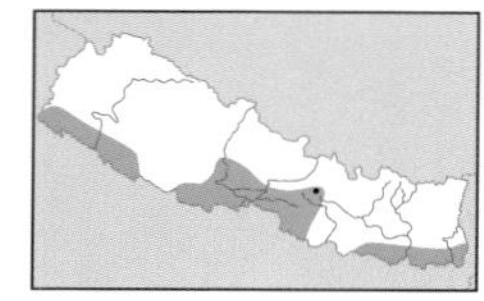

Short-toed Snake Eagle *Circaetus gallicus* 62–67 cm

Mainly a winter visitor; locally uncommon; 75–150 m (–2,130 m). **ID** Long broad wings, pinched-in at base, and rather long tail. Head broad and rounded. Soars with wings held flat or slightly raised; frequently hovers. When perched, appears large-headed, with wingtips reaching tail tip, and shows long unfeathered tarsus. Plumage variable, often has dark head and breast, barred underparts, dark trailing edge to underwing, and broad subterminal tail-band; can be very pale on head, underparts and underwing. On upperside, in all plumages, pale brown inner wing-coverts contrast with dark greater coverts and flight feathers. Juvenile similar to adult. **Voice** Mainly silent outside breeding season. **HH** Spends most of day soaring, searching for prey. Typically hunts 15–30 m above ground, occasionally higher, frequently hovering on gently beating wings; on spotting prey, plummets down almost vertically. Feeds mainly on snakes. Open country and wetlands.

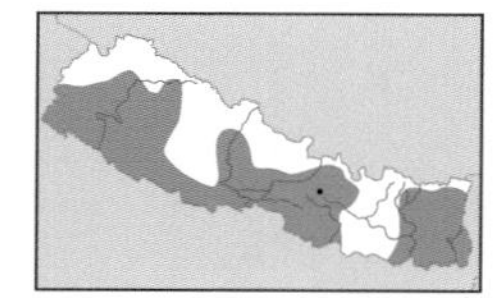

Crested Serpent Eagle *Spilornis cheela* 56–74 cm

Fairly common and widespread resident subject to some altitudinal movements; 75–2,100 m (–3,350 m) summer; 75–915 m (winter). **ID** Broad, rounded wings. Soars with wings held forward and in pronounced V. At rest has black-and-white crest, yellow cere and lores, and unfeathered yellow legs. Adult has broad white bands on wings and tail, and white spotting and barring on brown underparts. Juvenile has blackish ear-coverts, whitish head and underparts, narrower barring on tail (than adult), and largely white underwing with fine dark barring and dark trailing edge. **Voice** Varied loud, ringing, musical whistles or screams in flight. **HH** Often quite tame. Has characteristic habit of soaring over forest in pairs, the birds screaming to each other and sometimes rising to great heights. Spends hours perched very upright on forest tree. Raises crest if alarmed. Usually hunts by dropping almost vertically onto prey from a perch. Stick and twig nest is built high in a tree, often near stream. Forest and well-wooded country.

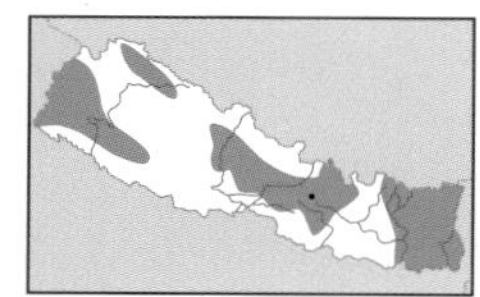

Black Eagle *Ictinaetus malayensis* 69–81 cm

Frequent resident, or locally fairly common; mainly 1,000–3,100 m (75–4,000 m). **ID** Distinctive wing shape and long tail. Flies with wings raised in V, primaries upturned. At rest, long wings extend to tip of tail. Adult dark brownish-black, with striking yellow cere and feet; in flight, whitish barring on uppertail-coverts, and faint greyish barring on tail and underside of remiges (cf. dark-morph Changeable Hawk Eagle, Plate 30). Juvenile as adult; may show indistinct pale streaking to head and underparts. **Voice** Normally silent, shrill yelping cries in aerial courtship. **HH** Invariably seen on wing. Frequently soars over forest, often reaching considerable heights. Hunts by flying buoyantly and slowly very low over forest. Nest a compact platform of sticks and twigs concealed in foliage of tall tree. Hill and mountain forests.

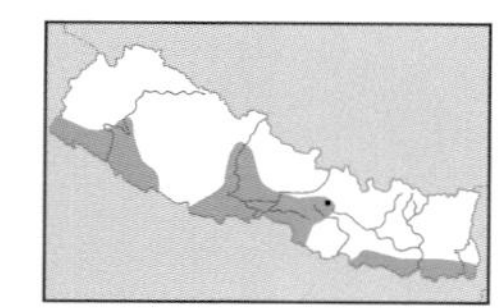

Eurasian Marsh Harrier *Circus aeruginosus* 42–54 cm

Local and uncommon winter visitor and passage migrant; 75–915 m (–1,525 m). **ID** A broad-winged, stout-bodied harrier; like other harriers, glides and soars with wings held in noticeable V, which helps separate it from Booted Eagle and Black Kite. Adult male distinguished from other male harriers by combination of chestnut-brown mantle and upperwing-coverts contrasting with grey secondaries/inner primaries and black outer primaries, pale head (variably streaked brown) and pale leading edge to wing, and brown streaking on breast and belly, becoming uniformly brown on lower belly and vent. Variable amount of brown on underwing-coverts, with rest of underwing being white except black tips to primaries. Melanistic morphs occur, being sooty-grey above and blackish below, with grey patch at base of underside of primaries. Adult female mainly dark brown except creamy crown, nape and throat, creamy leading edge to wing, and paler patch at base of underside of primaries. Melanistic female (and juvenile) much as melanistic male, but grey or whitish extends across underside of all flight feathers (not just primaries). Juvenile similar to female, but head and wing-coverts may be entirely dark. **Voice** Usually silent outside breeding season. **HH** Like other harriers has characteristic method of hunting. Systematically quarters the ground a few metres above it, gliding slowly on raised wings and occasionally flapping rather heavily several times; on locating prey, drops quickly with claws held out. Marshes, also grasslands and paddyfields. **AN** Western Marsh Harrier, if eastern form *spilonotus* is split.

soaring
ad pale
ad pale
ad dark
Short-toed Snake Eagle
juv
ad
juv
ad
soaring
Crested Serpent Eagle
ad
soaring
Black Eagle
♂
♂
♀
♀
aeruginosus
juv
Eurasian Marsh Harrier

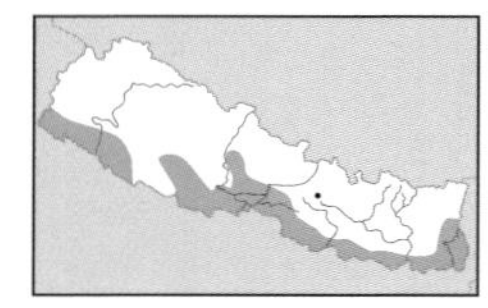

Pied Harrier *Circus melanoleucos* 41–46.5 cm

Local and very uncommon winter visitor; 75–350 m (–3,810 m). **ID** Usually looks broader in wings and body than Pallid and Montagu's Harriers (see below), and slightly heavier in flight, although male can appear dainty and buoyant. Most likely to be confused with Eastern form of Eurasian Marsh Harrier (potential vagrant to Nepal). Adult male distinctive, with black head and breast contrasting with white underparts, black upperbody, median coverts and primaries contrasting with grey rest of wing, and white leading edge to wing. Yellow cere and iris especially striking against black of head. Adult female superficially resembles female Pallid and Montagu's, but has paler underwing with narrow dark barring on flight feathers and sparse streaking on underwing-coverts, greyer primaries and secondaries with more pronounced dark banding, and greyer tail with narrower dark barring. Juvenile has dark brown head and upperparts, with white supercilium and patch below eye, white uppertail-coverts, rufous-brown underparts and underwing-coverts, and pale underside to primaries (with dark fingers) contrasting with dark underside to secondaries. **Voice** Usually silent. **HH** Habits like Eurasian Marsh, but more graceful on wing. Open grassland and cultivation.

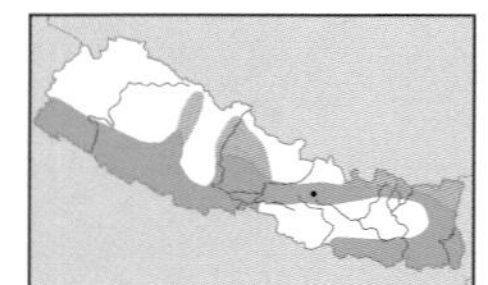

Hen Harrier *Circus cyaneus* 44–52 cm

Frequent winter visitor and passage migrant; 75–3,000 m (–5,400 m on passage). **ID** Compared to Pallid and Montagu's, has slower and more laboured flight, and is stockier, with broader wings and more rounded hand, normally with five (rather than four) visible primaries at tip. Adult male from male Pallid and Montagu's by combination of dark grey upperparts with extensive black wingtips, prominent white uppertail-coverts patch, absence of black banding on secondaries (although has variable dark trailing edge to underwing), and dark grey head and breast contrasting with white belly. Adult female has boldly streaked underparts. White band on uppertail-coverts broader than on Pallid and Montagu's. Has narrow pale neck-collar, but otherwise head is typically rather plain compared with those species, usually lacking dark ear-coverts patch. Juvenile recalls female but has rufous-brown wash to underparts and underwing-coverts, although these are noticeably streaked dark brown (unlike juvenile Pallid and Montagu's). **Voice** Usually silent. **HH** Flight and hunting behaviour like Eurasian Marsh, but more graceful on wing. Open grassland and cultivation.

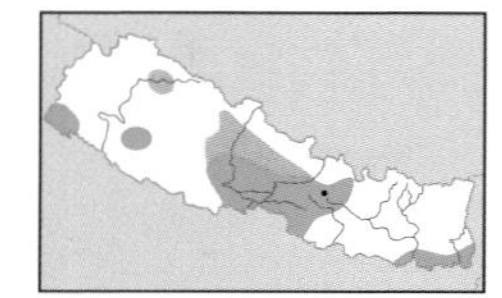

Pallid Harrier *Circus macrourus* 40–48 cm

Rare and local passage migrant and winter visitor; 75–2,200 m (–3,350 m). **ID** Slim-winged and fine-bodied, with buoyant flight. Folded wings fall short of tail tip, and legs longer than Montagu's. Male has pale grey upperparts, dark wedge on primaries, very pale grey head and underparts, and lacks black secondary bars. Immature male may show rusty breast-band and juvenile face markings. Female has distinctive underwing pattern: pale primaries, irregularly barred and lacking dark trailing edge, contrast with darker secondaries, which have narrower pale bands than in female Montagu's that taper towards body, and lacks prominent barring on axillaries. Typically, female has stronger head pattern than Montagu's, with more pronounced pale collar, dark ear-coverts and dark eye-stripe, and upperside of flight feathers darker and lacks banding; from female Hen by narrower wings with more pointed hand, stronger head pattern, and pattern on underside of primaries. Juvenile has unstreaked orange-buff underparts and underwing-coverts; on underwing, primaries evenly barred (lacking pronounced dark fingers), without dark trailing edge, and usually a pale crescent at base; head pattern more pronounced than Montagu's, with narrower white supercilium, more extensive dark ear-coverts patch, and broader pale collar contrasting strongly with dark neck-sides. **Voice** Usually silent. **HH** Habits like Hen Harrier, but lighter on wing. Open country, mainly grasslands.

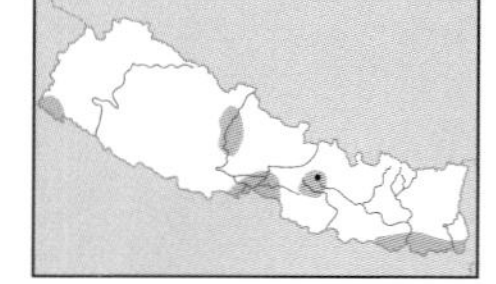

Montagu's Harrier *Circus pygargus* 43–47 cm

Very rare passage migrant; 75–2,630 m. **ID** Folded wings reach tail tip, and legs shorter than Pallid. Male has black band on secondaries, extensive black on underside of primaries, and rufous streaking on belly and underwing-coverts. Female differs from female Pallid in distinctly and evenly barred underside to primaries with dark trailing edge, broader and more pronounced pale bands on secondaries, barring on axillaries, less pronounced head pattern, and distinct dark banding on upperside of remiges. Juvenile has unstreaked rufous underparts and underwing-coverts, and darker secondaries than female; differs from juvenile Pallid in having broad dark fingers and dark trailing edge to hand on underwing, paler face with smaller dark ear-coverts patch and less distinct collar. **Voice** Usually silent. **HH** Habits like Hen Harrier, but lighter on wing. Open country including cultivation, grassland and marshes.

♂
♂
juv
juv
Pied Harrier
♀
♀
♂
♀
♀
♂
Hen Harrier
♂
♀
Pallid Harrier
♀
Pallid juv
Montagu's juv
juv
♂
juv
Montagu's Harrier
♂
♀
♀

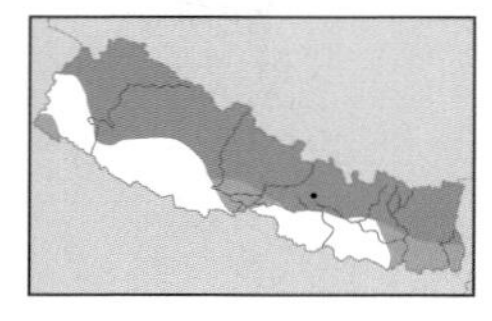

Northern Goshawk *Accipiter gentilis* 50–61 cm

Frequent resident; 75–4,880 (–6,100 m). **ID** Very large, with heavy, deep-chested appearance. Wings comparatively long, with bulging secondaries. Male has grey upperparts (greyer than female Eurasian Sparrowhawk), white supercilium and finely barred underparts. Female considerably larger with browner upperparts. Juvenile has heavily streaked buff underparts. **Voice** A shrill chatter and female gives disyllabic *hee-aa*. **HH** Hunting habits similar to Eurasian Sparrowhawk, but more powerful. Oak and coniferous forest, sometimes hunts above treeline.

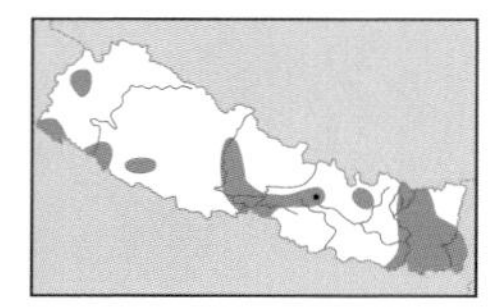

Crested Goshawk *Accipiter trivirgatus* 30–46 cm

Uncommon resident; 75–1,370 m (–2,100 m). **ID** Larger size and crest best distinctions from Besra. Short broad wings, pinched-in at base. Wingtips barely extend beyond tail base at rest. Yellow (rather than grey) cere and shorter crest help separate from Jerdon's Baza. Compared to other *Accipiter*, has short stoutish legs. Male has dark grey crown and paler grey ear-coverts, black submoustachial and gular stripes, and rufous-brown streaking on breast and barring on belly/flanks. Female larger with browner crown and ear-coverts, and browner streaking and barring on underparts. Juvenile has paler brown upperparts with pale fringes, buffish fringes to crown and crest, streaked ear-coverts, and buffish wash to underparts, which are mainly streaked brown (with barring on flanks and thighs). **Voice** Calls include passerine-like whistles, a shrill scream, *he he, hehehehe* and shrill whistle *chewee...chewee...* **HH** Broadleaved forest.

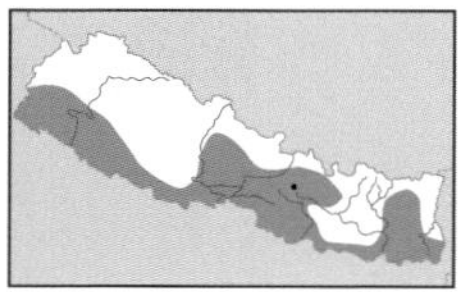

Shikra *Accipiter badius* 30–36 cm

Fairly common resident; 75–1,370 m (–2,250 m). **ID** Adult paler than Besra and Eurasian Sparrowhawk. Underwing pale, with finely barred remiges and slightly darker wingtips. Head cuckoo-like compared to other *Accipiter*. Male has pale blue-grey upperparts with contrasting dark primaries, indistinct grey gular stripe, fine brownish-orange barring on underparts, unbarred white thighs, and unbarred or lightly barred central tail feathers. Female upperparts more brownish-grey. Juvenile has pale brown upperparts; from juvenile Besra by paler upperparts and narrower tail barring, and from Eurasian by streaked underparts. **Voice** Loud, harsh *titu-titu* similar to Black Drongo call, also long drawn-out screams, *iheeya, iheeya*, and a constantly repeated *ti-tui* in display. **HH** Open wooded country and groves around villages and cultivation.

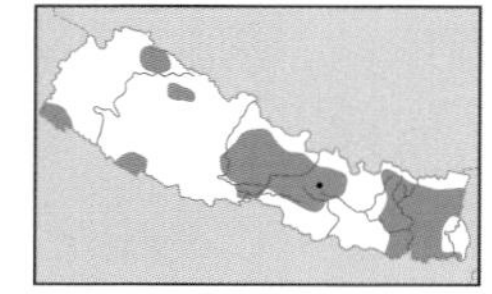

Besra *Accipiter virgatus* 29–36 cm

Uncommon resident; summers 1,350–2,800 m (–3,440 m), winters down to 250 m (–75 m). **ID** Small, with short primary projection (less than one-third of tail). Upperparts darker and underwing strongly barred compared to Shikra, while prominent gular stripe and streaked breast should separate it from Eurasian. In all plumages resembles Crested Goshawk, but considerably smaller, lacks crest, and has longer and narrower legs. Adult male has dark slate-grey upperparts lacking strong contrast with primaries, broad blackish gular stripe, bold rufous streaking on breast and barring on belly, flanks and thighs, and broader dark tail barring. Adult female has browner upperparts, with blackish crown and nape, and yellow iris (red in male). Juvenile has brown upperparts with rufous fringes, prominent gular stripe, and streaked/spotted underparts with barred flanks. From immature Shikra by darker, richer brown upperparts, broader gular stripe, and broader tail bars. From Eurasian by streaked underparts. **Voice** Loud squealing *ki-weeer* and a rapidly repeated *tchew-tchew-tchew* during displays. **HH** Dodges and twists at speed through dense forest. Breeds in dense forest; also open wooded country in winter.

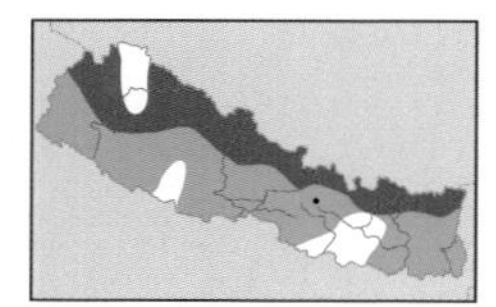

Eurasian Sparrowhawk *Accipiter nisus* 31–36 cm

Fairly common resident, winter visitor and passage migrant; summers 2,440–4,200 m (–5,180 m), winters 250–1,450 m (–100 m). **ID** Upperparts darker than Shikra, with prominently barred underparts and underwing. Uniform barring on underparts and absence of prominent gular stripe should separate from Besra. Male has dark slate-grey upperparts and reddish-orange barring on underparts. Female has dark brown upperparts, brown-barred underparts and yellow iris (red in some males). Juvenile has dark brown upperparts and barred underparts. Compared to migrant *nisosimilis*, male of resident Himalayan race *melaschistos* has darker grey upperparts, with almost black crown and mantle, and stronger rufous barring on underparts. **Voice** Two long notes followed by 3–4 very short notes, *tiu-tiu-tititi*. **HH** Captures prey in a short swift dash, relying on surprise, or by swift low flight between trees or patches of cover. Well-wooded country, open forest and groves in cultivation.

Northern Goshawk
juv
♀
Crested
Goshawk
♂
juv
juv
Shikra
♂
♀
juv
♀
♂
melaschistos
♀
♂
Eurasian
Sparrowhawk
juv
♀
♀
juv
Besra

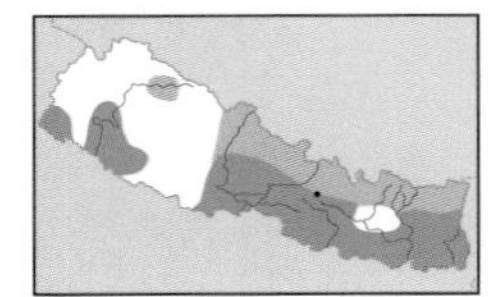

Oriental Honey-buzzard *Pernis ptilorhynchus* 57–60 cm

Local, fairly common resident and passage migrant; 75–1,700 m (–3,050 m). **ID** Tail long and broad, with narrow neck, small head and bill, and short bare tarsi. Soars on flat wings. Pronounced crest. Underparts and underwing-coverts range from dark brown through rufous to white, and unmarked, streaked or barred; often has dark moustachial stripe and gular stripe, and gorget of streaking on lower throat. Lacks dark carpal patch. Male has grey face, two black tail-bands, usually three black underwing bands, and brown iris. Female has browner face and upperparts, three black tail-bands, four narrower black underwing bands, and yellow iris. Juvenile has narrower underwing banding, three or more tail-bands, and extensive dark tips to primaries; cere yellow (grey in adult) and iris dark. **Voice** High-pitched screaming whistle, *wheeew*. **HH** Eats mainly honey and bee larvae. Spends long periods perched in foliage. Open woodland and groves near cultivation and villages. **AN** Crested Honey-buzzard.

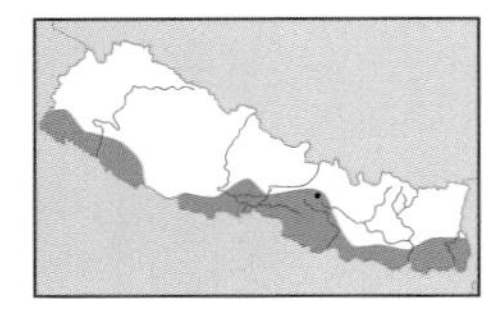

White-eyed Buzzard *Butastur teesa* 43 cm

Fairly common resident; 75–300 m (–1,500 m). **ID** Longish, rather slim wings, long tail, and buzzard-like head. Pale median coverts panel. Adult has black gular stripe, white nape patch, barred underparts, dark wingtips, and rufous tail; iris white and cere yellow. Juvenile has buffish head and breast streaked dark brown, with throat stripe indistinct or absent; rufous uppertail more strongly barred; iris brown. **Voice** Plaintive mewing *pit-weer, pit-weer* in breeding season. **HH** Sluggish, spends long periods perched very upright on vantage point such as an isolated tree, post or mound. Hunts chiefly by dropping to ground to seize prey. Dry open country.

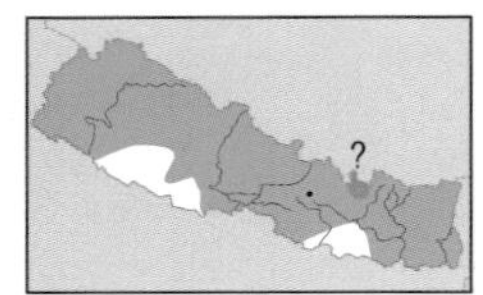

Himalayan Buzzard *Buteo* (*buteo*) *burmanicus* 51–57 cm

Fairly common winter visitor and passage migrant, some probably resident; 75–4,300 m. **ID** Stocky, with broad rounded wings and medium-length tail. Plumage very variable. Typically has variable streaking on breast and brown patches on sides of belly; tail grey-brown to greyish-white with diffuse dark barring (appearing pale and unbarred from below); prominent dark carpal patches on underwing; can be similar to larger Upland. Himalayan breeders ('*refectus*') are larger than migrants and very rufous. Their taxonomic status remains to be determined. Common Buzzard *B. buteo* (subspecies *vulpinus*) also reported recently. **Voice** Loud, repeated *mew peee-oo*. **HH** Rather sluggish. Hunts by pouncing on prey from a vantage point, or by soaring over open country and at times hovering with gently fanned wingtips. Mountains and open country. **TN** The name *burmanicus* is now considered to be antedated by *refectus*.

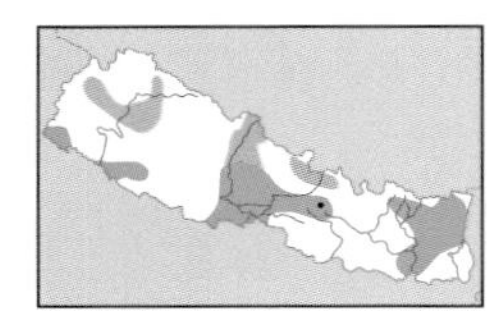

Long-legged Buzzard *Buteo rufinus* 61 cm

Uncommon to frequent winter visitor and passage migrant; 75–2,755 m (–5,000 m). **ID** Larger and longer-necked than Himalayan Buzzard, with longer wings and tail; soars with wings in deeper V. Most differ from Himalayan Buzzard in having combination of paler head and upper breast, rufous-brown lower breast and belly, more uniform rufous underwing-coverts, more extensive black carpal patches, larger pale primary patch on upperwing, and unbarred pale orange uppertail. Intermediate and dark morphs much like some plumages of Common Buzzard (subspecies *vulpinus* which may occur). Juvenile generally less rufous, with narrower and more diffuse trailing edge to wing, and lightly barred tail, which in many is pale greyish-brown. **Voice** Rather silent, in anxiety a short, sharp high-pitched *mew*. **HH** Habits similar to Himalayans. Open country.

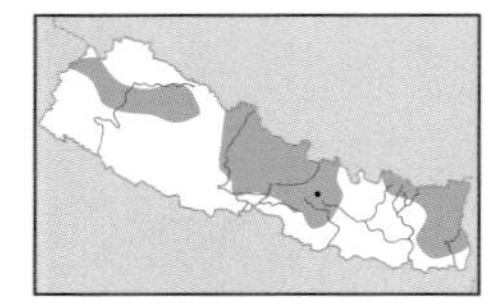

Upland Buzzard *Buteo hemilasius* 71 cm

Probably an uncommon winter visitor, possibly also a resident breeder; 1,370–4,420 m (–75 m). **ID** Larger, longer-winged and longer-tailed than Himalayan Buzzard; soars with wings in deeper V. Tarsus at least three-quarters feathered (half-feathered or less in other *Buteo*). 'Classic' pale morph has combination of large white primary patch on upperwing, greyish-white tail (with fine bars towards tip), whitish head and underparts with dark brown streaking, brown thighs, and extensive black carpal patches (some Himalayan Buzzards very similar); never has rufous tail or rufous thighs as in many Long-legged Buzzard. 'Blackish morph' indistinguishable by plumage from similar morph in Long-legged. Juvenile has less distinct trailing edge to wing than adult and more prominently barred tail. **Voice** Calls similar to Himalayan, but more nasal and prolonged. **HH** Habits similar to Himalayan. Open country in hills and mountains.

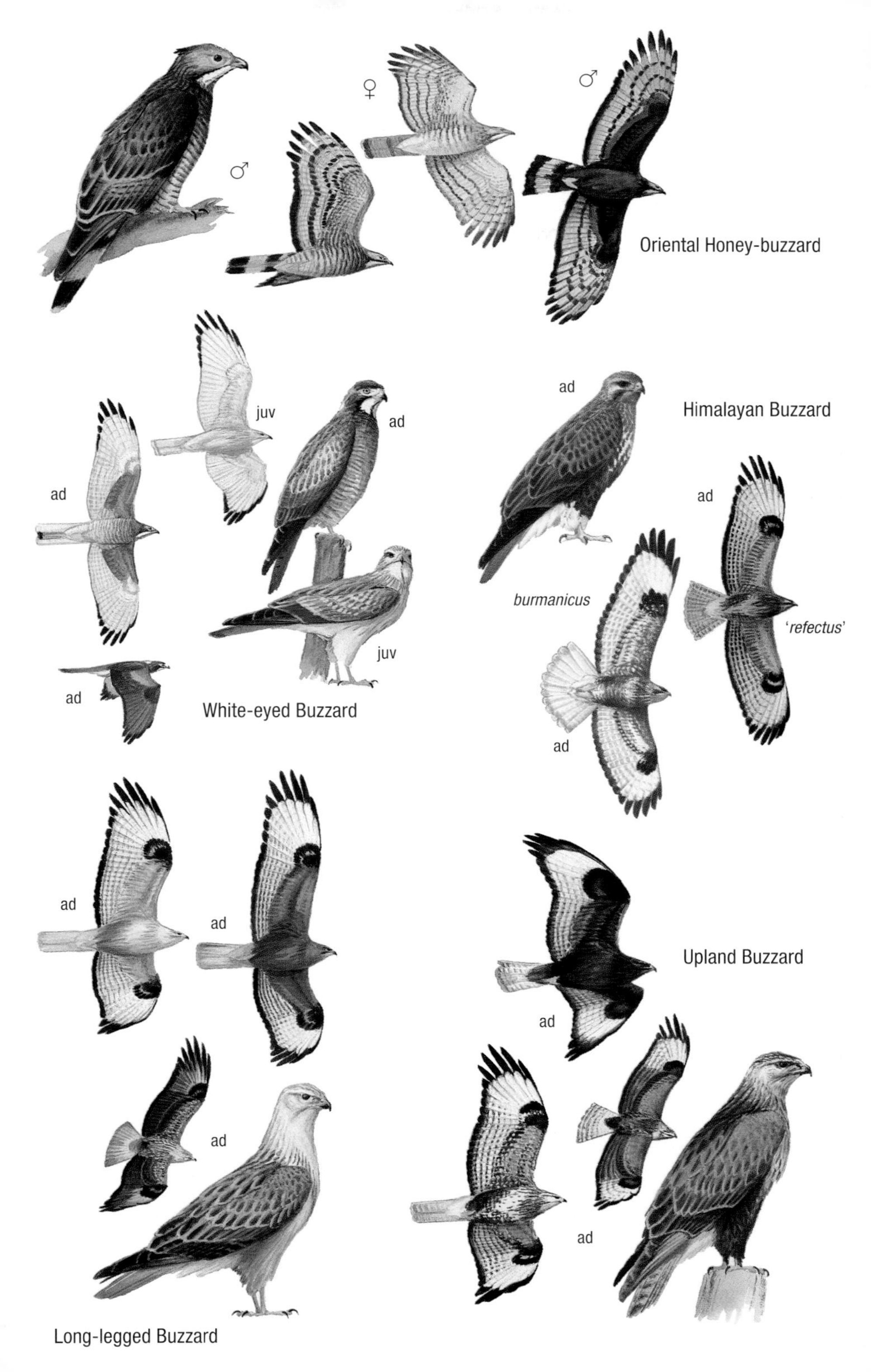

♀
♂
♂
Oriental Honey-buzzard
juv
ad
ad
ad
juv
White-eyed Buzzard
ad
Himalayan Buzzard
ad
burmanicus
'refectus'
ad
ad
ad
Upland Buzzard
ad
ad
ad
Long-legged Buzzard

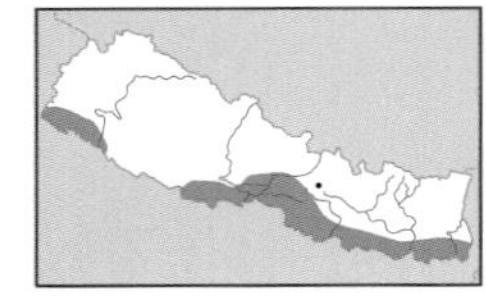

Indian Spotted Eagle *Aquila hastata* 59–67 cm

Rare resident; 75–350 m. **ID** As Greater Spotted, Indian Spotted is a stocky, medium-sized eagle with rather short broad wings, a buzzard-like head with comparatively fine bill (especially so in Indian Spotted), long and closely feathered tarsi, and a rather short tail. The wings are angled down at carpals when gliding and soaring. Adult similar in overall appearance to Greater Spotted Eagle, but warmer brown in coloration. Has wider gape than Greater Spotted, with thick 'lips' (gape flanges) visible at a distance, with gape-line extending to back of or behind eye (reaching level to centre of eye in Spotted). Lacks spiky nape feathers of Greater Spotted and has shorter thigh feathering. Underwing-coverts paler or same colour as flight feathers (darker in Greater Spotted). Juvenile more distinct from juvenile Greater Spotted. Spotting on upperwing-coverts less prominent, tertials pale brown with diffuse white tips (dark with bold white tips in Greater Spotted), uppertail-coverts pale brown with white barring (white in Greater Spotted), and underparts paler light yellowish-brown with dark streaking. In some plumages can resemble Steppe Eagle – differences mentioned for Greater Spotted should be helpful for separation (although gape-line is also long in Steppe). **Voice** Very high-pitched cackling laugh. **HH** Like other *Aquila*, an aggressive and powerful eagle which can soar well, often at considerable heights. Hunts by quartering with slow glides over areas within and near forest, usually flying above treetop level; seizes most prey on the ground. Takes wide range of prey, mainly small mammals, also amphibians, medium to small-sized birds, reptiles and insects. Nest is a large platform of sticks and twigs, some with leaves attached, in a large tree. Favours mixed habitats: hunts in open wetlands, marshes and river valleys within or near woodland or deep forest; breeds in lightly wooded areas. Globally threatened. **TN** Formerly included in Lesser Spotted Eagle *A. pomarina*.

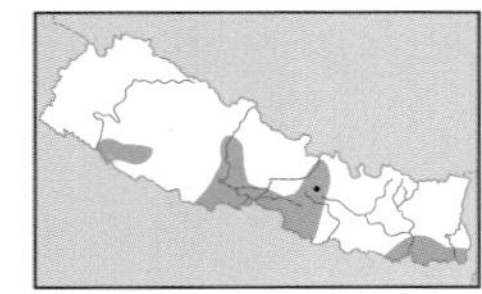

Greater Spotted Eagle *Aquila clanga* 65–72 cm

Rare winter visitor and passage migrant; 75–250 m (–3,840 m on passage). **ID** Medium-sized eagle with rather short and broad wings, stocky head, and short tail. Wings distinctly angled down at carpals when gliding, almost flat when soaring. See text for Indian Spotted for differences from that species. Compared to Steppe Eagle, has less protruding head in flight, with shorter wings and less deep-fingered wingtips; at rest, trousers less baggy, and bill smaller with rounded (rather than elongated) nostrils and shorter gape; lacks adult Steppe's barring on underside of flight and tail feathers, and dark trailing edge to wing, and has dark chin. Pale variant '*fulvescens*' distinguished from juvenile Imperial Eagle by structural differences, lack of prominent pale wedge on inner primaries of underwing, and unstreaked underparts. Juvenile has bold whitish tips to dark brown coverts. **Voice** A barking *kluck-kluck* or *tyuck-tyuck*. **HH** Often seen perched on a treetop, bush or bank near water. Hunts on wing or on ground, generally taking slow-moving prey, especially frogs. Also takes faster-moving waterbirds by swooping low and scattering flock, thereby isolating an individual; also eats stranded or dead fish and lizards. Well-wooded areas in vicinity of wetlands. Globally threatened.

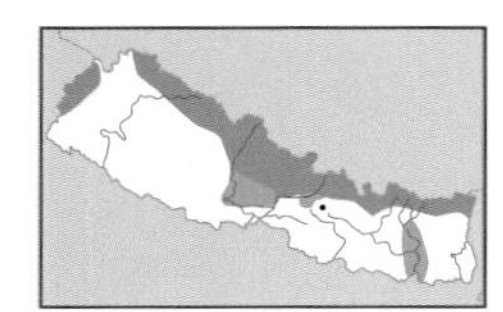

Golden Eagle *Aquila chrysaetos* 75–88 cm

Widespread but sparsely distributed resident; mainly 2,745–4,575 m (75–6,190 m). **ID** Large, with long broad wings (with pronounced curve to trailing edge), long tail, and distinctly protruding head and neck. Wings clearly pressed forward and raised (with upturned fingers) in pronounced V when soaring. Adult has pale panel on upperwing-coverts, golden crown and nape, and two-toned tail. Juvenile has white base to tail and white patch at base of flight feathers. **Voice** Generally silent; occasionally utters a loud, clear yelping *weee-o* in display or a thin, shrill *pleek*. **HH** Found alone or in pairs. Soars for hours over ridges. Usually hunts by quartering a slope, flying low and then striking with talons in a swift dash; occasionally pounces from a perch. Takes wide range of prey, mainly gamebirds and medium-sized mammals. Nest is an enormous platform of sticks, on a cliff ledge or in a tree overhanging or growing from a cliff. Inhabits high rugged mountains, usually well above the treeline.

ad
ad
juv
juv
juv
Indian Spotted Eagle
ad
ad
juv
juv
fulvescens
juv
Greater Spotted Eagle
sub-ad
ad
ad
juv
Golden Eagle

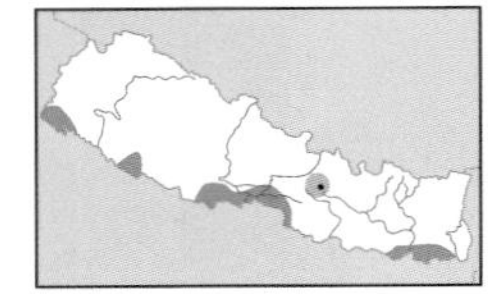

Tawny Eagle *Aquila rapax* 63–71 cm

Very rare, possibly resident; 75–250 m. **ID** Compared to Steppe Eagle, hand of wing does not appear as long and broad, tail slightly shorter, and looks smaller and weaker at rest; gape-line ends level with centre of eye (extends to rear of it in Steppe), and adult has yellowish iris (usually brown in Steppe). Differs from Greater and Indian Spotted in more protruding head and neck in flight, baggy trousers, yellow iris, and oval nostrils. Adult extremely variable, from dark brown through rufous to pale cream, and unstreaked or streaked rufous or dark brown. Dark morph very similar to adult Steppe (which shows much less variation); distinctions include less pronounced barring and dark trailing edge on underwing, dark nape, and dark throat. Tawny uniformly pale from uppertail-coverts to back, with undertail-coverts same colour as belly (contrast often apparent on similar species). Pale adults also lack prominent whitish trailing edge to wing, tip to tail and greater coverts bar (present on immatures of similar species). Characteristic, if present, is distinct pale inner-primary wedge on underwing. Juvenile also variable, with narrow white tips to unbarred secondaries; otherwise as similar-plumaged adult. Immature/sub-adult can show dark throat and breast contrasting with pale belly, and dark banding on underwing-coverts; whole head and breast may be dark. **Voice** A variety of loud, raucous cackles; a distinctive guttural *kra* while in pursuit, and a harsh grating *kekeke* in display flight. **HH** Spends much time perched on treetops in cultivation or near village rubbish dumps or slaughterhouses. Feeds on carrion and refuse; also small mammals, birds and reptiles, mainly stolen from smaller birds of prey, but also captures small mammals by pouncing on them from a small bush. Nest unrecorded in Nepal. In India a large platform of sticks and twigs in the top of a large isolated tree. Inhabits open dry country.

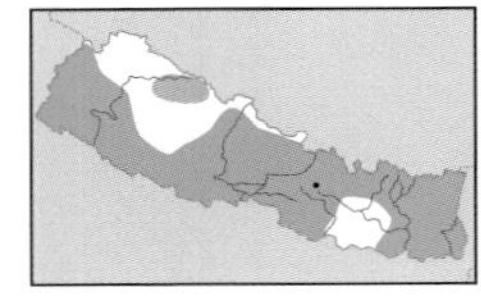

Steppe Eagle *Aquila nipalensis* 76–80 cm

Fairly common and widespread winter visitor and passage migrant; 75–2,200 m (–7,925 m). **ID** Broader and longer wings than Greater and Indian Spotted, with more pronounced and spread fingers, and more protruding head and neck; wings flatter when soaring, and less distinctly angled down at carpals when gliding. When perched, clearly larger and heavier, with heavier bill and baggy trousers. Adult separated from adult spotted eagles by underwing pattern (dark trailing edge, distinct barring on remiges, indistinct/non-existent pale crescents in carpal region), pale rufous nape patch and pale chin. Juvenile has broad white bar on underwing, double white bar on upperwing, and white crescent on uppertail-coverts; prominence of bars on upperwing and underwing much reduced in older immatures (more similar in appearance to Indian Spotted, which see). **Voice** Undescribed in region. **HH** A pirate and carrion eater, also takes easily available prey, such as injured birds, stranded fish, rodents, lizards and snakes. Frequents wooded hills, open country, lakes and large rivers. Globally threatened.

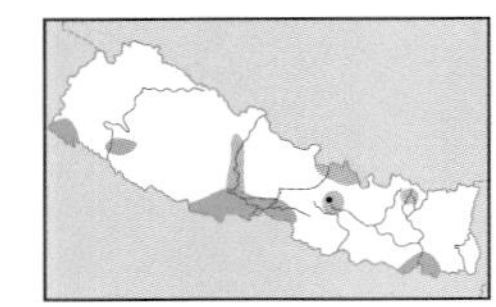

Eastern Imperial Eagle *Aquila heliaca* 72–83 cm

Rare winter visitor, mainly to lowlands, and passage migrant; winters 75–250 m (–3,900 m on passage). Large, stout-bodied eagle with long broad wings, longish tail, and distinctly protruding head and neck. Wings flat when soaring and gliding. Adult has almost uniform upperwing, small white scapular patches, golden-buff crown and nape, and two-toned tail. Juvenile has pronounced curve to trailing edge of wing, pale wedge on inner primaries, streaked buffish body and wing-coverts, uniform pale rump and back (lacking distinct pale crescent of other species, except Tawny), and white tips to median and greater upperwing-coverts. **Voice** A sonorous barking *owk-owk-owk*; rather silent outside breeding season. **HH** A solitary, majestic eagle. Rather inactive, typically spends most of day perched on a good vantage point, such as a tree, or often on ground. In the region, mainly feeds by robbing other birds of prey in flight; also eats carrion and small mammals, birds and reptiles taken on ground in very open country. Soars high overhead and can fly at great speeds. Vicinity of large waterbodies in winter. Globally threatened.

sub-ad
ad
ad
ad
ad
Tawny Eagle
juv
ad
ad
juv
juv
imm
imm
Steppe Eagle
ad
ad
juv
juv
Eastern Imperial Eagle

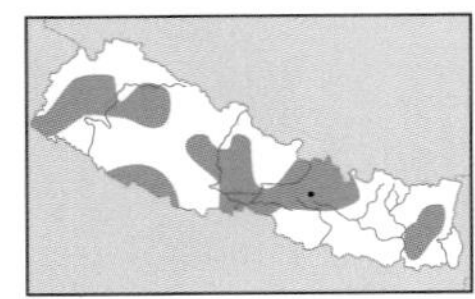

Bonelli's Eagle *Aquila fasciata* 65–72 cm

Resident, uncommon in centre and west, rare in east; mainly 1,400–2,600 m (100–1,350 m). **ID** Medium-sized eagle with long broad wings, distinctly protruding head, and long square-ended tail. Soars with wings flat. Adult has pale underparts and forewing, blackish carpals and band on underwing-coverts, greyish underside to flight feathers, whitish patch on mantle, and pale greyish tail with broad dark terminal band. Juvenile has pronounced curve to trailing edge of wing. Ginger-buff to reddish-brown underparts and underwing-coverts (with variable dark band on greater coverts), and narrow greyish barring on underside of wings and tail, which both lack dark trailing edge. Also shows pale inner primaries on underwing, uniform upperwing and pale crescent across uppertail-coverts and patch on back. **Voice** Silent; mellow fluting calls, *klu-klu-klu-kluee* or *kluu-klu-klu-klu-klu* in display and near nest. **HH** Forested hills and mountains. **TN** Formerly placed in *Hieraaetus*.

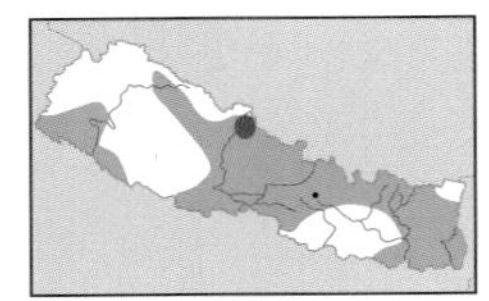

Booted Eagle *Hieraaetus pennatus* 45–53 cm

Chiefly uncommon winter visitor and passage migrant; has bred; 75–4,000 m. **ID** Smallish, rather kite-like eagle. Wings comparatively long and narrow, and tail long and square-ended. When gliding and soaring, wings held slightly forward and flat or slightly angled down at carpal. Twists tail in kite-like fashion. In all plumages, shows small white shoulder patches, pale panel on median coverts, pale wedge on inner primaries, pale scapulars, white crescent on uppertail-coverts, and greyish underside to tail. Head, underparts and underwing-coverts whitish, brown or rufous respectively in pale, dark and rufous morphs. Juvenile as adult, but has broad white trailing edge to wings and tail. **Voice** High-pitched double whistle, *ki-keee*. In display very noisy, *pi-peee, pi-pi-pi-pi-peee*. **HH** Well-wooded country in hills, mountains and lowlands.

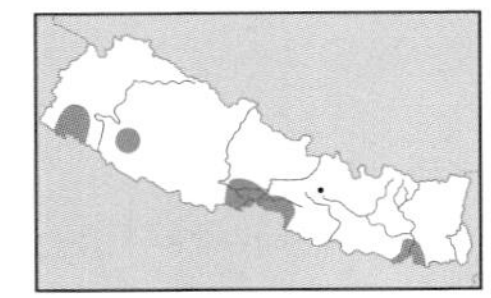

Rufous-bellied Eagle *Lophotriorchis kienerii* 53–61 cm

Very rare, possibly only a visitor; 75–300 m. **ID** Smallish, with buzzard-like wings and tail. Glides and soars on flat wings. Adult has blackish hood and upperparts, white throat and breast, and rufous rest of underparts. Upperwing uniformly dark except pale patches at base of primaries. Juvenile has white underparts and underwing-coverts, dark mask and white supercilium, dark patch on sides of upper breast, and dark patch on flanks. Head-on, shows striking white leading edge to wing; upperwing dark with paler base to primaries. **Voice** Silent except in breeding season, when utters a piercing or plaintive scream. **HH** Frequents broadleaved evergreen, semi-evergreen and moist deciduous broadleaved forest. **TN** Formerly placed in *Hieraaetus*.

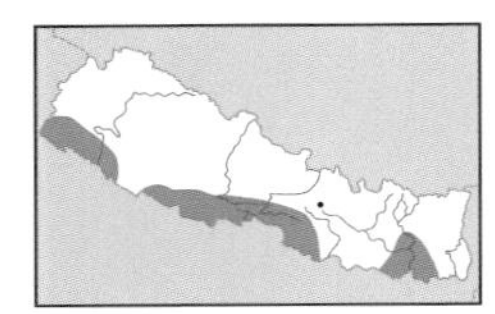

Changeable Hawk Eagle *Nisaetus* (*cirrhatus*) *limnaeetus* 63–77 cm

Local resident, mainly in lowland protected areas; 75–360 m (–1,050 m). **ID** Wings slightly narrower and more parallel-edged, compared to Mountain Hawk Eagle, with proportionately longer tail. Soars with wings held flat (except in display, when both wings and tail raised). Best separated from Mountain by lack of prominent crest, paler sides to head, boldly streaked underparts (any barring is confined to flanks, thighs and vent), and narrower dark tail barring. Dark morph confusable with Black Eagle; best told by structural differences, greyish undertail and greyish bases to underside of flight feathers. Juvenile has pale fringes to upperparts (some have white forewing), buff underparts and underwing-coverts, and narrower and more numerous tail-bands than adult. Difficult to separate from juvenile Mountain but has paler head and lacks crest. Has pale crescent across outer primaries, lacking in Mountain. **Voice** Silent except in breeding season; gives an ascending series of shrill whistles, *kri-kri-kri-kri-kree-ah* and *kreeee-krit* and a loud high-pitched *ki-ki-ki-ki-ki-ki-ki-ki-kee* rising in crescendo and ending in a long, drawn-out scream. **HH** Fairly open sal and mixed broadleaved forest.

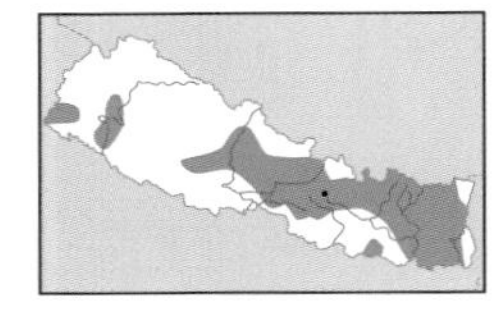

Mountain Hawk Eagle *Nisaetus nipalensis* 72 cm

Frequent resident; 1,500–2,835 m (–75 m). **ID** Wings broader than Changeable, with squarer wingtips and more pronounced curve to trailing edge, and has proportionately shorter tail. Soars with wings in shallow V. Distinguished from Changeable by prominent crest, dark crown and ear-coverts (with variable pale supercilium), heavily barred underparts and underwing-coverts, whitish-barred uppertail-coverts, and stronger dark barring on tail. Juvenile from juvenile Changeable by more extensive dark streaking on crown and sides of head, white-tipped black crest, and fewer, more prominent tail-bands. Has patch of light and dark barring on inner primaries, lacking in Changeable. **Voice** Three shrill notes, *tlueet-weet-weet*; also a rapid, bubbling call in display. **HH** Forested hills and mountains. **TN** Formerly placed in *Spizaetus*.

ad
juv
ad
juv
Bonelli's Eagle
ad
pale morph
dark morph
pale morph
Booted Eagle
dark morph
ad
juv
ad
Rufous-bellied Eagle
ark morph
juv
juv
dark morph
pale morph
ad
Mountain Hawk Eagle
pale morph
juv
juv
Changeable Hawk Eagle

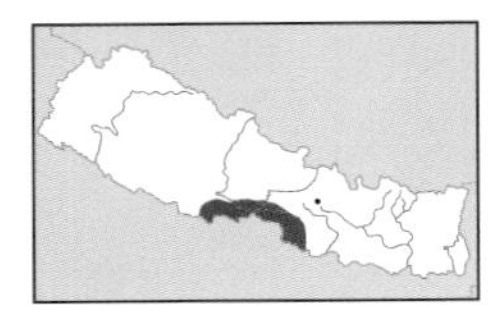

Slaty-legged Crake *Rallina eurizonoides* 25 cm

Probably a very rare and very local summer visitor; 75–375 m. **ID** Greenish or grey legs and feet and more extensive black-and-white barring on underparts are best features from adult Ruddy-breasted. Also olive-brown mantle contrasts with rufous neck and breast. Juvenile has dark olive-brown head, breast and upperparts. More prominent and extensive black-and-white barring on underparts are best features from juvenile Ruddy-breasted. **Voice** A *kek-kek, kek-kek*, a loud drumming croak, a subdued *kok*, and a *krrrr* alarm call. **HH** A typical rail though partially nocturnal. Very skulking, emerging from thick cover in early morning and at dusk like other rails. Nests in dense vegetation, sometimes away from water. Marshes in forest and well-wooded country.

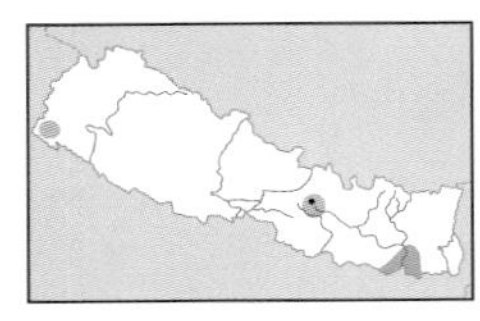

Brown-cheeked Rail *Rallus indicus* 23–28 cm

Rare and very local winter visitor and passage migrant; 100–120 m (–1,340 m). **ID** Similar to Water Rail (see Appendix 1), but has more pronounced supercilium (due to darker eye-stripe), browner wash to breast, and barred undertail-coverts. Juvenile similar to juvenile Water Rail but has barred undertail-coverts. **Voice** A metallic, strident *skrink, skrink* beginning explosively and repeated after a few seconds. **HH** Like other rails, secretive, keeping within thick cover, except during early morning and dusk; typically bolts into cover with head and tail lowered at least sign of danger. Marshes and reedbeds. **AN** Eastern Water Rail.

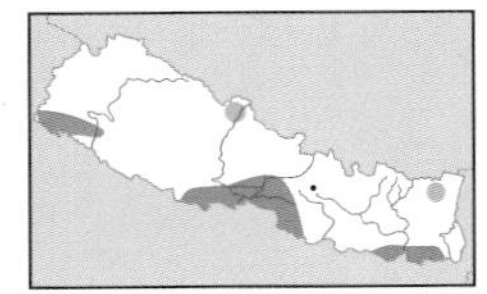

Brown Crake *Amaurornis akool* 28 cm

Local resident, fairly common in Chitwan National Park, rare elsewhere; 75–365 m. **ID** Olive-brown upperparts, grey face and breast, and olive-brown flanks and undertail-coverts; underparts lack barring. Has red iris, greenish bill and pinkish-brown to purple legs. Juvenile similar to adult, but has dull iris and paler grey underparts. **Voice** Calls include a shrill rattle, a long drawn-out vibrating whistle and a short plaintive note. **HH** A typical rail though less secretive than most species. Like other rails, flies reluctantly and feebly, with legs dangling, for a short distance and then drops into cover again. Nests in or near marshes. Inhabits reedy marshes and vegetation bordering watercourses.

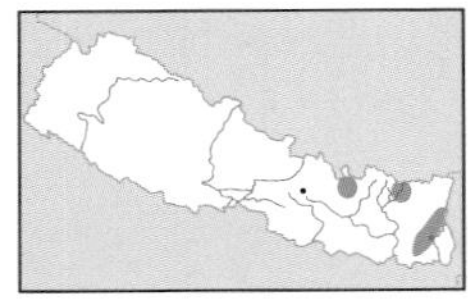

Black-tailed Crake *Porzana bicolor* 25 cm

Very rare and very local, presumably resident; 1,925–2,600 m. **ID** Red legs, iris and eye-ring. Bill greenish with variable red at base. Sooty-grey head and underparts, rufous-brown upperparts, and sooty-black tail and undertail-coverts. Juvenile has dark brown upperparts and brownish-olive underparts with some white mottling on breast and belly; iris brown, and bill and legs duller/browner. **Voice** Quite harsh rasping notes, often followed by a prolonged trill that is more obviously descending than Ruddy-breasted Crake. **HH** Poorly known, very skulking, probably like those of other rails. Upland marshes.

Baillon's Crake *Porzana pusilla* 17–19 cm

Rare winter visitor and passage migrant; has bred; 75–1,350 m. **ID** Adult has rufous-brown upperparts (extensively marked white), grey underparts, and black-and-white barring on flanks. Legs and bill green. Juvenile has buff underparts. **Voice** Song may be heard on migration: a rattling rasp, *trrrrr-trrrrr*, also a *tyiuk* alarm call. **HH** Typical rail. Secretive, though not particularly shy, mainly crepuscular. Nest usually has a partial canopy of plant stems. Reedy edges of lakes and large rivers, marshes and wet paddyfields.

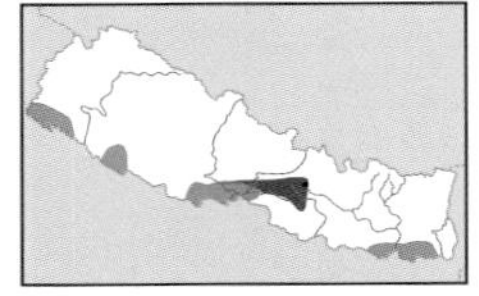

Ruddy-breasted Crake *Porzana fusca* 22 cm

Locally fairly common resident in lowlands; summer visitor to mid-hills; 75–1,280 m. **ID** From other crakes by combination of dull chestnut underparts, unmarked dark olive-brown upperparts, indistinct dark brown and white barring on rear flanks and undertail-coverts (much more restricted than Slaty-legged Crake) and red legs. Juvenile dark olive-brown, with white-barred undertail-coverts and fine greyish-white mottling/barring over rest of underparts. Legs duller and iris brown (rather than red as in adult). **Voice** Single soft *crake* at considerable intervals; a loud metallic *twek* repeated at 2–3-second intervals, often followed by a squeaky trilling. **HH** Typical rail. Skulking. Like other rails, walks with rhythmic movements of head and neck, and jerks tail. Nests on marshy ground among grass, reeds or rice plants; stalks sometimes bent over to form a canopy. Reedy lake edges, marshes and wet fields.

ad
juv
Slaty-legged Crake
ad
Brown-cheeked Rail
ad
Brown Crake
ad
juv
Black-tailed Crake
ad
juv
Baillon's Crake
ad
juv
Ruddy-breasted Crake

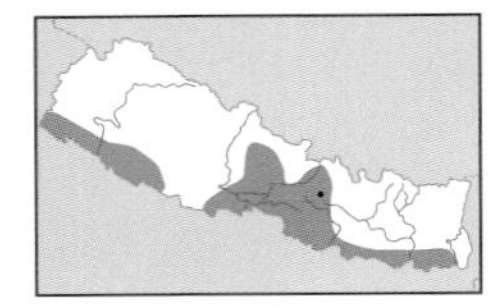

White-breasted Waterhen *Amaurornis phoenicurus* 32 cm

Fairly common resident; 75–1,370 m (–3,800 m). **ID** Adult has grey upperparts and white face, foreneck and breast; undertail-coverts rufous-cinnamon. Bill and legs greenish or yellowish, with swollen reddish base to upper mandible. Juvenile has greyish face, foreneck and breast, and olive-brown upperparts; bill and legs darker. **Voice** Highly vocal when breeding; calls include a metallic *krr-kwaak-kwaak* and a *kook... kook... kook* often following loud roars, croaks and chuckles. **HH** Habits similar to other rails, but less shy than most species; often feeds in open and on quite dry land. Clambers among bushes with ease. Nests on ground in dense undergrowth or thick bushes at water's edge. Inhabits thick cover close to all wetland types.

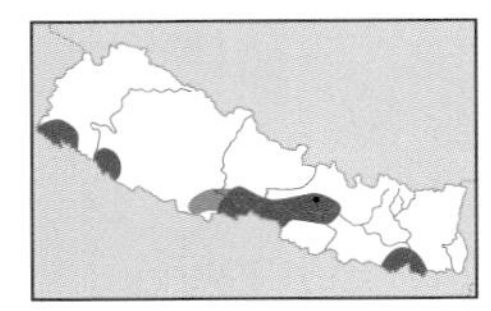

Watercock *Gallicrex cinerea* ♂ 43 cm, ♀ 36 cm

Locally frequent monsoon visitor; 75–1,280 m. **ID** Breeding male mainly greyish-black, with yellow-tipped red bill and red shield and horn. Upperparts fringed grey and buff. Legs bright red. First-summer male has broad rufous-buff fringes to plumage. Non-breeding male and female have buff underparts with fine barring, and buff fringes to dark brown upperparts. Legs greenish. Juvenile has uniform rufous-buff underparts, and rufous-buff fringes to upperparts. Male much larger than female with heavier bill. **Voice** Series of 10–12 *kok-kok-kok* notes followed by a deep, hollow *utumb-utumb-utumb* repeated 10–12 times, and then by 5–6 *kluck-kluck-kluck* notes. **HH** Very similar to other rails. Mainly crepuscular, also emerging from cover in cloudy weather. In breeding season, in the monsoon, males call persistently. Nest a deep cup or dome of vegetation in reedbeds or paddyfields. Marshes and flooded fields.

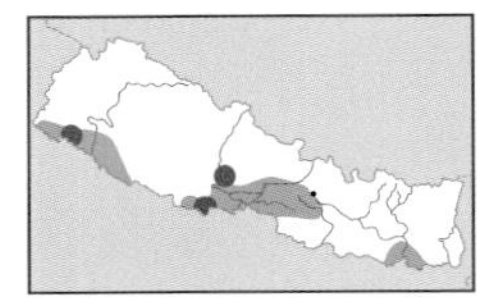

Purple Swamphen *Porphyrio porphyrio* 45–50 cm

Chiefly a winter visitor and passage migrant, also breeds locally; 75–915 m (–1,370 m). **ID** Large size, purplish-blue coloration with variable greyish head, and huge red bill and frontal shield. Female smaller than male. Juvenile duller than adult, with greyer neck and underparts, more olive-brown above, duller red bill (blackish at first), and duller legs and feet. **Voice** An explosive, nasal, rising *quinquinkrrkrr* in alarm; also a soft *chuck-chuck*. **HH** Typically in small parties and in larger numbers in extensive marshes. Diurnal, and where undisturbed not shy. Forages mainly in reedbeds and readily clambers on reeds. Nest is a bulky mass of vegetation on a floating islet or just above water in reeds. Inhabits dense reedbeds at pool edges and in marshes. **TN** The Indian subspecies is sometimes treated as a separate species, Grey-headed Swamphen *P.* (*porphyrio*) *poliocephalus*.

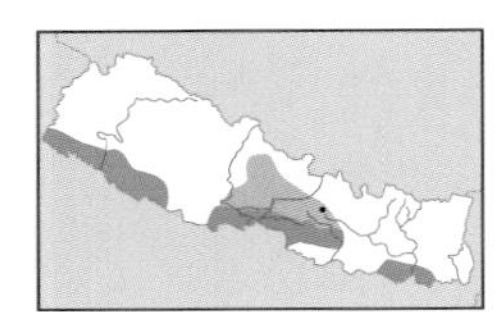

Common Moorhen *Gallinula chloropus* 32–35 cm

Locally common resident and winter visitor; 75–250 m (–4,575 m on passage). **ID** White lateral undertail-coverts, with black central stripe, and usually shows white line on flanks. Breeding adult has blackish head and neck, slate-grey underparts, and dark olive-brown upperparts; red bill with yellow tip and red frontal shield. Non-breeding adult has duller bill and legs. Juvenile has dull green bill, and is mainly brown with whitish throat, grey wash to breast and flanks, and variable whitish patch on belly. Immature resembles non-breeding adult, but has duller body plumage, with grey of underparts washed brown and buff, whitish throat, and less prominent pale border to flanks. **Voice** A *cuk cuk cuk* and *kekuk* in alarm; also a loud explosive *kurr-ik* and *kark*, and soft muttering *kook... kook*. **HH** Usually forages in open; spends most of day swimming among aquatic plants; also feeds on land. Generally escapes danger by swimming into vegetation or running quickly. Slow and heavy in flight, and patters along the water before becoming airborne. Nest is a large mass of vegetation in reedbeds, built a little above water level. Lakes, pools and marshes with emergent vegetation.

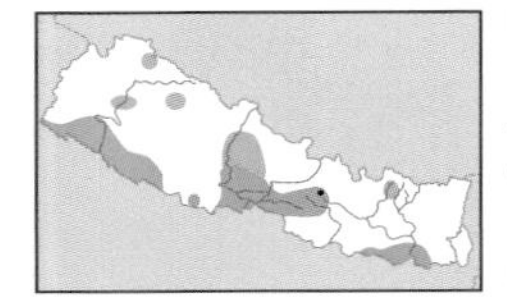

Eurasian Coot *Fulica atra* 36–38 cm

Regular winter visitor and passage migrant, locally common; 75–3,500 m (–5,000 m). **ID** Blackish, with white bill and frontal shield. Paler trailing edge to secondaries in flight. Immature duller than adult with whitish throat. Juvenile grey-brown, with whitish throat and breast. **Voice** Calls include a high-pitched *pyee* and a series of long, soft *dp... dp* notes. **HH** Gregarious, diurnal and not shy. Forages chiefly for aquatic vegetation in open water, mainly by diving; also by sieving plant material from surface. When disturbed, prefers to skitter away along water surface rather than fly. Takes to air with difficulty, after prolonged pattering across water. Large areas of open water with emergent vegetation.

White-breasted Waterhen
ad
juv
♀
♀ juv
♂ br
Watercock
♂ non-br
(transitional ad)
Purple Swamphen
ad
juv
Eurasian Coot
ad
ad
juv
juv
Common Moorhen

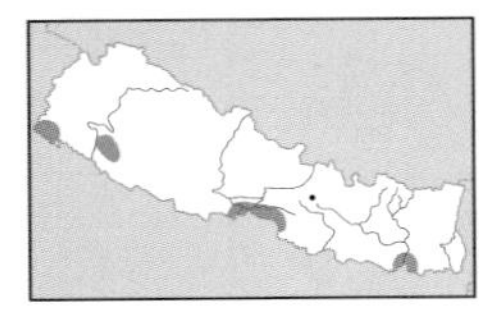

Bengal Florican *Houbaropsis bengalensis* 66 cm

Rare and local resident; 75–305 m. **ID** Larger and stockier than Lesser Florican, with broader head and thicker neck. Breeding male has black head, neck and underparts, and white wing-coverts. In flight, wings are entirely white except black tips. Non-breeding male similar to female but has white wing-coverts and retains some black on belly. Female larger than male and has buffish neck and underparts, and pale buff coverts with dark flight feathers. Immature similar to female but has banding on flight feathers and wing-coverts are more heavily marked. **Voice** Shrill, metallic *chik-chik-chik* when disturbed. **HH** Very shy. Forages in short-grass areas. In breeding season males perform striking displays, leaping above grassland in early mornings and evenings. Grassland. Globally threatened.

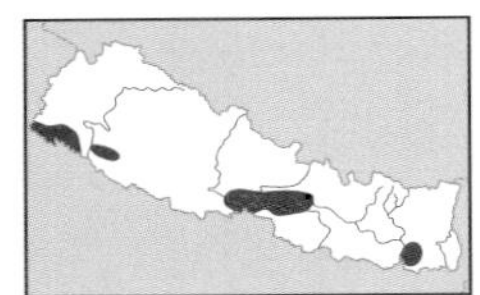

Lesser Florican *Sypheotides indicus* 46–51 cm

Very rare summer visitor; 250–300 m (–1,310 m). **ID** Smaller and slimmer than Bengal, with smaller head, finer neck and more rapid wingbeats. Male breeding has black head/neck and underparts, and white wing-coverts. Differs in white throat, spatulate-tipped head plumes, and white 'hind collar'; also white of wing more restricted, and has rufous banding on dark flight feathers. Non-breeding male similar to female, but has whiter wing-coverts and black underwing-coverts. Female and immature from Bengal by dark crescent below eye, dark stripes on foreneck, and rufous-barred primaries. **Voice** Short whistle when disturbed. **HH** When flushed, flies off with fast wingbeats, covering some distance before landing and running into cover. Dry grassland. Globally threatened.

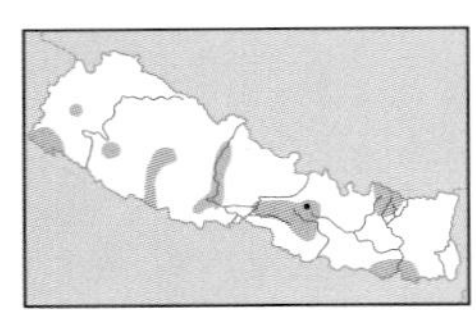

Demoiselle Crane *Grus virgo* 90–100 cm

Fairly common passage migrant; 75–5,185 m. **ID** Small crane, with short, fine bill. Adult has black head and neck with white tuft behind eye, and grey crown; black neck feathers extend as point beyond breast, and elongated tertials project in shallow arc beyond body, giving rise to distinctive shape. Immature initially almost entirely grey, with slate-grey on foreneck and shorter all-grey tertials. By first winter is similar to adult, but head and neck dark grey and less contrasting, tuft behind eye grey and less prominent, has brown cast to upperparts, and elongated feathers of foreneck and tertials are shorter. In flight, black breast helps separate from Common Crane at distance; also legs and neck appear relatively shorter, and wings shorter and broader-based. **Voice** Flight call *garrooo*, higher pitched than Common. **HH** Highly gregarious and like other cranes flocks often fly in V-formation. Cultivation and large rivers.

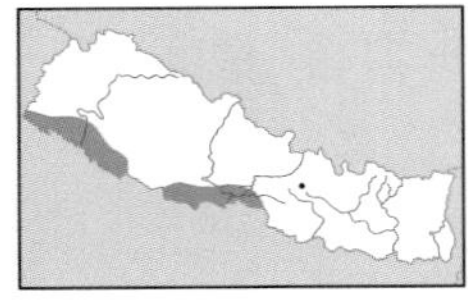

Sarus Crane *Grus antigone* 156 cm

Uncommon and local resident; 150–300 m. **ID** A huge, mainly pale grey crane with reddish legs and very large bill. Adult grey, with bare red head and upper neck, and bare ashy-green crown. In flight, black primaries contrast with rest of wing. Immature has rusty-buff head and neck, and upperparts marked with brown; older immature similar to adult but has dull red head and upper neck, and lacks greenish crown. **Voice** Very loud trumpeting, usually a duet by pairs. **HH** Like other cranes has powerful flight with head and neck extended and legs and feet outstretched behind. Cultivation in well-watered country. Globally threatened.

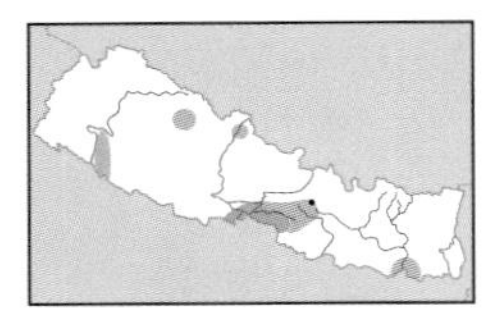

Common Crane *Grus grus* 110–120 cm

Irregular winter visitor and passage migrant; 75–3,050 m. **ID** Adult has mainly black head and foreneck, with white stripe behind eye extending down side of neck; red patch on crown visible at close range. Immature has brown markings on upperparts with buff or grey head and neck; adult head pattern apparent on some by first winter and as adult by second winter. In flight, both adult and immature show black primaries and secondaries contrasting with grey wing-coverts. **Voice** Loud, trumpeting *krrooah* flight call; bugling duet on ground. **HH** Winter crops and large rivers, often with Demoiselles.

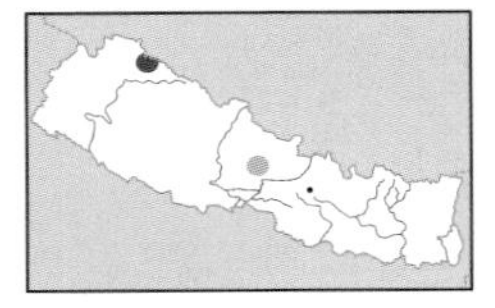

Black-necked Crane *Grus nigricollis* 139 cm

Very rare and very local summer visitor to Humla, in far north-west; 4,800 m. **ID** Large, stocky crane, with comparatively short neck and legs. Adult pale grey with contrasting black head, upper neck and bunched tertials; more contrast between black flight feathers and grey coverts than Common, and has black tail-band. Immature has buff or brownish head, neck, mantle and mottling to wing-coverts. **Voice** Variety of calls, most slightly higher pitched than Common. **HH** Breeds in high-altitude moist pasture. Globally threatened.

imm
♀
♂
♀
♂ non-br
♀
♂ br
♂ br
Bengal Florican
Lesser Florican
Sarus Crane
ad
juv
ad
juv
ad
ad
Demoiselle Crane
juv
ad
ad
juv
ad
juv
ad
ad
Common Crane
Black-necked Crane

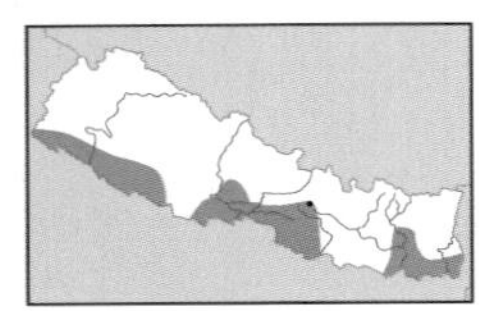

Indian Thick-knee *Burhinus (oedicnemus) indicus* 36–39 cm

Frequent and locally common resident; 75–375 m (–1,310 m). **ID** Sandy-brown and streaked. Short yellow-and-black bill, striking yellow eye, and long yellow legs. In flight has black flight feathers with patches of white in primaries. Treated as separate species from Eurasian Thick-knee by some authorities. Smaller, darker and more heavily streaked than Eurasian, with larger bill, shorter tail and longer tarsi. Bill mainly black, and shows more pronounced barring on wing-coverts (with broader pale panel). **Voice** Piercing calls recalling Great. **HH** Typically spends day in shade. Very wary; if suspicious runs off furtively with its head low, then squats and flattens itself on ground. When foraging, makes short runs, stopping to capture prey with swift snatch. Sandy or stony riverbeds and open dry fields. **AN** Indian Stone-curlew.

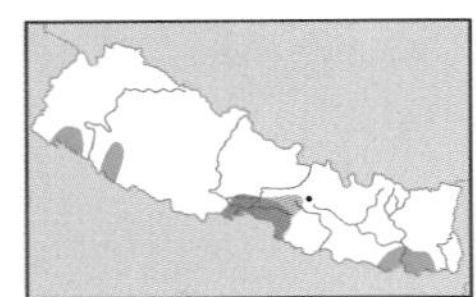

Great Thick-knee *Esacus recurvirostris* 49–54 cm

Rare, now mainly recorded at Koshi in far east; 75–250 m. **ID** Large, slightly upturned black-and-yellow bill, and yellow eye. At rest, most striking features are white forehead and 'spectacles' contrasting with black ear-coverts, and blackish and whitish bands on wing-coverts. In flight, grey panel on wings and white patches on primaries. **Voice** A rising, wailing whistle of two or more syllables; a loud, harsh *see-eek* alarm call. **HH** Habits similar to Eurasian but usually rests in full sun during day. Stony, shingle or sandy beds and banks of large rivers. **AN** Great Stone-curlew.

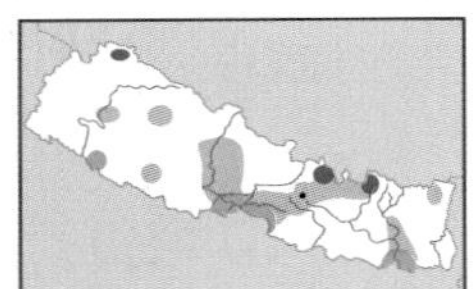

Ibisbill *Ibidorhyncha struthersii* 38–41 cm

Very uncommon resident, breeds locally; 3,800–4,200 m (breeding season), 100–915 m (winter). **ID** Adult has black face, downcurved dark red bill, and black and white breast-bands. In flight, white patch at base of inner primaries and blackish tail-band. Juvenile has brownish upperparts with buff fringes, faint breast-band, and dull legs and bill. **Voice** A ringing *klew-klew* and rapid *tee-tee-tee-tee*. **HH** Singles, pairs or small groups forage inconspicuously and quietly in mountain rivers and streams. Usually quite wary; if alarmed nervously bobs head and wags tail. Seeks aquatic invertebrates by walking slowly through water, using its long decurved bill to probe around and under stones; frequently wades belly-deep. Fast-flowing mountain streams and rivers with shingle beds.

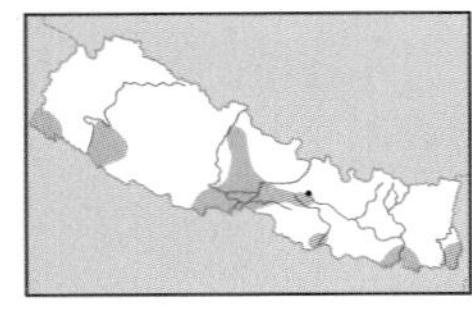

Black-winged Stilt *Himantopus himantopus* 35–40 cm

Uncommon passage migrant; 75–1,380 m (–3,355 m). **ID** Slender appearance, with long pinkish legs and a fine straight bill. Black upperwing strongly contrasts with white V on back in flight. Adult at rest has mainly white head, neck and underparts, contrasting with upperparts, and reddish-pink legs. Both sexes can show variable amounts of black and/or dusky grey on crown and hindneck. Juvenile has browner upperparts with buff fringes. **Voice** A noisy wader, readily agitated; calls include *kek... kek* and a rather anxious *kikikikiki*. **HH** Gregarious throughout year. Graceful, walking slowly and deliberately. Forages on dry mud and by wading in shallows, sometimes belly-deep in water. Picks prey from surface, probes in soft mud and sweeps bill from side to side; sometimes immerses head and neck in water. Shallow marshes, pools and lakes.

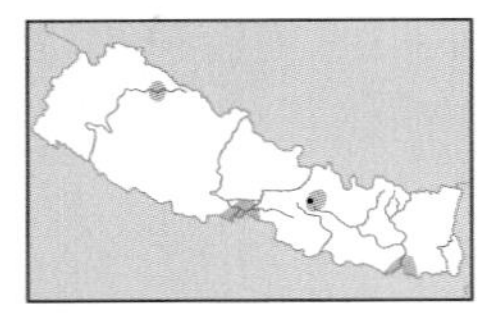

Pied Avocet *Recurvirostra avosetta* 42–45 cm

Rare passage migrant; 75–275 m. **ID** Upward kink to black bill. Distinctive black-and-white pattern. Juvenile has brown and buff mottling on mantle and scapulars. **Voice** A throaty *quib... quib*. **HH** Characteristically feeds by sweeping bill and head from side to side in shallow water; often swims and upends like a dabbling duck, and picks prey from surface of water or mud. Marshes, lagoons and mudflats.

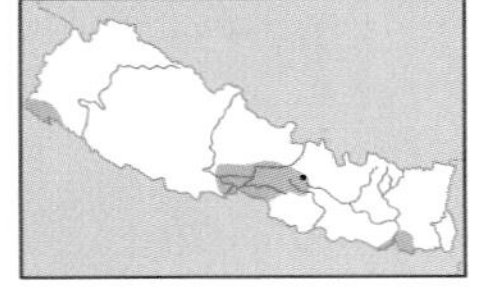

Ruff *Philomachus pugnax* ♂ 26–32 cm, ♀ 20–25 cm

Very uncommon passage migrant; 75–1,310 m. **ID** Distinctive shape, with long neck, small head, short and slightly downcurved bill, and long yellowish or orangey legs. In all plumages, lacks prominent supercilium and, in flight, shows narrow white wing-bar and prominent white sides to uppertail-coverts. Male considerably larger than female. Non-breeding and juvenile have neatly fringed upperparts, juvenile has buff underparts. Breeding birds typically have black and chestnut markings on upperparts, male with striking ruff. **Voice** Generally silent. **HH** Forages by walking purposefully, picking prey from surface and vegetation, and probing in mud. When stationary, has upright posture with head raised. Marshes, wet fields and mudbanks of rivers and lakes.

Indian Thick-knee
ad
ad
ad
Great Thick-knee
ad
Ibisbill
ad
ad
Black-winged
Stilt
ad
imm
♂ br
♂ br
♂ br
♀ br
Pied Avocet
ad
♀ juv
Ruff
♂ non-br

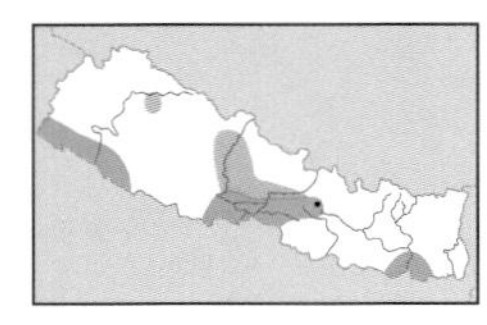

Northern Lapwing *Vanellus vanellus* 28–31 cm

Uncommon winter visitor; 75–1,380 m (–2,700 m). **ID** Black crest, white (or buff) and black face pattern, black breast-band, and dark green upperparts. Juvenile has prominent buff fringes to mantle, scapulars and wing-coverts. Very broad, rounded wingtips. Whitish rump and blackish tail-band in flight. **Voice** Mournful *eu-whit*. **HH** In pairs or small flocks. Feeding behaviour similar to Red-wattled. Wet grassland, marshes, lake margins, riverbanks; sometimes fallow fields and dry stubble.

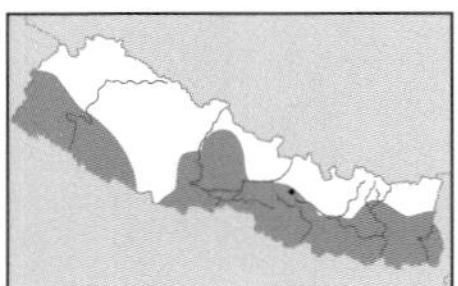

River Lapwing *Vanellus duvaucelii* 29–32 cm

Locally fairly common resident in lowlands, frequent in mid-hills; 75–915 m (–1,380 m). **ID** Black crest, face and throat, grey sides to neck, and black bill and legs. Black patch on belly. In flight, broad white greater coverts wing-bar contrasting with black flight feathers, and black tail. Juvenile similar to adult, but black of head partly obscured by white tips, and has buff fringes and dark subterminal marks to feathers of upperparts. **Voice** Sharp insistent, high-pitched *did, did, did*, sometimes ending with *did-did-do-weet*. **HH** Usually occurs singly, in pairs or small groups. Feeding behaviour similar to Red-wattled. Often has a hunched posture with head drawn in. Sandbanks and shingle banks of rivers.

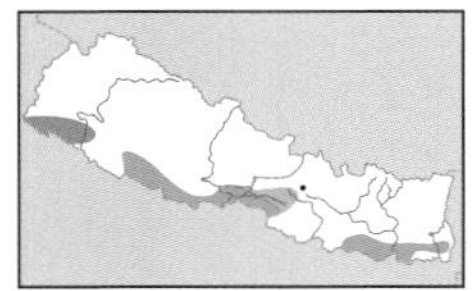

Yellow-wattled Lapwing *Vanellus malabaricus* 26–28 cm

Local resident and visitor; 100–185 m **ID** Yellow wattles and legs. White eye-stripes joining on nape, dark cap, and brown breast-band. In flight, white greater coverts wing-bar contrasting with black flight feathers, and white tail with black subterminal band. Juvenile has small and dull yellow wattles, white chin, brown cap, and prominent buff fringes and dark subterminal bars to feathers of upperparts. **Voice** Strident *chee-eet* and a hard *tit-tit-tit*. **HH** Occurs singly, in pairs, and sometimes in small flocks in non-breeding season. Foraging behaviour similar to Red-wattled Lapwing. Flight buoyant with rather slow wingbeats. Dry fields, open dry country and dry riverbeds.

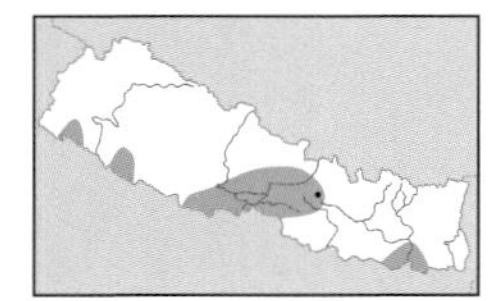

Grey-headed Lapwing *Vanellus cinereus* 34–37 cm

Uncommon winter visitor and passage migrant 1,310 m in Kathmandu Valley and uncommon elsewhere below 275 m. **ID** Yellow bill with black tip, and yellow legs. Grey head, neck and breast, latter with diffuse black border, and black tail-band. Secondaries white. Juvenile has brownish head and neck, lacks dark breast-band, and has prominent buff fringes to feathers of upperparts. **Voice** Plaintive *chee-it, chee-it*. Usually found in small parties or flocks of up to 50 birds. **HH** Behaviour similar to Red-wattled Lapwing, with which it often feeds. Riverbanks and wet fields.

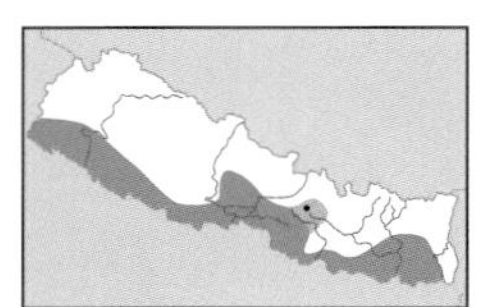

Red-wattled Lapwing *Vanellus indicus* 32–35 cm

Common and widespread resident; 75–1,050 m. **ID** Black cap and breast, red bill with black tip, and yellow legs. In flight, white greater coverts wing-bar and black tail-band. Juvenile duller than adult, with whitish throat. **Voice** Agitated and penetrating *did he do it, did he do it* and a less intrusive *did did did*. **HH** In pairs or small flocks of up to *c*.12. A vigilant and noisy bird; when alarmed calls loudly and frantically while circling overhead. Forages by walking or running in short spurts, then stops and probes with body tilted forward; also vibrates its foot rapidly on surface to flush invertebrates. Feeds mainly at night and in early mornings and evenings. Usually flies slowly with deep flaps. Open ground near water.

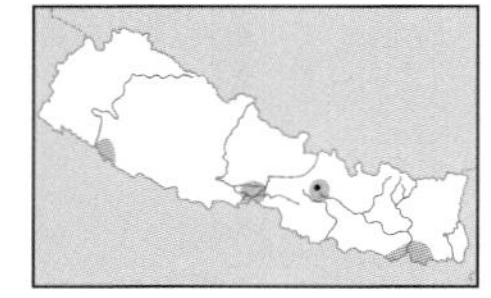

Black-tailed Godwit *Limosa limosa* 36–44 cm

Rare and local passage migrant; 75–1,525 m. **ID** White wing-bars and white rump with black tail-band. Long straight bill mainly pinkish with darker tip, and long dark legs. In breeding plumage, male has rufous-orange neck and breast, with blackish barring on underparts and white belly; breeding female larger and duller than male. In non-breeding plumage, is uniform grey on neck, upperparts and breast. Juvenile has cinnamon underparts and cinnamon fringes to dark-centred upperparts. **Voice** Yapping *kek-kek* occasionally uttered in flight. **HH** Feeds mainly by walking slowly and probing in open mud or shallows; also by picking prey from surface. Feeds in deeper water than most waders; sometimes wades up to the belly and probes with head and neck almost completely submerged. Mainly shallows and mudbanks of large rivers and lakes. **AN** Western Black-tailed Godwit, if eastern form *melanuroides* is split.

Northern Lapwing
ad
ad
River Lapwing
ad
Yellow-wattled Lapwing
non-br
ad
Grey-headed Lapwing
Red-wattled Lapwing
non-br
non-br
♂ br
juv
Black-tailed Godwit

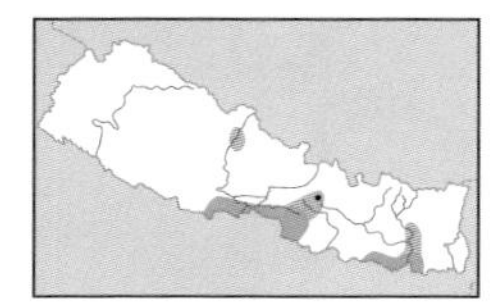

Pacific Golden Plover *Pluvialis fulva* 23–26 cm

Uncommon, mainly a passage migrant, and winter visitor; 75–250 m (–2,950 m). **ID** Slim-bodied, long-necked and long-legged plover with short dark bill. In all plumages, has golden-yellow markings on upperparts, and dusky-grey underwing-coverts and axillaries. In flight shows narrower white wing-bar and dark rump. Adult breeding has black on face, foreneck, breast and belly strikingly bordered by white. Adult non-breeding and juvenile have yellowish-buff wash to supercilium, cheeks and neck; usually show pronounced supercilium, which often curves down as diffuse crescent behind ear-coverts, and a dark patch on rear ear-coverts. See Appendix 1 for comparison with Grey Plover. **Voice** An abrupt disyllabic *chi-vit* recalling Spotted Redshank and a more plaintive *tu-weep* or *chew you.* **HH** Associates in flocks which disperse when feeding. Flight is swift and direct. Very wary and, if disturbed, birds rise almost simultaneously in a compact flock, twisting and turning rapidly in unison. Typical plover feeding behaviour: runs in short spurts, pausing and standing erect, then stooping to pick up prey. Mudbanks of wetlands and ploughed fields.

Long-billed Plover *Charadrius placidus* 19–21 cm

Very rare winter visitor and passage migrant; 75–1,380 m. **ID** Like a large Little Ringed, but has longer tail with clearer dark subterminal bar and more prominent white wing-bar; ear-coverts never black, and has less distinct eye-ring than Little Ringed. White forehead and supercilium more prominent in non-breeding plumage compared with Little Ringed. Habits similar to those of Little Ringed, but usually solitary. **Voice** A clear, penetrating *piwee* in flight. **HH** Very well camouflaged in its usual habitat of shingle banks of large rivers.

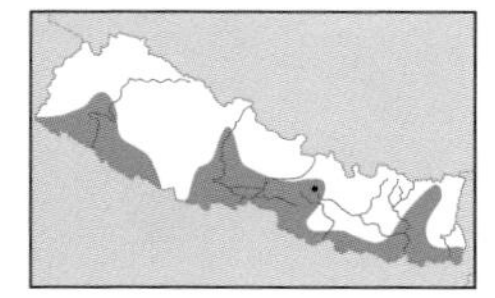

Little Ringed Plover *Charadrius dubius* 14–17 cm

Common and widespread resident and winter visitor; 75–1,500 m (–2,745 m). **ID** Small, elongated and small-headed appearance, and uniform upperwing with only a very narrow wing-bar. Bill small and mainly dark. Legs yellowish or pinkish. Adult breeding has striking yellow eye-ring. **Voice** Clear, descending *pee-oo* or shorter *peeu* in flight. **HH** In pairs or small flocks that scatter over wide area when feeding; often mixed with other waders. If disturbed, they rise in a compact flock and fly off rapidly, low over ground, veering and swerving in unison. Carries head low and drawn into shoulders. Has typical plover feeding action: makes short runs, then pauses and stoops stiffly without bending legs to pick up small invertebrates. During pauses has habit of vibrating one foot on mud, probably to attract prey to surface. Shingle and mudbanks of rivers, pools and lakes.

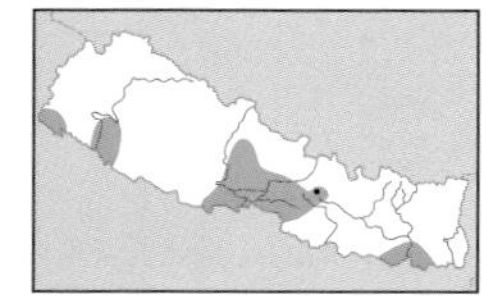

Kentish Plover *Charadrius alexandrinus* 15–17 cm

Winter visitor and passage migrant; fairly common at Koshi but mainly uncommon elsewhere; 75–1,380 m. **ID** Small size and stocky appearance. White hind collar and usually small, well-defined patches on sides of breast. Upperparts paler, more sandy, than Little Ringed. Legs usually appear blackish, but may be tinged brown or olive-yellow. Male has rufous cap and black eye-stripe and forecrown. **Voice** Flight call a soft *pi... pi... pi* recalling Little Stint, or a rattling trill *prrr* or *prrtut* (harsher than similar call of Lesser Sand); and a plaintive *whoheet.* **HH** Typical plover gait and feeding behaviour (see Little Ringed), but runs more rapidly. Shingle and sandy riverbeds.

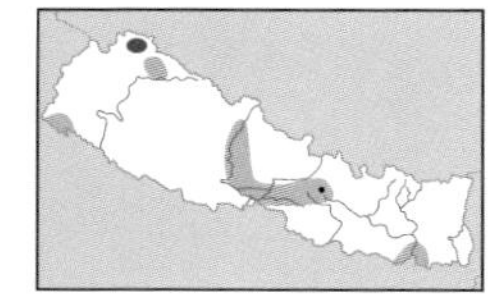

Lesser Sand Plover *Charadrius mongolus* 19–21 cm

Rare winter visitor and spring migrant; possibly breeds in Humla district in far north-west; 75–4,195 m. **ID** Larger and longer-legged than Kentish, lacking white hind collar. Very difficult to identify from Greater Sand Plover (see Appendix 1) although is smaller and has stouter bill (equal to or shorter than distance between bill base and rear of eye, with blunt tip), and shorter dark grey or dark greenish legs (with tibia shorter than tarsus). In flight, feet do not usually extend beyond tail and white wing-bar is narrower on primaries. Breeding male typically has full black mask and forehead, and more extensive rufous on breast compared to Greater (although variation in these characters). **Voice** A hard *chitik, chi-chi-chi,* and *kruit-kruit* in flight, which are rather short and sharp. **HH** Gait and feeding behaviour typical of plovers. Breeds in high-altitude semi-desert; on banks of lakes and large rivers on passage and in winter.

Pacific Golden Plover
br
non-br
non-br
br
br
br
non-br
non-br
Long-billed Plover
non-br
non-br
Little Ringed Plover
non-br
♂ br
Kentish Plover
non-br
♀ br
non-br
non-br
♂ br
non-br
Lesser Sand Plover

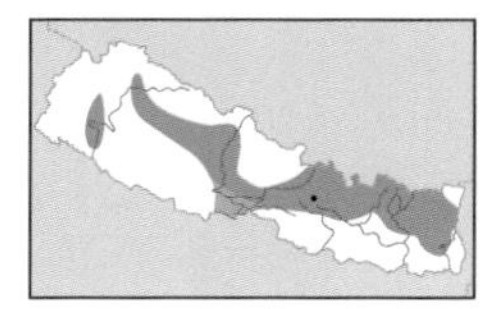

Eurasian Woodcock *Scolopax rusticola* 33–35 cm

Widespread and locally fairly common resident; breeds 1,980–3,900 m, non-breeding season 1,350–2,100 m (–100 m). **ID** Bulky and rufous-brown with broad, rounded wings. Crown and nape banded black and buff; lacks sharply defined mantle and scapular stripes. **Voice** Usually silent when flushed, sometimes gives harsh *schaap*. **HH** Solitary, crepuscular and nocturnal, passing day in thick cover. When flushed flies off without calling, with wings making a swishing sound; zigzags away with wavering wingbeats and quickly drops into cover again. Male has roding display flight in breeding season when gives sharp, regularly repeated *chiwich*. Moist forest with dense undergrowth and marshy glades.

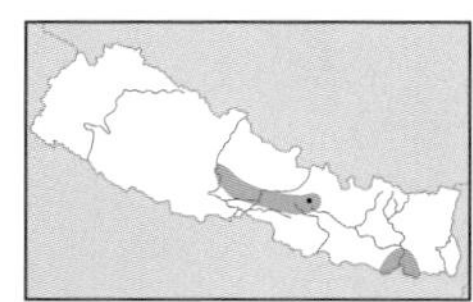

Jack Snipe *Lymnocryptes minimus* 17–19 cm

Very uncommon winter visitor and passage migrant; 75–1,500 m. **ID** Small, with short bill. Flight weaker and slower than that of Common Snipe, with rounded wingtips. Has divided supercilium but lacks pale crown-stripe. Mantle and scapular stripes very prominent. **Voice** Invariably silent when flushed. **HH** When feeding has characteristic habit of bobbing its body. Usually crepuscular and nocturnal like other snipes. Marshes and wet fields.

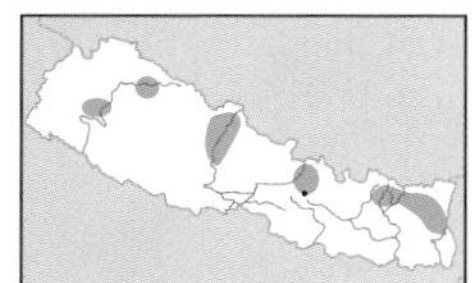

Solitary Snipe *Gallinago solitaria* 29–31 cm

Uncommon and local winter visitor and passage migrant, probably also resident; 2,135–4,000 m (–915 m). **ID** Large, dull-coloured snipe with long bill. Compared to Wood Snipe is colder-coloured and less boldly marked, with less striking head pattern and narrower white mantle and scapular stripes. Furthermore, has gingery-brown breast finely spotted and barred white, and rufous barring on mantle and scapulars. Wings longer and narrower than in Wood. Legs yellowish. **Voice** Harsh *kensh*, deeper than Common's if flushed. Male has aerial 'drumming' breeding display when it utters deep *chok-achock-a* call, combined with mechanical bleating produced by outer tail feathers. **HH** If flushed, zigzags away, flying more heavily and slowly than Common. High-altitude marshes and streams.

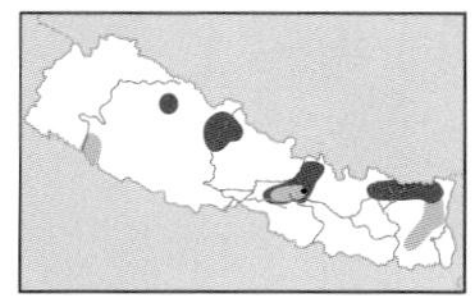

Wood Snipe *Gallinago nemoricola* 28–32 cm

Rare and sparsely distributed, possibly partial resident; breeds 3,650–4,520 m (–4,900 m); non-breeding 75–3,050 m. **ID** Large, with heavy and direct flight on broad wings. Bill relatively short and broad-based. More boldly marked than Solitary, with buff and blackish head-stripes, broad buff stripes on blackish mantle and scapulars (white in juvenile), and warm buff neck and breast with brown streaking. Legs greenish. **Voice** Long series of nasal notes, *check-check-check*, from ground on breeding area. In display flight nasal *che-dep, che-dep, ip-ip-ip, ock, ock*; when flushed *che-dep, che-dep*. **HH** If flushed has slow, heavy, wavering flight and soon settles. Breeds in marshy grassland with patchy dwarf scrub; winters in forest marshes. Globally threatened.

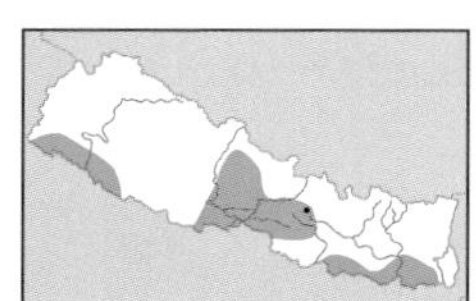

Pin-tailed Snipe *Gallinago stenura* 25–27 cm

Fairly common to frequent winter visitor and passage migrant; 75–1,370 m. **ID** Compared to Common, has more rounded wings, and slower and more direct flight. Lacks well-defined white trailing edge to secondaries, and has densely barred underwing-coverts and pale upperwing-coverts panel. Feet project beyond tail in flight. At rest, shows little or no tail projection beyond wings. Usually has bulging supercilium in front of eye, with little contrast between buff supercilium and cheeks, and eye-stripe often narrow in front of eye and poorly defined behind it. Width and colour of edges to lower large scapulars similar on inner and outer webs, creating scalloped appearance. **Voice** If flushed gives short, rasping *tetch*, deeper than Common. **HH** Flushes with little or no zigzagging; usually drops into cover more quickly than Common. Marshy pool edges, wet paddyfields; sometimes dry ground unlike Common.

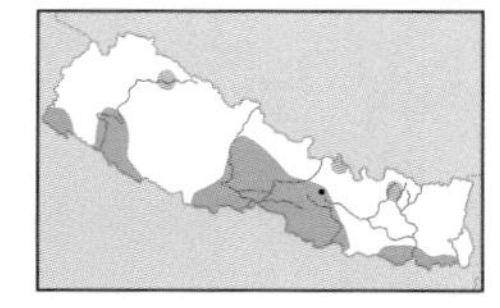

Common Snipe *Gallinago gallinago* 25–27 cm

Locally fairly common winter visitor and passage migrant; 75–1,500 m (–4,700 m on passage). **ID** Compared to Pin-tailed, wings more pointed, faster and more erratic flight. In flight, prominent white trailing edge to wing, white banding on underwing-coverts, and more extensive white belly patch. At rest, shows noticeable projection of tail beyond wings, poorly defined median coverts panel, buff supercilium contrasts with white cheek-stripe, and broad buff edges to outer webs of lower scapulars contrast with narrower, browner inner webs. **Voice** If flushed, gives an anxious, rising, grating *scaaap* which usually sounds higher pitched and more urgent than Pintail. **HH** When flushed rises steeply with rapid zigzagging, circles high and lands some distance away. Marshes, wet fields and muddy edges to rivers and pools.

ad
Eurasian Woodcock
ad
ad
Jack Snipe
Solitary Snipe
ad
ad
ad
Wood Snipe
Pin-tailed Snipe
ad
ad
Common Snipe

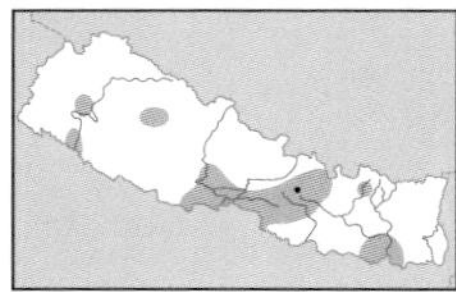

Eurasian Curlew *Numenius arquata* 50–60 cm

Rare passage migrant; 75–3,050 m. **ID** From Whimbrel (see Appendix 1) by larger size, much longer bill, more uniform head pattern (lacking prominent supercilium and crown-stripe of Whimbrel). Juvenile has shorter bill. **Voice** Distinctive mournful rising *cur-lew* call and an anxious *were-up* in alarm. Its song, a sequence of bubbling phrases, accelerating and rising in pitch is often heard in winter and on passage. Frequently calls in flight. **HH** Usually singly or in small groups. Generally wary and difficult to approach. Feeds by walking on mud or grassland and probing deeply, also by picking from surface. Muddy riverbanks and wet grass fields.

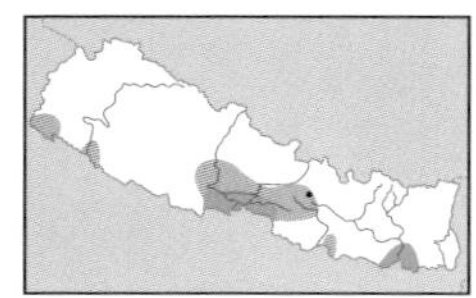

Spotted Redshank *Tringa erythropus* 29–32 cm

Uncommon passage migrant; 75–250 m (–1,350 m). **ID** Red at base of bill and red legs. Longer bill and legs than Common Redshank, lacking broad white trailing edge to wing. Non-breeding plumage paler grey above and whiter below than Common with more prominent white supercilium. Underparts black in breeding plumage. In first-summer plumage has dark barring on underparts and dark-mottled upperparts; legs can be black. Juvenile similar to non-breeding adult, but has darker grey upperparts more heavily spotted white, and underparts finely barred with grey. **Voice** A distinctive *tu-ick* in flight and a shorter *chip* alarm call. **HH** Usually solitary or in small groups. Feeds by picking from surface, often after a short dash; frequently forages in water, either sweeping bill from side to side or probing rapidly, often submerging the head and neck completely. Swims readily and upends like a surface-feeding duck to reach bottom mud. Muddy banks and shallow water of rivers and lakes.

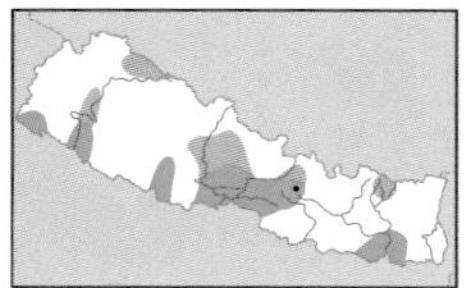

Common Redshank *Tringa totanus* 27–29 cm

Locally frequent winter visitor and passage migrant; 75–915 m (–4,270 m). **ID** Orange-red at base of bill, orange-red legs, and broad white trailing edge to wing. Non-breeding plumage grey-brown above, with grey breast. Neck and underparts heavily streaked in breeding plumage; upperparts with variable dark brown and cinnamon markings. Juvenile quite different from juvenile Spotted, with brown upperparts entirely fringed and spotted buff, underparts heavily streaked dark brown, and dull orange legs and base to bill. **Voice** Very noisy, often giving alarm call, an anxious *teu-hu-hu* in flight; also a mournful *tyuuu* on ground. **HH** Mainly found singly or in small groups. Generally wary. Feeds by walking briskly and picking from surface; also probes and wades in shallow water. Marshes, muddy edges of rivers and lakes.

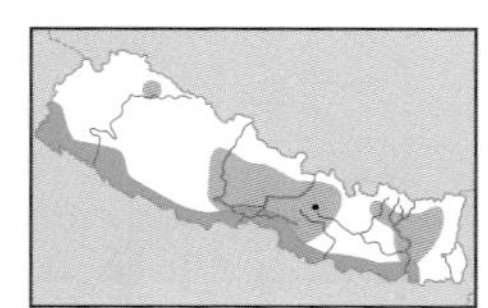

Common Greenshank *Tringa nebularia* 30–34 cm

Mainly a common winter visitor 75–1,500 m and passage migrant up to 4,800 m. **ID** Stocky, with long, stout (and slightly upturned) bill and long, stout greenish legs. Upperparts grey and foreneck and underparts white in non-breeding plumage. In breeding plumage, foreneck and breast streaked, upperparts untidily streaked. Juvenile has dark-streaked upperparts with fine buff or whitish fringes. **Voice** Loud, ringing *tu-tu-tu* flight call; sometimes a throatier *kyoup-kyoup-kyoup*. **HH** Usually forages singly. Generally wary and, when alarmed, bobs head and body nervously up and down. Feeds actively, chiefly in shallow water or at water's edge; detects prey mainly by sight, and makes frequent rapid runs to seize fast-moving prey. Flies strongly and sometimes erratically. Wide range of wetlands.

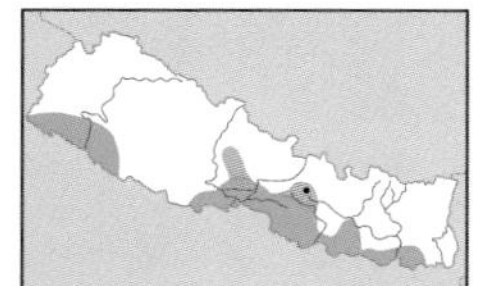

Marsh Sandpiper *Tringa stagnatilis* 22–25 cm

Uncommon passage migrant and winter visitor; 75–1,300 m. **ID** Smaller and daintier than Common Greenshank, with proportionately longer legs and finer bill. Legs greenish or yellowish. Upperparts grey and foreneck and underparts white in non-breeding plumage, when pale lores, forehead and chin create a pale-faced appearance. In breeding plumage, foreneck and breast streaked and upperparts blotched and barred. Juvenile upperparts appear streaked blackish, with feathers notched and fringed buff, and head-sides, hindneck and upper mantle are streaked dark grey and white. **Voice** An abrupt, dull *yup* flight call; also rapid, excitable series of *kiu-kiu-kiu* notes. **HH** A particularly graceful wader. Mainly found singly or in small groups. Forages actively, often in water or at water's edge, picks delicately from surface, making frequent rapid darts to seize prey, probes occasionally. Pools, lakes and marshes.

Eurasian Curlew
ad
non-br
br
juv
non-br
Spotted Redshank
non-br
br
non-br
juv
Common Redshank
br
br
Marsh Sandpiper
non-br
non-br
Common Greenshank

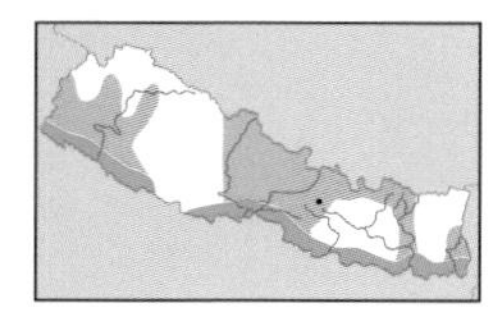

Green Sandpiper *Tringa ochropus* 21–24 cm

Common winter visitor 75–1,380 m; common passage migrant 75–4,250 m. **ID** From Wood Sandpiper by shorter greenish legs and stockier appearance, darker and less heavily spotted upperparts, and supercilium indistinct or absent behind eye. In flight, very dark underwing, strongly contrasting with white belly and vent, and striking white rump is distinctive. Adult breeding has white streaking on crown and neck, heavily streaked breast, and prominent whitish spotting on upperparts. Adult non-breeding more uniform on head and breast, and less distinctly spotted on upperparts. Juvenile has browner upperparts with buff spotting. **Voice** Ringing *tluee-tueet* and *tuee-weet-weet* calls. **HH** Favours small waters: ditches, small pools and streams; also marshes, wet fields and banks of rivers and lakes.

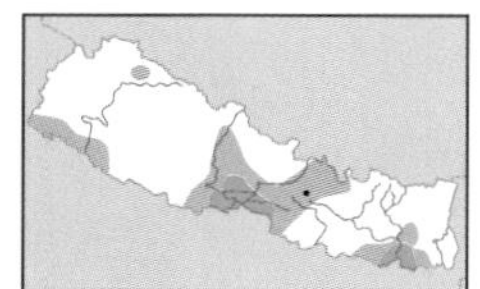

Wood Sandpiper *Tringa glareola* 18–21 cm

Uncommon winter visitor, frequent on passage; 75–3,780 m. **ID** From Green by longer, yellowish legs and slimmer appearance, heavily speckled upperparts, and prominent supercilium behind eye; in flight by call, slimmer body and narrower wings, toes projecting clearly beyond tail, paler underwing contrasting less with white underparts, and paler brown upperparts contrasting less with smaller white rump. Adult breeding has heavily streaked breast and barred flanks; upperparts barred and spotted pale grey-brown and white. Adult non-breeding has more uniform grey-brown upperparts, spotted whitish, and breast brownish and lightly streaked. Juvenile has warm brown upperparts speckled warm buff and lightly streaked buff breast. **Voice** Soft *chiff-if* or *chiff-if-if* flight call. **HH** Favours shallow water with emergent vegetation: marshes, pools, lakes and wet paddyfields.

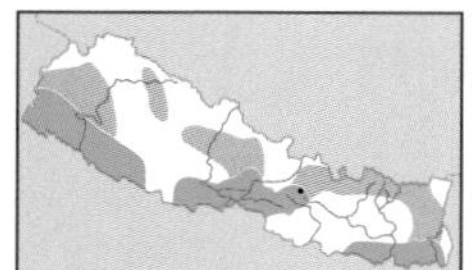

Common Sandpiper *Actitis hypoleucos* 19–21 cm

Common and widespread; winter visitor 75–1,370 m, passage migrant 75–5,400 m; **ID** Horizontal stance, long tail projecting well beyond closed wings. White wing-bar and brown rump and centre of tail in flight. In breeding plumage has irregular dark streaking and barring on upperparts, lacking in non-breeding. Juvenile has buff fringes and dark subterminal crescents to upperparts. **Voice** Anxious *wee-wee-wee* when flushed or alarmed. **HH** Characteristically rocks rear body and bobs head constantly when feeding. Runs along water's edge and picks prey from ground. Flies low over water, with rapid, shallow wingbeats alternating with brief glides on stiff downcurved wings. Often on small waterbodies: pools and streams, rivers, lakes and marshes.

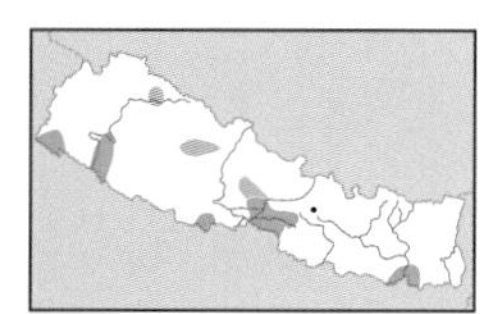

Little Stint *Calidris minuta* 13–15 cm

Uncommon winter visitor and passage migrant; 75–1,370 m (–3,050 m). **ID** More rotund and upright than Temminck's, with dark legs. In flight, grey sides to tail. Adult breeding has pale mantle V, rufous wash to face, neck-sides and breast, and rufous fringes to upperparts feathers. Non-breeding has untidy, mottled/streaked appearance (Temminck's more uniform), with grey breast-sides. Juvenile has whitish mantle V, greyish nape, prominent white supercilium typically split above eye (not shown in plate), and rufous fringes to upperparts. **Voice** Flight call a weak *pi, pi, pi*. **HH** An active wader, rapidly picks at surface, and frequently darts about to catch very tiny prey. Muddy edges of lakes, streams and rivers, also wet paddyfields.

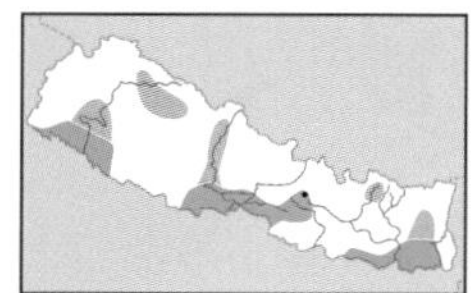

Temminck's Stint *Calidris temminckii* 13–15 cm

Fairly common winter visitor and passage migrant; 75–1,370 m (–4,710 m on passage). **ID** More elongated than Little, with more horizontal stance, tail extending noticeably beyond closed wings at rest. In flight, white sides to tail. Legs yellowish. In all plumages, lacks mantle V and is usually rather uniform, with complete breast-band and indistinct supercilium. Adult breeding has irregular dark markings on upperparts and juvenile regular buff fringes (pattern very different from Little). See Appendix 1 for Long-toed Stint. **Voice** A trilling, cicada-like *trrrrrit*. **HH** Unobtrusive. Forages more among vegetation at wetland edges than other stints. Favours vegetated freshwater habitats: marshes, pools, lakes, wet fields and riverbanks.

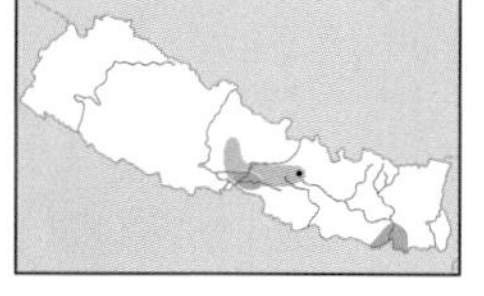

Dunlin *Calidris alpina* 16–22 cm

Very uncommon, mainly a passage migrant, also rare winter visitor; 75–1,370 m. **ID** Shorter legs and bill compared to Curlew Sandpiper (see Appendix 1), with dark centre to rump. Adult breeding has black belly. Adult non-breeding darker grey-brown than Curlew Sandpiper, with less distinct supercilium. Juvenile has streaked belly, rufous fringes to mantle and scapulars, and buff mantle V. **Voice** Flight call a distinctive slurred *screet*. **HH** Makes short runs over wet mud and wades near water's edge. Mud and sandbanks of rivers.

non-br
non-br
Green Sandpiper
non-br
juv
Wood Sandpiper
non-br
ad
juv
ad
Common Sandpiper
br
juv
non-br
juv
non-br
Little Stint
br
non-br
juv
non-br
br
Temminck's Stint
br
non-br
juv
Dunlin

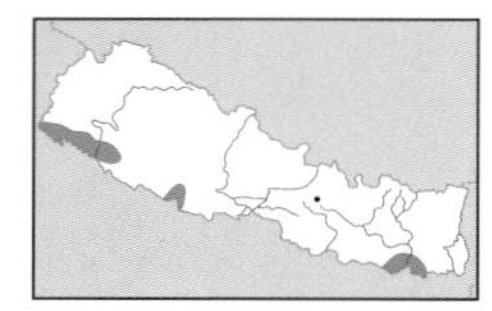

Indian Courser *Cursorius coromandelicus* 23 cm

Local resident, mainly in Koshi area; 75–275 m. **ID** Adult has rich orange underparts contrasting with grey-brown upperparts, and blackish centre of belly, chestnut crown and black lores. In flight, white band on uppertail-coverts and very dark underwing. Juvenile has dark brown crown, pale lores, strong brown-and-cream barring and blotching on upperparts, and brown markings on pale chestnut-brown underparts. Initially lacks dark patch on belly. **Voice** Generally silent. A low *gwut* or *wut*. **HH** Usually shy and readily takes flight. When alert has very alert posture. Flight direct and rapid, with powerful regular wingbeats and legs extending slightly beyond tail. Diurnal. When foraging, runs rapidly in short bursts and pauses to bend and pick up prey like a plover; also often digs in dry ground. Open country: dry areas with scattered scrub, stony ground and dry riverbeds.

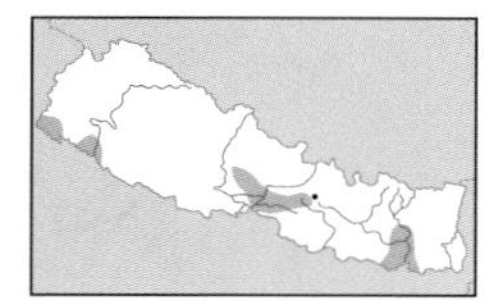

Oriental Pratincole *Glareola maldivarum* 23–24 cm

Rare passage migrant; 75–1,310 m. **ID** Adult breeding has black-bordered creamy-yellow throat and peachy-orange wash to underparts (pattern much reduced in non-breeding plumage). Red underwing-coverts in flight. **Voice** Sharp *kyik*, *chik-chik* or *chet* calls. **HH** Usually crepuscular, also active in overcast conditions; rests during hot part of day, squatting on ground. Hawks insects with mouth wide open in powerful swallow-like flights; also feeds on ground like a plover, making short dashes to capture prey. Favours dried-out bare flats by larger rivers and marshes; also low-lying pastures and fields, often near water.

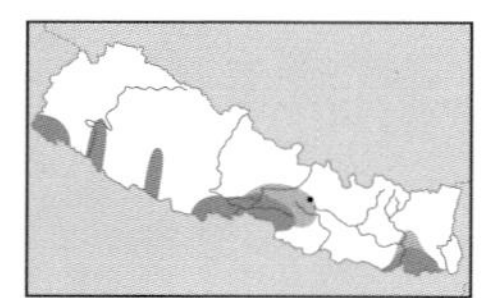

Small Pratincole *Glareola lactea* 16–19 cm

Local resident and partial migrant; 75–250 m. **ID** Small size, with sandy-grey coloration and square-ended tail (or has shallow fork). White panel on secondaries, blackish underwing-coverts and black tail-band in flight. Adult breeding has black lores and buff wash to throat; non-breeding plumage lacks these features and has streaked throat. Juvenile has indistinct buff fringes and brown subterminal marks to upperparts. **Voice** A high-pitched, rattling *tiririt*. **HH** Habits similar to Oriental's but often hawks insects later in evening. Breeds colonially. Large rivers with sand or shingle banks.

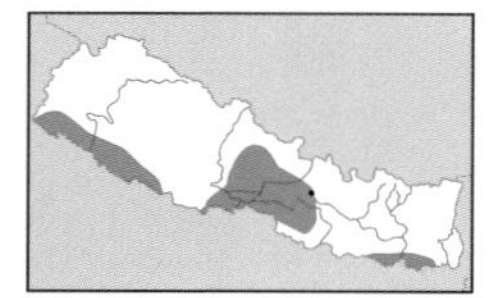

Greater Painted-snipe *Rostratula benghalensis* 25 cm

Uncommon resident; 75–350 m (–1,370 m). **ID** Rail-like wader, with broad, rounded wings and longish, downcurved bill. White or buff 'spectacles' and 'braces'. Adult female has maroon head and neck, and dark greenish wing-coverts. Adult male and juvenile duller, with buff spotting on wing-coverts; juvenile lacks dark breast-band, and throat and breast finely streaked. **Voice** Occasionally an explosive *kek* when flushed; female has a soft *koh koh* in display. **HH** Found singly and in small flocks. Chiefly crepuscular or nocturnal. Skulking and reluctant to fly if approached; rises heavily with legs trailing and lands in cover again a short distance away. Feeds by probing in mud, or by sweeping bill from side to side in shallow water while bobbing rear body. Movements slow and deliberate. Has roding display flight like Eurasian Woodcock. Marshes, pools and ditches, thickly vegetated, with mud patches.

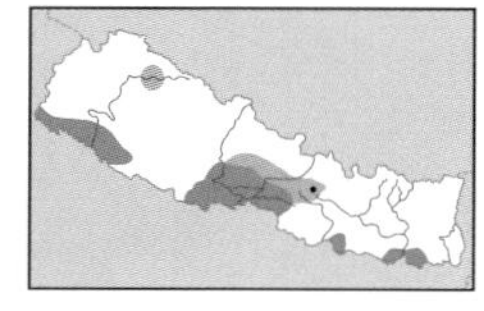

Pheasant-tailed Jacana *Hydrophasianus chirurgus* 31 cm

Uncommon resident subject to local movements depending on water conditions; 75–1,525 m (–3,050 m). **ID** Extensive white on upperwing, and white underwing. Yellowish patch on sides of neck. Adult breeding has brown underparts and long, curved tail. Adult non-breeding and juvenile lack elongated tail and have white underparts, with dark line down side of neck and dark breast-band. Juvenile has buff fringes to upperparts and barring on breast. **Voice** A distinctive *me-e-ou* or *me-oup* in breeding season, and a nasal *tewu*. **HH** Gregarious outside breeding season. Walks or rests on floating vegetation; usually in open. Forages actively throughout day. Swims well and floats buoyantly. Flight slow and flapping, with its large feet dangling behind. Lakes and pools with floating aquatic plants.

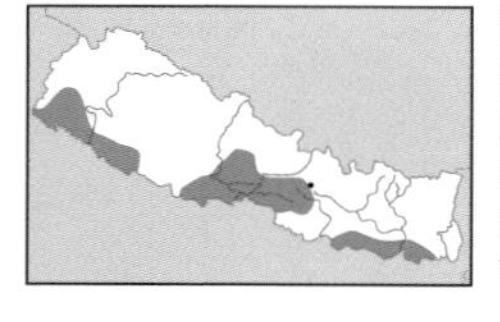

Bronze-winged Jacana *Metopidius indicus* 28–31 cm

Resident, fairly common in lowlands, uncommon up to 915 m. **ID** Dark upperwing and underwing. Adult has white supercilium, bronze-green upperparts and blackish underparts. Juvenile has orange-buff wash on breast, short white supercilium and yellowish bill. **Voice** A short, harsh grunt and a wheezy, piping *seek-seek-seek*. **HH** Habits and habitat similar to those of Pheasant-tailed. If alarmed will partially submerge its body among aquatic vegetation.

ad
br
juv
non-br
br
br
Indian Courser
Oriental Pratincole
br
Small
Pratincole
br
br
non-br
♂
juv
♀
Greater Painted-snipe
br
non-br
Pheasant-tailed Jacana
ad
ad
imm
Bronze-winged Jacana

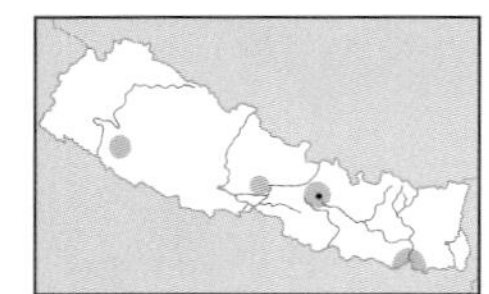

Steppe Gull *Larus* (*heuglini*) *barabensis* 59 cm

Rare passage migrant, possibly a winter visitor in very small numbers. **ID** Due to uncertainty about the status of large gulls in Nepal, comparison is made here (and in account for Heuglin's) with Caspian Gull (although there are no confirmed records). Appears smaller, slighter-bodied, with smaller more rounded head and smaller bill than these species. Adult similar to Heuglin's, but has paler grey upperparts (similar to *L. h. taimyrensis*). Upperparts darker grey, with fuller black wingtip than Caspian (similar to Heuglin's). Eyes yellow but may appear dark (resembling Caspian). Adult breeding has deeper yellow legs and bill than Caspian (latter with more extensive red gonys-spot variably marked with black). Bill paler in winter with dark subterminal marks, and pale tip. Immature plumages as Heuglin's (except, like Caspian, moults juvenile mantle feathers by autumn). **Voice** Typical long call is braying *ka-yaow-owowow-ow-ow ow* etc., which is similar to that of most other large white-headed gulls and probably indistinguishable in the field. **HH** Lakes and large rivers. **TN** Sometimes considered to be a race of Caspian Gull *L. cachinnans*.

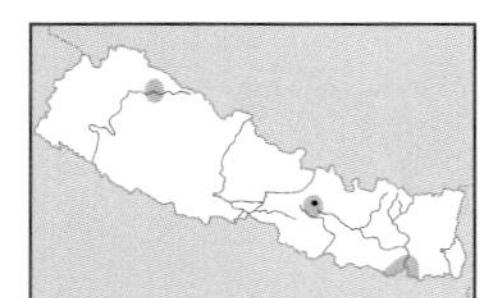

Heuglin's Gull *Larus heuglini* 58–65 cm

Probably a rare passage migrant and possibly a winter visitor in very small numbers; 75–3,050 m. **ID** Generally stockier and squarer-headed than Steppe and Caspian. Adult has darker grey upperparts than Steppe and especially Caspian (although *L. h. taimyrensis* is more similar to Steppe); head more heavily streaked in non-breeding plumage than in other species. Eyes yellow (appear small and dark in Caspian). Adult shows more black on wingtips than Caspian, typically with smaller white 'mirror' at tip of longest primary, and no white 'mirror' on second longest primary (latter present in *taimyrensis*). Juvenile and first-winter from Caspian by darker inner primaries, greater coverts and underwing-coverts. Retains neat pale-fringed juvenile mantle feathers for longer than Caspian (to December). Heuglin's attains adult plumage one year ahead of Caspian. First-summer has grey feathers in mantle, coverts and tertials. Second-winter has mantle and coverts largely grey (coverts of Caspian retain many immature feathers). *L. h. taimyrensis*, which may occur, is bulkier and broader-winged than *L. h. heuglini*, and upperparts of adult are a shade paler. **Voice** See Steppe Gull. **HH** Lakes and large rivers.

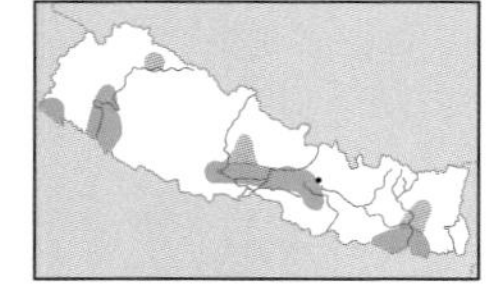

Pallas's Gull *Ichthyaetus ichthyaetus* 69 cm

Frequent winter visitor and passage migrant at Koshi, uncommon elsewhere; 75–3,050 m. **ID** Larger than Heuglin's. Head more angular, with sloping forehead, and crown peaks behind eye; bill longer and strikingly dark-tipped (except in juvenile), with bulging gonys. Eyes always dark. Adult breeding has black hood, white eye-crescents and yellow bill with black and red at tip. White primary tips contrast with black subterminal marks, and white patch on outer wing. Adult non-breeding has white head with black mask (and white eye-crescents). Juvenile has brown mantle and scapulars with pale fringes. First-winter/first-summer has grey mantle (unlike Heuglin's), pronounced dark mask and streaked hindcrown, and clear-cut dark tail-band. May acquire partial hood as first-summer. Second-winter has largely grey upperwing, with dark lesser coverts bar and extensive black on primaries. Third-winter as adult non-breeding, but more black on primaries. **Voice** Corvid-like *kra-ah*. **HH** Lakes and large rivers. **TN** Formerly placed in genus *Larus*. **AN** Great Black-headed Gull.

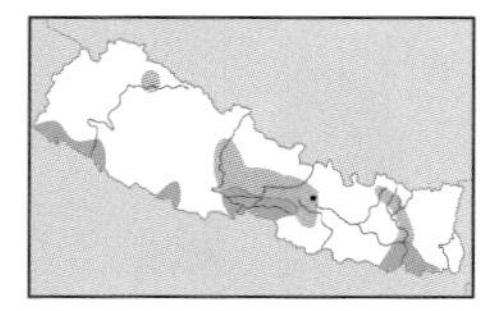

Brown-headed Gull *Chroicocephalus brunnicephalus* 42 cm

Rare, mainly a passage migrant, also winter visitor; 75–5,490 m. **ID** Slightly larger than Black-headed, with more rounded wingtips, and broader bill that is dark-tipped in all ages. Adult has broad black wingtips (broken by white 'mirrors') and white patch on outer primaries and primary-coverts; underside to primaries largely black; iris pale yellow (brown in adult Black-headed). In breeding plumage, hood paler brown than Black-headed. Juvenile and first-winter have broad black wingtips contrasting with white patch on primary-coverts and base of primaries. **Voice** As Black-headed, but deeper and gruffer. **HH** Lakes and large rivers.

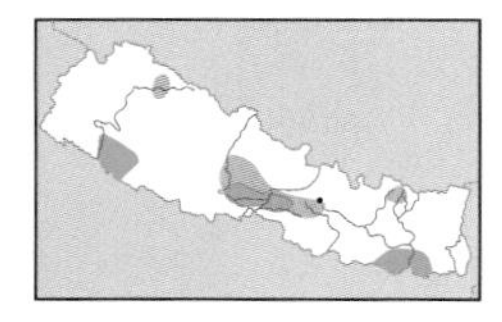

Black-headed Gull *Chroicocephalus ridibundus* 38 cm

Rare, mainly a passage migrant, also winter visitor; 75–5,490 m. **ID** Smaller than Brown-headed, with finer bill and narrower, more pointed wings. In all plumages, has distinctive white 'flash', and less black on wingtips than Brown-headed. Bill blackish-red and hood dark brown in breeding plumage. In non-breeding and first-winter plumages, bill tipped black and head white with dark ear-coverts patch. **Voice** Nasal *kyaaar*, short *keck* and deeper *kuk*. **HH** Lakes and large rivers. **TN** Brown- and Black-headed Gulls formerly placed in genus *Larus*.

non-br
1st-win
Steppe Gull
non-br
1st-win
Heuglin's Gull
br
non-br
non-br
Pallas's Gull
1st-year
non-br
1st-win
br
Brown-headed Gull
non-br
non-br
1st-win
br
non-br
Black-headed Gull

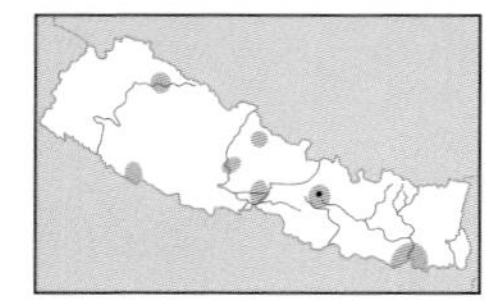

Gull-billed Tern *Gelochelidon nilotica* 35–38 cm

Very rare and very local passage migrant; 75–100 m (–3,100 m). **ID** Stout gull-like black bill, and broader-based, less pointed wings compared to other terns (except Caspian, which has a huge red bill). Flight steady and more gull-like, less graceful, with shallower wingbeats. Rump and tail grey and concolorous with back in all plumages. Adult breeding has black cap and darker grey upperparts than other plumages. Black of head reduced to black mask in non-breeding and immature plumages. Juvenile has sandy tinge to crown and mantle, and indistinct dark fringes to tertials and some wing-coverts, but is less heavily marked on upperparts than other juvenile terns. **Voice** Guttural *gek-gek-gek* or *gir-vit*. **HH** Frequently feeds by hawking, swooping or dipping to take food, often skimming close to surface, or seizing insects in mid-air. Lakes and large rivers.

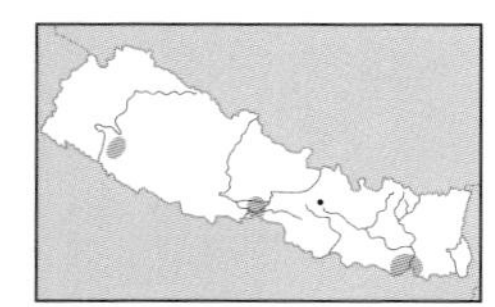

Caspian Tern *Hydroprogne caspia* 47–54 cm

Very rare and very local visitor; 75–250 m. **ID** Large size and broad-winged/short-tailed appearance. Huge red bill and black underside to primaries. Adult breeding has complete black cap. Adult non-breeding has black-streaked crown and black mask; bill duller, with more black at tip. First-winter and first-summer are similar to adult non-breeding, but show faint dark lesser-covert and secondary bars and dark-tipped tail. Juvenile has narrow dark subterminal bars to scapulars and wing-coverts; forehead and crown more heavily marked, almost forming dark cap. **Voice** Loud and far-carrying, hoarse *kretch*. **HH** Fishes by patrolling high above water, bill pointing downwards, hovering occasionally, before plunging to seize prey. Lakes and large rivers. **TN** Formerly placed in genus *Sterna*.

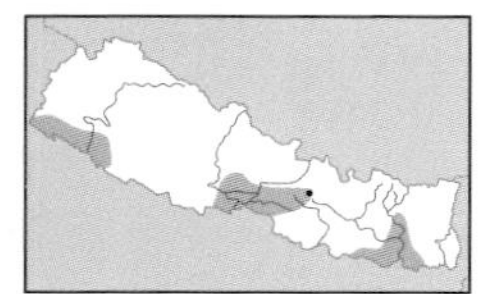

River Tern *Sterna aurantia* 38–46 cm

Rare and very local visitor; 75–610 m (–1,310 m). **ID** Adult breeding has orange-yellow bill, black cap, greyish-white underparts, and long greyish-white outer tail feathers; whitish primaries contrast with otherwise grey wing to form striking 'flash' on outer wing in flight. Non-breeding plumage lacks elongated outer tail feathers, and has blackish mask and mainly grey crown. Large size, stocky appearance and stout yellow bill (with dark tip) help separate adult non-breeding and immature from Black-bellied. Juvenile has dark fringes to upperparts, black-streaked crown and nape, whitish supercilium, and dark mask extending as dark streaking onto ear-coverts and sides of throat. **Voice** Fairly short, shrill, staccato *kiuk-kiuk* in flight, quite high-pitched and melodious. **HH** Mainly feeds by plunge-diving from air; also by dipping to surface and picking up prey. Lakes and large rivers.

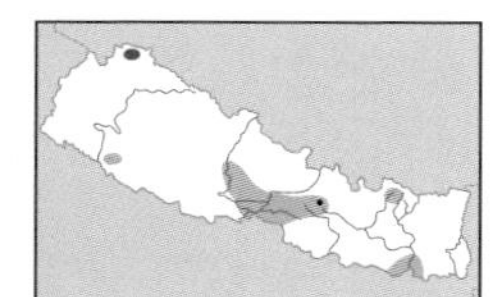

Common Tern *Sterna hirundo* 31–35 cm

Very rare passage migrant, probably also breeds in far north-west; 75–4,000 m. **ID** In breeding plumage, orange-red bill with black tip, orange-red legs, pale grey wash to underparts, and elongated outer tail feathers that reach tail tip at rest. Non-breeding and first-winter have mainly dark bill, darker grey upperparts and shorter tail. Juvenile has orange legs and bill base (bill becoming black with age). *S. h. tibetana* breeding in north-west subcontinent has been recorded; adult breeding has darker grey upperparts than widespread nominate, with a shorter bill with more extensive black tip. *S. h. longipennis* also reported; in breeding plumage, mostly black bill, with greyer upperparts and underparts than nominate and a more distinct white cheek-stripe; legs dark reddish-brown. **Voice** Drawn-out and harsh *krri-aaah* and a short *kik*. **HH** Fishes mainly by plunge-diving from air. Lakes and large rivers.

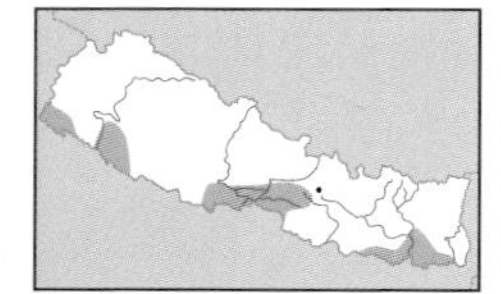

Black-bellied Tern *Sterna acuticauda* 33 cm

Very rare and very local visitor; 750–730 m. **ID** Smaller than River Tern, with orange bill (and variable black tip) in all plumages. Adult breeding has grey breast, black belly and vent, and long outer tail feathers. Like River Tern, whitish primaries contrast with grey rest of wing to form striking 'flash' on outer wing in flight. Long orange bill and deeply forked tail are best features from Whiskered Tern. Adult non-breeding and immature have white underparts, shorter tail, and black mask and streaking on crown. Confusingly, can have black cap and white underparts, when most similar to River, but structural differences and orange bill are diagnostic. Juvenile has dark mask and streaking on crown and nape, sandy coloration to head and mantle, and brown fringes to upperparts. **Voice** Clear, piping *peuo*. **HH** Feeds by plunge-diving from air, also by dipping to surface and picking up prey. Lakes, reservoirs and large rivers. Globally threatened.

non-br
br
Gull-billed Tern
non-br
br
Caspian Tern
br
non-br
River Tern
juv
br
hirundo
non-br
juv
Common Tern
non-br
juv
br
br
Black-bellied Tern

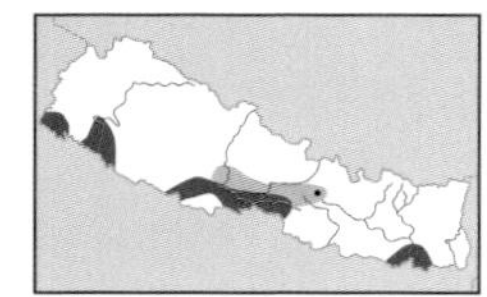

Little Tern *Sternula albifrons* 22–24 cm

Local summer visitor, fairly common in Koshi area, uncommon elsewhere; 75–120 m (–1,280 m). **ID** Fast flight with rapid wingbeats from narrow-based wings. Adult breeding has white forehead and black lores, black-tipped yellow bill, orange legs and feet, and black outer primaries. Adult non-breeding and immature have blackish bill, black mask and nape-band, dark lesser coverts bar, and dark legs. Juvenile has dark subterminal marks to upperparts feathers, and whitish secondaries form a broad pale trailing edge to wing. **Voice** Gives *ket* or *ket-ket*. **HH** Hovers more frequently and for longer periods than other terns, with faster-fluttering beats before plunge-diving steeply onto prey; also dips steeply to surface, skimming prey from water. Lakes and rivers. **TN** Formerly placed in genus *Sterna*.

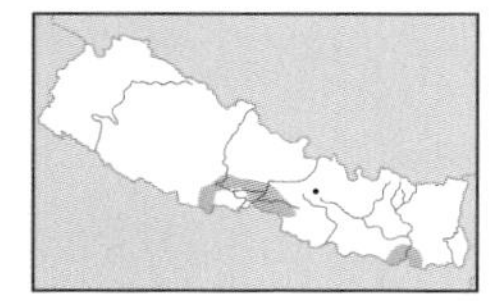

Whiskered Tern *Chlidonias hybrida* 23–25 cm

Irregular, uncommon passage migrant; 75–915 m. **ID** In breeding plumage, white cheeks contrast with black cap and grey underparts. In non-breeding and juvenile plumage, from White-winged by larger bill, grey rump concolorous with back and tail, and different head pattern (see White-winged). Head markings can be limited to dark mask recalling small Gull-billed. Compared to White-winged Tern, juvenile generally lacks pronounced dark lesser coverts and secondary bars and has black and buff markings on mantle/scapulars that appear more checkered (more uniformly dark in White-winged Tern). **Voice** Hoarse *eirchk* or *kreep*. **HH** Feeds mainly on insects by hawking or picking from water surface; unlike White-winged occasionally plunge-dives for fish. Marshes, lakes, rivers and rivers.

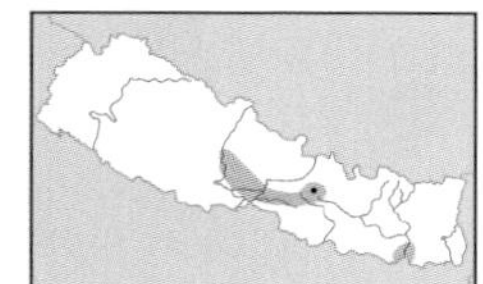

White-winged Tern *Chlidonias leucopterus* 20–23 cm

Rare spring passage migrant; 75–1,350 m. **ID** In breeding plumage, black head and body contrast with pale upperwing-coverts, and has black underwing-coverts. Black underwing-coverts are last feathers to be lost during moult into non-breeding plumage (always white in Whiskered). In non-breeding and juvenile plumage, smaller bill, whitish rump contrasting with grey tail, and different head pattern distinguish from Whiskered. Black ear-coverts patch is bold and reaches below eye, and usually has well-defined black line on nape. First-year has dark lesser coverts and secondary bars, and by late winter these contrast strongly with pale (worn) median and greater coverts, which form pale panel in wing, while mantle also can appear noticeably darker than pale coverts, giving rise to 'saddled' appearance like juvenile; in this plumage distinct from non-breeding and first-year Whiskered, which have more uniform mantle and wings. **Voice** Hoarse, dry *kirsch*. **HH** Hawks insects or swoops down to pick small prey from water surface; flies with great agility. Marshes, large rivers, flooded paddyfields, pools and lakes.

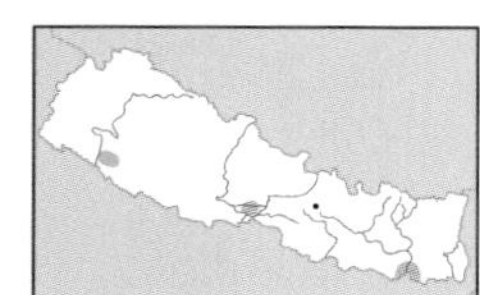

Indian Skimmer *Rynchops albicollis* 40 cm

Irregular, very rare non-breeding visitor; 75–300 m. **ID** Adult has large, drooping orange-red bill (with lower mandible projecting noticeably beyond upper), black cap, and black mantle and wings contrasting with white underparts. In flight, broad white trailing edge to upperwing, white underwing with blackish primaries, and white rump and tail with black central tail feathers. Juvenile has whitish fringes to browner mantle and upperwing-coverts, diffuse cap, and dull orange bill with black tip. **Voice** Nasal *kap kap*. **HH** Feeds chiefly in early morning or near dusk. Flight fast, powerful and graceful, resembling a *Sterna* tern. Systematically quarters water surface and has characteristic skimming method of foraging. Larger rivers with sandbanks. Globally threatened.

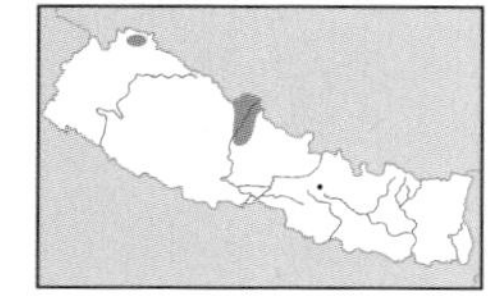

Tibetan Sandgrouse *Syrrhaptes tibetanus* 48 cm

Rare and local resident in upper Mustang, Annapurna Conservation Area and in Humla district, far north-west; 4,800–5,540 m. **ID** Large and pin-tailed. In flight, black flight feathers contrast with sandy coverts on upperwing, while underwing is mainly black except for white lesser coverts and trailing edge to primaries. Both sexes distinctive, with pale orange face and throat, fine black barring on crown and breast, sandy upperparts with bold black spotting on scapulars, and white lower breast and belly. Male has unbarred sandy mantle and wing-coverts and fawn wash on lower breast. Female similar, but has fine black barring on mantle, coverts and tertials, and more extensive black barring on breast. Immature similar to female, but has only faint traces of pale orange on throat and lacks pin-tail. **Voice** Deep *guk-guk* or *caga-caga* calls. **HH** Terrestrial. Seen in small groups. When disturbed rises with clatter of wings, flying off rapidly and directly with fast wingbeats. Walks and runs well, foraging mainly for small seeds on ground. High-altitude Tibetan steppe grassland, stony pasture, and open alpine meadow habitats.

juv
br
non-br
Little Tern
non-br
br
non-br
br
Whiskered Tern
non-br
br
juv
non-br
juv
White-winged Tern
ad
juv
Indian Skimmer
♀
♂
Tibetan Sandgrouse

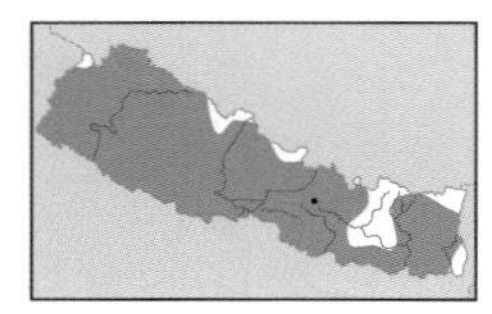

Common Pigeon *Columba livia* 33 cm

Common and widespread resident; breeds 75–4,270 m (–5,150 m), winters up to at least 2,810 m. **ID** Grey tail with blackish terminal band, and broad black bars on greater coverts and tertials/secondaries. Darker grey and lacks whitish tail-band of Hill Pigeon. Northern subspecies *neglecta* has pale grey to whitish back. Feral populations differ considerably in coloration and patterning. **Voice** Deep, repeated *gootr-goo, gootr-goo*. **HH** Lives in colonies and nests in hole or crevice of cliffs and ruined walls. Feeds chiefly in cultivation, mainly on seeds, also green shoots. Feral birds live in villages and towns; wild birds around cliffs and ruins. **AN** Rock Dove.

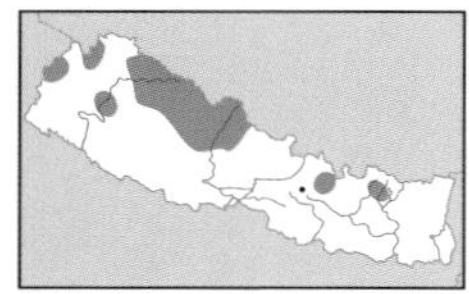

Hill Pigeon *Columba rupestris* 33 cm

Common resident in Trans-Himalayas in north-west; breeds 2,900–5,490 m, winters above 1,650 m. **ID** Similar to Common Pigeon, but paler grey, with white back and white band across tail contrasting with blackish terminal band. Juvenile browner and lacks iridescence on neck; feathers of neck and breast are fringed rusty-buff, and coverts fringed creamy-buff. **Voice** High-pitched quickly repeated *gut-gut-gut-gut*. **HH** Habits similar to wild Common Pigeon. Often very tame. Feeds on grain in cultivation and along mule tracks. Nests in crevice or hole in cliff or wall. Inhabits high-altitude villages and cliffs, mainly in Tibetan plateau country.

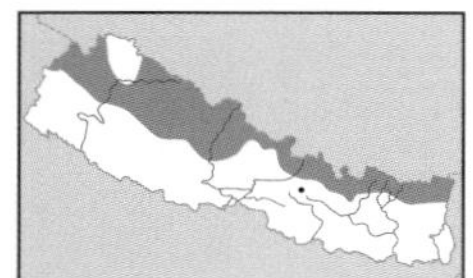

Snow Pigeon *Columba leuconota* 34 cm

Common and widespread resident, breeds 3,000–5,200 m (–5,700 m). **ID** Adult has slate-grey head, creamy-white collar and underparts, fawn-brown mantle contrasting with pale grey wing-coverts, white back contrasting with blackish rump and uppertail-coverts, and white band on blackish tail. Juvenile has greyish-buff wash to neck, breast and underparts, and fine whitish fringes to coverts and scapulars. **Voice** A prolonged, high-pitched *coo-ooo-ooo*. **HH** In pairs and small parties in summer and large flocks in winter. Forages on grass, rocky slopes and cultivation; in summer also at edge of melting snow-fields. Eats mainly seeds and small bulbs. Nests colonially on ledges, in cliff fissures and caves. Inhabits cliffs and gorges in mountains with plentiful rainfall; absent in dry steppe regions.

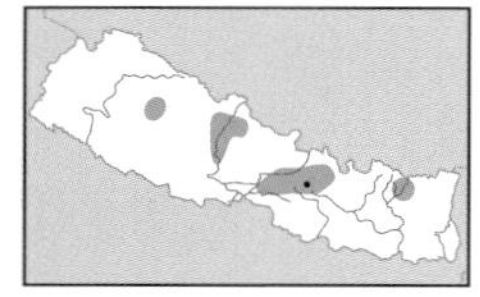

Common Wood Pigeon *Columba palumbus* 43 cm

Erratic winter visitor; 950–2,275 m. **ID** White wing patch and dark tail-band, buff neck patch and deep vinous underparts. In flight, from below, greyish-white band across tail and grey undertail-coverts concolorous with base of tail. Juvenile duller and browner, and lacks green gloss and buff patch on neck. **Voice** Deep and throaty repeated *kookooo-koo...kookoo*. **HH** Usually in small parties; sometimes large flocks. Feeds on acorns, berries and buds plucked from bushes and trees; also grain. Clambers among foliage while feeding; may hang upside-down to reach food items. Wooded hillsides.

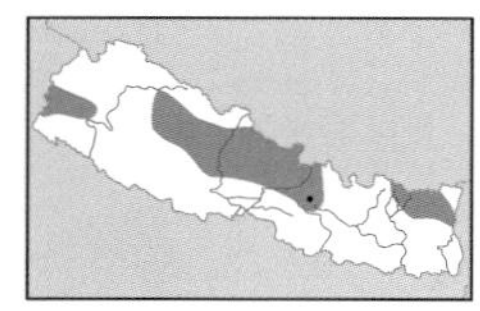

Speckled Wood Pigeon *Columba hodgsonii* 38 cm

Frequent and quite widespread resident; 1,500–3,050 m. **ID** From Ashy Wood Pigeon by lack of buff patch on neck, white spotting on wing-coverts, 'speckled' underparts, and dark grey vent and undertail-coverts concolorous with undertail. Like Ashy, has very dark underwing and blackish uppertail and undertail in flight. Male has maroon mantle and maroon on underparts, replaced by grey in female. Juvenile similar to female, but neck pattern less distinct; has finer white tips to coverts, and underparts more diffusely patterned. **Voice** Deep *whock-whr-ooooo whroo*. **HH** Usually in pairs or small flocks. Mainly frugivorous and arboreal; sometimes feeds on grain in crop stubbles. Clambers about trees when feeding. Nests in forest trees. Mainly inhabits oak–rhododendron forest.

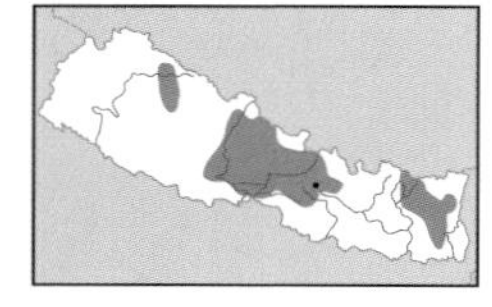

Ashy Wood Pigeon *Columba pulchricollis* 36 cm

Frequent resident, mainly from west-central Nepal eastwards; 1,100–2,440 m. **ID** From Speckled by combination of dark slate-grey upperparts, no white spotting on wing-coverts, uniform dark slate-grey breast without 'speckling', and creamy-buff belly and undertail-coverts that contrast with dark undertail. Has buff collar (at close range, rufous-buff tips to neck feathers contrast with dark bases, forming checkerboard pattern), and metallic green-and-purple sheen to lower neck and back. Juvenile has browner upperparts, with less distinct pattern on neck, and rufous fringes to feathers of breast and belly. **Voice** A deep, slightly booming, repeated *whuoo... whuoo... whuoo*. **HH** Alone, in pairs or small flocks. Chiefly arboreal and frugivorous; wanders in search of fruiting trees. Typically perches very quietly, concealed among foliage in canopy. Nests in trees. Inhabits dense broadleaved forest.

ad
Common Pigeon
ad
Hill Pigeon
ad
Snow Pigeon
ad
Common Wood Pigeon
♀
♂
Speckled Wood Pigeon
ad
Ashy Wood Pigeon

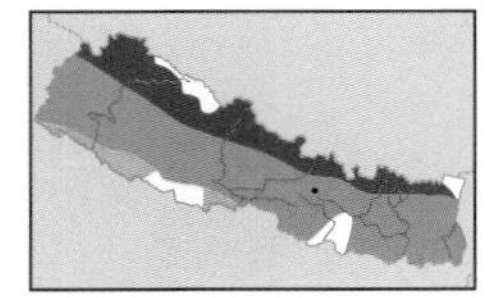

Oriental Turtle Dove *Streptopelia orientalis* 33 cm

Common and widespread resident and winter visitor; breeds 365–4,570 m; winters mainly below 1,370 m. **ID** Rufous-scaled scapulars and wing-coverts, dusky underparts and barred neck. In flight, has dusky-grey underwing. Juvenile lacks neck bars, and has buffish-grey head and underparts, and pale buff fringes to dark-centred feathers of upperparts. In western Nepal, *S. o. meena* has white rather than grey sides and tip to tail, and white rather than grey undertail-coverts. **Voice** A hoarse, mournful repeated *goor... gur-grugroo.* **HH** Found alone or in pairs when breeding, and in small parties in winter; may form flocks on migration. A ground-feeder, gleaning grain from cultivation; also eats grass, bamboo and weed seeds, and often forages on dusty tracks. Flight direct with fast wingbeats, and on take-off and alighting the tail is widely fanned. Nests in a bush, young tree or bamboo clump. Open forest, especially near cultivation.

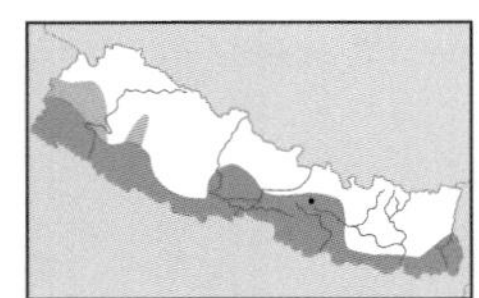

Eurasian Collared Dove *Streptopelia decaocto* 32 cm

Fairly common and widespread resident; 75–400 m (–2,440 m). **ID** Sandy-brown with black half-collar, white sides to tail, and white underwing-coverts. Juvenile lacks neck-collar, and feathers of upperparts are fringed buff. **Voice** A repeated cooing *kukkoo... kook.* **HH** Habits similar to those of Oriental Turtle Dove. Breeds all year, varying locally. Nests low down in a bush or small tree. Open dry country with cultivation and groves.

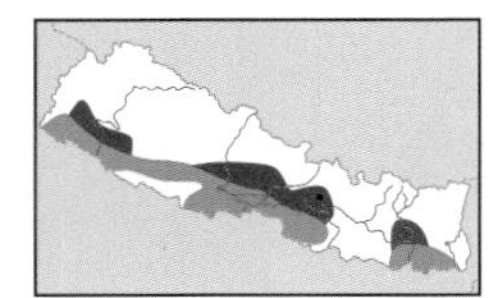

Red Collared Dove *Streptopelia tranquebarica* 23 cm

Fairly common and quite widespread resident; 75–300 m (–1,370 m). **ID** Male has blue-grey head with black half-collar, pinkish-maroon upperparts and pink underparts. Compared to Eurasian Collared, female has darker buffish-grey underparts, darker fawn-brown upperparts, greyer underwing-coverts, white (rather than grey) vent, and is smaller with shorter tail. Juvenile lacks neck-collar, and feathers of upperparts and breast are fringed buff. **Voice** A harsh, rolling, repeated *groo-gurr-goo.* **HH** Habits similar to those of Oriental Turtle Dove. Breeds all year, varying locally. Nests in trees 3–8 m above ground. Light woodland and trees in open country. **AN** Red Turtle Dove.

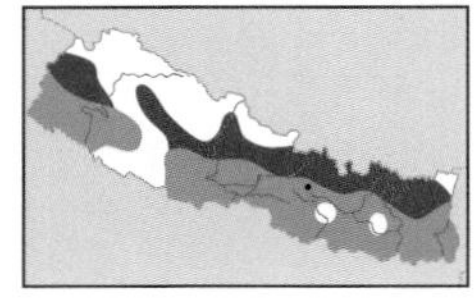

Spotted Dove *Stigmatopelia chinensis* 30 cm

Common and widespread resident; 75–1,500 m. **ID** Upperparts scaled or spotted pinkish-buff. Has extensive black-and-white checkered patches on sides of neck, vinaceous-pink-tinged neck and breast, and dark grey-brown rump and tail with blackish base to outer tail feathers. Juvenile paler and browner, lacks checkered patch on sides of neck, has faintly barred mantle and scapulars, and narrow rufous fringes to wing-coverts. **Voice** A soft, mournful *krookruk-krukroo... kroo-kroo-kroo.* **HH** Habits similar to those of Oriental Turtle Dove. When disturbed, bursts upwards with noisy clatter of wings, then glides down to settle nearby. Breeds almost all year, varying locally. Nests fairly low down in tree, thorn bush or bamboo clump. Cultivation, habitation and open forest. **TN** Formerly placed in *Streptopelia.*

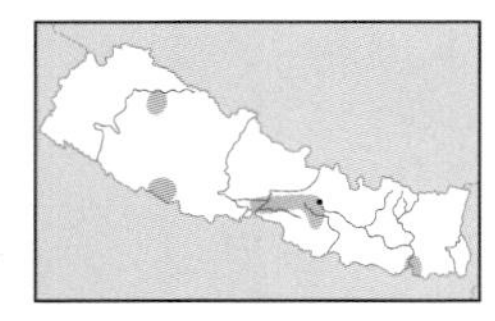

Laughing Dove *Stigmatopelia senegalensis* 27 cm

Rare, status and movements uncertain; 610–2,440 m. **ID** Slim, small, with fairly long tail. Brownish-pink head and underparts, uniform upperparts, and black stippling on upper breast. Juvenile duller, lacks black stippling and has whitish fringes to scapulars and coverts. **Voice** A soft *coo-rooroo-rooroo* or *cru-do-do-do-do.* **HH** Similar to those of Oriental Turtle Dove. Dry cultivation and scrub-covered hills. **TN** Formerly placed in *Streptopelia.*

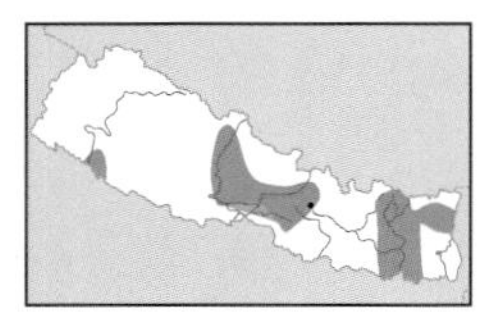

Barred Cuckoo Dove *Macropygia unchall* 41 cm

Rare resident; 250–2,800 m. **ID** Long, graduated tail, slim body and small head. Face, belly and vent pale. Upperparts and tail rufous, barred dark brown. Male has unbarred head and neck with extensive purple-and-green gloss. Female heavily barred on head, neck and underparts, with gloss restricted to nape and sides of neck. Juvenile more uniformly dark and heavily barred. **Voice** A very deep *croo-umm,* the second syllable a booming note, audible at long range; heard in distance as a low, muffled single *umm* repeated at short intervals. **HH** Usually in pairs or small flocks. Feeds on berries, acorns and shoots of forest trees, clambering about and sometimes swinging upside-down to reach food. Nests 2–8 m up in small forest tree. Dense broadleaved evergreen forest.

ad
meena
ad
agricola
Oriental Turtle Dove
ad
Eurasian Collared Dove
♀
♂
Red Collared Dove
ad
Spotted Dove
ad
Laughing Dove
♀
♂
Barred Cuckoo Dove

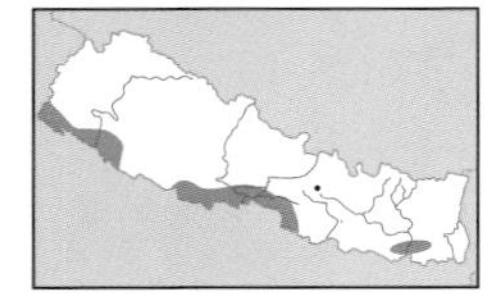

Orange-breasted Green Pigeon *Treron bicinctus* 29 cm

Local resident, common in Chitwan National Park; 75–305 m. **ID** Smaller than other green pigeons, with grey central tail feathers in both sexes (at rest, tail appears grey rather than green). Male from other green pigeons by orange breast bordered above by lilac band and yellowish-green forehead merging into pale blue-grey hindcrown and nape. Mantle uniform green. Female has yellow cast to breast and belly, and grey hindcrown and nape. **Voice** Very similar to Thick-billed. **HH** Often in mixed feeding flocks with other fruit-eating birds. Usually at tops of tall trees. Clambers about with great agility to reach fruit, sometimes hanging upside-down. Keeps well concealed in foliage; when approached 'freezes'. Wings make a loud clatter as birds burst out of a tree. Broadleaved evergreen and moist deciduous forest.

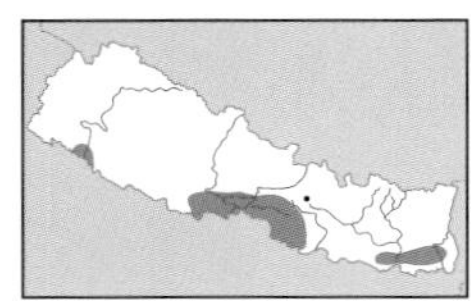

Ashy-headed Green Pigeon *Treron (pompadora) phayrei* 27 cm

Local resident; fairly common in Chitwan National Park, generally uncommon elsewhere; 75–250 m. **ID** Both sexes from Thick-billed by thin blue-grey bill (lacking red base) and lack of prominent greenish orbital skin. Male has maroon mantle; further differences from male Thick-billed are diffuse orange patch on breast, greenish-yellow throat, and uniform dark chestnut undertail-coverts. Female lacks maroon mantle. Green central tail feathers, greyish cap, yellowish throat and white undertail-covers separate from female Orange-breasted. Tail shape and paler green coloration separate from female Wedge-tailed Green Pigeon. **Voice** Very similar to Thick-billed. **HH** Habits like Orange-breasted. Broadleaved evergreen and moist deciduous forest.

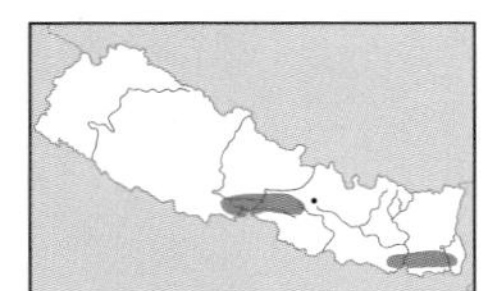

Thick-billed Green Pigeon *Treron curvirostra* 27 cm

Rare and local resident; 75–455 m. **ID** Both sexes from Ashy-headed (in 'Pompadour group') by thick bill with red base, prominent greenish orbital skin, whitish scaling on vent. Male has maroon mantle and green breast. **Voice** Pleasant wandering whistles and song a quiet, warbling series of whistling and cooing notes rising and falling in pitch. **HH** Habits like Orange-breasted. Dense evergreen or mixed broadleaved forests.

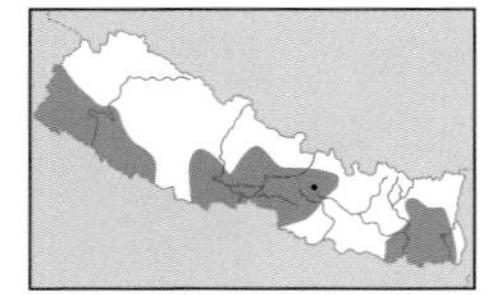

Yellow-footed Green Pigeon *Treron phoenicopterus* 33 cm

Widespread and locally common resident; 75–250 m, uncommon to 1,400 m. **ID** Large size, grey cap and greenish-yellow forehead and throat, broad olive-yellow collar, pale greyish-green upperparts, mauve shoulder patch, yellowish band at base of tail, and yellow legs and feet. Sexes similar, although female is duller. **Voice** Similar to Thick-billed. **HH** Habits like Orange-breasted. Deciduous forest and fruiting trees around villages and cultivation.

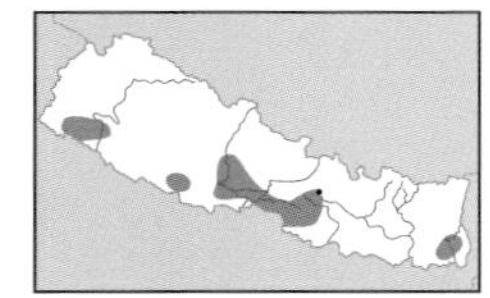

Pin-tailed Green Pigeon *Treron apicauda* 42 cm

Local resident locally frequent; 75–305 m (–915 m). **ID** Large green pigeon with extended and pointed central tail feathers. Grey tail (with greenish tip to central feathers), contrasting with lime-green rump and uppertail-coverts are additional features from female Wedge-tailed. Green crown, wing-coverts and back are additional differences from male Wedge-tailed. Has blue cere and bill base and naked blue lores. Male has longer central tail feathers, pale orange wash to breast, and more pronounced grey cast to upper mantle compared with female. **Voice** Distinctive, deep, tuneful short melody: *oou...ou-ruu...oo-ru...ou-rooou.* **HH** Habits like Orange-breasted. Tall broadleaved forest, especially evergreen.

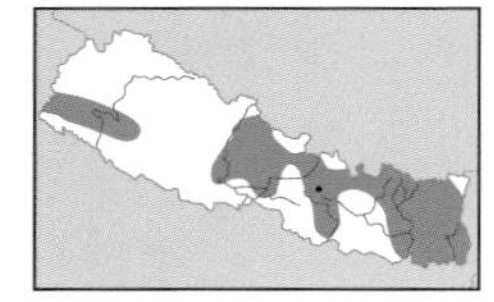

Wedge-tailed Green Pigeon *Treron sphenurus* 33 cm

Locally fairly common resident; 1,525–2,000 m (–2,800 m). **ID** Male from male Ashy-headed by larger size, long and wedge-shaped tail, less extensive maroon patch on upperparts (confined to inner wing-coverts, scapulars and lower mantle), pale cinnamon undertail-coverts, darker olive-green back and rump, and only indistinct, fine yellow edges to greater coverts and tertials. In flight, more uniform tail, lacking pale grey terminal band of Ashy-headed. Orange wash to crown, and very long pale cinnamon undertail-coverts are further differences. Female mainly green, lacking orange on crown and breast and maroon on upperparts of male. Undertail-coverts yellowish-white with grey-green centres. From female Ashy-headed by differences in tail shape and colour, lack of prominent yellow in wing, uniform green head (lacking grey crown). Tail shape, maroon on upperparts (male), and dull olive-green rump, uppertail-coverts and central tail feathers are best distinctions from Pin-tailed. **Voice** Series of mellow whistles like other green pigeons, but deeper with more hooting than Thick-billed. **HH** Habits like Orange-breasted, but less gregarious than other *Treron* species. Broadleaved forest.

♀
Orange-breasted Green Pigeon
♂
♀
Ashy-headed Green Pigeon
♂
♀
♂
Thick-billed Green Pigeon
ad
Yellow-footed Green Pigeon
♂
♀
Pin-tailed Green Pigeon
♀
♂
Wedge-tailed Green Pigeon

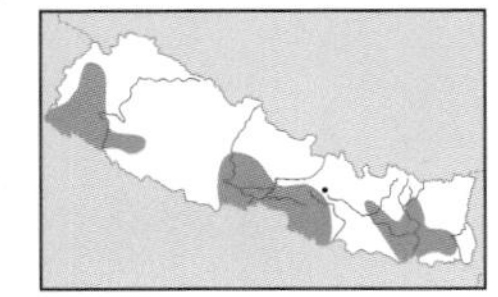

Emerald Dove *Chalcophaps indica* 27 cm

Locally common resident; 75–365 m (–1,370 m). **ID** Stocky, broad-winged, short-tailed pigeon with emerald-green upperparts. Typically very rapid in flight, and allows only brief views on being flushed from forest floor, when shows black-and-white banding on back. Male has grey crown, white forehead and supercilium, and deep vinaceous-pink head-sides and underparts; white shoulder patch. Female has warm brown crown, neck and underparts, forehead and supercilium suffused with grey, and shoulder patch generally warm brown. Juvenile resembles female, but has dark grey barring on buffy-white forehead, narrow buff fringes (and some dark subterminal bars) to neck and underparts, dark brown tertials with chestnut tips, brown primaries with chestnut edges and tips, rufous-brown rump, and brownish bill. **Voice** Mournful booming *tk-hoon... tk-hoon*. **HH** Singly or in pairs. Feeds on ground, often on forest tracks. Frequently seen flying away rapidly and directly through forest, only a few metres above ground. Moist broadleaved forest.

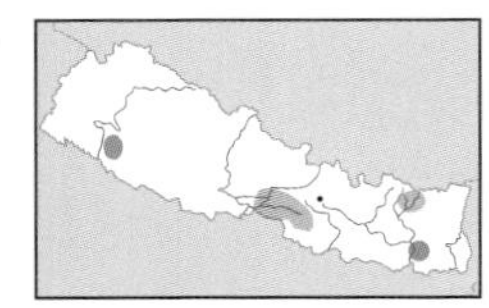

Mountain Imperial Pigeon *Ducula badia* 43–51 cm

Extremely rare, no recent records; 100–1,250 m. **ID** Very large pigeon, with pinkish-grey head and underparts, brownish upperparts, and pale terminal band to tail. Juvenile has rufous fringes to mantle and wing-coverts, with chestnut leading edge to wing, and tail pattern is less well defined. **Voice** Deep, resonant, booming double note preceded by a click heard at close range. Singly or in pairs; in small flocks at fruiting trees, often associates with other frugivorous birds. Flies swiftly high above treetops when moving between feeding grounds. **HH** Tall broadleaved evergreen forest.

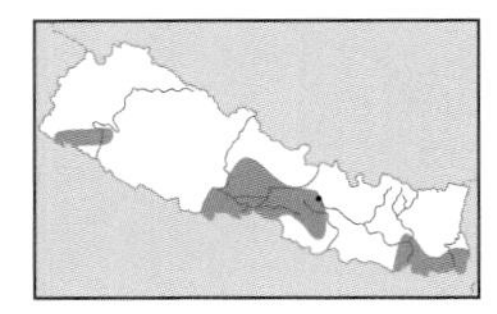

Long-tailed Broadbill *Psarisomus dalhousiae* 28 cm

Local resident; frequent in Chitwan National Park and Pokhara Valley, uncommon elsewhere; 275–1,340 m. **ID** 'Dopey-looking' with big head, large eyes, stout lime-green bill, long thin tail, and upright stance. Mainly green with black cap, blue crown, and yellow 'ear' spot, throat and collar. White patch at base of primaries in flight. Juvenile has green cap and lacks blue crown spot. **Voice** Loud, piercing *pieu-wieuw-wieuw-wieuw* call. **HH** Arboreal, keeps in forest canopy or midstorey in flocks of up to 20. Very upright when perched. Unobtrusive and lethargic when not feeding. Broadleaved evergreen and semi-evergreen forest.

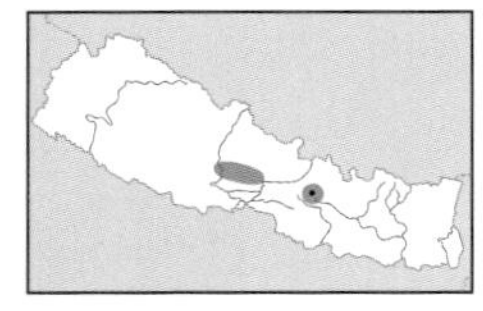

Blue-naped Pitta *Pitta nipalensis* 25 cm

Very rare; 915–1,525 m. **ID** Large pitta with fulvous sides of head and underparts, and uniform green upperparts. Male has glistening blue hindcrown and nape. Female has rufous-brown crown and smaller, greenish-blue patch on nape. Juvenile mainly brown, streaked and spotted buff, with brownish-white supercilium; wings and tail brownish-green. **Voice** Powerful double whistle. **HH** Habits like Indian. Broadleaved evergreen forest and moist shaded ravines.

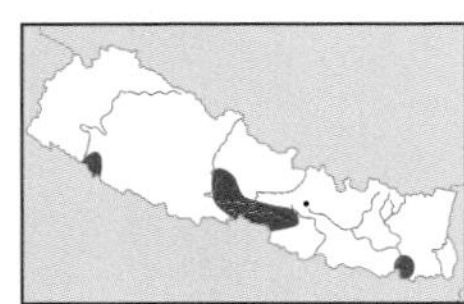

Hooded Pitta *Pitta sordida* 19 cm

Summer visitor, frequent in Chitwan National Park and buffer zone, rare elsewhere; 100–305 m (–1,400 m). **ID** Largely black head with chestnut crown and nape, glistening blue forewing and uppertail-coverts, green breast and flanks, and scarlet belly and vent with black abdominal patch. Larger white wing patch in flight than Indian. Juvenile duller with black scaling on crown, white patch on median coverts, brownish chin and dirty white throat, and brownish underparts with dull pink belly and vent. **Voice** Song a loud explosive double whistle, *wieuw-wieuw*; a *skyeew* contact call and a bleating, whining note. **HH** Habits like Indian Pitta. Broadleaved evergreen forest, often near water.

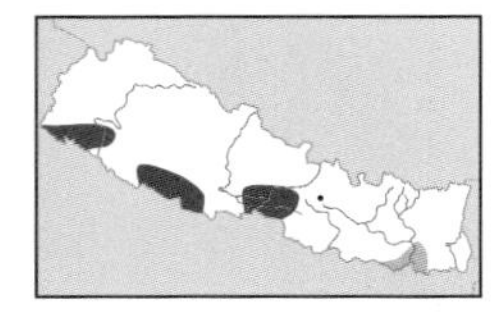

Indian Pitta *Pitta brachyura* 19 cm

Local summer visitor; 100–245 m (–1,360 m); fairly common in central Nepal, less common elsewhere. **ID** Bold black stripe through eye contrasting with white throat and supercilium, and buff lateral crown-stripes separated by black centre to crown. Underparts buff, with reddish-pink lower belly and vent. Upperparts green, with shining blue uppertail-coverts and forewing. In flight, small white patch on wing. Juvenile much duller with lateral crown-stripes scaled with black. **Voice** Sharp two-noted whistle, second note descending, *pree-treer*. **HH** Usually seen singly or in pairs. Keeps mainly to forest floor. Forages by flicking leaves and other vegetation, and probing leaf litter and damp earth. Usually moves in long hopping bounds. Skulking and often most easily located by call. Sings and roosts in trees or bushes, sometimes high up. Broadleaved forest with dense undergrowth.

♀
♂
♀
Emerald Dove
ad
Mountain Imperial Pigeon
♂
Long-tailed
Broadbill
ad
♀
Blue-naped Pitta
ad
ad
Hooded Pitta
Indian Pitta

PLATE 48: PARAKEETS

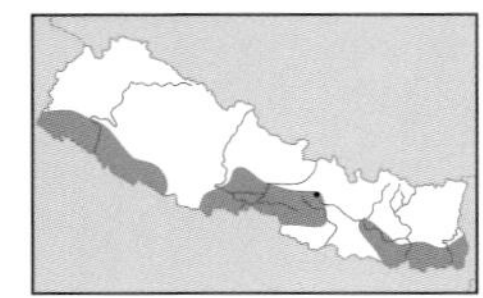

Alexandrine Parakeet *Psittacula eupatria* 53 cm

Widespread resident, common in some protected areas and generally frequent or uncommon elsewhere; 75–365 m (–1,380 m). **ID** From Rose-ringed by combination of larger size, maroon shoulder patch and massive bill. Deeper, more raucous call and slower and more laboured flight are additional pointers. Male has black chin-stripe joining pink and turquoise hind collar, both of which are lacking on female and immature. **Voice** Loud guttural *keeak* or *kee-ah* call, deeper and more raucous than Rose-ringed. **HH** Quite wary. Small parties clamber about in tall fruiting trees. Flies with deliberate wingbeats uttering a harsh, loud scream. Roosts communally. Sal and riverine forests.

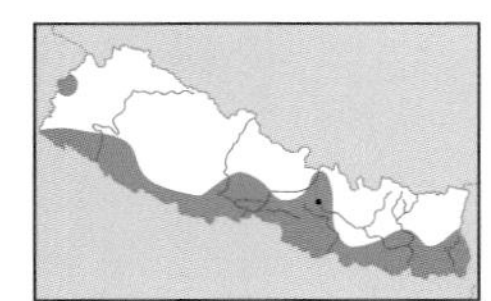

Rose-ringed Parakeet *Psittacula krameri* 42 cm

Abundant and widespread resident in lowlands; only occasionally seen higher up except in Pokhara and Kathmandu Valleys; 75–1,600 m. **ID** From Alexandrine by smaller size, lack of maroon shoulder patch and smaller bill. Dark blue-green (rather than pale yellowish) dorsal aspect to tail is a further feature. Male has black chin-stripe joining pink hind collar. Female lacks chin-stripe and collar, and is all green (with indistinct pale green collar). **Voice** Shrill, loud and variable *kee-ah*, higher pitched and less guttural than Alexandrine. **HH** Flocks raid orchards and crops in large numbers. Can form enormous communal roosts, often with crows and mynas. Constantly screeches and squabbles. Often associated with habitation and cultivation; also open woodland and secondary growth. **AN** Ring-necked Parakeet.

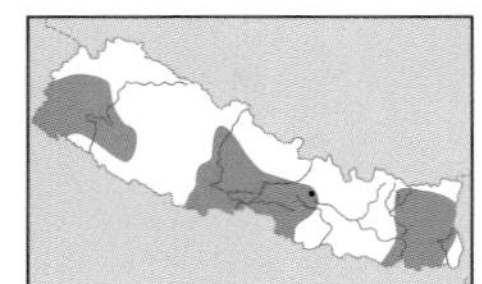

Slaty-headed Parakeet *Psittacula himalayana* 41 cm

Resident, locally fairly common in some protected areas, frequent elsewhere; breeds up to 2,135 m (–3,260 m), non-breeding season down to 250 m (–75 m). **ID** Adult has grey head, stout red bill and yellow-tipped tail. Underside of tail is strikingly yellow. Male has maroon shoulder patch, lacking in female. Larger than female Plum-headed; head is darker slate-grey with black chin-stripe and half-collar, has a stouter red bill with pale yellow lower mandible, lacks yellowish collar, and has yellow (rather than white) tip to tail. Immature similar to female and immature Rose-ringed but has darker, dull green head, yellow tip to tail (may not be apparent on younger birds). **Voice** Shrill *tooi... tooi* call, deeper and harsher than Plum-headed. **HH** Feeding habits like Alexandrine. Agile in flight, in compact flocks twisting through trees in unison. Broadleaved forests and well-wooded areas.

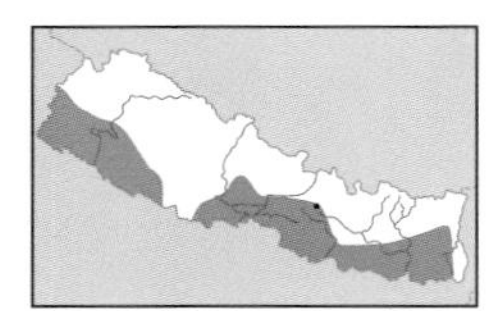

Plum-headed Parakeet *Psittacula cyanocephala* 36 cm

Resident, common in some protected areas and locally common elsewhere; 75–500 m (–1,525 m). **ID** Male has plum-red and purplish-blue head, yellow upper mandible and white-tipped blue-green tail. Female has greyish head; smaller-bodied and with daintier head and bill than Slaty-headed, with lilac cast to paler grey head (lacking black chin-stripe and half-collar), yellow upper mandible, yellowish collar and upper breast, and white tip to tail. Juvenile has green head with buffish forehead, lores and cheeks. **Voice** Shrill *tooi-tooi*, higher pitched and less harsh than Slaty-headed. **HH** Less associated with people than Rose-ringed. Roosts communally in large numbers. In flight weaves through forest trees with great agility. Can be destructive to crops and orchards. Broadleaved forests and well-wooded areas.

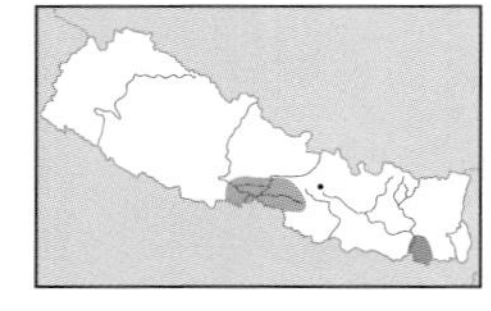

Blossom-headed Parakeet *Psittacula roseata* 36 cm

Very local; winter visitor to Chitwan National Park, resident or visitor to Koshi area; 75–250 m. **ID** Male from male Plum-headed by paler pink and lilac-blue on head. Also lacks turquoise collar, and has pale yellow tip to tail. Female similar to female Plum-headed, but has paler greyish-blue head, less distinct collar, maroon shoulder patch, pale yellow tail tip. **Voice** As Plum-headed. **HH** Habits similar to Plum-headed. Open forest and well-wooded areas.

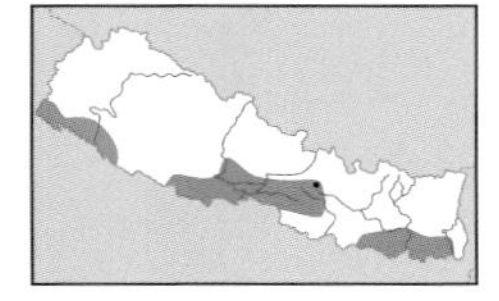

Red-breasted Parakeet *Psittacula alexandri* 38 cm

Resident, frequent in Chitwan, uncommon elsewhere; 75–365 m (–1,800 m). **ID** Male has lilac-grey crown and ear-coverts (with variable pinkish wash), broad black chin-stripe, deep lilac-pink breast and belly, greenish-yellow lesser wing-coverts, and yellowish-tipped blue-green tail. Female similar, but has blue-green tinge to head, purer peach-pink breast, and black upper mandible. Immature duller, with green underparts and orange-red bill. **Voice** Short, sharp nasal *kaink*. **HH** Usually quiet while feeding in treetops. Forms large communal roosts and can attack crops. In winter found in riverine forests near settlements, sometimes in large flocks; moves deeper into broadleaved forests and in smaller flocks in breeding season.

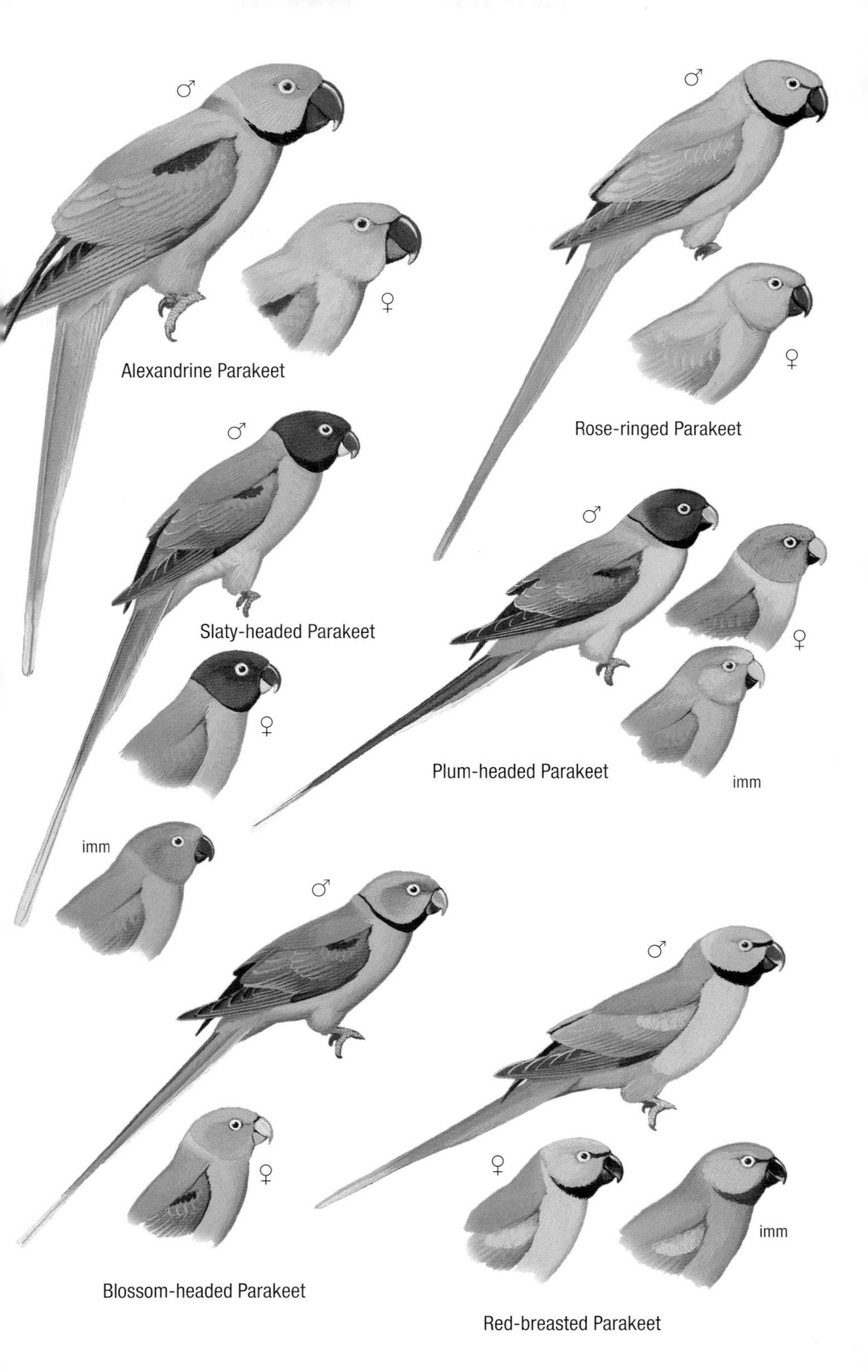
♂
♀
Alexandrine Parakeet
♂
♀
Rose-ringed Parakeet
♂
Slaty-headed Parakeet
♀
imm
♂
♀
Plum-headed Parakeet
imm
♂
♀
Blossom-headed Parakeet
♂
♀
imm
Red-breasted Parakeet

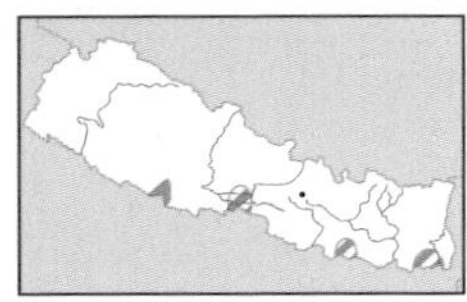

Vernal Hanging Parrot *Loriculus vernalis* 14 cm

Very rare resident or visitor; 75–275 m. **ID** Small (sparrow-sized), stocky green parrot with red rump and uppertail-coverts, and red bill. Adult has yellowish-white iris. Male has turquoise throat patch, lacking or much reduced in female. **Voice** Distinctive di- or trisyllabic rasping flight call, *de-zeez-zeet*, occasionally given at rest. **HH** Mainly keeps in tall treetops. Short, rapid wingbeats and undulating flight. Broadleaved evergreen and moist deciduous forest.

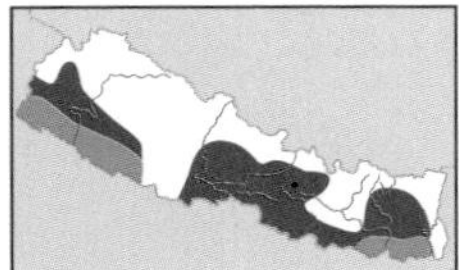

Asian Koel *Eudynamys scolopaceus* 43 cm

Common and widespread resident and summer visitor; 75–1,370 m (–1,800 m). **ID** Large, with long broad tail. Male glossy black (with green iridescence), with a dull lime-green bill and brilliant red eye. Female brown above (with faint green gloss), spotted and barred white and buff, and white below, strongly barred dark brown. Also has striking red eye. Juvenile blackish, with white or buff tips to wing-coverts and tertials, and variable white barring on underparts; tail black, although shows pronounced rufous barring in some. **Voice** Loud, rising and increasingly anxious, repeated *ko-el...ko-el...ko-el* and a bubbling more rapid repeated *koel... koel*. **HH** Typically concealed in dense foliage when not feeding. Open woodland, groves, gardens and cultivation.

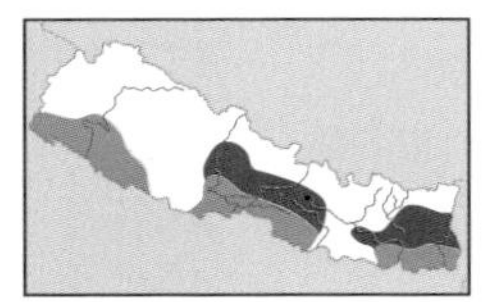

Green-billed Malkoha *Rhopodytes tristis* 38 cm

Locally fairly common resident, 75–700 m; uncommon summer visitor higher up to 2,000 m. **ID** Large and very long-tailed. Greyish-green coloration, with lime-green bill, red eye-patch, white-streaked supercilium and broad white tips to tail feathers. **Voice** Low croaking *ko... ko... ko*, and a chuckle when flushed. **HH** Very skulking, creeps and clambers unobtrusively through branches low down in thick vegetation. Dense broadleaved forest and thickets. **TN** Formerly placed in genus *Phaenicophaeus*.

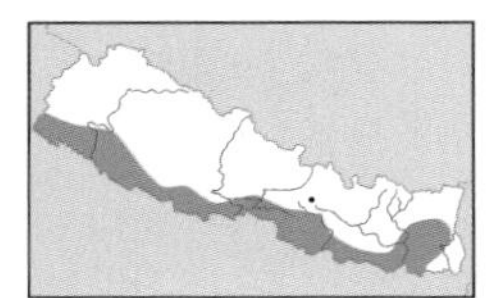

Sirkeer Malkoha *Taccocua leschenaultii* 42 cm

Quite widespread resident; uncommon in far west, rare further east; 75–365 m (–1,370 m). **ID** Adult mainly sandy grey-brown, with yellow-tipped red bill and white-edged dark facial skin giving masked appearance. Black shaft-streaking on crown, mantle and breast, throat buff, belly rufous-buff. Long, graduated, white-tipped tail. Immature very similar, but has indistinct buff fringes to wing-coverts, scapulars and tertials. Juvenile has broad dark brown streaking on head, mantle, throat and breast, and buff fringes to mantle, wing-coverts and tertials. **Voice** Normally silent, but can give parakeet-like *kek-kek-kerek-kerek-kerek*. **HH** Largely terrestrial; sometimes clambers among shrubs and small trees. Thorn scrub and *Acacia* bushes; also bushy rocky places. **TN** Formerly placed in genus *Phaenicophaeus*.

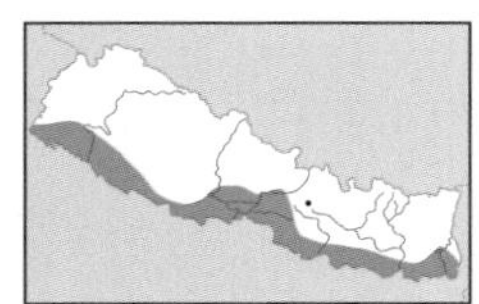

Greater Coucal *Centropus sinensis* 48 cm

Common and widespread resident; 75–365 m (–915 m). **ID** Adult from adult breeding Lesser Coucal by much larger size, black underwing-coverts (difficult to see in the field), and brighter and more uniform chestnut wings. Juvenile has brownish-black head and body, with chestnut spotting on crown and nape (becoming barred on mantle), and diffuse whitish barring over entire underparts; chestnut-brown coverts and flight feathers are barred dark brown, and tail is narrowly barred buff or greyish-white. Immature resembles adult, but head and body are duller black and has barred (juvenile) flight feathers and tail. **Voice** Deep, resonant and primate-like *hoop-hoop-hoop-hoop-hoop-hoop*, descending and then rising at end of series. **HH** Walks sedately with tail held horizontal, or skulks in dense vegetation. Tall grass and thickets near cultivation.

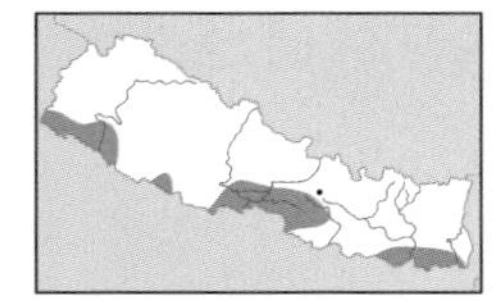

Lesser Coucal *Centropus bengalensis* 33 cm

Locally fairly common resident, mainly in protected areas; 75–365 m (–1,400 m). **ID** Smaller than Greater, with stouter bill, duller chestnut mantle and wings (including browner tertials and primary tips), and chestnut underwing-coverts. Eyes dark (red in Greater). Often shows buff streaking on some scapulars and wing-coverts (unlike Greater). Adult non-breeding has dark brown head and mantle with prominent buff shaft-streaks, and dark brown and rufous barring on rump and very long uppertail-coverts. Wings and tail as adult breeding. Juvenile similar to adult non-breeding, but has less distinct pale shaft-streaking on upperparts, dark barring on crown, mantle and back, dark brown barring on wings, and narrow rufous barring on tail. Immature has head and body as adult non-breeding, but wings and tail barred like juvenile. **Voice** Series of deep resonant *pwoop-pwoop-pwoop* notes, very similar to Greater, but usually slightly faster and more interrogative, initially ascending, then descending and decelerating. **HH** Habits similar to Greater. Tall grassland, reedbeds and shrubberies.

♂
imm
Vernal Hanging Parrot
♂ imm
♂
♀ imm
Asian Koel
♀
ad
Sirkeer Malkoha
juv
ad
Green-billed
Malkoha
imm
juv
br
non-br
ad
Lesser Coucal
Greater Coucal

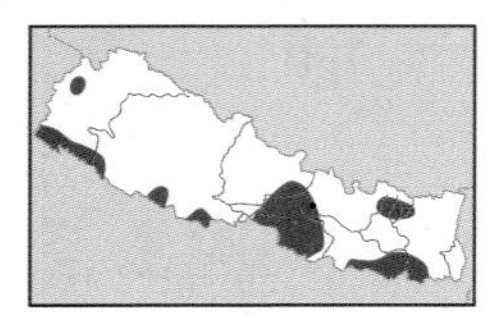

Jacobin Cuckoo *Clamator jacobinus* 33 cm

Summer visitor; frequent or uncommon in some protected areas; very uncommon elsewhere; 305–365 m (–3,875 m). **ID** Black and white with crest. White patch at base of primaries, and prominent white tips to tail feathers. Juvenile has browner upperparts, grey wash to throat and upper breast, and buffish wash to rest of underparts. Smaller crest than adult, with smaller white wing patch, and paler bill. **Voice** Metallic *piu... piu... pee-pee piu, pee-pee piu* or *piu...piu... piu* uttered frequently day and night. **HH** Conspicuous, often perching in open. Chiefly arboreal, but may forage in low bushes. Broadleaved forest and well-wooded areas. **AN** Pied Cuckoo.

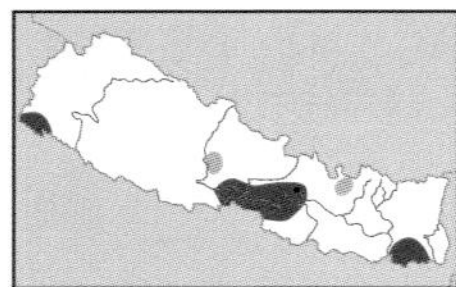

Chestnut-winged Cuckoo *Clamator coromandus* 47 cm

Local summer visitor; 150 m (–75 m) to 365 m (–1,370 m). **ID** Prominent crest, whitish collar, chestnut wings, and orange wash to throat and breast. Long black tail has narrow greyish-white feather tips. Lacks white in wing. Juvenile has shorter crest, rufous fringes to upperparts, buff collar, whitish throat and breast, broad buff tips to rectrices, and paler bill. Immature similar to adult, but retains some buff tips to scapulars, coverts and tail feathers. **Voice** Series of double metallic whistles, *breep breep*; also a harsh, grating scream. **HH** Arboreal and rather retiring, favours the canopy, but may forage in bushes. Broadleaved forest.

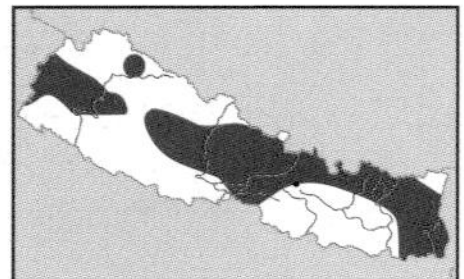

Large Hawk Cuckoo *Hierococcyx sparverioides* 38 cm

Fairly common and widespread summer visitor, very rare in winter; 1,830–3,000 m (–150 m on passage). **ID** Larger than Common Hawk Cuckoo, with browner mantle (contrasting with slate-grey head), blackish chin, grey streaking on throat and breast, irregular rufous breast-band, broad dark brown barring on underparts, and broader and stronger dark tail-bands. Underwing-coverts white, barred dark brown. Juvenile has strongly barred rather than spotted underparts, and broader tail-bands. Immature has darker slate-grey head than immature Common, with blackish chin and grey throat streaking. **Voice** Shrill repeated *pee-pee-ah... pee-pee-ah* rising in pitch to a hysterical crescendo. Often calls throughout night. **HH** Usually keeps well hidden among foliage of forest canopy, even when calling. *Accipiter*-like in flight; low with a few fast wingbeats followed by a glide. Broadleaved forest.

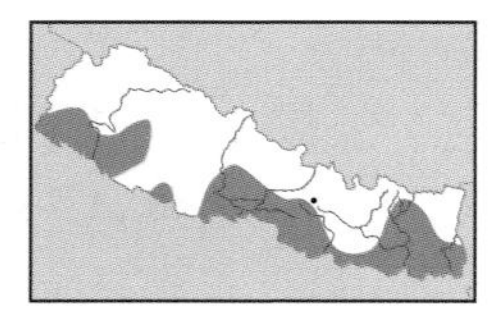

Common Hawk Cuckoo *Hierococcyx varius* 34 cm

Common and widespread resident; 75–1,000 m (–1,500 m). **ID** Smaller than Large, with whitish or greyish chin and throat, uniform grey upperparts, more rufous on underparts, indistinct barring on belly and flanks, and narrower tail-bands. Underwing-coverts rufous and only faintly barred. In juvenile, flanks typically less heavily marked than in Large, with spots or chevrons rather than bars (although some very similar), while rufous tail-bands and tail tip are typically brighter and more clearly defined. **Voice** Call as Large, but more shrill and manic, becoming more vocal during hot weather; often calls throughout night. **HH** Habits similar to Large, but more often seen. Lightly wooded areas.

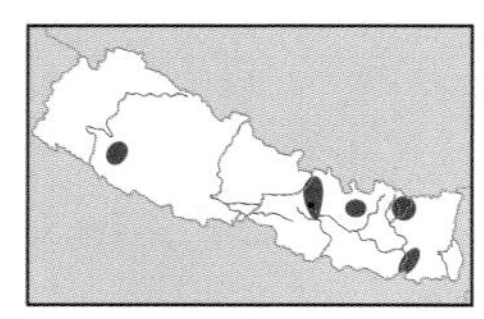

Hodgson's Hawk Cuckoo *Hierococcyx fugax* 29 cm

Rare and local summer visitor, mainly in east. **ID** Smaller than Common, with stouter bill. Upperparts darker slate-grey, with slate-grey chin, more extensive rufous on underparts, and unbarred white belly and flanks. Throat and breast can show dark grey streaking. Has more pronounced rufous tip to tail, and frequently shows a single pale inner tertial (on both wings on some; not present in Common). Juvenile has darker brown and more uniform upperparts than juvenile Common, and broader (squarer) spots on underparts. Immature has dark grey chin, ear-coverts and crown, rufous barring to upperparts, and strongly streaked underparts. **Voice** Shrill repeated *gee-whiz*, becoming more frantic and high-pitched. **HH** Usually in low trees or bushes, but moves higher when calling. Broadleaved evergreen and moist deciduous forest. **AN** Whistling Hawk Cuckoo, if local form *nisicolor* is split.

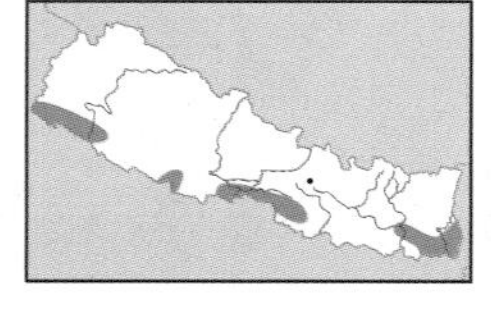

Banded Bay Cuckoo *Cacomantis sonneratii* 24 cm

Uncommon resident in Chitwan National Park, rare resident or summer visitor elsewhere; 150–250 m (–2,440 m). **ID** White supercilium (finely barred black and encircles brown ear-coverts), finely barred white underparts, and fine and regular dark barring on upperparts. Juvenile has broader (and more diffuse) barring on underparts, and crown and nape have some buff barring. **Voice** A shrill, whistled *pi-pi-pew-pew*, first two notes on same pitch, the last two descending. **HH** Calls from bare branches of treetops, usually holding tail depressed, wings drooped and rump feathers fluffed out. Dense broadleaved forest.

juv
Chestnut-winged
Cuckoo
ad
ad
juv
Jacobin Cuckoo
ad
ad
imm
imm
Common Hawk Cuckoo
juv
Large Hawk Cuckoo
juv
Hodgson's
Hawk Cuckoo
ad
ad
juv
imm
juv
Banded Bay Cuckoo

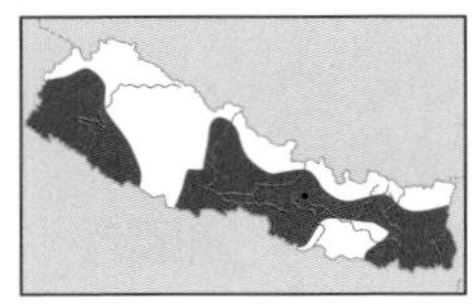

Indian Cuckoo *Cuculus micropterus* 33 cm

Locally common summer visitor; 75–2,100 m. **ID** From Eurasian and Himalayan by browner mantle, and broader, more widely spaced black barring on underparts. Tail has broader (diffuse) dark subterminal band, broader white barring on outer tail feathers and larger white spots on central feathers. Eyes reddish-brown (yellow in Eurasian Cuckoo; yellow or brown in Himalayan Cuckoo). Female has rufous-buff wash to base of grey breast, and rufous suffusion to whitish barring of lower breast. No hepatic female morph. Juvenile distinctive with broad white tips to feathers of crown, nape, scapulars and wing-coverts; throat and breast creamy-white with irregular brown markings, and barring on underparts is broader and more irregular than on Eurasian and Himalayan. **Voice** Descending whistle, *kwer-kwah... kwah-kurh.* **HH** Frequents treetops and canopy; sometimes flies hawk-like above forest. Often calls at night. Forest and well-wooded country.

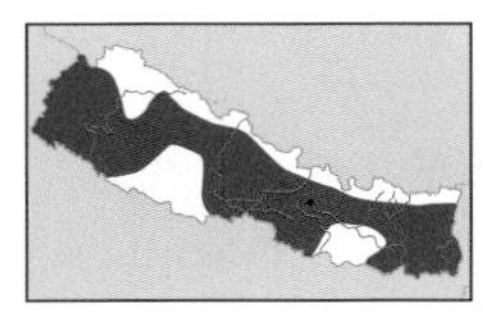

Eurasian Cuckoo *Cuculus canorus* 32–34 cm

Common and widespread summer visitor; 915–3,800 m. **ID** Best distinguished from Himalayan by song; see that species. Non-hepatic female has rufous wash to lower border of grey breast. Hepatic female is rufous-brown above and whitish below, and strongly barred all over with dark brown. Juvenile very variable, some superficially resembling grey adult, others hepatic female, with whitish fringes to upperparts and white nape patch. **Voice** *Cuck-oo... cuck-oo* call. **HH** Habits similar to Indian, less vocal at night. Often perches in open when calling. Forest and well-wooded country. **AN** Common Cuckoo.

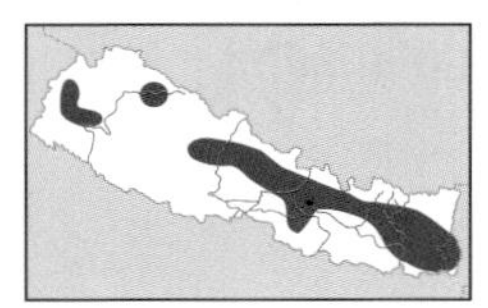

Himalayan Cuckoo *Cuculus saturatus* 30–32 cm

Locally fairly common summer visitor; 1,525–3,355 m. **ID** Compared to Eurasian, has broader black barring on buffish-white (rather than pure white) underparts, and darker grey upperparts, sometimes contrasting with paler head, and can be distinctly smaller. Female and juvenile, like Eurasian, occur as grey and rufous morphs, and have broader black barring, especially on breast, back, rump and tail than Eurasian. **Voice** Resonant *ho... ho... ho... ho*, very similar to Hoopoe. **HH** Habits similar to Indian, but usually hidden among foliage and less noisy at night. Forest and well-wooded country. **TN** Formerly considered conspecific with Oriental Cuckoo *C. optatus*.

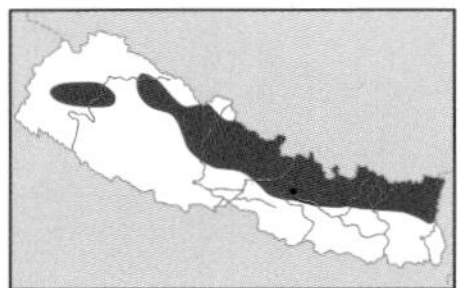

Lesser Cuckoo *Cuculus poliocephalus* 25 cm

Locally fairly common summer visitor; 1,500–3,700 m. **ID** Smaller than Himalayan (although Himalayan can be similar-sized) with finer bill. Plumage almost identical, but has darker rump and uppertail-coverts, contrasting less with tail. Hepatic morph of female prevails, and is typically more rufous than hepatic Himalayan (some with almost unmarked rufous crown, nape, rump and uppertail-coverts). Juvenile like juvenile Himalayan but has dark grey-brown upperparts and whiter (broadly barred) underparts. **Voice** Strong, cheerful *pretty-peel-lay-ka-beet.* **HH** Habits similar to Indian. Often calls noisily in flight. Forest and well-wooded country.

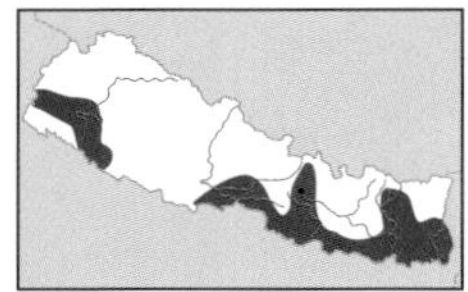

Grey-bellied Cuckoo *Cacomantis passerinus* 23 cm

Mainly a summer visitor; 305–1,400 m. **ID** Grey adult is grey with white vent and undertail-coverts. In hepatic female, base colour of underparts is mainly white, upperparts are bright rufous with crown and nape only sparsely barred, and tail is unbarred. Juvenile varies. Some have brownish-black upperparts (without distinct barring), dusky-grey underparts with indistinct buffish-grey barring mainly on belly and flanks (some more heavily barred on underparts), and tail grey-brown with whitish notching. Others are barred rufous on upperparts, and underparts are similar to hepatic female; tail dark brown, barred rufous (and is similar to Plaintive). Intermediates occur, e.g. with uniform grey upperparts and strong rufous barring on tail. **Voice** Clear *pee-pipee-pee... pipee-pee* ascending in scale and higher pitched with each repetition. **HH** Keeps mainly to leafy tops of trees and bushes. Very active and restless. Open forest and groves.

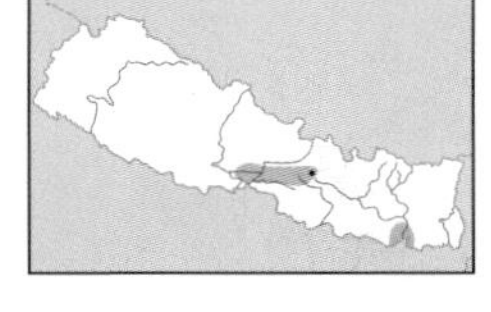

Plaintive Cuckoo *Cacomantis merulinus* 23 cm

Extremely rare winter visitor and passage migrant; 75–1,340 m. **ID** Adult has orange underparts. In hepatic female, compared to Grey-bellied, base colour of underparts is pale rufous, upperparts duller rufous-brown with more regular dark barring, and tail strongly barred. Juvenile has bold streaking on rufous-orange head and breast, and is distinct from hepatic and juvenile Grey-bellied. **Voice** Mournful whistle *tay... ta... tee* second note lower and third note higher than first; also *tay... ta... ta... tay* repeated with increasing speed and ascending pitch. **HH** Habits like Grey-bellied. Open forest and groves.

Indian Cuckoo
♂
♀
juv
♂
Eurasian Cuckoo
♂
♀
hepatic
♀
hepatic
Himalayan Cuckoo
♀
♂
Lesser Cuckoo
♀
hepatic
juv
♂
♀
hepatic
♂
juv
rufous juv
grey juv
♀
hepatic
Plaintive Cuckoo
Grey-bellied Cuckoo

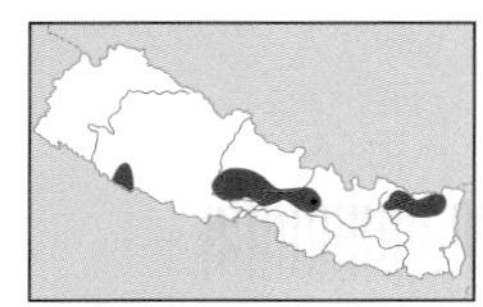

Asian Emerald Cuckoo *Chrysococcyx maculatus* 18 cm

Rare summer visitor; mainly 1,280–1,800 m (150–2,700 m). **ID** Male has emerald-green upperparts. Female has rufous-orange crown and nape and unbarred bronze-green mantle and wings. Yellow bill with dark tip. Juvenile similar to female but has rufous fringes to mantle and wing-coverts, and rufous-orange wash to barred throat and breast. **Voice** Clear, loud three-noted whistle, a sharp *chweek* flight call and loud descending *kee-kee-kee-kee*. **HH** Usually keeps to leafy canopy of tall trees, but on arrival in spring often flies about conspicuously. Very active, moving rapidly between branches and making sallies to capture flying insects. Habitually perches along a branch, rather than across it. Flight fast and direct. Broadleaved evergreen forest.

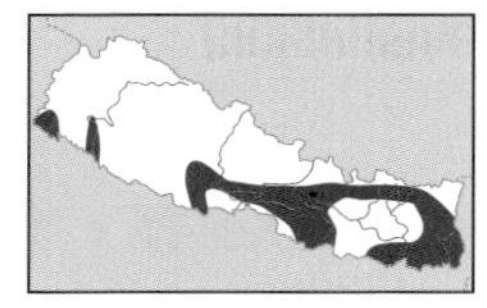

Drongo Cuckoo *Surniculus lugubris* 25 cm

Locally fairly common or frequent summer visitor; 75–1,500 m (–2,000 m). **ID** Adult glossy black, except for fine white barring on very long undertail-coverts, white thighs, and tiny white patch on nape (difficult to see in field). From a drongo by fine, downcurved black bill and white-barred undertail-coverts. Juvenile similar, but dull black, spotted white. **Voice** Distinctive series of ascending whistles, *pee-pee-pee-pee-pee-pee*, broken off and quickly repeated. **HH** Resembles a drongo, especially when perched upright, but less active and flies like a cuckoo. Perches on a bare branch when calling, but otherwise usually keeps in canopy foliage of trees and bushes. Forest edges and clearings, also groves. **TN** Birds with forked tails are often treated as a separate species, Fork-tailed Drongo Cuckoo, *S. l. dicruroides*, but there is much confusion over vocalisations and distribution.

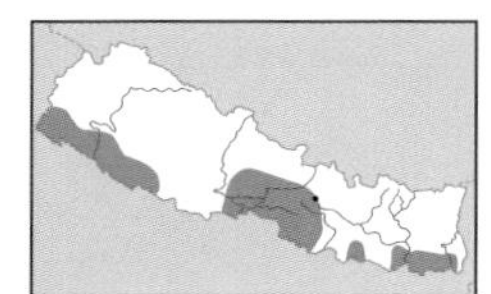

Blue-bearded Bee-eater *Nyctyornis athertoni* 31–34 cm

Uncommon resident; 150–365 m (–2,440 m). **ID** Large green bee-eater with a broad square-ended tail. Adult has blue forehead and 'beard', green upperparts, broad greenish streaking on yellowish-buff belly and flanks, and yellowish-buff undertail-coverts and undertail. Yellowish-buff underwing-coverts in flight. Juvenile similar to adult and has blue 'beard' even when very young. **Voice** A gruff *gga gga ggr gr* or *kor-r-r kor-r-r.* **HH** Spends much time perched inactively and inconspicuously among foliage in middle or upper storey; sometimes in open on treetops. Typically perches in hunched posture with its tail hanging vertically. Makes aerial sallies after insects from a vantage point like other bee-eaters. Flight laboured and deeply undulating. Edges and clearings of dense broadleaved forest and open forest.

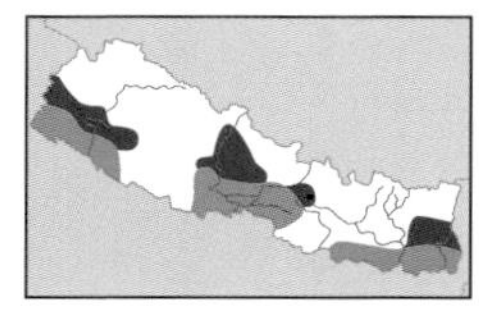

Green Bee-eater *Merops orientalis* 16–18 cm

Common and widespread resident and summer visitor; 75–620 m (–2,800 m). **ID** Small with elongated central tail feathers, blue or green throat with black gorget, variable golden-brown to rufous crown and nape, and green tail. Juvenile has square-ended tail; crown and mantle green, lacks black gorget, and throat pale yellowish- or bluish-green. **Voice** Pleasant throaty trill, *tree-tree-tree.* **HH** Often in small, loose parties. Perches on vantage point such as small tree, dead branch or post, sometimes on backs of cattle; then launches sallies after insects and circles gracefully back to base. Open country with scattered trees.

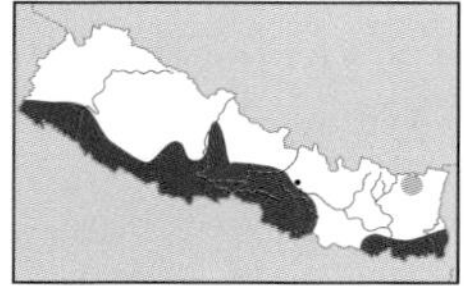

Blue-tailed Bee-eater *Merops philippinus* 23–26 cm

Locally common summer visitor; rarely over-winters; 75–300 m (–1,525 m). **ID** Mainly green bee-eater with blue rump and tail. Green crown and nape (with hint of blue on supercilium in front of eye), Throat and ear-coverts chestnut (with hint of blue below eye), and upperparts and underparts washed rufous and turquoise. Juvenile similar to adult but duller. Strong blue cast to rump, uppertail-coverts and tail. **Voice** A rolling *diririp.* **HH** In loose flocks when foraging. Darts out from exposed perches to seize prey; also hawks insects in continuous flight. Sometimes hunts from treetops at forest edges and clearings. Near water in open wooded country.

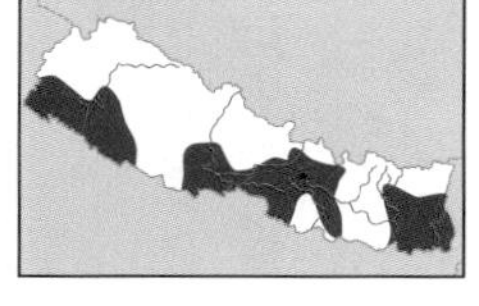

Chestnut-headed Bee-eater *Merops leschenaulti* 18–20 cm

Mainly a summer visitor, some resident at lower altitudes; mainly 75–680 m, locally to 1,525 m (–2,135 m). From Green by combination of bright chestnut crown, nape and mantle, yellow throat, turquoise rump, and broad tail with shallow fork. Juvenile duller, with chestnut of upperparts absent or reduced to a wash on crown (crown and nape uniform dark green in some). **Voice** A *pruik* or *churit*, briefer or less melodious than calls of larger bee-eaters. **HH** Similar to Blue-tailed, but a forest bird; often sallies from treetops or feeds high above canopy. Open broadleaved forest, often near water.

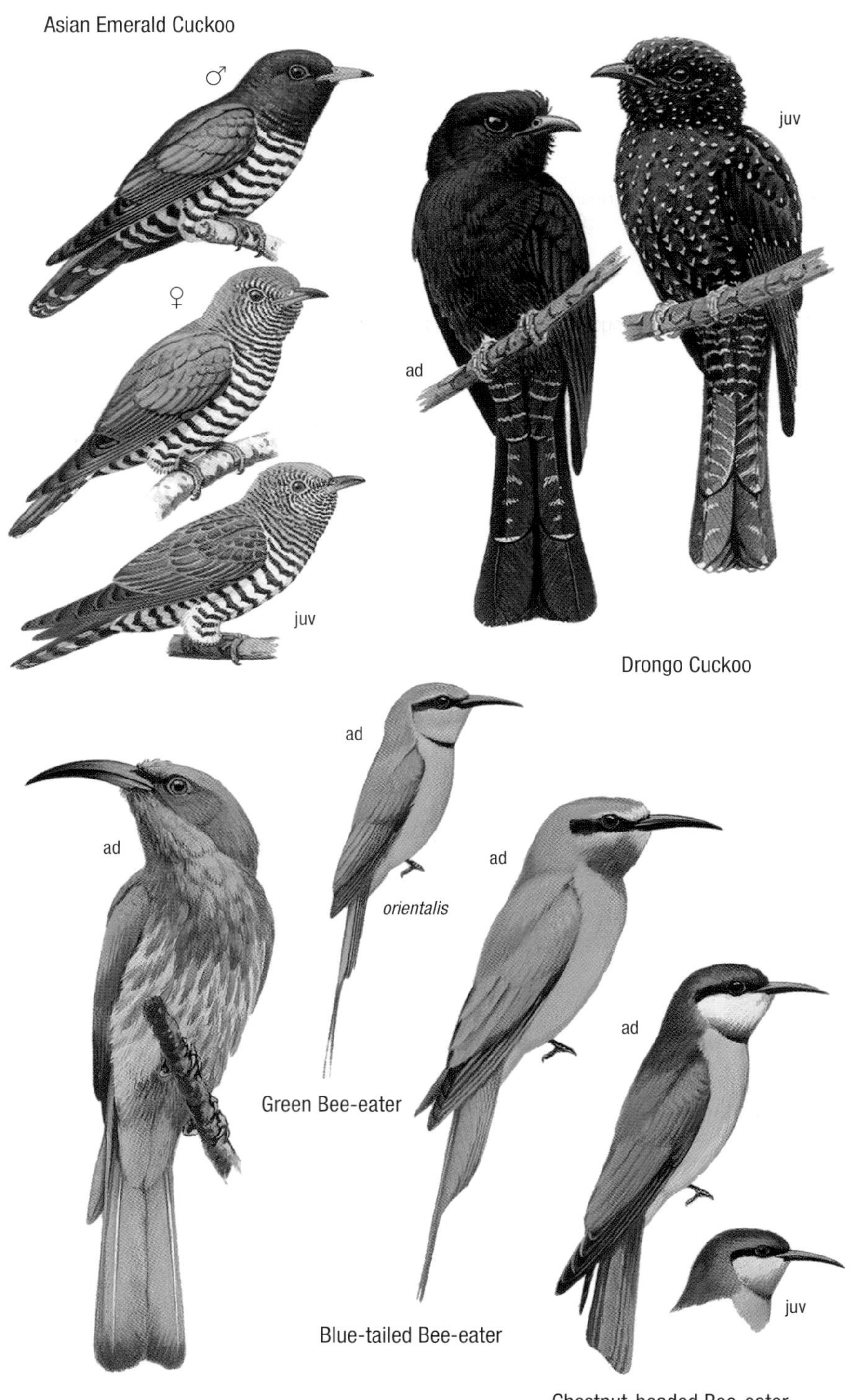
Asian Emerald Cuckoo
♂
♀
juv
ad
juv
Drongo Cuckoo
ad
ad
orientalis
Green Bee-eater
ad
ad
juv
Blue-tailed Bee-eater
Blue-bearded Bee-eater
Chestnut-headed Bee-eater

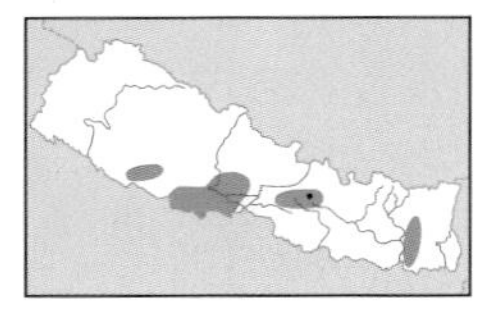

Common Barn Owl *Tyto alba* 36 cm

Local resident; 75–1,320 m. **ID** Readily identified by combination of unmarked white face and contrasting black eyes, white to golden-buff underparts finely spotted black, and golden-buff and grey upperparts finely spotted black and white. Wings and tail appear very uniform in flight, lacking any prominent barring or patches. **Voice** Variety of eerie screeching and hissing notes. **HH** Mainly nocturnal. Hunts by quartering open country a few metres above ground. Roosts and nests in large, old buildings in cities, towns and villages, also in ruins.

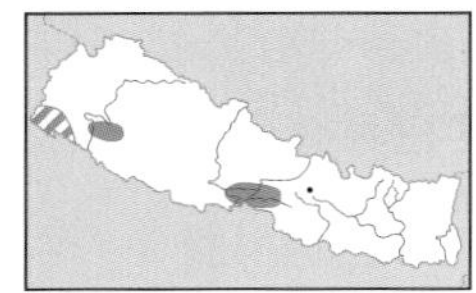

Eastern Grass Owl *Tyto longimembris* 36 cm

Very rare and very local resident; 150–225 m. **ID** Similar to Barn in size and structure, with dark eyes, and whitish face and underparts. Upperparts darker and contrast more with underparts than Barn, being more heavily marked dark brown (especially on crown and scapulars) and golden-buff (particularly on nape). Further, has dark barring on flight feathers, with prominent golden-buff patch at base of primaries contrasting with dark carpal patch and primary tips, and has dark-barred white or buff tail, which usually contrasts with dark uppertail-coverts. Legs longer and feathered only halfway down tarsus (to feet in Barn). Mottled, rather than streaked upperparts, lack of prominent streaking on breast, white facial discs, pale bill and black eyes are useful features from Short-eared Owl, which may occur in similar habitats. **Voice** Like Barn Owl. **HH** Nocturnal; hunting behaviour similar to Barn Owl. Tall grassland. **TN** Formerly placed within Grass Owl *T. capensis*.

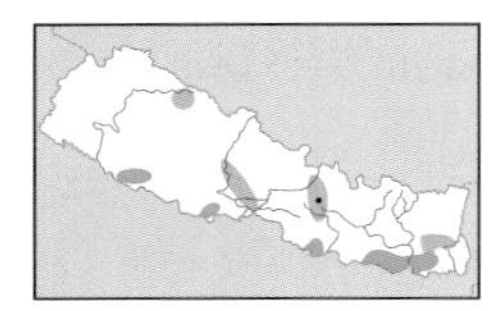

Short-eared Owl *Asio flammeus* 37–39 cm

Winter visitor and passage migrant, uncommon at Koshi, rare elsewhere; 75–3,350 m. **ID** From Long-eared (see Appendix 1) at rest by short or apparently absent ear-tufts, buffish-white coloration to facial discs, yellow (rather than orange) eyes, buff background to upperparts, and non-existent or indistinct streaking on belly and flanks. Further differences in flight are buffish-white background to primaries and tail, with less prominent dark barring, pronounced dark carpal patch, noticeable white trailing edge to wing, and narrower wings and longer tail. **Voice** Usually silent. **HH** Diurnal and crepuscular. Hunts by quartering low over ground. Open country.

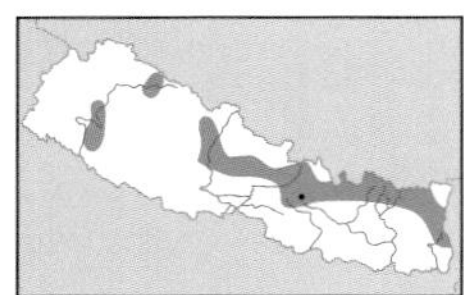

Mountain Scops Owl *Otus spilocephalus* 20 cm

Frequent resident; 1,525–2,745 m (–970 m). **ID** From similar species by unstreaked underparts, which are indistinctly spotted buff and barred brown, and by unstreaked upperparts mottled buff and brown (crown and nape usually most heavily so). Poorly defined facial disc and stubby ear-tufts. Bill and claws pale. Often shows a paler band on upper mantle (forming diffuse 'collar'), and head can appear slightly paler than rest of upperparts. In W Himalayas (*O. s. huttoni*) grey or fulvous-brown, while those in E Himalayas (*O. s. spilocephalus*) are rufous. It is probable that both races intergrade in Nepal. **Voice** Repeated double whistle, *toot-too*. **HH** Nocturnal like other scops owls. Roosts by day in tree hollow. Calls intermittently at night. Dense broadleaved forest.

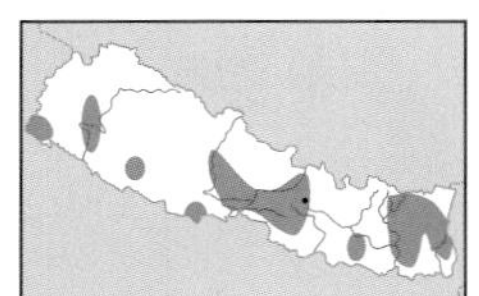

Collared Scops Owl *Otus bakkamoena* 23–25 cm

Locally fairly common resident; 185–1,525 m. **ID** From Oriental by larger size, prominent buff nuchal collar edged dark brown, more sparsely streaked underparts, and buffish (less distinct) scapular spots. Eyes dark orange or brown (yellow in Oriental). The two forms, Indian (*O. b. gangeticus*) and Collared (*O. b. lettia*) are separated mainly by call, although this is not considered diagnostic. Ear-tufts are longer and spotted (Collared), or shorter and barred (Indian). Upperparts are more heavily and irregularly marked with short dark streaks and cross-bars (Collared); or finer, longer streaks (Indian). Yellowish bill lacks black tip (Collared) or has black tip (Indian). **Voice** Call is a subdued frog-like *whuk* with rising inflection repeated at irregular intervals (Indian); or softer and less staccato than Indian with falling inflection (Collared). **HH** Nocturnal. Open forest and well-wooded areas. **TN** The two subspecies are not separated or mapped here because vocal differences are considered unreliable and plumage differences are small and variable.

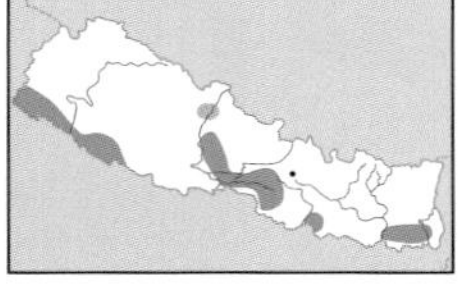

Oriental Scops Owl *Otus sunia* 19 cm

Locally fairly common resident; 100–365 m (–1,525 m). **ID** Highly variable, with grey, brown and rufous morphs. Smaller than Collared with prominent white scapular spots, profusely streaked underparts, and lacks prominent nuchal collar. Rufous morph distinct from Collared. Eyes yellow (dark in Collared). **Voice** Repeated, resonant, rhythmic, frog-like *wut-chu-chraaii*, first note musical, last two more rasping. **HH** Hides by day in dense foliage. Broadleaved forest.

ad
Common Barn Owl
ad
Eastern Grass Owl
ad
Short-eared Owl
lettia
ad
grey morph
ad
rufous morph
huttoni
spilocephalus
ad
ad
Collared Scops Owl
ad
ad
Mountain Scops Owl
Oriental Scops Owl

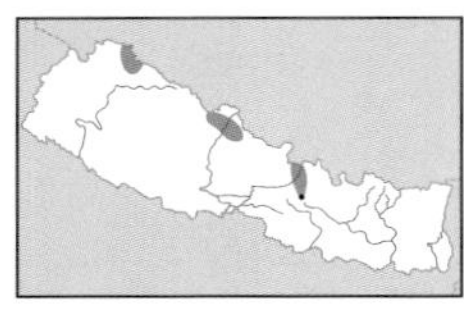

Eurasian Eagle Owl *Bubo bubo* 56–66 cm

Rare resident; 3,700–5,000 m. **ID** Very large, with pronounced upright ear-tufts. Upperparts mottled dark brown and greyish-buff; underparts heavily streaked. Larger, paler and greyer than Indian Eagle Owl, with less heavily marked upperparts, plainer facial discs (lacking pronounced dark border) and less heavily barred tail. **Voice** Call a resonant *whooh-tu*, the first note stressed and longer. **HH** Habits like Indian Eagle. Cliffs and open rocky areas.

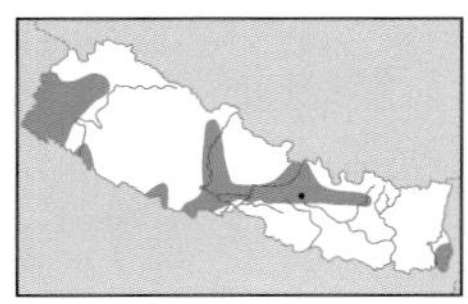

Indian Eagle Owl *Bubo (bubo) bengalensis* 48.5 cm

Very uncommon resident; 1,500–1,800 m (–2,100 m). **ID** From Eurasian by darker more heavily marked upperparts, pronounced dark border to facial discs, buff scapular spots, and heavily barred tail. From Brown Fish Owl by more upright ear-tufts, pronounced facial discs, broader breast streaking, and entirely feathered legs. **Voice** Call a deep, resonant *woo-hoooo*, the second note stressed. **HH** Mainly nocturnal, roosting by day in a cleft or ledge of a rocky cliff. Usually perches on a rock pinnacle or similar exposed situation well before sunset and until long after sunrise. Detects prey chiefly by scanning from a perch; also by making short flights over open country. Often detected by its call, usually given after emerging from roost. Chiefly nests on a cliff ledge. Frequents earth or rock cliffs and nearby open country.

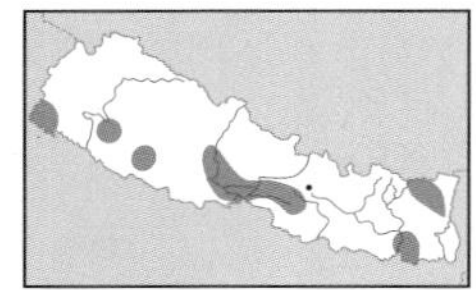

Spot-bellied Eagle Owl *Bubo nipalensis* 63 cm

Rare and local resident; 250–2,150 m. **ID** Very large, with bold chevron-shaped spots on whitish underparts, whitish facial discs, buff-barred dark brown upperparts, large pale bill, and brown eyes. Juvenile very distinctive: crown, mantle, coverts, rump and underparts white and buff, with brown spotting and barring, and face off-white. **Voice** Low, deep, moaning hoot, and a far-carrying mournful scream. **HH** Largely nocturnal, usually hiding by day among dense foliage in deep forest. Very bold and powerful owl, able to overcome large prey such as Kalij Pheasant. Hunts in forest. Nests in a tree hollow, deserted raptor nest or cliff cave or fissure. Inhabits dense broadleaved forest.

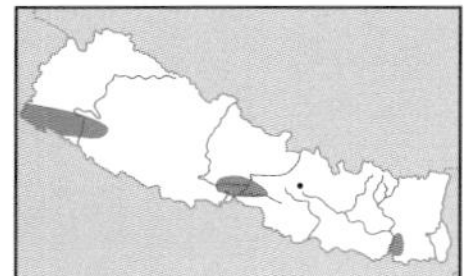

Dusky Eagle Owl *Bubo coromandus* 58 cm

Very rare and very local resident; 75–300 m. **ID** Upperparts greyish-brown, finely vermiculated whitish, with diffuse darker brown streaking; underparts greyish-white, finely vermiculated and more strongly streaked brown. Greyer and much less heavily marked than Indian. From Brown Fish Owl by more upright ear-tufts, more pronounced facial disc, more uniform grey-brown upperparts, and lack of any rufous tones. Legs feathered to toes (largely unfeathered in Brown Fish Owl). **Voice** Deep, resonant *wo, wo, wo, wo-o-o-o-o*. **HH** Crepuscular. Generally roosts by day in a shady tree, sometimes a thicket, emerging about an hour before sunset. May hunt in daylight, especially in cloudy weather. Frequently begins calling in early afternoon, then continues intermittently, but can call at any time. Usually nests in a deserted raptor nest high in a tree near water. Inhabits well-watered areas with extensive cover of well-foliaged trees.

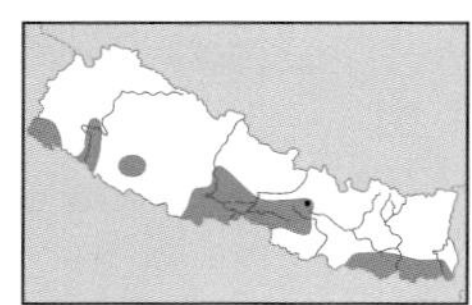

Brown Fish Owl *Ketupa zeylonensis* 56 cm

Local and uncommon resident; 75–1,525 m. **ID** From Tawny Fish Owl by combination of duller brown upperparts, finer dark brown streaking on crown, mantle and scapulars, finer streaking on dull buff underparts (with close cross-barring, lacking in other species), and absence of white above bill. **Voice** Calls include a soft, rapid, deep *hup-hup-hu*, maniacal laugh *hu-hu-hu-hu... hu ha* and mournful scream similar to Spot-bellied Eagle Owl. **HH** Usually in pairs. Generally roosts in densely foliaged trees. Partly diurnal, emerging from roost long before sunset, when pair members start to call. Sometimes hunts by day, in dull weather. Nests in a tree, on a ledge of steep riverbank or in a deserted raptor nest. Inhabits forest and well-wooded areas near water.

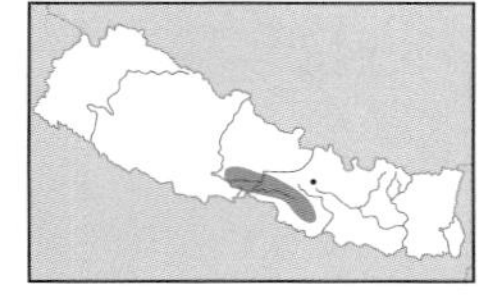

Tawny Fish Owl *Ketupa flavipes* 61 cm

Very rare and very local resident; 250–305 m. **ID** From smaller Brown Fish Owl by pale orange upperparts (much more richly coloured than Brown) with bolder and more distinct black streaking, bold orange-buff barring on wing-coverts and flight feathers, and broader and more prominent black streaking on pale rufous-orange underparts, which lack black cross-bars, and often shows prominent whitish patch on forehead. **Voice** Deep *whoo-hoo* and a cat-like mewing. **HH** Habits similar to those of Brown Fish Owl. Banks of streams and rivers and ravines in dense broadleaved forest.

Eurasian Eagle Owl
ad
Indian Eagle Owl
ad
Spot-bellied Eagle Owl
juv
ad
ad
Dusky Eagle Owl
ad
ad
Tawny Fish Owl
Brown Fish Owl

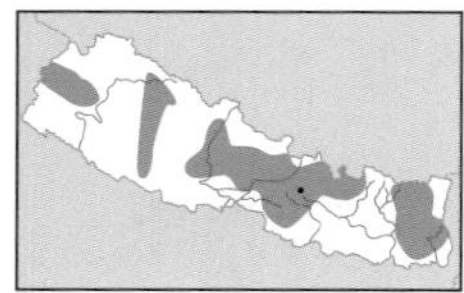

Collared Owlet *Glaucidium brodiei* 17 cm

Fairly common resident; mainly 610–3,050 m (250–3,500 m). **ID** Most similar to Jungle Owlet, but much smaller, and has distinct buff (or rufous) 'spectacles' on upper mantle, which frame blackish patches (creating pattern resembling an owl's face). Further, crown appears spotted rather than neatly and finely barred, and is barred rather than diffusely streaked on flanks, with broader white 'eyebrows'. Occurs as rufous, grey and brown morphs. **Voice** Pleasant four-noted bell-like whistle *toot... tootoot... toot*, uttered in runs of three or four and repeated at intervals. **HH** Diurnal and crepuscular, often hunting in daylight. Usually seeks prey by watching and listening from a prominent perch. Calls persistently by day and night in breeding season. Broadleaved and coniferous forest.

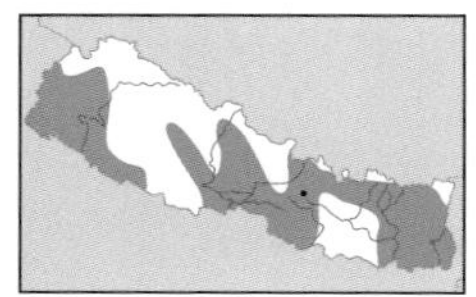

Asian Barred Owlet *Glaucidium cuculoides* 23 cm

Fairly common and widespread resident; 245–2,000 m (–2,745 m). **ID** From Jungle by larger size and longer, fuller-looking tail, buff barring on wing-coverts and flight feathers (wings of Jungle barred rufous and contrast with buff-barred mantle), and less finely barred upperparts and underparts, with lower flanks usually noticeably streaked (barring continues onto lower flanks on Jungle). Lower flanks pattern varies and may appear barred (but more broadly and diffusely than Jungle). Juvenile has buff spotting on crown, nape and mantle; breast barring and flanks streaking more diffuse. **Voice** Crescendo of harsh squawks. In breeding season a continuous bubbling whistle that lasts up to seven seconds. **HH** Mainly diurnal, scans and listens for prey from a prominent perch. Puffs itself into a ball before starting its bubbling call, and then gradually subsides to its normal size. Broadleaved forests including open woodland.

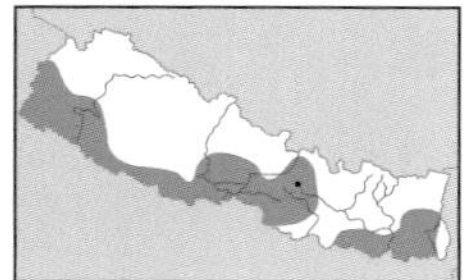

Jungle Owlet *Glaucidium radiatum* 20 cm

Common and widespread resident; 915–1,600 m. **ID** From Asian Barred by smaller size, bright rufous barring on wing-coverts and flight feathers contrasting with buff barring on mantle (wings of Asian Barred are barred buff and therefore concolorous with mantle), more closely barred upperparts and underparts, with bars continuing onto lower flanks. **Voice** Loud *kao... kao... kao* followed by a *kao... kuk* then a *kao..kuk*, which is repeated at an increasing rate for several seconds; also notes similar to Asian Barred. **HH** Mainly crepuscular, sometimes active in daytime, especially in dull weather. Spends day in foliage or tree hollow. Open forests and second growth.

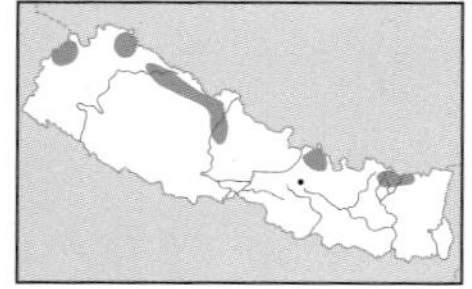

Little Owl *Athene noctua* 23 cm

Rare resident; 2,715–4,950 m (–2,300 m). **ID** From Spotted Owlet by streaked rather than spotted breast and flanks, neatly streaked crown with white streaks arranged in lines (rather than irregularly scattered spots). **Voice** Plaintive *quew* repeated every few seconds and a soft barking *werro-werro*. **HH** Crepuscular and partly diurnal; when feeding young can hunt at any time. Mainly hunts by pouncing from vantage point, also by running on ground. When alarmed, stretches upwards and bobs its head in curious fashion. Cliffs and ruins.

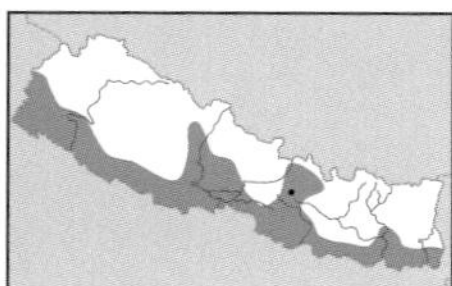

Spotted Owlet *Athene brama* 21 cm

Common and widespread resident; 75–1,525 m (–2,745 m). **ID** From Jungle and Asian Barred by spotted rather than barred appearance (with prominent white spots on crown, mantle and wing-coverts, and brown spotting rather than close dark barring on underparts). In addition, has pale facial discs and pale hind collar. From similar Little Owl by spotted or barred rather than streaked breast and flanks. **Voice** Harsh screechy *chirurr-chirurr-chirurr*, followed by or alternated with *cheevak, cheevak, cheevak* and a variety of other discordant screeches and chuckles. **HH** Mainly crepuscular and nocturnal. Hides by day in tree hollow, chimney or roof. Frequently hunts around streetlights for insects. Usually in pairs or family groups. Around habitation and cultivation.

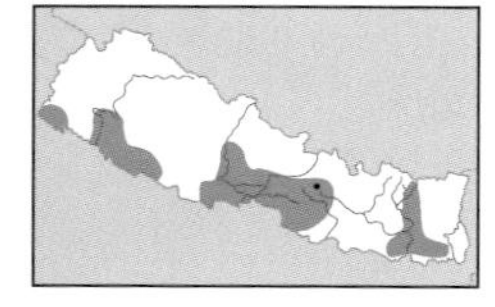

Brown Hawk Owl *Ninox scutulata* 32 cm

Locally fairly common resident; 75–1,500 m. **ID** Hawk-like profile (with slim body, long tail and narrow head). Uniform brown upperparts with variable amounts of white spotting on scapulars, all-dark face except variable white patch above bill (lacks pale facial disc shown by many owl species), and bold rufous-brown streaking and spotting on underparts. **Voice** A soft *oo... ok, oo... ok, oo... ok* in runs of 6–20 calls. **HH** Crepuscular and nocturnal, typically spending day concealed in top of forest tree. Flight hawk-like: a series of rapid wingbeats followed by a glide. Broadleaved forest.

ad
turning
away
ad
facing
Collared Owlet
juv
ad
Asian Barred Owlet
ad
Jungle Owlet
ad
bactriana
Little Owl
ad
indica
Spotted Owlet
ad
lugubris
Brown Hawk Owl

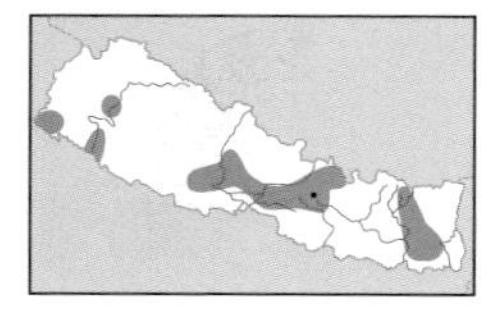

Brown Wood Owl ***Strix leptogrammica*** 47–53 cm

Rare and local resident; 150–2,700 m (–3,300 m). **ID** From Himalayan Wood Owl by uniform dark brown crown and upperparts (with patch of white barring on scapulars), and greyish-white underparts finely barred brown; flight feathers dark brown, narrowly barred paler brown. Also has dark brown face, with prominent white eyebrows, and striking white band across foreneck. **Voice** Calls include a double hoot, *tu-whoo*, a sonorous squawk, *hoo-hoohoohoo (hoo)*, and a variety of eerie shrieks and chuckles. **HH** Nocturnal. Roosts by day in large trees. Very shy; if disturbed will compress itself to resemble a stump or fly off a long distance through forest. Dense broadleaved forest.

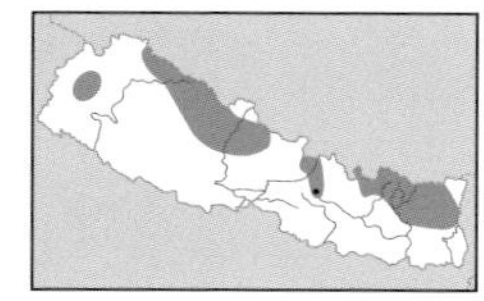

Himalayan Wood Owl ***Strix (aluco) nivicolum*** 41 cm

Locally frequent resident; 2,000–4,000 m. **ID** Rufous to dark brown, with heavily streaked underparts, white markings on scapulars, dark centre to crown, and pale forecrown stripes. Larger and greyer Tawny Owl *Strix aluco* (race *biddulphi*) possibly occurs. From Brown Wood Owl by boldly marked crown and mantle, and heavily streaked underparts. Eyes dark. **Voice** Trisyllabic *too-tu-whoo*, sometimes shortened to a two-noted *turr-whooh*; also low-pitched *ku-wack-ku-wack*. **HH** Nocturnal; seldom emerges before dusk. Typically roosts in a tree, perched close to trunk, partly concealed by leaves and resembling a dead stump. Hunts mainly by looking and listening from a perch. Oak/rhododendron and coniferous forests. **TN** Calls are very different from those of Tawny Owl and the two are often treated as separate species. However, the calls of other subspecies in the complex are still not well known, so they are retained as conspecific here.

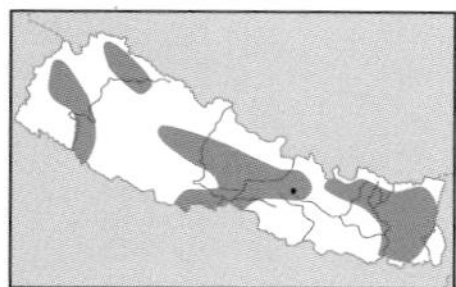

Grey Nightjar ***Caprimulgus (indicus) jotaka*** 32 cm

Fairly common and quite widespread resident and partial migrant; breeds 610–2,895 m (–3,500 m); winters 180–915 m. **ID** Dark grey-brown heavily marked with black. Colder coloured and less strongly patterned than Large-tailed Nightjar, with greyer upperparts and lacks diffuse warm rufous-brown nuchal collar. Breast dark grey-brown, lacking warm buff or brown tones. Further, has bold, irregular black markings on scapulars, usually lacking prominent pale edges, variable but rather poorly defined rufous-buff spotting on coverts, and broader dark bands on tail (less white at tip than Large-tailed). **Voice** Song a rapid series of loud, knocking *tuck* or *SCHurk* notes. **HH** A typical nightjar. Crepuscular and nocturnal. By day perches on ground or lengthwise on branch, and is difficult to detect. Feeds on insects caught in flight. Typically flies erratically to and fro over and among vegetation, occasionally wheeling, gliding and hovering to pick insects from foliage. Most easily located by their calls and mainly heard between sunset and dark and in earliest dawn. Forest clearings, open forest and scrub-covered slopes.

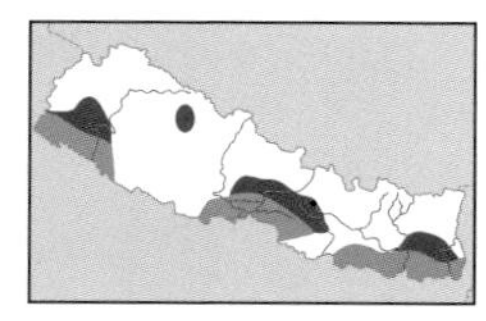

Large-tailed Nightjar ***Caprimulgus macrurus*** 33 cm

Locally fairly common, probably resident; 100–915 m (–3,100 m). **ID** Larger, longer tailed and more warmly coloured and strongly patterned than Grey, with pale rufous-brown nuchal collar, complete white throat, well-defined buff edges and bold wedge-shaped black centres to scapulars, broad buff tips to coverts forming wing-bars, and more extensive white or buff in outer tail feathers. **Voice** Series of loud, resonant calls: *chaunk-chaunk-chaunk*, repeated at rate of *c.*100 per minute. **HH** Habits like Grey. Forest edges and clearings; on forest paths and roads at night.

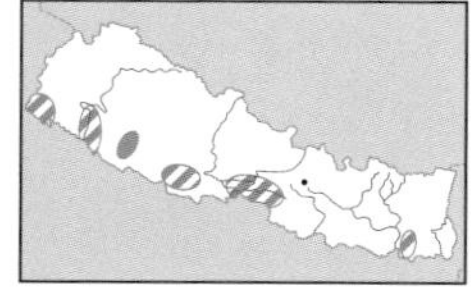

Indian Nightjar ***Caprimulgus asiaticus*** 24 cm

Very rare resident or visitor; 75–250 m. **ID** Grey, sandy-grey to brownish-grey. Best identified by combination of small size and relatively short wings and tail, boldly streaked crown, rufous-buff markings on nape forming distinct collar, bold black centres and broad buff edges to scapulars, prominent buff or rufous-buff spotting on wing-coverts, and pale, relatively unmarked central tail feathers. **Voice** Far-carrying *chuk-chuk-chuk-chuk-tukaroo*; likened to a ping-pong ball bouncing to rest on a hard surface; male gives short sharp *qwit-qwit* in flight. **HH** Habits like Grey. Dry, fallow cultivation and dry scrub; on dusty tracks near cultivation at night.

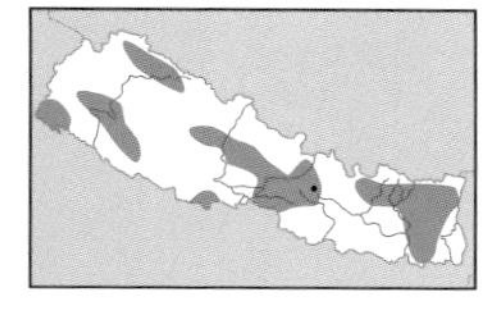

Savanna Nightjar ***Caprimulgus affinis*** 23 cm

Locally fairly common resident; 75–250 m (–915 m). **ID** Medium-sized, dark brownish-grey nightjar. Less strikingly marked than other nightjars; crown and mantle finely vermiculated, and lack bold dark streaking; has more uniform coverts with fine dark vermiculations and irregular rufous-buff markings, and scapulars usually edged rufous-buff. Male has largely white outer tail feathers. **Voice** Strident and shrill *dheet*. **HH** Habits like Grey. Open, short grassland and scrub.

ad
newarensis
Brown Wood Owl
Himalayan Wood Owl
ad
brown morph
ad
rufous morph
♂
♀
♂
Grey Nightjar
♀
♂
♂
Large-tailed Nightjar
ad
Indian Nightjar
♂
♀
♂
Savanna Nightjar

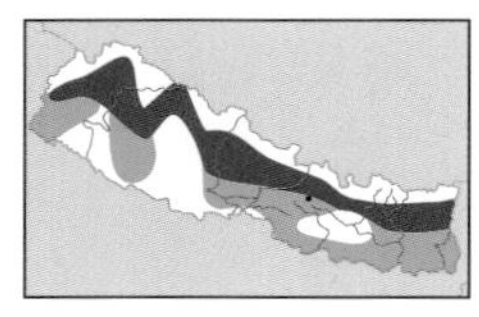

Himalayan Swiftlet *Collocalia brevirostris* 14 cm

Fairly common and widespread resident; summers up to 4,575 m; winters 915–2,745 m. **ID** A stocky brown swiftlet with slight gloss to upperparts, and pronounced indentation to tail. Paler grey-brown underparts than upperparts, and distinct pale grey rump-band. **Voice** Low, rattling call and twittering *chit-chit* at roost. **HH** Gregarious. Wanders erratically over large distances to feed. Frequently hunts over open country, including agricultural fields, often close to ground above meadows, fields or rivers. Roosts in caves.

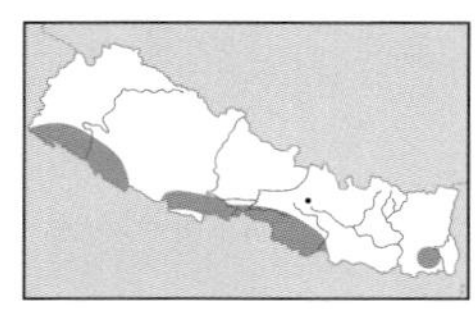

White-rumped Spinetail *Zoonavena sylvatica* 11 cm

Local resident; 150–250 m. **ID** Small and stocky, with broad wings, pinched-in at base and pointed at tip. Flight fast with rapid wingbeats, banking from side to side, interspersed by short glides on slightly bowed wings. Upperparts mainly blue-black with contrasting white rump; throat and breast grey-brown, merging into whitish lower belly and undertail-coverts. Long white undertail-coverts contrast with black of sides and tip of undertail. 'Spines' at tip of tail visible at close range. Wing shape and flight action differ from House Swift, and lacks white throat. **Voice** Twittering *chick-chick* in flight. **HH** In flocks of up to 50 birds. Hawks over forest with great manoeuvrability. Broadleaved forest.

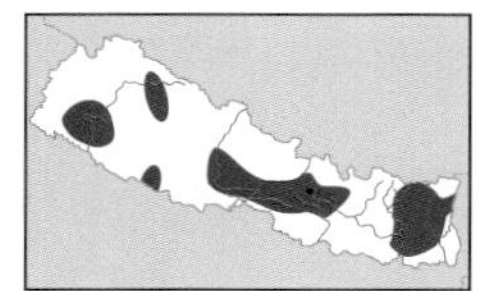

White-throated Needletail *Hirundapus caudacutus* 20 cm

Locally frequent summer visitor; 150–3,200 m. **ID** As other needletails, a magnificent flier combining very strong flapping with swooping, gliding and soaring, often at very high speeds. Like Silver-backed has pale 'saddle' on upperparts and striking white 'horseshoe' crescent at rear end. From Silver-backed by clearly demarcated white throat, and white inner webs to tertials (showing as white patch, but may be obscured). Juvenile has less clear-cut white throat (and is much more similar to Silver-backed); has black streaking and spotting on white of rear flanks, and dark fringes to white undertail-coverts. **Voice** Feeble, rapid metallic chittering, audible only at close range. **HH** Usually occurs singly or in loose parties. Dashes around crags with amazing adroitness. Birds may cover huge distances in a day's foraging. Roosts colonially on cliffs and trees. Often seen over ridges, cliffs, forest, upland grassland and river valleys.

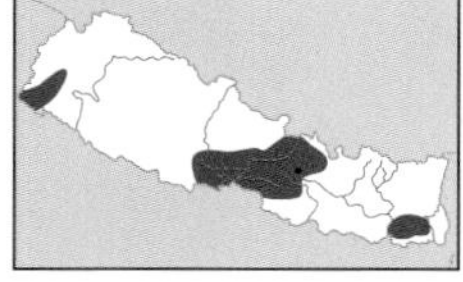

Silver-backed Needletail *Hirundapus cochinchinensis* 20 cm

Uncommon in Chitwan National Park, rare elsewhere, possibly a summer visitor; 250–1,540 m (–2,440 m). **ID** Throat pale brown or grey and can appear distinctly pale greyish-white, but never pure white and sharply divided from breast as in White-throated. Tertials have a pale grey inner web that may be visible in the field (strikingly white in White-throated). Juvenile has dark fringes to white undertail-coverts. **Voice** Soft rippling *trp-trp-trp-trp-trp*. **HH** Habits very similar to White-throated. Mainly hawks over forest and forested hills.

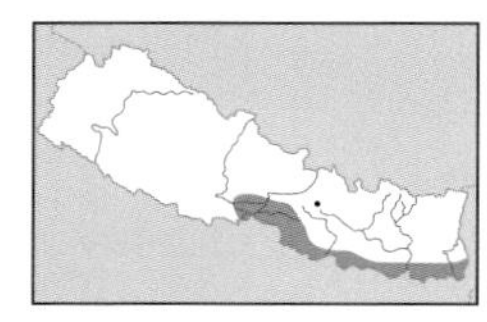

Asian Palm Swift *Cypsiurus balasiensis* 13 cm

Generally uncommon resident; more frequent in east; 75–120 m. **ID** Small and very slim with fine scythe-shaped wings and deeply forked tail (usually held closed). Rapid fluttering wingbeats interspersed by short glides. Throat slightly paler than rest of underparts, and may also show a marginally paler rump. Much smaller than Crested Treeswift, with weaker, more fluttering flight, and is browner (also does not show whiter belly and undertail-coverts). **Voice** A trilling *te-he-he-he-he*. **HH** Usually hawks insects around palms. Particularly active foraging in evening and readily joins mixed flocks with other swifts and hirundines. Twists and turns in air with great agility. Open country; closely associated with palms.

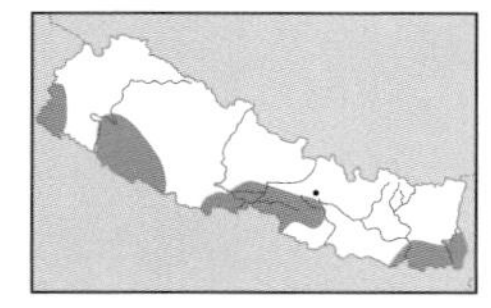

Crested Treeswift *Hemiprocne coronata* 23 cm

Frequent resident, uncommon in west; 75–365 m (–1,280 m). **ID** Large size with sickle-shaped wings and long, deeply forked tail usually held closed and pointed in flight. Typically flies above tree canopy with mixture of rapid and rather heavy fluttering, and periods of banking and gliding. In flight, appears mainly blue-grey with darker upperwing and tail, and whitish abdomen and undertail-coverts. At rest, both sexes show prominent dark green-blue crest, and wing-coverts are glossed blue. Male has dull orange ear-coverts. Female has dark grey ear-coverts, forming mask, bordered below by whitish moustachial stripe. Juvenile has extensive white fringes to upperparts (especially noticeable on lower back and rump), and feathers of underparts are fringed white with grey-brown subterminal bands. **Voice** Harsh *whit-tucck... whit-tuck* in flight. **HH** Unlike other swifts, perches readily in trees. Forages over well-wooded areas and forest, usually deciduous.

ad
ad
ad
White-rumped Spinetail
White-throated Needletail
Himalayan Swiftlet
ad
ad
ad
Asian Palm Swift
Silver-backed Needletail
♂
Crested Treeswift
♀

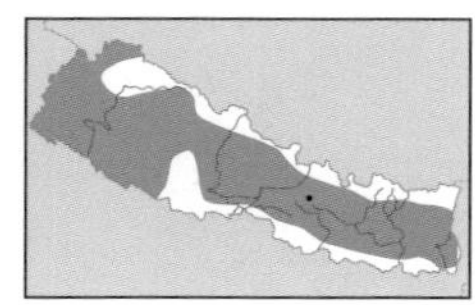

Alpine Swift *Tachymarptis melba* 22 cm

Fairly common and widespread, possibly resident; breeds 75–2,200 m (–3,700 m); non-breeding season 75–915 m (–2,200 m). **ID** Larger and more powerful than *Apus* swifts, with deeper and slower wingbeats. Best identified by white throat with brown breast-band, and white breast and belly contrasting with brown underwings, flanks and vent. **Voice** A high-pitched trilling *tri-hi-hi-hi-hi*. **HH** Gregarious, usually in scattered flocks. Like other swifts feeds entirely in air, drinks and bathes by swooping low over water. Eats mainly tiny insects caught by flying back and forth among aerial concentrations with its large mouth open; also pursues individual insects. Roosts in clefts in rock faces. Builds a typical swift nest, a half-cup of plant material glued together and attached to a cleft or rocky cliff with the birds' saliva. Skims over hills and mountains with cliffs and river gorges; may occur briefly over any habitat.

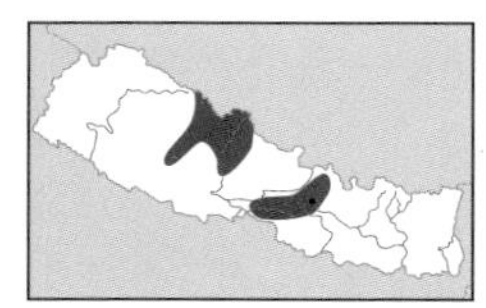

Common Swift *Apus apus* 17 cm

Local summer visitor; 2,300–4,200 m (–1,300 m). **ID** From Fork-tailed Swift by uniform brown upperparts (lacking white rump) . Juvenile has whiter forehead, more extensive white throat and extensive pale scaling on underparts. **Voice** A high-pitched screaming *screee... screee... screee*. **HH** Habits similar to Alpine. Nest undescribed in Indian Subcontinent. Chiefly dry mountainous areas, but can occur briefly over any habitat.

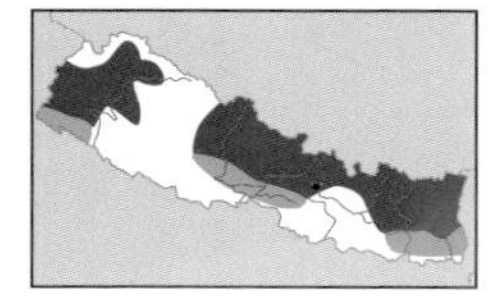

Fork-tailed Swift *Apus pacificus* 15–18 cm

Chiefly a summer visitor, few winter records; 75–3,800 m. **ID** Dark swift with prominent white rump and deeply forked tail. From House Swift by longer, deeply forked tail, and slimmer-bodied and longer-winged appearance. **Voice** Call less wheezy and softer than Common, a *sreee*. **HH** Habits similar to Alpine. Builds a typical swift nest (see Alpine) in small colonies inside fissures in cliff faces. Favours hawking over open ridges or hilltops. **TN** Race that occurs, *leuconyx* (Blyth's Swift), sometimes considered a separate species; underparts are less prominently scaled than nominate.

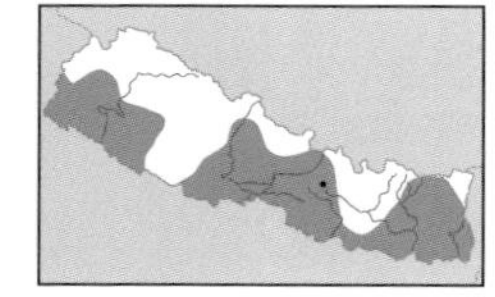

House Swift *Apus affinis* 15 cm

Common and widespread resident; summers 75–2,200 m; winters 75–1,350 m. **ID** A small, stocky swift with prominent white throat and rump-band. From Fork-tailed by smaller size, shorter and broader wings, stout body and rather big head, and much shallower tail fork. Flight is weaker then Fork-tailed. **Voice** Rapid and shrill *sik-sik-sik-sik... sik-sik-sik-sik-sik-sik*. **HH** Usually in large scattered flocks and keeps within wider vicinity of nest when breeding. Breeds and roosts communally. Depends entirely on habitations for nesting; only nests in larger settlements. Builds typical swift nest, each placed haphazardly one upon another, usually under eaves or verandas of buildings, under bridges or arches. Inhabits cities, towns and larger villages. **TN** *Apus nipalensis*, sometimes treated as a separate species, is no longer considered a valid split.

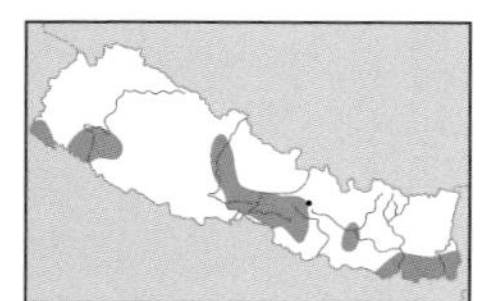

Ashy Woodswallow *Artamus fuscus* 19 cm

Local resident; frequent or uncommon; 75–365 m (–2,560 m). **ID** Adult has stout blue-grey bill, uniform slate-grey head, greyish-maroon mantle and pinkish-grey underparts. In flight, white-tipped tail and greyish-white band on uppertail-coverts. Juvenile has browner upperparts with buff fringes, paler grey throat (with indistinct brownish barring). **Voice** Harsh *chek-chek-chek*. Song a drawn-out pleasant twittering, starting and finishing with harsh *chack* notes. **HH** In flocks of up to 30. Spends much time hawking insect prey on wing. Perches on dead branches near treetops, telegraph wires or other vantage points and makes frequent aerial sallies. Flies in wide circle with rapid wingbeats alternating with glides. Has distinctive habit of wagging its stumpy tail when perched. Nest is a shallow cup of fine fibres, placed on horizontal branch. Open wooded country.

ad
Common Swift
ad
Alpine Swift
ad
Fork-tailed Swift
ad
House Swift
ad
Ashy Woodswallow

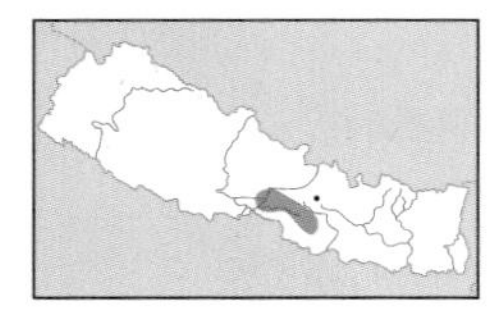

Ruddy Kingfisher *Halcyon coromanda* 25 cm

Very rare and very local, probably resident, Chitwan National Park, mainly in Churia Hills; 200–500 m. **ID** Medium-sized forest-dwelling kingfisher, with large coral-red bill, rufous-orange upperparts with brilliant violet gloss, and paler rufous underparts. In flight, striking bluish-white rump. Juvenile darker and browner on upperparts, with bluer rump, faint blackish barring on rufous underparts, and blackish bill. **Voice** Call a descending, high-pitched *titititititititi*, rather like White-throated, but more musical. Song a soft, trilling *tyuur-rrrr* incessantly repeated at one-second intervals from a perch. **HH** Secretive, shy and most easily detected by its call. Pools and streams in dense broadleaved tropical and subtropical evergreen forest.

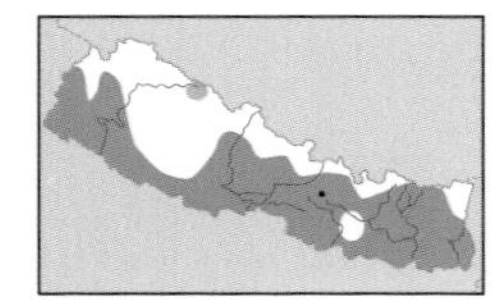

White-throated Kingfisher *Halcyon smyrnensis* 27–28 cm

Widespread resident, common 75–1,000 m, rare above 1,800 m (–3,050 m). **ID** Large kingfisher with large red bill, chocolate-brown head and underparts, white throat and centre of breast, and brilliant turquoise-blue upperparts including rump and tail. In flight, prominent white patches at base of black primaries. Juvenile duller, with brown bill and dark scalloping on breast. **Voice** Call a loud, rattling laugh. Song a drawn-out musical whistle *kilillili*. **HH** Bold and noisy. Like most kingfishers, spends long periods perched alone or in well-separated pairs, watching intently for prey. Typically perches on fence posts, telegraph wires or branches. Occasionally jerks its tail, and on spotting quarry often bobs head and body to help judge distance before dropping swiftly downwards to seize prey in its bill. Can be found far from water: cultivation, forest edges, gardens and wetlands.

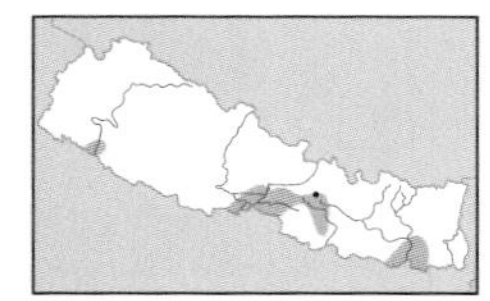

Black-capped Kingfisher *Halcyon pileata* 28 cm

Rare and irregular visitor; 75–300 m (–1,800 m). **ID** Large, mainly coastal kingfisher with coral-red bill. Black cap, white collar, deep purplish-blue upperparts, black coverts contrasting with blue secondaries, white breast, and pale orange-buff belly and flanks. In flight, bright blue rump and prominent white patches at base of primaries. Juvenile has dusky scalloping on collar and breast. **Voice** Distinctive ringing cackle, *kikikikikiki*, similar to but higher pitched than White-throated. **HH** Perches in open at forest edges or on telegraph wires. Dives down obliquely to catch prey; rarely plunges into water. Lakes and rivers.

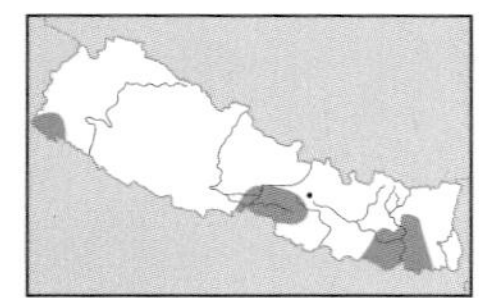

Blue-eared Kingfisher *Alcedo meninting* 17 cm

Rare and very local resident; 150–350 m. **ID** From Common by blue ear-coverts (except juvenile), darker blue upperparts (lacking greenish tones to crown, scapulars and wings), darker brilliant blue back and rump, and deeper orange underparts. Female has red on lower bill. Juvenile has rufous-orange ear-coverts as in Common, but darker blue upperparts (similar to adult) and lacks broad blue moustachial stripe of latter species (although can show short black moustachial that does not extend beyond ear-coverts). **Voice** Call a shrill, single *seet* or *tsit*, higher pitched and less strident than Common. Contact calls a thin, shrill *striiiiit, trrrrrt tit... trrrreu* etc. **HH** Habits very similar to Common. Mainly streams in dense broadleaved evergreen forest.

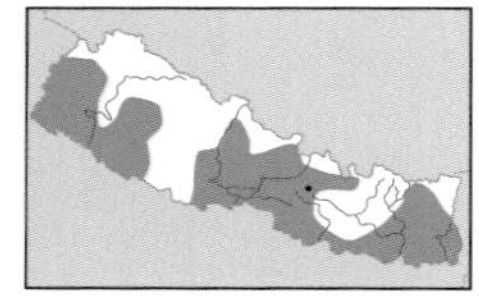

Common Kingfisher *Alcedo atthis* 16 cm

Locally common and widespread resident; 75–1,000 m (–5,350 m). **ID** From similar forest-dwelling Blue-eared by orange ear-coverts, paler greenish-blue upperparts and paler orange underparts. However, juvenile Blue-eared has rufous ear-coverts. Female has red on lower mandible. Juvenile similar to adult, but duller and greener above, with dusky scaling on breast. **Voice** Call a high-pitched, shrill *chee*, usually repeated, and *chit-it-it* alarm call. **HH** Uses post, reed or bank at water's edge as a vantage point, perching 1–2 m above surface. Plunges headlong into water to catch prey; may submerge completely, and sometimes hovers before diving. Wetlands in open country.

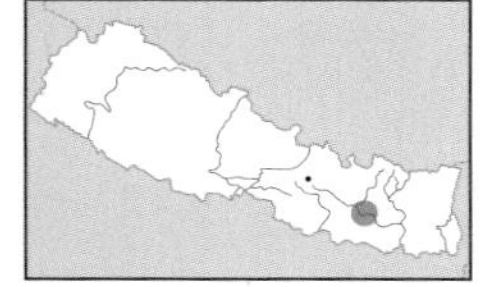

Blyth's Kingfisher *Alcedo hercules* 22 cm

Very rare and very local, possibly resident; 250 m. **ID** From Common and Blue-eared by considerably larger size and larger and longer bill, darker greenish-blue (almost brownish-black) scapulars and wings (with prominent turquoise spotting on lesser and median coverts), almost blackish crown with sharply contrasting turquoise spotting, and less distinct orange loral spot. Female has red on lower mandible. In flight dark mantle and wings contrast with brilliant blue back and rump. **Voice** Flight call a shrill, single note, more powerful and deeper than other *Alcedo* kingfishers. **HH** Habits similar to Common, but shy and poorly known. Shaded streams in dense broadleaved forest.

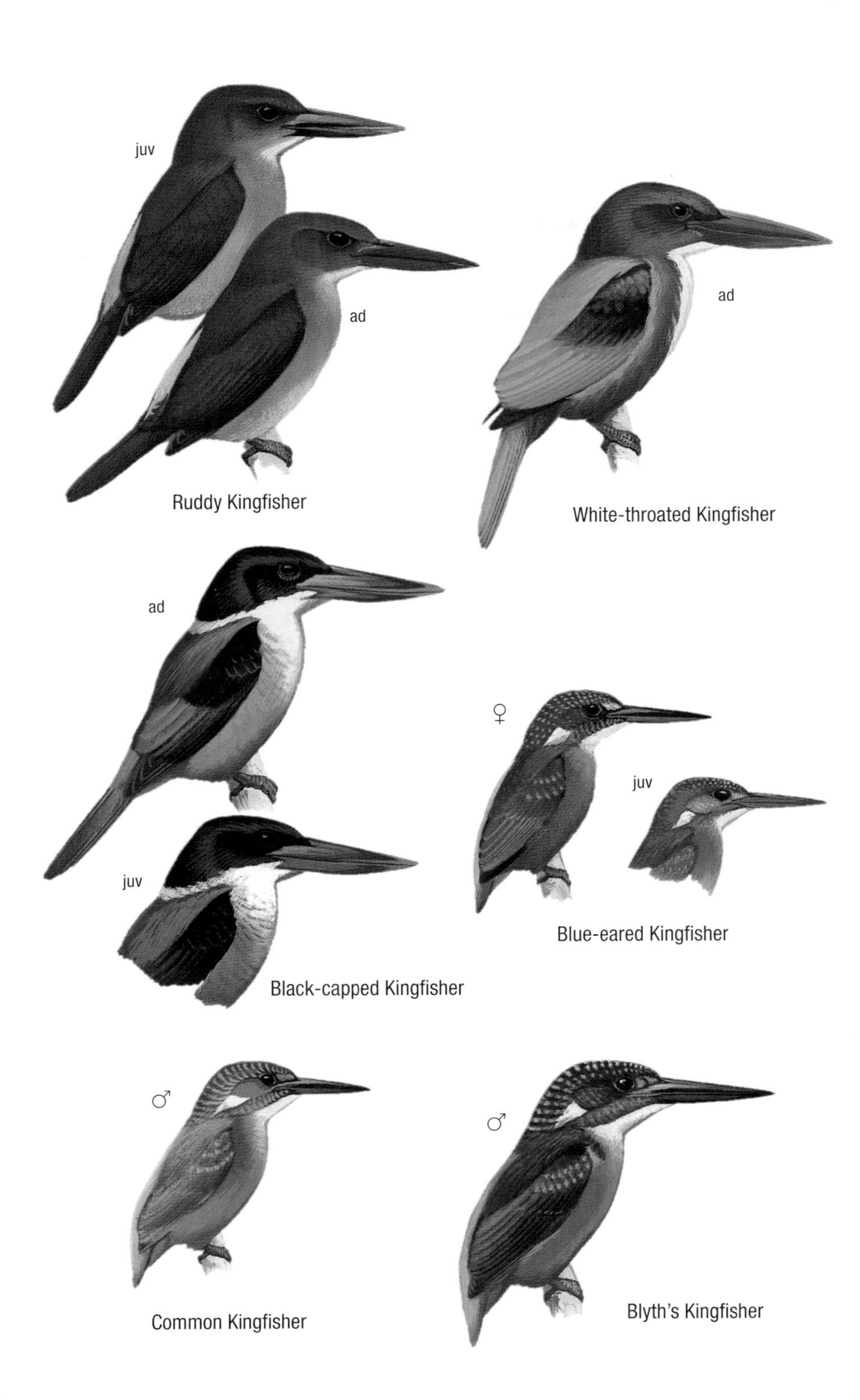
juv
ad
Ruddy Kingfisher
ad
White-throated Kingfisher
ad
juv
Black-capped Kingfisher
♀
juv
Blue-eared Kingfisher
♂
Common Kingfisher
♂
Blyth's Kingfisher

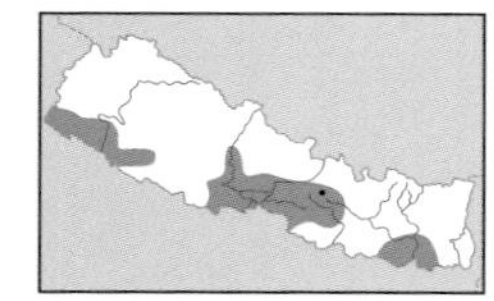

Stork-billed Kingfisher *Pelargopsis capensis* 35 cm

Locally fairly common and quite widespread resident; 75–760 m (–1,830 m). **ID** Very large with huge coral-red bill, brownish cap, pale orange-buff collar and underparts, and blue-green upperparts. In flight, turquoise rump and lower back. Juvenile has dusky barring on underparts, especially breast (forming broad band). **Voice** Explosive, shrieking *ke-ke-ke-ke-ke* and a pleasant *peer, peer, pur.* Song a long series of often paired melancholy whistles, *iuu-iuu...iuu-iuu...iuu-iuu* etc. **HH** Rather sluggish and heard more often than seen. Perches, often half-hidden, on a branch overhanging water, occasionally diving down to catch prey. Flight strong, direct and steady like other kingfishers. Shaded pools, slow-moving rivers and streams in well-wooded country.

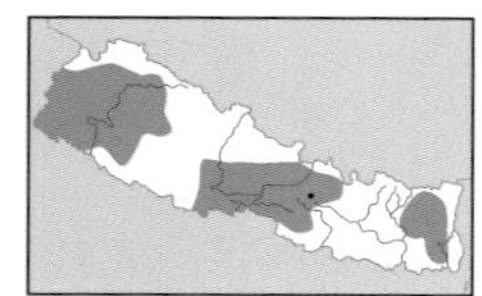

Crested Kingfisher *Megaceryle lugubris* 41–43 cm

Frequent and widespread resident; 250–1,800 m (–3,050 m). **ID** Very large with prominent crest, often held open. From much smaller Pied Kingfisher by lack of white supercilium, complete white neck-collar, finely spotted breast-band (sometimes mixed rufous), and dark grey wings and tail finely barred white (lacking Pied's prominent white patches in wings). Female and juvenile similar to male, but have pale rufous underwing-coverts. **Voice** Usually silent, although has squeaky *aik* flight call and a loud, hoarse, grating trill. **HH** Keeps to favourite stretches of river; spends most of day perched on branches close to and overhanging water or on rocks at river's edge or in the middle, where current is swift. Unlike Pied does not hover, but dives from a perch to catch fish. Rocky, fast-flowing mountain rivers and larger rivers in foothills; rarely by lakes.

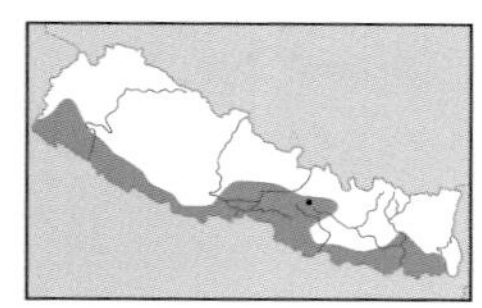

Pied Kingfisher *Ceryle rudis* 25 cm

Common and widespread resident; 75–915 m. **ID** Crested black-and-white kingfisher. White-streaked black crown and crest, white supercilium contrasting with broad black eye-stripe, white underparts with black breast-band, and black-and-white wings and tail. Male has double breast-band, female a single, usually broken, breast-band. **Voice** Sharp *chirruk chirruk.* **HH** Characteristically hunts by hovering over water with bill pointing down and fast-beating wings, then plunging vertically when it sees a fish. Still fresh waters, slow-moving rivers and streams, and lakes and pools in open country.

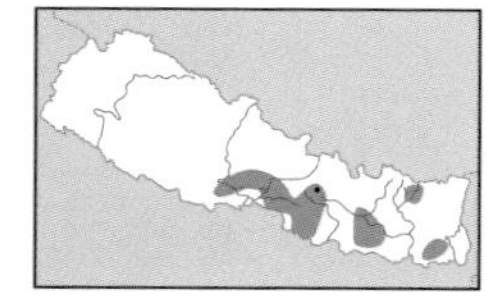

Red-headed Trogon *Harpactes erythrocephalus* 35 cm

Local and uncommon resident; 250–1,000 m (–1,830 m). **ID** Eye-ring and bill can be bright blue. Male has crimson head and breast, white breast-band, pinkish-red underparts, and black-and-grey vermiculated wing-coverts. Female similar but has dark cinnamon head and breast, and brown-and-buff vermiculated coverts. Immature resembles female. **Voice** Descending *tyaup, tyaup, tyaup, tyaup, tyaup.* **HH** Usually singly or in widely separated pairs. Perches almost motionless in upright posture for long periods. Captures flying insects on wing, twisting and turning like a flycatcher, or snatches prey while hovering in front of foliage and occasionally by swooping to ground. Mid-storey of dense broadleaved forest.

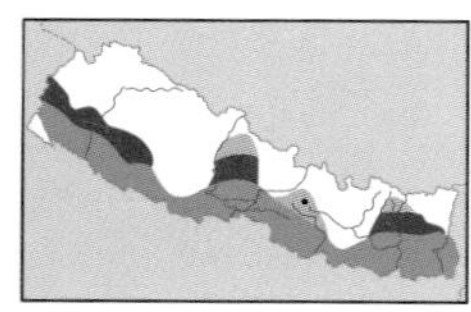

Indian Roller *Coracias benghalensis* 33 cm

Common and widespread resident; 75–1,050 m (–3,655 m). **ID** Rufous-brown on nape and underparts, white streaking on ear-coverts and throat, and greenish mantle. Turquoise band across primaries and dark blue terminal band to tail. Juvenile duller with more heavily streaked throat and breast. *C. b. affinis* in east has purplish-brown face and underparts, blue streaking on throat, and dark corners to tail. **Voice** Raucous *chack-chack-chack* and discordant screeches. **HH** Spends most of day on prominent perch such as dead tree, post or telegraph wire in open country, scanning for prey. Swoops leisurely down to capture prey on ground. Often hunts until dusk well advanced. Aggressively territorial throughout most of year. Cultivation, open woodland, groves and gardens.

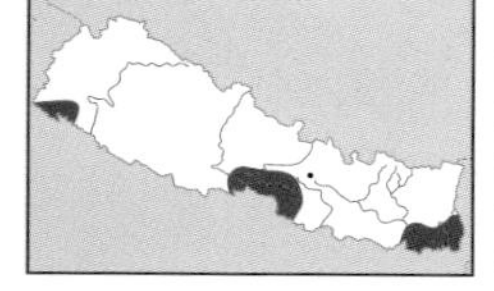

Dollarbird *Eurystomus orientalis* 28 cm

Local summer visitor, common in Chitwan National Park, uncommon elsewhere; 75–365 m (–1,300 m). **ID** Dark greenish to bluish, appearing black at distance, with red bill and eye-ring. In flight, turquoise patch on primaries. Flight buoyant, on broad wings. Juvenile similar, but has dull pinkish bill. **Voice** Raucous *check check.* **HH** Spends long periods in daytime perched inactively, often on tops of dead trees, only occasionally making short flights after insects. Feeds chiefly in late afternoon and evening until dark. Hawks insects on wing; small parties gather and hawk in circles where insects are swarming. Agile flight when chasing prey. Territorial when breeding. Forest and forest clearings. **AN** Oriental Dollarbird.

ad
Stork-billed Kingfisher
ad
Crested Kingfisher
♂
♀
Pied Kingfisher
♀
♂
Red-headed Trogon
ad
ad
benghalensis
affinis
ad
benghalensis
Indian Roller
ad
Dollarbird

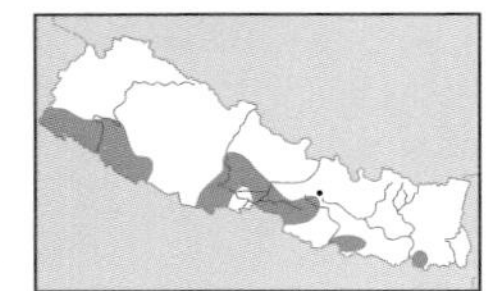

Brown-headed Barbet *Megalaima zeylanica* 27 cm

Resident, fairly common in far west, rare further east; 75–300 m. **ID** From similar Lineated Barbet by much finer whitish streaking on head and breast (which look more uniformly brown), brown chin and throat concolorous with breast, and virtual absence of streaking on belly and flanks. In addition, whitish-tipped wing-coverts, more extensive bare orange patch around eye that invariably extends to bill (eye-patch becoming yellow in non-breeding season), and deeper reddish-orange bill (orange-brown in non-breeding season). **Voice** Monotonous *kutroo, kutroo, kutroo* or *kutruk, kutruk, kutruk* uttered throughout day. **HH** Very noisy in hot weather, often calling in chorus. Found singly or in small groups and in parties of up to 20+, sometimes with other frugivorous birds, such as green pigeons, bulbuls and mynas, in favoured fruiting trees. Especially fond of figs. Broadleaved forest, wooded areas and trees near habitation.

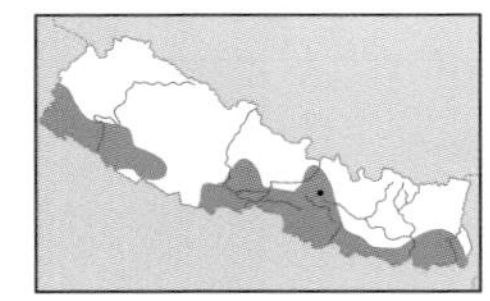

Lineated Barbet *Megalaima lineata* 28 cm

Widespread and locally common resident; 75–915 m. **ID** From similar Brown-headed by bold white streaking on head, upper mantle and breast (extending onto centre of belly), with usually whitish chin and throat (can be dusky-brown in some). Underparts can look white, streaked brown. In addition, has less extensive naked yellowish patch around eye, which (unlike Brown-headed) is usually separated from yellowish (not orange) bill, and uniform unspotted wing-coverts. Juvenile is more like Brown-headed, but orbital skin as adult and has unspotted wing-coverts. **Voice** Monotonous *kotur, kotur, kotur*, slightly mellower and softer than Brown-headed. **HH** Habits very similar to Brown-headed. Open deciduous forest, well-wooded areas and roadside avenues with fruiting trees.

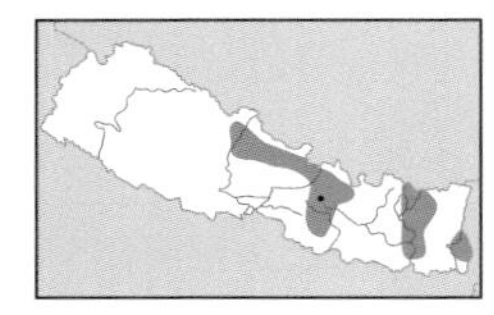

Golden-throated Barbet *Megalaima franklinii* 23 cm

Frequent resident; 1,500–2,400 m. **ID** Medium-sized barbet. From Blue-throated by broad black stripe through eye, greyish-white (not blue) cheeks, and yellow crown centre, chin and upper throat. From smaller Coppersmith Barbet by uniform green breast and dark legs and feet. Juvenile duller and yellow is less prominent. **Voice** Wailing, repetitive *peeyu, peeyu*, recalling Great, but higher pitched; also a monotonous *pukwowk, pukwowk, pukwowk*. **HH** Habits very similar to Brown-headed. Moist, broadleaved subtropical and temperate forest.

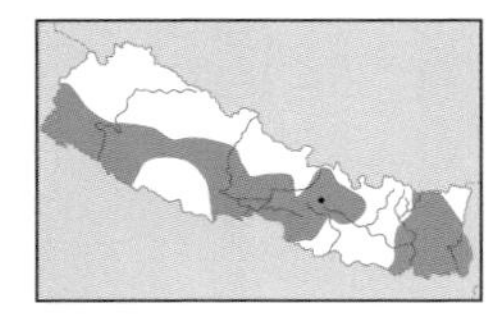

Blue-throated Barbet *Megalaima asiatica* 23 cm

Widespread resident, common 75–1,500 m, frequent to 2,100 m. **ID** Medium-sized barbet, with red forehead, black band across centre of crown, red hindcrown, and blue 'face', throat and upper breast. Possibly confusable with Blue-eared Barbet, but is larger, has red on crown, uniform blue head-sides, and larger pale bill (with variable dark culmen and tip). Juvenile similar to adult, but head pattern duller and poorly defined (with red of crown intermixed green and black). **Voice** Loud, harsh *took-a-rook, took-a-rook* uttered very rapidly. Habits similar to Brown-headed. Evergreen and deciduous trees, especially figs; open forest, groves near habitation and gardens.

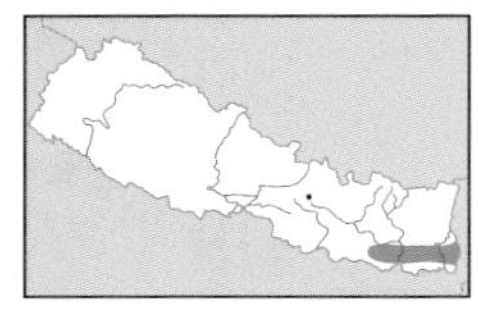

Blue-eared Barbet *Megalaima australis* 17 cm

Very rare and local resident in far east; 120–305 m. **ID** Small barbet with black forehead, blue throat and complex black, blue, red and yellow pattern on sides of head. Easily separated from larger Blue-throated by blackish forehead and blue crown, multi-coloured face and smaller all-dark bill. Juvenile mainly green, with blue wash to 'face' and throat, and lacks black and red head markings. **Voice** Disyllabic, repetitive *tk-trrt* repeated *c.*120 times a minute, and a throaty whistle. **HH** Habits similar to Coppersmith. Usually found singly perched in treetops. Dense, broadleaved evergreen forest.

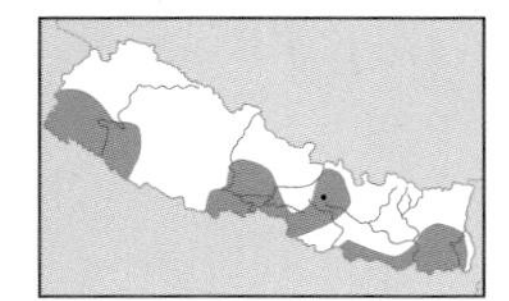

Coppersmith Barbet *Megalaima haemacephala* 17 cm

Widespread resident, common 75–915 m, frequent to 1,830 m. **ID** Small, brightly coloured barbet, with crimson forehead and patch on breast, yellow patches above and below eye contrasting with blackish hindcrown and sides of head, yellow throat, dark streaking on belly and flanks, and bright red legs and feet. Juvenile lacks red on forehead and breast, has prominent pale yellow patches above and below eye (surrounded by dark olive head-sides and moustachial stripe), whitish throat, olive-green breast-band, and broad olive-green streaking on belly and flanks. **Voice** Call a loud, metallic, monotonous, repetitive *tuk, tuk, tuk* etc. **HH** Usually singly or in small parties. Particularly vocal in heat of day. Deciduous biotope, especially with figs; open wooded country, groves and wooded gardens.

ad
Brown-headed Barbet
ad
Lineated Barbet
ad
Golden-throated Barbet
ad
Blue-throated Barbet
ad
juv
Blue-eared Barbet
ad
juv
Coppersmith Barbet

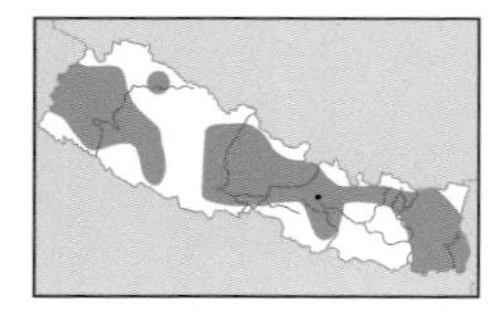

Great Barbet *Megalaima virens* 33 cm

Common, widespread resident; mainly 900–2,200 m (305–2,600 m). **ID** Largest barbet; unmistakable, with large pale yellow bill, violet-blue head, brown breast and mantle, olive-streaked yellowish underparts, and red undertail-coverts. Juvenile duller with greener head. **Voice** Monotonous, incessant and far-reaching *piho piho* uttered throughout day; also repetitious *tuk tuk tuk* often given in duet (presumably with female). **HH** Rather silent in winter, but noisy in hot weather. Usually singly or in groups of up to 5–6, but congregates in large parties in fruit-laden trees. Mainly moist, broadleaved forest; also well-wooded country.

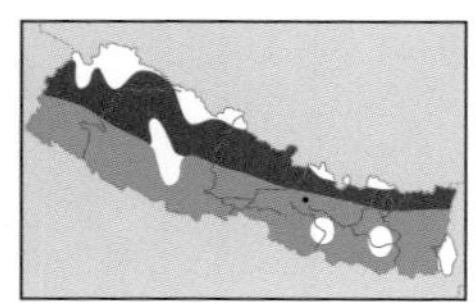

Common Hoopoe *Upupa epops* 31 cm

Fairly common, widespread resident 75–1,500 m; breeds to 4,400 m (–5,900 m on passage). **ID** Mainly rufous-orange to orange-buff, with striking black-and-white wings and tail, black-tipped fan-like crest is usually held flat, and downcurved bill. Broad, rounded wings recall giant butterfly in flight. **Voice** Repetitive *poop, poop, poop*; similar to Oriental Cuckoo, which usually has four notes instead of 2–3. **HH** Usually singly or in pairs. Searches for insects on ground, running and walking about, probing and pecking. Flight undulating, slow and butterfly-like, with irregular wingbeats. Open country, cultivation, around villages and grassy alpines slopes. **AN** Eurasian Hoopoe.

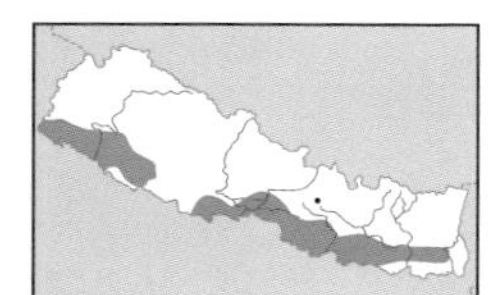

Indian Grey Hornbill *Ocyceros birostris* 50 cm

Local, frequent resident; 75–305 m (–760 m). **ID** Small hornbill with sandy-grey upperparts. Broad greyish-white supercilium with dark grey ear-coverts. Bill has prominent blackish casque and extensive black at base. Tail has white tips, dark grey subterminal band and elongated central feathers. In flight, white tips to primaries and secondaries. Female similar to male, but has smaller casque with less pronounced tip. Immature has bill as female, but smaller with smaller casque; lacks white wingtips. **Voice** Calls include loud cackling *k-k-k-ka-e*, rapid piping *pi-pi-pi-pi-pipipieu-pipipieu-pipipieu* and kite-like *chee-oowww*. **HH** Feeds mainly in fruiting trees, often with other frugivores. Occasionally descends to ground to take fallen fruit, large insects or to dustbathe. Favours large old trees. Open deciduous forest and wooded areas with fruiting trees.

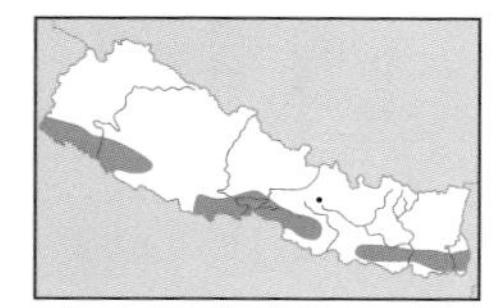

Oriental Pied Hornbill *Anthracoceros albirostris* 55–60 cm

Local, frequent resident; 75–250 m. **ID** Much smaller than Great Hornbill, with black head and neck, and mainly black tail with white tips to outer feathers. Upperwing black except narrow white trailing edge. Sexes similar, but female smaller with less convex casque lacking projecting tip, and has black at tip of bill and casque. Both sexes have pale blue circumorbital skin. Immature has smaller bill and casque with less black; orbital skin and throat patch whitish. **Voice** Varied loud, shrill, nasal squeals and raucous chucks, including loud cackling. **HH** Habits like Indian Grey. Mature broadleaved forest with fruiting trees.

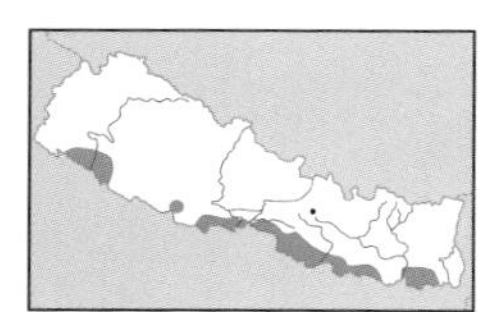

Great Hornbill *Buceros bicornis* 95–105 cm

Rare and local resident; 75–250 m (–500 m). **ID** Huge size, massive yellow casque and bill, and white tail with black subterminal band. Black face-band contrasting with yellowish-white nape, neck and upper breast. In flight, broad whitish wing-bar and trailing edge to wing. Neck, breast, white wing-bars and base of tail typically stained yellow with preen-gland oils. Male has red iris, black circumorbital skin and black at each end of casque. Female smaller, with smaller bill, white iris and red circumorbital skin, and lacks black on casque. Immature lacks casque until at least six months old. **Voice** Loud, deep, retching calls, often in short series and frequently in duet. Loud *ger-onk* flight call. **HH** Quite wary. Keeps to regular schedule of feeding circuits and roosting flights. Often flies high above forest for long distances. Chiefly arboreal, but sometimes descends to ground to feed. Mature broadleaved forest with fruiting trees.

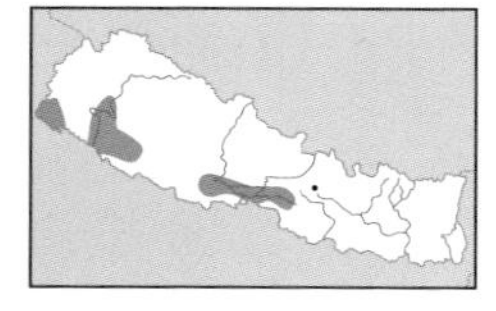

Great Slaty Woodpecker *Mulleripicus pulverulentus* 51 cm

Rare and local resident; 175–350 m. **ID** Giant slate-grey woodpecker with huge pale bill, long neck and long tail. Male has pinkish-red moustachial patch and pink on lower throat. **Voice** Loud goat-like bleating; very loud cackle in flight. **HH** Forages chiefly on trunks and large branches; birds follow each other in flight with slow deliberate wingbeats. Has large home range. Sal and other broadleaved forests; dependent on mature trees. Globally threatened.

ad
Great Barbet
ad
Common Hoopoe
♂
imm
Indian Grey Hornbill

♂
♂
Great Hornbill
♂
♀
Oriental Pied Hornbill
♀
Great Slaty Woodpecker

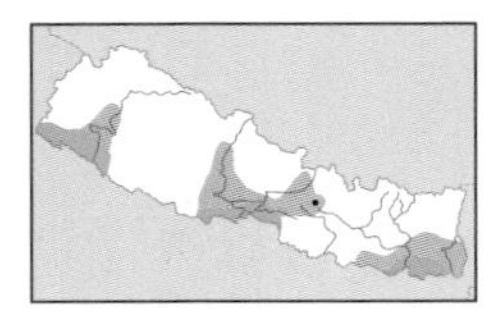

Eurasian Wryneck *Jynx torquilla* 16–17 cm

Frequent passage migrant and winter visitor; 75–915 m (–3,445 m). **ID** Cryptically patterned in grey, buff and brown. Dark stripe through eye, irregular dark stripe on nape and mantle, buff to pale rufous throat and breast finely barred with black, and long, barred tail. **Voice** Piping *kek-kek-kek*. **HH** Rather sluggish and unobtrusive. Feeds mainly on ground, hopping with tail slightly raised, picking up insects. Can cling to trunks. Scrub, second growth and edges of cultivation and marshes.

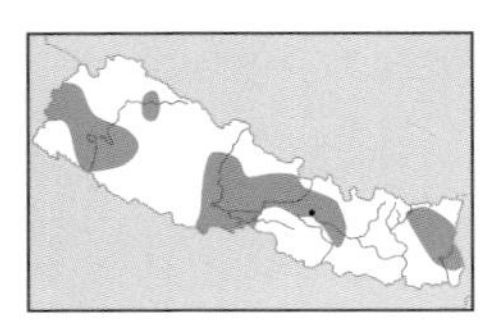

Speckled Piculet *Picumnus innominatus* 10 cm

Locally fairly common resident; 275–1,830 m. **ID** Tiny size. Whitish face broken by blackish ear-coverts patch and malar stripe, and white to yellowish-white underparts heavily spotted black. Also greyish crown, yellowish-green upperparts, and short, square-ended blackish tail with white on central and outer feathers. Male has dull orange forehead and forecrown, barred black. Female has uniform forehead and crown. **Voice** Sharp *tsick*, high-pitched *sik-sik-sik*, a *ti-ti-ti* and loud drumming. **HH** Usually with mixed feeding flocks of insectivores. Energetic and agile when feeding; small enough to forage on slender twigs, often on bushes close to ground. Creeps along branches, clings upside-down and perches sideways across branches. Bushes and bamboo in forest and second growth.

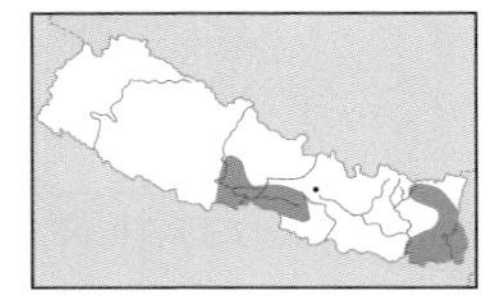

White-browed Piculet *Sasia ochracea* 9–10 cm

Rare resident; 250–2,135 m. **ID** Tiny size and tail-less appearance. Greenish-olive upperparts (variably washed rufous) and rufous underparts, very short black tail, fine white supercilium behind eye, and red iris and orbital skin. Male has golden-yellow forehead, rufous in female. **Voice** Short, sharp *chi*, sometimes followed by a rapid trill. Rapid, tinny drumming; also taps loudly on bamboo. **HH** Habits similar to Speckled. Sometimes hops among leaf litter on forest floor, cocking its tail like a wren. Bamboo and dense low vegetation in broadleaved forest and second growth.

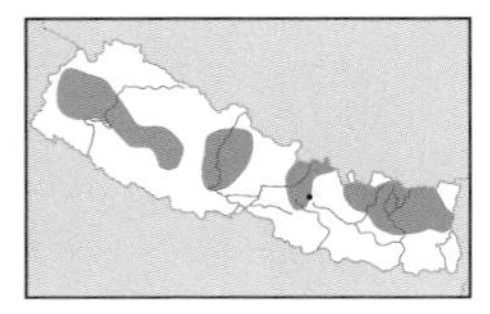

Rufous-bellied Woodpecker *Dendrocopos hyperythrus* 20 cm

Locally fairly common resident; 1,830–3,400 m (–3,500 m). **ID** Distinctive with white-barred mantle and wings, whitish face, and uniform rufous underparts. Male has red crown and nape. Female has white-spotted black crown and nape. Immature has blackish streaking on buff throat and blackish barring on rufous-buff underparts, and both sexes show scarlet feather tips on blackish crown. **Voice** Rapid, high *tik-tik-tik-tik* and *titi-r-r-r-r* alarm call. **HH** Like other woodpeckers, works up trunks and along branches in jerky spurts, directly or in spirals. Usually forages in upper storey. Often feeds by probing and pecking loose bark and in crevices; also bores holes into oak bark and drinks sap. Broadleaved forests.

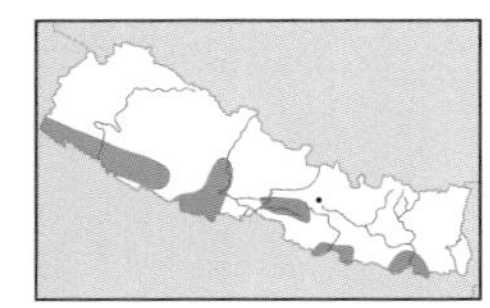

Brown-capped Pygmy Woodpecker *Dendrocopos nanus* 13 cm

Resident, locally fairly common in west, rare from central Nepal east; 75–275 m. **ID** From Grey-capped by brown crown (warmer than mantle), browner upperparts, and greyish-white to brownish-white underparts streaked (sometimes faintly) brown. Further, throat is evenly mottled and streaked brown, and stripe behind eye is brown and concolorous with crown. White spotting on central tail feathers is a further feature. Male has small crimson patch on sides of hindcrown. **Voice** Rather feeble, unobtrusive drum; also powerful rapid, metallic rattle probably indistinguishable from Grey-capped but, unlike that species, rarely gives *kik* calls. **HH** Very similar to Grey-capped. Light forest, second growth and trees in cultivation.

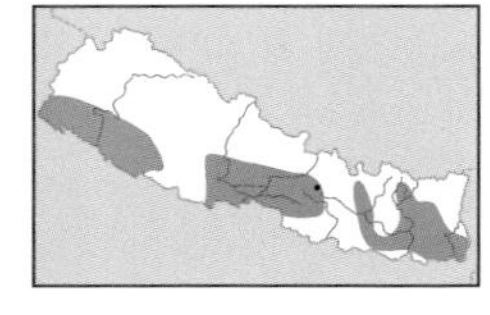

Grey-capped Pygmy Woodpecker *Dendrocopos canicapillus* 14 cm

Resident, fairly common/common from central areas east, uncommon in west; 75–1,370 m. **ID** From Brown-capped by grey crown (blackish on sides and towards nape), blackish stripe behind eye, and by deeper fulvous underparts with heavier black streaking. Further, tends to show diffuse blackish malar and whiter throat. Upperparts blacker, but difference is only very slight. Lacks white spotting on central tail feathers. Male has crimson on nape. Two intergrading races occur: male *D. c. semicoronatus* (in east) has more extensive red on sides of crown, meeting in narrow band on nape, than *D. c. mitchellii* (in west). **Voice** High, quickly repeated *tit-tit-errrrr* rattle like Brown-capped; also weak, repeated *kik*. **HH** Often with foraging mixed parties of insectivores. Typically in canopy; its small size enables it to forage on outermost twigs. Often perches crosswise and hangs upside-down. Broadleaved forest, second growth and trees in cultivation.

Eurasian Wryneck

Speckled Piculet

White-browed Piculet

Rufous-bellied Woodpecker

Brown-capped Pygmy Woodpecker

Grey-capped Pygmy Woodpecker

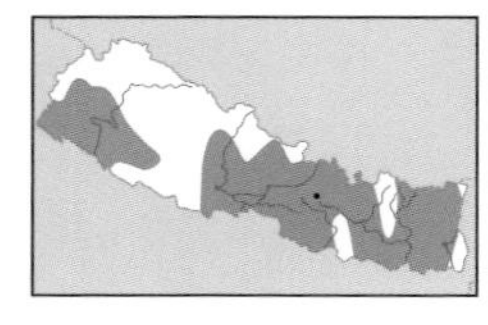

Fulvous-breasted Woodpecker *Dendrocopos macei* 18–19 cm

Resident, common from west-central areas east, frequent further west; 75–1,830 m (–2,745 m). **ID** Medium-sized woodpecker with white-barred black mantle and wings, and lightly streaked dirty buff underparts. Male has red on crown, black in female. **Voice** An explosive *tchick*, rapid *pik-pipipipipipipipi* and soft chattering *chik-a-chik-a-chit*. **HH** Often with mixed-species feeding flocks. Forages chiefly on tree trunks and larger high branches. Forests, edges, open wooded country, and gardens with trees.

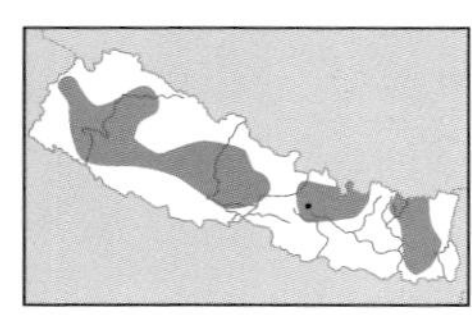

Brown-fronted Woodpecker *Dendrocopos auriceps* 19–20 cm

Locally fairly common resident; 1,065–2,440 m (500–2,895 m). **ID** From Yellow-crowned Woodpecker by white-barred mantle, brownish forehead and forecrown, and prominent black moustachial stripe and patch on sides of breast. Also smaller bill, well-defined black streaking on underparts, pinkish undertail-coverts and all-black central tail feathers. Male has red nape and yellow hindcrown. Female has dull yellow hindcrown and nape. **Voice** A series of rapidly repeated cries, *chitter-chitter-chitter-r-rh*; also short, sharp *chik-chik*. **HH** Often with roving mixed-species feeding flocks. Forages mainly in trees, sometimes on bushes in forest understorey, very rarely on ground. Coniferous and dry broadleaved forests.

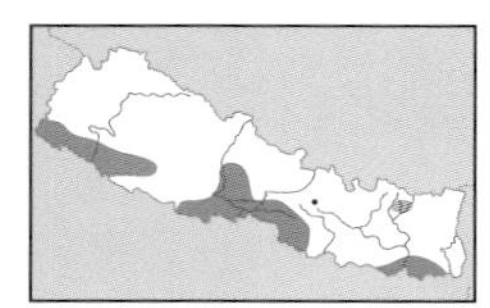

Yellow-crowned Woodpecker *Dendrocopos mahrattensis* 17–18 cm

Resident, fairly uncommon in west, very uncommon further east; 75–275 m (–1,700 m). **ID** From Brown-fronted by yellowish forehead and forecrown, white-spotted mantle and wing-coverts, bold white barring on central tail feathers, whitish rump, and diffuse brown moustachial stripe and patch on sides of neck. Underparts rather dirty grey, with fairly heavy brown streaking, and small red patch on lower belly. Male has red hindcrown and nape, female brownish hindcrown and nape. **Voice** Shrill *peek-peek* calls and a rapidly repeated *kik-kik-kik-r-r-r-h*. **HH** Often with mixed bands of insectivores. In canopy and on trunks. Often excavates bark and dead wood. Open woodland and dry country with scattered trees.

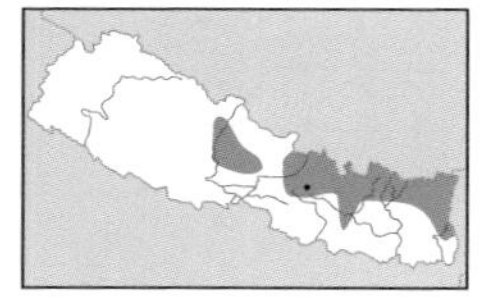

Crimson-breasted Woodpecker *Dendrocopos cathpharius* 18 cm

Locally frequent resident; 1,500–2,745 m (915–3,050 m). **ID** Similar to Darjeeling Woodpecker with white 'shoulder' patch. Smaller than Darjeeling with smaller bill; also, male has red of nape extending to rear and sides of neck, while female has diffuse, but fairly distinct, red patch at rear of ear-coverts. In addition, compared to Darjeeling, both sexes have a diffuse red patch on breast, lack heavy barring on flanks and thighs, and have indistinct red streaking on undertail-coverts (wholly pale red on Darjeeling). **Voice** A loud, monotonous *tchick*, higher pitched and less sharp than Darjeeling, and a shrill *kee-kee-kee*. **HH** Frequently feeds in lower parts of trees and on bushes; favours dead trees. Singly or in pairs and is frequently restless. Heavy moist broadleaved forest; mixed broadleaved forest with dense undergrowth.

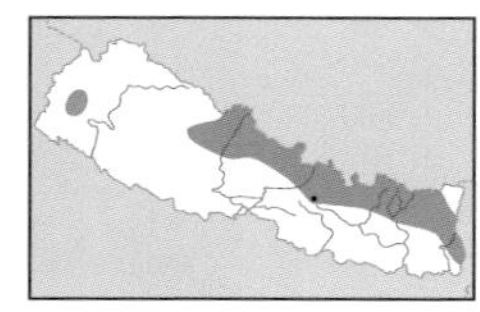

Darjeeling Woodpecker *Dendrocopos darjellensis* 25 cm

Resident, fairly common from west-central areas east, uncommon in west; 1,830–3,500 m (–3,750 m). **ID** Medium-sized black-and-white woodpecker, with white 'shoulder' patch. From Himalayan by black streaking on yellowish-buff underparts, yellowish-buff to pale orange neck-sides, absence of black rear border to ear-coverts, and male has black crown and red nape. From Crimson-breasted by larger size and bill, and yellowish-buff patch on neck. **Voice** Calls include a single *tsik*, rattling *di-di-di-d-dddddt* and *tchew-tchew-tchew-tchew*. **HH** Often in pairs or with mixed-species flocks. Forages in trees at varying heights, often high on moss-covered trunks and in canopy, also on dead trees and mossy logs on forest floor. Oak/rhododendron and mixed broadleaved/coniferous forests, also hemlock forests.

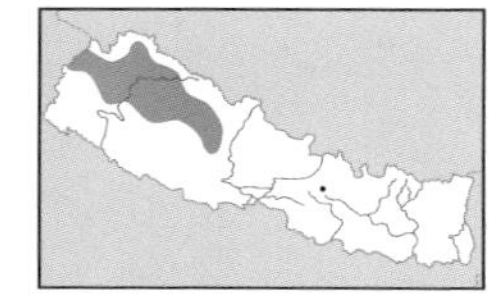

Himalayan Woodpecker *Dendrocopos himalayensis* 23–25 cm

Fairly common resident in west; 1,830–3,050 m. **ID** Medium-sized black-and-white woodpecker, with yellowish or buffish underparts, and white 'shoulder' patch. From Darjeeling by unstreaked underparts, black rear edge to ear-coverts, and red crown of male (although juvenile male Darjeeling has reddish crown). **Voice** A single *kit*, rapid *tri-tri-tri-tri* and rapid high-pitched *chisik-chisik*. **HH** Usually in pairs, searching trunks and larger branches; sometimes fallen logs. Searches for beetle larvae by boring holes along branches; also opens pine cones by wedging them in clefts in bark, or between trunks and branches, hammering them to extract the seeds. Mainly coniferous forest, also mixed coniferous/broadleaved forest.

Fulvous-breasted Woodpecker

Brown-fronted Woodpecker

Yellow-crowned Woodpecker

Crimson-breasted Woodpecker

Darjeeling Woodpecker

Himalayan Woodpecker

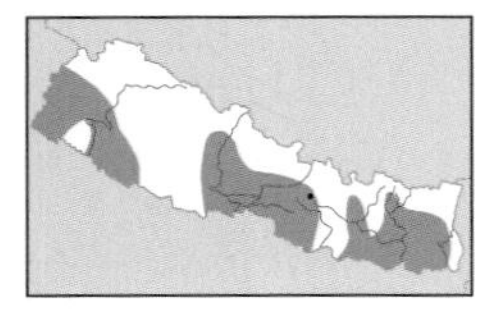

Lesser Yellownape *Picus chlorolophus* 27 cm

Locally fairly common resident; 75–1,750 m (–2,135 m). **ID** From Greater Yellownape by smaller size and bill, red and white head markings, rufous panel in wing, white barring on primaries and white barring on underparts. Male has red moustachial and line above eye. **Voice** A buzzard-like, drawn-out *pee-oow* and a descending series of shrill *kee kee kee kee kee* notes. **HH** Often accompanies mixed parties of insectivores or other woodpecker species. Feeds chiefly in smaller trees, occasionally in understorey shrubs or on fallen logs. Noisy and conspicuous. Broadleaved forest and second growth.

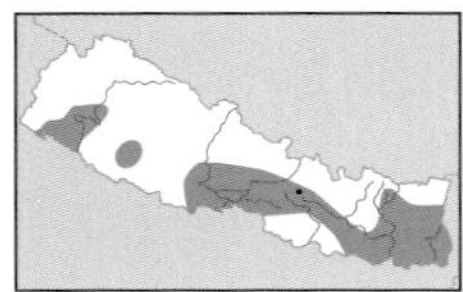

Greater Yellownape *Picus flavinucha* 33 cm

Locally frequent resident; 305–915 m (150–2,135 m). From Lesser by larger size and bill, brown on crown, and lack of red and white markings on head. Further differences include dark olive neck-sides adjoining dark-spotted white foreneck, uniform underparts, black barring on primaries, and yellow (male) or rufous-brown (female) throat. **Voice** A plaintive, descending *pee-u... pee-u* and single metallic *chenk*. **HH** Often with mixed-species feeding flocks and other woodpeckers. Feeds at all levels, from forest floor to highest branches. Shy and restless. Broadleaved forest and forest edges.

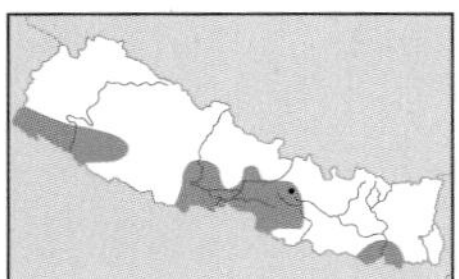

Streak-throated Woodpecker *Picus xanthopygaeus* 30 cm

Locally common resident; 75–465 m (–915 m). **ID** From similar Scaly-bellied Woodpecker by olive streaking on throat and upper breast. Also smaller size, smaller bill and usually darker upper mandible, indistinct moustachial stripe (often obscured by pale streaking) and comparatively uniform tail. Male has red crown, which is black (streaked grey) in female. Juvenile has grey bases to feathers of mantle and scapulars, creating mottled appearance. **Voice** Rather silent, although can give sharp, single *queemp*. **HH** Mainly feeds on ground on ants and termites; also pecks at cattle dung for beetle larvae and takes flower nectar. Open broadleaved forest and second growth.

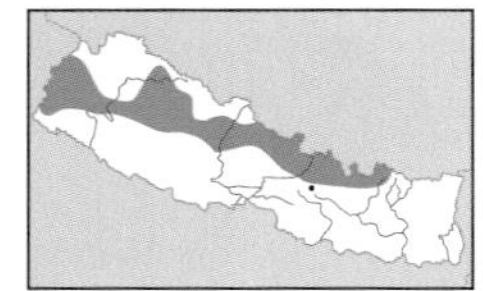

Scaly-bellied Woodpecker *Picus squamatus* 35 cm

Locally common resident; 1,850–3,700 m. **ID** From similar Streak-throated by larger size, larger yellowish bill, more prominent dark moustachial stripe, unstreaked throat and upper breast, more boldly scaled underparts, and prominent whitish barring on tail. Male has red crown, which is black (streaked grey) in female. Juvenile has mottled upperparts, and dark spotting and scaling on throat and breast, which could cause confusion with Streak-throated (larger and has white-barred tail). **Voice** Flight or contact call a *kuik-kui-kuik* usually rapidly repeated 3–4 times; advertising call a melodious, quavering, rapidly repeated *klee-gu-kleeguh*. **HH** Searches tree trunks for wood-boring beetles and ground for ants and termites. Noisy. Coniferous forest with large trees; also coniferous/oak forests.

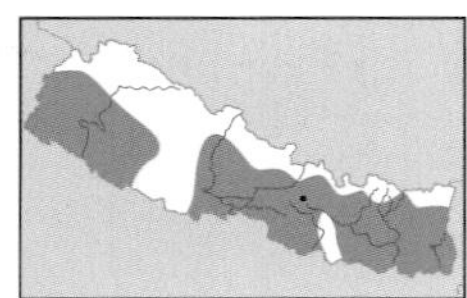

Grey-headed Woodpecker *Picus canus* 32 cm

Fairly common resident; mainly 150–2,200 m (75–2,600 m). **ID** Plain grey face, black nape and moustachial, dark bill, and uniform greyish-green underparts. Male has red forehead and forecrown (black in female). Juvenile duller, with greyer upperparts, less pronounced moustachial, and whitish barring on underparts. Two races. In west, *P. c. sanguiniceps* is darker and greener on upperparts and underparts than *P. c. hessei* in east, which has bronze sheen to upperparts and yellower underparts. **Voice** A high-pitched *peeek, peeek, peeek, peeek*, usually in runs of 4–5 notes and fading at end; also chattering alarm. **HH** Usually solitary and somewhat shy; edging up and down tree trunks out of sight. Often feeds on ground. Partial to ants, beetles, berries and nectar. Open to closed broadleaved forests and edges; groves in farmland and visits lone trees.

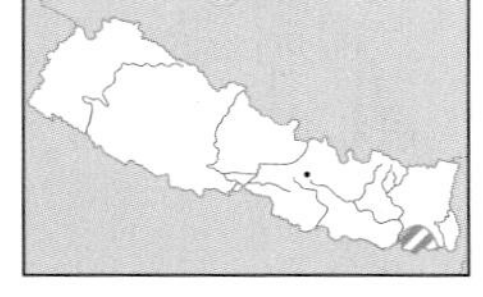

Pale-headed Woodpecker *Gecinulus grantia* 25 cm

Probably a former resident; no records since 1981, formerly recorded very locally in far east at 305 m. **ID** A smallish, mainly unbarred woodpecker with small, pale bill. Easily distinguished by golden-olive head and neck, dull crimson to crimson-brown upperparts, brown primaries barred buffish-pink, and dark olive underparts. Male has crimson-pink crown. **Voice** Territorial call reminiscent of Bay's, a loud, strident *yi-wee-wee-wee*; also a harsh, high-pitched and quickly repeated *grrritj-trrit-grrit* etc. Drumming is loud, evenly pitched and of fairly short duration. **HH** Forages noisily and often low down on large bamboos in moist broadleaved forest.

♂
♀
Lesser Yellownape
♂
♀
Greater Yellownape
♂
♀
Streak-throated Woodpecker
♂
♀
Scaly-bellied Woodpecker
♂
♀
Grey-headed Woodpecker
♂
♀
Pale-headed Woodpecker

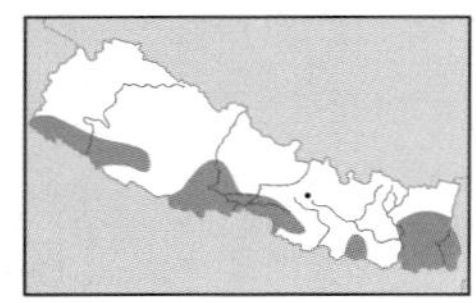

Himalayan Goldenback *Dinopium shorii* 30–32 cm

Locally fairly common resident in a few protected areas; frequent or uncommon elsewhere; 75–275 m (–650 m). **ID** Smaller size and bill than Greater Goldenback, with black hindneck, and brownish-buff centre of throat (and breast on some) with black spotting forming irregular border. Indistinctly broken moustachial stripe (centre is brownish-buff, with touch of red in male). Further, has reddish or brown eyes, and three toes. Breast irregularly streaked and scaled black, but sometimes almost unmarked. Female has white-streaked black crest (white-spotted in female Greater). **Voice** Rapid, repeated tinny *klak-klak-klak-klak-klak*, slower and softer than Greater. **HH** Often associates with other woodpeckers and small birds. Tall, mature broadleaved forest. **AN** Himalayan Flameback.

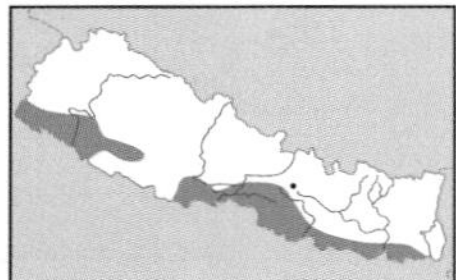

Lesser Goldenback *Dinopium benghalense* 26–29 cm

Common resident; 75–365 m. **ID** Smaller than other goldenbacks. Best identified by combination of black lower back and rump, different head pattern (white-spotted black throat and black stripe through eye, but lacks moustachial stripe), barred primaries and (variable) white or buff spotting on blackish lesser wing-coverts. Further, female has red hindcrown and crest. **Voice** Single strident *klerk* contact call and a whinnying *kyi-kyi-kyi*. **HH** Fairly bold and noisy. Often found in small parties with other species. Forages at all levels in trees and on ground. Chiefly eats ants. Light forests, groves and trees in open country. **AN** Black-rumped Flameback.

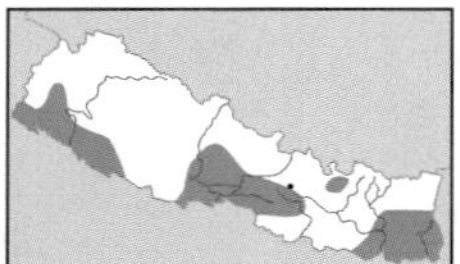

Greater Goldenback *Chrysocolaptes lucidus* 33 cm

Locally fairly common resident; 75–915 m (–1,700 m). **ID** From Himalayan by larger size and longer S-shaped neck, longer bill, white or black-and-white spotted hindneck, pale eyes, and four (not three) toes. In addition, has clearly broken moustachial stripe (with obvious white oval centre), clean single black line on centre of throat, and white spotting on black breast. Female has white-spotted crown (white-streaked in Himalayan). **Voice** Sharp, metallic, monotone *di-di-di-di-di-di-di* like a large cicada, and a lower-pitched, more rapid *di-i-i-i-i-t* in flight. **HH** Frequently associates with other woodpeckers, drongos and small insectivores. Chiefly visits large trees. Forages at all levels, especially on dead wood, sometimes on ground. Mainly eats insects and grubs, also nectar. Broadleaved forests and groves. **AN** Greater Flameback.

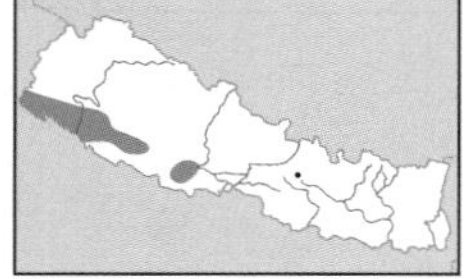

White-naped Woodpecker *Chrysocolaptes festivus* 29 cm

Uncommon resident in west; 75–245 m. **ID** Large, with large bill and divided moustachial stripe. Best identified by white hindneck and mantle contrasting with black scapulars and back (which form black V). Rump also black. Female has yellow hindcrown and crest. **Voice** Tinnier, higher-pitched and more hesitant rattle than goldenbacks; all phrases drop noticeably in pitch, reminiscent of Crested Kingfisher. **HH** Singly, in loose pairs and with other species. Forages on lower tree trunks and occasionally on ground. Light forest and scattered trees in open country.

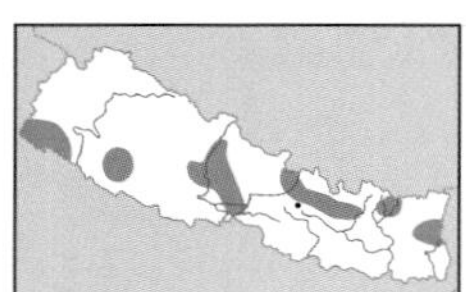

Bay Woodpecker *Blythipicus pyrrhotis* 27 cm

Local and uncommon resident from west-central Nepal east, rare further west; 1,525–2,500 m (–75 m). **ID** From Rufous by long yellowish-white bill. Also larger, with more angular head, more broadly barred and brighter rufous upperparts, diffuse streaking on forehead and crown, darker brown underparts, and largely unbarred tail. Male has prominent scarlet patch on sides of neck, extending to nape. Female lacks scarlet patch. Juvenile has more prominent barring on mantle, diffuse rufous and dark brown barring on underparts, and more prominent pale streaking on head. **Voice** Loud descending laughter, *keek, keek-keek-keek-keek-kerere-kerere*; also a loud chattering *kerere-kerere-kerere*. **HH** Forages mostly within a few metres of ground on moss-covered trunks, dead stumps and fallen logs; also on ground among roots; sometimes higher up, keeping close to or on trunk. Shy and elusive, heard much more often than seen. Eats ants and beetle larvae. Dense broadleaved forests.

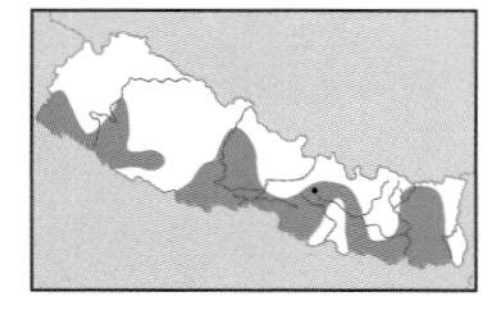

Rufous Woodpecker *Micropternus brachyurus* 25 cm

Locally frequent resident; 75–305 m (–2,300 m). **ID** Medium-sized, rufous-brown woodpecker with shaggy crest and short black bill. Heavily barred black on mantle, wings, flanks and tail. Male has small scarlet flash on ear-coverts. Female has pale buff ear-coverts. **Voice** A high-pitched, nasal *keenk keenk kenk*. Diagnostic drumming like stalling engine, *bdddd-d-d-d-dt*. **HH** Shy, forages in trees at all heights; often seen digging into tree-ant nests; also feeds on fallen logs, termite nests and cow dung. Mainly broadleaved forest and second growth. **TN** Formerly placed in genus *Celeus*.

Himalayan Goldenback

Lesser Goldenback

Greater Goldenback

White-naped Woodpecker

Bay Woodpecker

Rufous Woodpecker

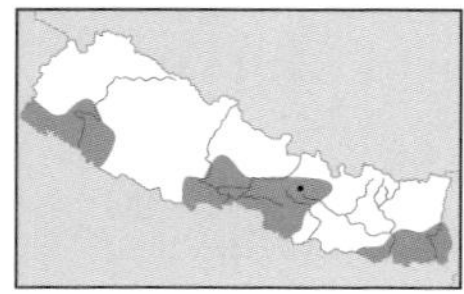

Bar-winged Flycatcher-shrike *Hemipus picatus* 15 cm

Common and widespread resident; 75–1,830 m. **ID** Dark cap contrasts with white sides of throat; has white patch on wing and white rump. Male has blackish cap and brown mantle. Female has brown cap and brown mantle. **Voice** Continuous *tsit-ti-ti-ti-ti-ti* or *whiriri-whiriri-whiriri*; high-pitched trilling *sisisisisisi* and insistent tit-like *chip*. **HH** In forest canopy, usually in pairs or small parties; often joins itinerant foraging flocks of mixed insectivores. Hunts insects among foliage, and by making frequent aerial sallies like a flycatcher. Broadleaved forest and forest edges.

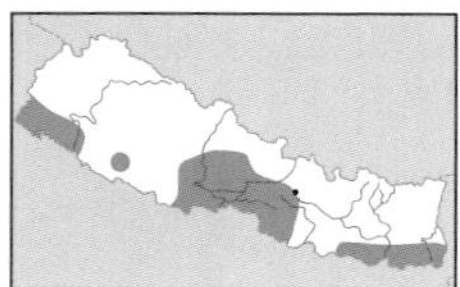

Large Woodshrike *Tephrodornis virgatus* 23 cm

Resident, locally frequent 75–365 m, uncommon to 1,450 m. **ID** From Common Woodshrike by larger size and bill, lack of pale supercilium, striking white lower back and rump, and uniform grey-brown tail. Female has poorly defined brown mask compared to male, with paler bill, darker eye, and brown crown and nape concolorous with mantle. Juvenile is scaled buff and brown on upperparts, and tertials and tail feathers are diffusely barred, with buff fringes and dark subterminal crescents. **Voice** A musical *kew-kew-kew-kew*; also harsh shrike-like calls. **HH** Quiet and unobtrusive. Seeks insects in foliage, on trunks and branches, usually high in trees. Usually in pairs or small parties, often joins other insectivores. Broadleaved forest and well-wooded areas, prefers moister habitats than Common. **TN** Formerly *T. gularis*.

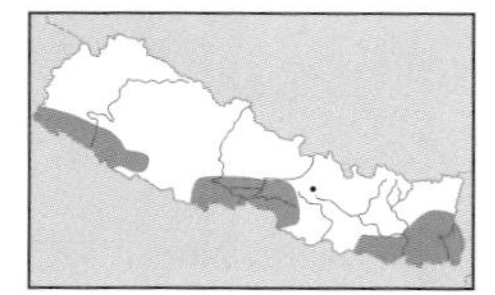

Common Woodshrike *Tephrodornis pondicerianus* 18 cm

Locally fairly common and widespread resident; 75–455 m. **ID** From Large by smaller size, white supercilium above dark mask, and dark brown tail with white sides. Iris brown. Sexes similar. Juvenile has buffish-white supercilium, whitish-spotted crown and mantle, pale and dark fringes to tertials and tail feathers, and indistinct brown streaking on breast. **Voice** A plaintive whistling *weet-weet*, followed by a quick interrogative *whi-whi-whi-wheee* and an accelerating trill, *pi-pi-i-i-i-i-i*. **HH** Dry country, where it replaces Large Woodshrike. Habits similar to Large. Open broadleaved forest, second growth and well-wooded areas.

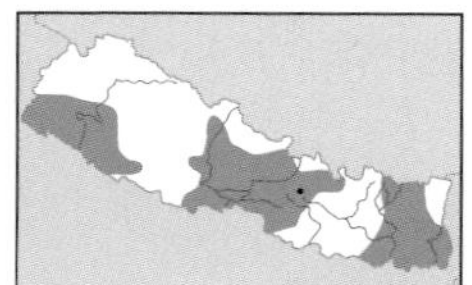

Large Cuckooshrike *Coracina macei* 30 cm

Common and widespread resident; breeds 75–2,135 m (–2,440 m), winters 75–1,525 m at least. **ID** Male has grey upperparts and underparts (latter lack barring) with blackish mask. Female has less pronounced dark mask and barring on whitish belly and flanks. Both sexes have broad whitish fringes to wing feathers. Juvenile heavily scaled brown and white. **Voice** Song a rich fluty *pi-io-io*; also gives a loud wheezy *jee-eet*. **HH** Singly, in scattered pairs or loose flocks. Noisy. Usually perches in topmost branches of trees, although sometimes descends to bushes and ground to feed. Has distinctive habit of flicking each wing slightly one after the other, when it alights. Open woodland, groves and trees in cultivation.

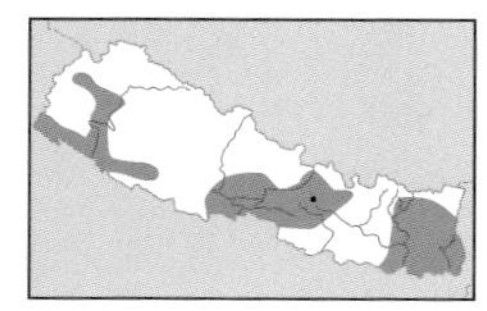

Black-winged Cuckooshrike *Coracina melaschistos* 24 cm

Frequent, widespread resident; 75–915 m all year, to 2,400 m in summer. **ID** Male has dark slate-grey head and body, black wings, fine white tips to undertail-coverts, and bold white tips to long tail. Female similar, but paler grey, with wings less contrastingly black, and faint barring on belly and vent. Juvenile has head and body boldly scaled white, with white tips to wing-coverts and tertials. **Voice** A descending monotonous *pity-to-be*. **HH** Generally singly or in pairs, often with minivets, drongos and other insectivores. Gleans invertebrates from foliage, usually high in trees, sometimes in undergrowth. Has undulating flight. Open forest, forest edges and groves.

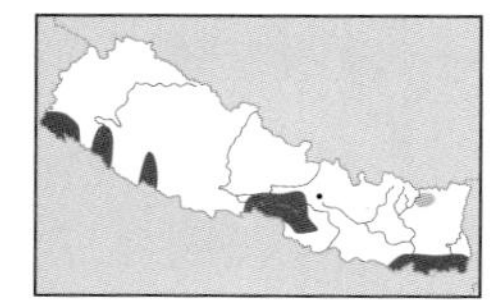

Black-headed Cuckooshrike *Coracina melanoptera* 18 cm

Very rare visitor, mainly in spring and summer; 75–275 m (–1,430 m). **ID** Male has dark slate-grey head, neck and upper breast, contrasting with pale grey mantle and rest of underparts; wings darker grey than mantle, with broad pale fringes to coverts and tertials. Female from female Black-winged by prominent supercilium, stronger dark grey and white barring on underparts, pale grey back and rump contrasting noticeably with shorter and squarer blackish tail, and broader white fringes to coverts and tertials. Juvenile has upperparts barred white. **Voice** Clear, mellow whistling notes, followed by a quick repeated *pit-pit-pit*. **HH** Habits similar to Black-winged. Open broadleaved forest, groves and second growth.

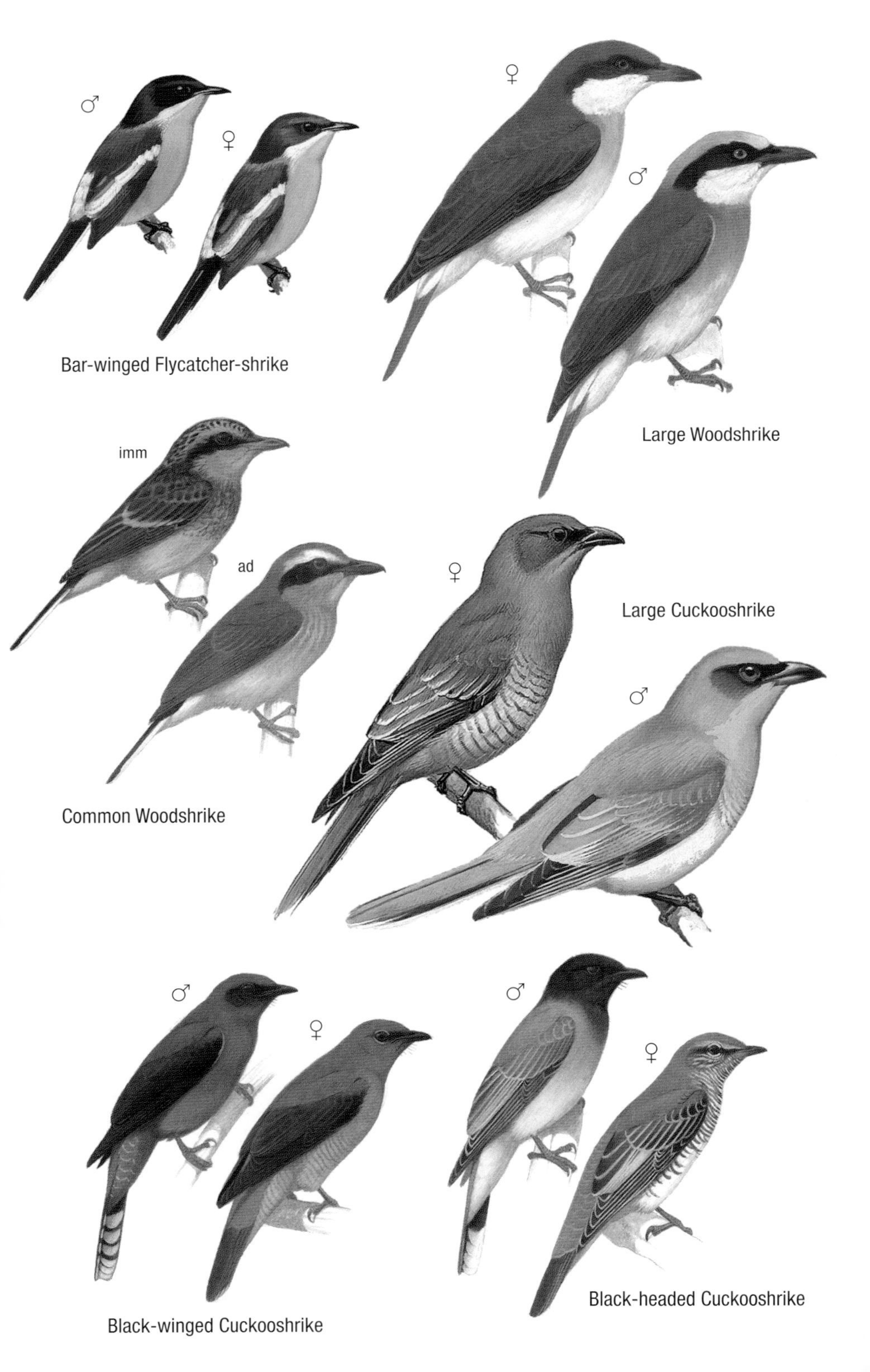
♂
♀
Bar-winged Flycatcher-shrike
♀
♂
Large Woodshrike
imm
ad
Common Woodshrike
♀
Large Cuckooshrike
♂
♂
♀
Black-winged Cuckooshrike
♂
♀
Black-headed Cuckooshrike

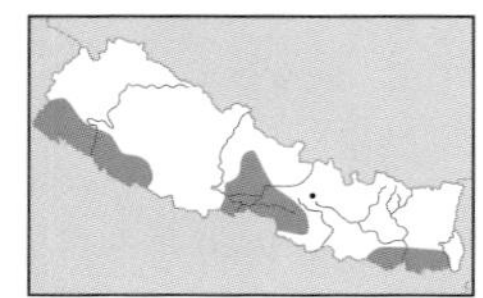

Rosy Minivet *Pericrocotus roseus* 20 cm

Local, uncommon resident; 75–245 m (–1,370 m). **ID** Male has grey-brown upperparts, white throat, pinkish-red edges to tertials, and pinkish underparts and rump. Female from other female minivets by combination of grey to grey-brown upperparts, with indistinct greyish-white forehead and supercilium, dull and indistinct olive-yellow rump, whitish throat, and pale yellow underparts. Juvenile has upperparts scaled yellow. **Voice** Whirring trill. **HH** Upright stance when perched. Arboreal and forages for insects in canopy and midstorey by flitting through foliage, sometimes hovering in front of a sprig or by making short sallies. Usually in pairs in breeding season, in small parties at other times. Similar habits to other minivets but rather less active. Broadleaved forests.

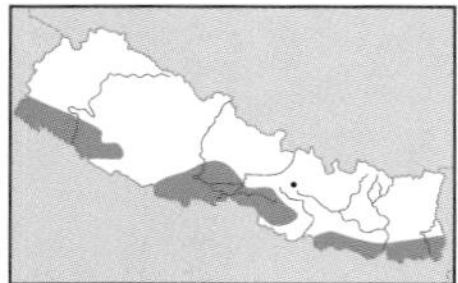

Small Minivet *Pericrocotus cinnamomeus* 16 cm

Widespread resident, common in far west, fairly common in central Nepal and rare in far east. **ID** Small size. Male has grey upperparts, dark grey throat, orange wing panel and orange underparts. Female has pale throat and pale underparts with orange wash; wing panel orange. Juvenile has upperparts scaled buff. **Voice** Continuous high-pitched *swee-swee* etc. **HH** Habits like Rosy, but more active and forages in canopy. More open wooded areas than those preferred by other minivets.

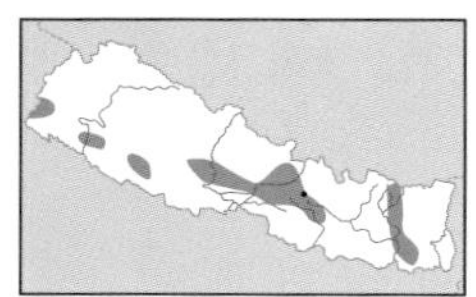

Grey-chinned Minivet *Pericrocotus solaris* 18 cm

Rare resident from west-central Nepal east; 250–2,075 m. **ID** Male has grey chin and pale orange throat, grey ear-coverts, slate-grey upperparts, orange-red underparts and rump, and orange-red wing patch and outer tail feathers. Female has grey forehead and supercilium, grey ear-coverts, whitish chin and sides to yellow throat. Juvenile has pale yellow fringes to feathers of upperparts. **Voice** Distinctive rasping *tsee-sip*. **HH** Habits like Rosy. Moist deciduous and evergreen broadleaved forests.

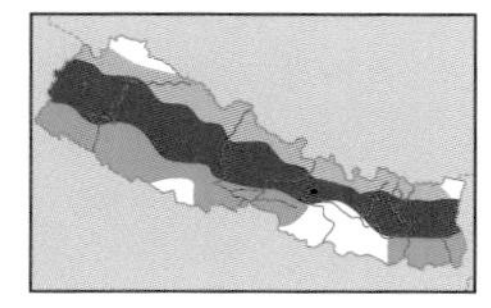

Long-tailed Minivet *Pericrocotus ethologus* 20 cm

Common and widespread resident; breeds 1,200–3,500 m (–3,965 m), winters 245–2,135 m. **ID** From very similar Short-billed by different shape of red wing patch (extending as narrow panel on tertials and secondaries). Also underparts are more scarlet-red, and black throat is less extensive and duller. Female best identified by narrow yellow forehead and supercilium. Ear-coverts more uniformly grey than female Short-billed Minivet, and has distinctly paler yellow throat (than breast); where ranges overlap brighter rump and uppertail-coverts than that species. **Voice** Distinctive, sweet double whistle *pi-ru*, the second note lower. **HH** Habits like Rosy, though highly gregarious except when nesting and forages chiefly in treetops. Broadleaved and coniferous forests; also well-wooded areas in winter.

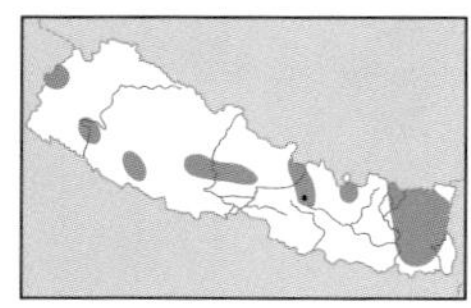

Short-billed Minivet *Pericrocotus brevirostris* 19 cm

Rare resident from mid-west Nepal east; 250–2,745 m. **ID** Male lacks extension to red wing patch on secondaries. Female has yellow forehead and yellow cast to ear-coverts, and deep yellow throat concolorous with rest of underparts. See Long-tailed Minivet for further differences from that species. **Voice** Distinctive loud, high-pitched, monotone whistle, often coupled with dry *tup* contact notes. **HH** Habits like Rosy. Broadleaved forest and forest edges.

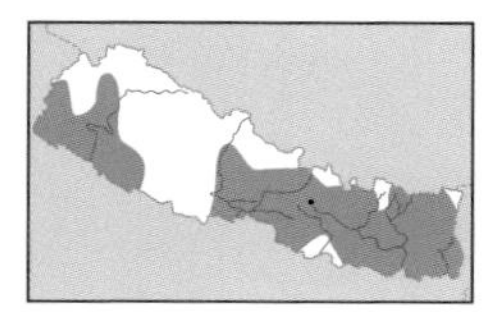

Scarlet Minivet *Pericrocotus flammeus* 22–23 cm

Common and widespread resident; 75–2,200 m. **ID** From Long-tailed and Short-billed by large size and isolated patch of colour on secondaries, red in male and yellow in female. Male is more orange-red than Long-tailed and Short-billed. Head pattern of female closest to female Short-billed, with extensive area of yellow on forehead and ear-coverts. **Voice** Piercing, loud *twee-twee-tweetywee-tweetyweetywee*. **HH** Habits like Rosy, but highly gregarious in non-breeding season. Usually broadleaved forest. **TN** Extralimital Orange Minivet *P. flammeus* is considered conspecific with Scarlet Minivet *P. speciosus*, together known as Scarlet Minivet *P. flammeus*.

♂
♀
Rosy Minivet
♂
♀
Small Minivet
♂
♀
Grey-chinned Minivet
♂
♀
Long-tailed Minivet
♂
♀
Short-billed Minivet
♂
♀
speciosus
Scarlet Minivet

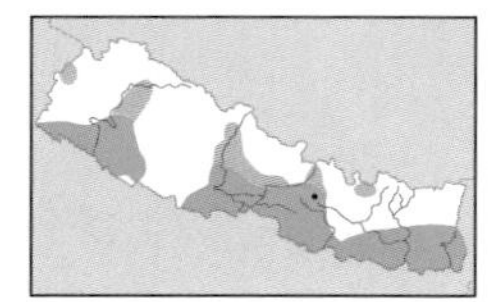

Brown Shrike ***Lanius cristatus*** 18–19 cm

Widespread winter visitor, fairly common in east, uncommon further west; 75–1,525 m (–2,700 m). **ID** Compared to Isabelline has darker and more uniform rufous-brown upperparts, with much less contrasting rufous tail, and warmer rufous flanks. Also thicker bill, more graduated tail and lacks white patch at base of primaries (apparent in male Isabelline). Female has paler lores and faint dark scaling on breast and flanks compared to male. Darker mask than female Isabelline with more prominent white supercilium. Juvenile has dark scaling on upperparts and underparts (mantle uniform by first-winter). **Voice** Rich varied chattering song; grating call. **HH** Watches for prey from top of bush, tree, post or telegraph wire. Swoops down to catch prey from ground or in flight. Forest edges and secondary scrub.

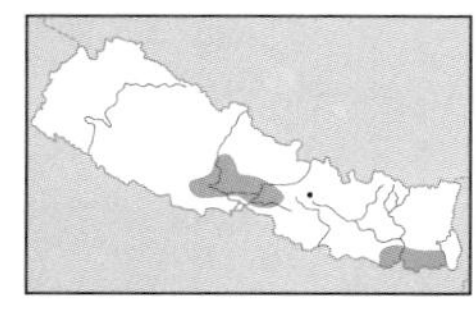

Isabelline Shrike ***Lanius isabellinus*** 17 cm

Very rare winter visitor; 75–1,340 m (–4,050 m). **ID** Pale sandy-brown mantle, contrasting with rufous rump and tail. Supercilium buffish and mask incomplete in front of eye. Female has paler ear-coverts than male, and usually has faint dark scaling on breast and flanks. Juvenile has dark scaling on upperparts and underparts (mantle uniform by first-winter). **Voice** Grating call. **HH** Habits like Brown. Open dry scrub.

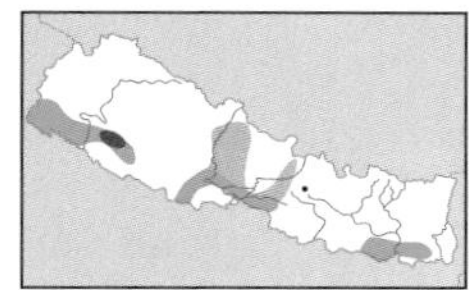

Bay-backed Shrike ***Lanius vittatus*** 17 cm

Mainly uncommon passage migrant, also rare summer visitor and winter visitor; 75–335 m (–3,965 m). **ID** Adult has black forehead, pale grey crown and nape, deep maroon mantle, whitish rump, and white patch at base of primaries. Juvenile from juvenile Long-tailed by smaller size, shorter tail, more uniform greyish/buffish base colour to upperparts, pale rump, more intricately patterned wing-coverts and tertials (buff fringes and dark subterminal crescents and central marks), and primary-coverts prominently tipped buff. First-year like washed-out version of adult; lacks black forehead. **Voice** Pleasant, rambling warbling song with much mimicry; harsh churring call. **HH** Habits like Brown. Open dry bushy areas, and bushes in cultivation.

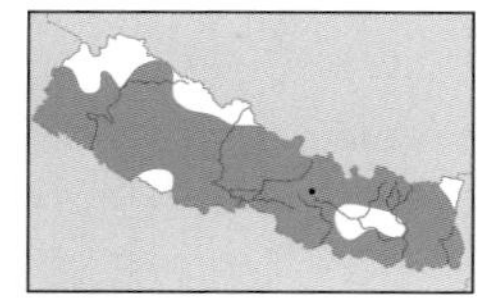

Long-tailed Shrike ***Lanius schach*** 25 cm

Common, widespread resident; breeds 300–3,100 m, chiefly 1,500–2,700 m. **ID** Two intergrading races occur: *L. s. tricolor* chiefly in centre and east; *L. s. erythronotus* in west. Adult *L. s. erythronotus* has grey mantle, rufous scapulars and upper back, narrow black forehead, rufous sides to black tail and small white patch on primaries. Juvenile has (dark-barred) rufous-brown scapulars, back and rump, dark greater coverts and tertials fringed rufous. *L. s. tricolor* has black crown and nape. **Voice** Pleasant, subdued, rambling warbling song with much mimicry; harsh grating call. **HH** Habits like Brown. Bushes in cultivation, open forest and grasslands.

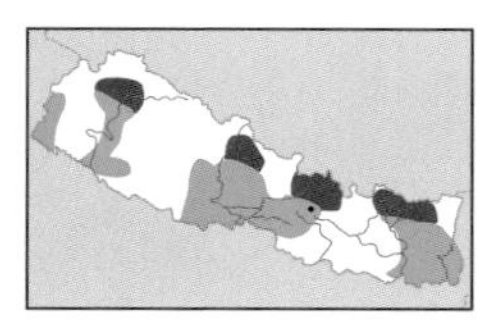

Grey-backed Shrike ***Lanius tephronotus*** 25 cm

Fairly common, widespread resident; breeds 2,240–4,000 m (–4,570 m); winters 275–2,560 m at least. **ID** Adult from Long-tailed by uniform (darker) grey upperparts, with rufous uppertail-coverts (no rufous on scapulars and upper back). Also lacks or has very narrow black forehead, and lacks (or has only very indistinct) white patch at base of primaries. Juvenile has brown ear-coverts, cold grey upperparts (except rufous-brown uppertail-coverts), with indistinct black subterminal crescents and buff fringes to feathers, rufous fringes and indistinct black subterminal borders to coverts and tertials, and dark scaling on breast and flanks. Uniform cold grey base coloration to upperparts is best distinction from juvenile Long-tailed. **Voice** Harsh grating call, mimics other birds. **HH** Habits like Brown. Bushes in cultivation and on hillsides, open scrub and second growth.

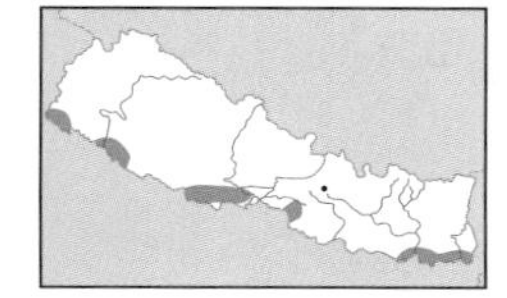

Southern Grey Shrike ***Lanius meridionalis*** 24 cm

Rare, local resident; 75–250 m. **ID** Adult from Grey-backed and Long-tailed by paler grey mantle and white scapulars, bold white markings on black wings and tail, greyish rump, and white breast and flanks. Black of mask extends over forehead and on sides of neck. Extensive white patch at base of primaries and, with inner webs of secondaries and tips of outer webs also largely white, shows much white in wing even at rest. Juvenile has sandy cast to grey crown and mantle, with very indistinct barring on crown, buff tips to tertials and median and greater coverts, and faint buffish wash to underparts; mask grey, and does not extend over forehead as in adult. **Voice** Chattering song with harsh notes; mimics other birds; drawn-out *kwiet* whistle. **HH** Habits like Brown. Dry country, open thorn scrub, cultivation edges.

1st-win
♂
1st-win
♀
♂
Brown Shrike
Isabelline Shrike
ad
juv
juv
erythronotus
imm
Bay-backed Shrike
ad
ad
Long-tailed Shrike
tricolor
juv
ad
ad
juv
Grey-backed Shrike
Southern Grey Shrike

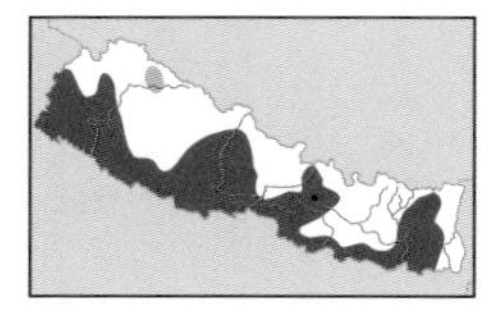

Indian Golden Oriole *Oriolus* (*oriolus*) *kundoo* 25 cm

Widespread summer visitor, fairly common in lowlands, frequent elsewhere; 75–1,830 m (–2,600 m). **ID** Adult male has small black eye-patch, golden-yellow head and body, largely black wings with yellow carpal patch and prominent tips to tertials/secondaries, and yellow-and-black tail. Adult female has yellowish-green upperparts, blackish streaking on whitish underparts (with variable yellow on sides of breast), brownish-olive wings, yellow rump, and brownish-olive tail with yellow corners. Immature like adult female, but bill and eyes dark; initially duller on upperparts and more heavily streaked on underparts; suggestion of dark stripe behind eye recalls Slender-billed. **Voice** Loud, fluty *weela-whee-oh* song; harsh, nasal *kaach* call. **HH** Usually keeps hidden in leafy canopy. Flight powerful and undulating with fast wingbeats. Open deciduous woodland. **TN** Sometimes considered race of Eurasian Golden Oriole *O. oriolus*.

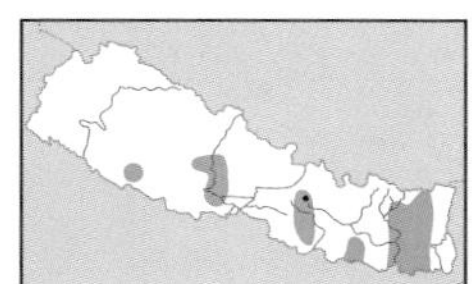

Slender-billed Oriole *Oriolus tenuirostris* 27 cm

Very uncommon winter visitor; 75–2,285 m. **ID** Long, slender, slightly downcurved bill, and black nape band. Male from male Indian Golden by olive-yellow mantle and olive-yellow wing panels. Female similar, but has dull black nape-band, duller yellow head and underparts, and duller olive-yellow upperparts. Immature initially has dark bill and eye, more uniformly olive wings, and whitish underparts with black streaking; nape-band and eye-stripe can be very diffuse. Diffuse nape-band if apparent, and longer bill, are best features from immature Indian Golden. See Appendix 1 for differences from Black-naped. **Voice** Mellow, fluty song, *wheeow* or *chuck, tarry-you*; diagnostic, high-pitched woodpecker-like *kick* call. **HH** Habits like Indian Golden. Well-wooded areas.

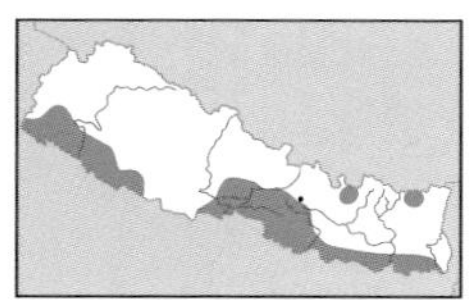

Black-hooded Oriole *Oriolus xanthornus* 25 cm

Common, widespread resident; mainly 75–365 m, uncommon to 1,370 m. **ID** Adult male has black head contrasting with golden-yellow body, yellow edges to black tertials and secondaries, and mostly yellow tail. Adult female similar, but has olive-yellow mantle. Immature has dark bill, yellow forehead, yellow-streaked head, black-streaked white throat, diffuse black streaking on yellow breast, and duller wings with narrow yellowish edges to flight feathers. **Voice** Song a mixture of mellow, fluty and harsh notes; harsh, nasal *kwaak* call. **HH** Habits like Indian Golden, but not shy. Frequently seen flying from tree to tree, sometimes with itinerant bands of insectivores. Open broadleaved forest and well-wooded areas.

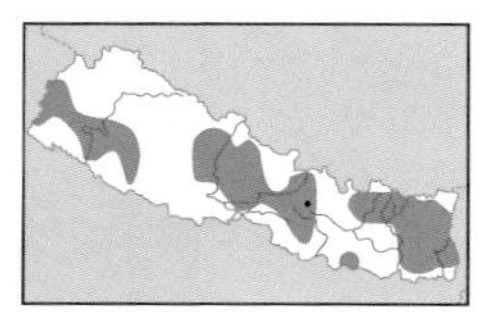

Maroon Oriole *Oriolus traillii* 27 cm

Locally fairly common and widespread resident; breeds mainly 1,500–2,400 m, winters 1,200–1,800 m (–75 m). **ID** Maroon rump, vent and tail in all plumages. Adult has blue-grey bill and pale yellow iris. Male has black head, breast and wings contrasting with glossy maroon body. Female similar but duller maroon mantle, whitish belly and flanks with diffuse maroon-grey streaking. Immature has uniform brown upperparts, including wings, whitish underparts streaked dark brown, brown iris. Juvenile has orange-buff fringes to upperparts and tips to coverts. **Voice** Rich fluty *pi-io-io* song; nasal squawking call. **HH** Habits like Indian Golden. Often with mixed-species foraging flocks. Broadleaved forest.

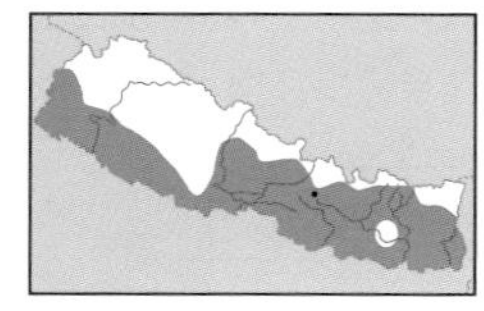

Black Drongo *Dicrurus macrocercus* 28 cm

Common, widespread resident; 75–1,525 m, sometimes to 2,000 m. **ID** From Ashy Drongo, which can appear dark and glossy, by blacker upperparts and shiny blue-black throat and breast, merging into black of rest of underparts; also usually shows white rictal spot and eye is duller. First-winter has whitish fringes to black underparts. Juvenile uniform dark brown. **Voice** Harsh *ti-tiu* and *cheece-cheece-chichuk*; pairs duet during breeding season; a good mimic. **HH** Often associates with grazing cattle. Crepuscular. Makes frequent sallies to seize insects in mid-air or on ground from vantage point. More open country than other drongos: cultivation, around villages and suburbs of towns and cities.

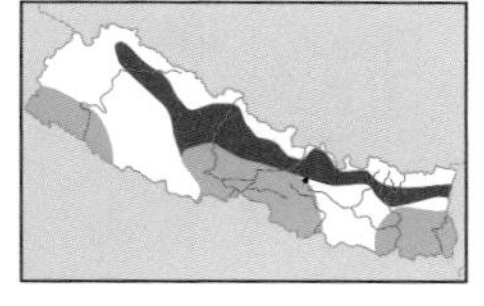

Ashy Drongo *Dicrurus leucophaeus* 29 cm

Common, widespread resident and partial migrant; breeds 1,220–2,745 m, winters 1,065–1,525 m; also locally common resident in lowlands. **ID** Adult has dark grey underparts and slate-grey upperparts with blue-grey gloss; iris bright red. First-winter has brownish-grey underparts with indistinct pale fringes (underparts blacker and whitish fringes more distinct in Black). Juvenile as juvenile Black. **Voice** Like Black but more varied and a good mimic, includes a whistling *kil-ki-kil*. **HH** Usually uses bare branches near treetop as vantage point. Very agile in pursuit of insects like other drongos. Breeds in forest; winters in well-wooded areas.

♂
♀
imm
Indian Golden Oriole
♂
♀
Slender-billed Oriole
imm
♂
♀
imm
Black-hooded Oriole
imm
♂
♀
Maroon Oriole
Black Drongo
ad
imm
imm
ad
Ashy Drongo

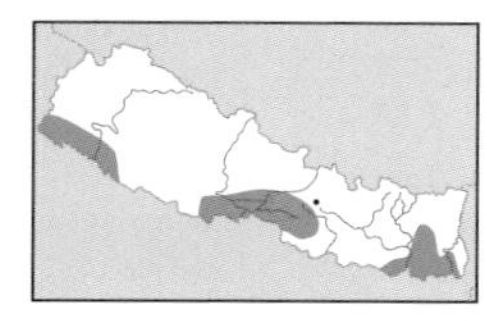

White-bellied Drongo *Dicrurus caerulescens* 24 cm

Locally fairly common and quite widespread resident; 75–305 m. **ID** White belly and vent. Smaller than Black and Ashy, and tail shorter and fork typically shallower. Upperparts glossy slate-grey, much as Ashy (and thus less black than Black). Throat and breast browner in first-winter compared to adult and border between breast and white belly less clearly defined. **Voice** Typical drongo song, with melodious and jarring notes, frequent changes of tempo and much mimicry. **HH** Typical drongo. Hawks insects from exposed perches. Swift, strong, undulating flight. Bold and pugnacious. Highly crepuscular. Open forest and well-wooded areas.

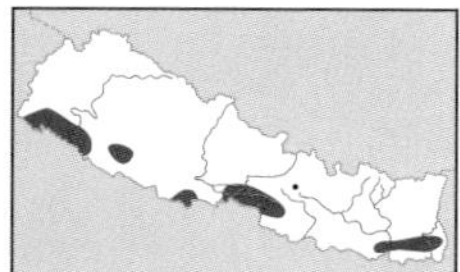

Crow-billed Drongo *Dicrurus annectans* 28 cm

Local and uncommon summer visitor, rare resident; 75–250 m. **ID** From Black by much stouter bill, more extensive area of rictal bristles (resulting in tufted forehead not unlike Lesser Racket-tailed), and shorter, broader tail that is widely splayed at tip but not deeply forked (with outer feathers more noticeably curving outwards). Note habitat differences from Black. First-winter has white spotting from breast to undertail-coverts (recalling immature Greater Racket-tailed). Juvenile has uniform brownish-black upperparts and underparts lack gloss. **Voice** Loud, musical whistles and churrs; characteristic descending series of harp-like notes. **HH** Usually hunts in mid-storey or lower canopy. Associates in loose flocks on migration. Moist broadleaved forest.

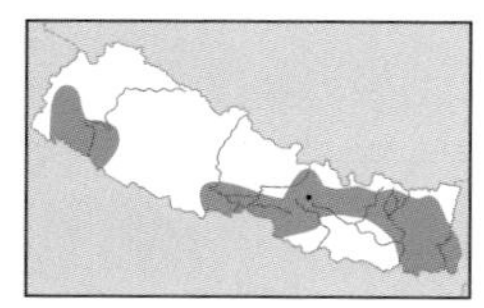

Bronzed Drongo *Dicrurus aeneus* 24 cm

Frequent resident; 75–1,600 m, sometimes to 2,000 m. Small size, with flatter bill compared to Black and less deeply forked tail (can be almost square-ended on moulting or juvenile birds). Note also habitat differences. Adult strongly glossed metallic blue-green. Juvenile has brown underparts, and duller and less heavily spangled upperparts. **Voice** Loud, varied musical whistles and churrs. **HH** Usually singly or in pairs. A regular member of mixed feeding parties of insectivores. Hunts in shady areas of forest in overgrown forest clearings and along paths. Moist broadleaved forest.

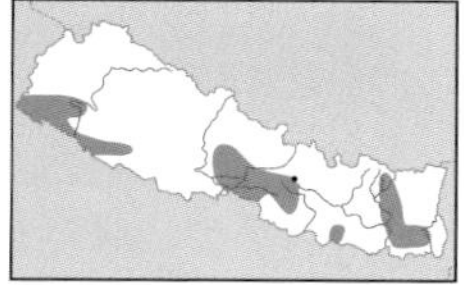

Lesser Racket-tailed Drongo *Dicrurus remifer* 25 cm

Locally frequent resident; mainly 915–1,800 m (150–2,440 m). **ID** Tufted forehead without crest (giving rise to rectangular head shape), square-ended tail, and smaller size and bill than Greater Racket-tailed; has smaller flattened rackets. As in Greater, tail streamers and rackets can be missing or broken in adult, and are missing in immature. **Voice** Loud, varied, musical whistles, screeches and churrs, with much mimicry. **HH** Singly or in pairs, often with roving mixed flocks of other forest species. Normally keeps to leafy canopy in clearings or forest edges and along streams. Dense, moist broadleaved forest.

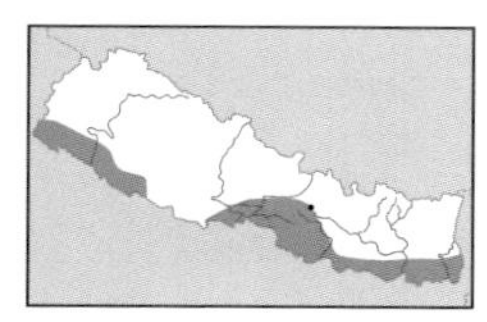

Greater Racket-tailed Drongo *Dicrurus paradiseus* 32 cm

Locally frequent resident; 75–365 m. **ID** Adult from Lesser Racket-tailed (where ranges overlap) by larger size and less tidy appearance, larger bill, crested head, forked tail and longer, twisted tail-rackets. Tail-streamers and rackets can be missing or broken, and tail can appear almost square-ended when in moult (or juvenile plumage). Juvenile initially lacks rackets, is less heavily glossed than adult and has much-reduced crest. White fringes to belly and vent in first-winter plumage. **Voice** Loud, varied musical whistles, screeches and churrs, with much mimicry. **HH** Similar to other drongos, but more sociable, often in small groups and joins mixed flocks of other species. Frequently crepuscular. Forages in lower and mid-storeys of forest. Broadleaved forest.

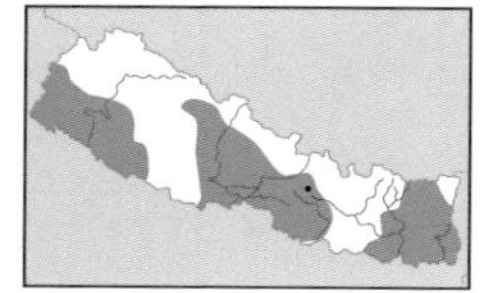

Spangled Drongo *Dicrurus hottentottus* 32 cm

Quite widespread resident, fairly common 75–1,050 m, uncommon to 1,525 m (–4,115 m). **ID** Broad tail with upward-twisted corners, and long downcurved bill. Adult has extensive spangling and hair-like crest. Juvenile browner and lacks spangling; also lacks crest and has square-ended tail with less pronounced upward twist to outer feathers. **Voice** A loud *chi-wiii*, the first note stressed and second rising, or sometimes the *wiii* note is given singly. **HH** Singly or in small parties. Frequently joins flocks of insectivores. Feeds mainly on flower nectar, also insects, located chiefly by searching flowers, leaves and trunks, although also catches them on wing. Moist broadleaved forest. **TN** Extralimital Spangled Drongo *D. bracteatus* is considered conspecific with Hair-crested Drongo *D. hottentottus*, together known as Spangled Drongo.

ad
imm
ad
imm
Crow-billed Drongo
ad
Bronzed Drongo
White-bellied Drongo
ad
imm
Lesser Racket-tailed Drongo
ad
ad
imm
Greater Racket-tailed Drongo
juv
Spangled Drongo

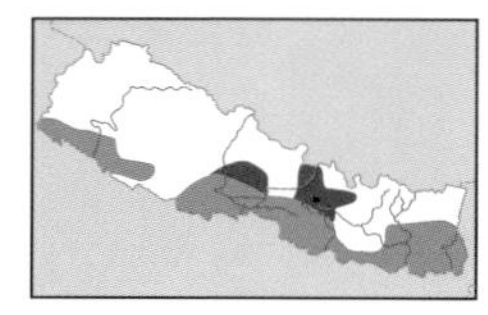

Common Iora *Aegithina tiphia* 14 cm

Widespread, locally common resident; 75–365 m; frequent to 1,900 m and a summer visitor at higher altitudes. **ID** Greenish upperparts, yellow underparts and prominent white wing-bars. Male has blackish wings and tail, and can have some black markings on crown, nape and mantle in breeding plumage. Female has greenish-grey wings and greenish tail. Bill stout and pointed and has pale eye. **Voice** Song a long, drawn-out two-toned whistle; also a piping *tu-tu-tu-tu*. **HH** Singly or in pairs, each bird calling frequently. Arboreal. Forages methodically amongst foliage of trees and bushes. Hops on branches and sometimes hangs upside-down. Open forest and well-wooded areas.

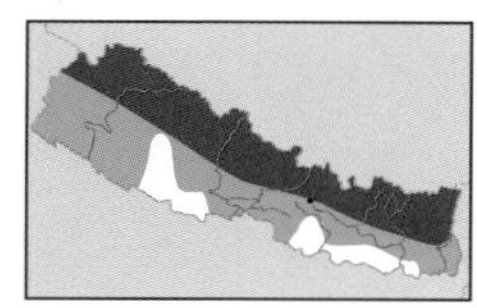

Yellow-bellied Fantail *Chelidorhynx hypoxantha* 13 cm

Locally common resident; breeds 2,440–4,000 m, winters 1,200–1,800 m (150–2,560 m). **ID** From other fantails by yellow forehead and supercilium, dark mask, yellow underparts, greyish-olive upperparts, and blackish tail boldly tipped white. Long, fanned tail is best feature from Black-faced Warbler. Male has black mask, dark olive-brown in female. Juvenile resembles female, but lacks yellow on forehead, has duller and paler supercilium and underparts, and greyer upperparts. **Voice** Constant *sip sip* notes comprise a trilling song. **HH** Very lively and active. Like other fantails, characteristically erects and spreads tail and turns whole body from side to side. Broadleaved and coniferous forests and shrubberies above treeline. **TN** Formerly placed in genus *Rhipidura*.

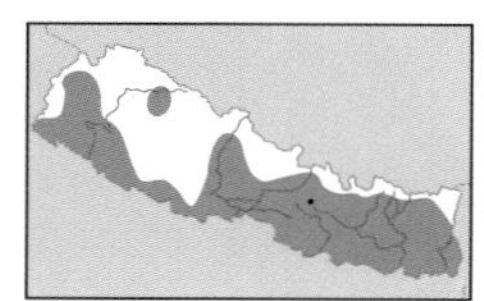

White-throated Fantail *Rhipidura albicollis* 19 cm

Fairly common resident; 75–2,440 m (–3,100 m). **ID** From White-browed Fantail by narrow white supercilium and white throat, lack of spotting on wing-coverts, slate-grey underparts, and smaller white tips to tail. Juvenile browner, with body feathers, wing-coverts and tertials tipped rufous; throat dark. **Voice** Song a descending series of weak whistles, *tri... riri... riri... riri... riri*; squeaky *cheek* call. **HH** Habits like Yellow-bellied but less active. Usually forages in mid-storey and often near main trunk of trees. Breeds in ravines and shady areas in broadleaved forests; also second growth and groves in winter.

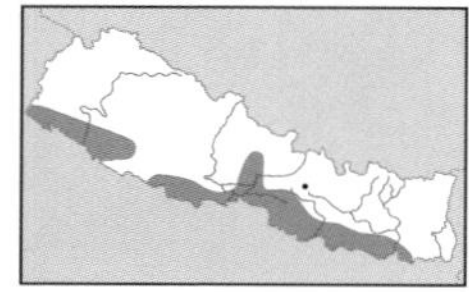

White-browed Fantail *Rhipidura aureola* 18 cm

Rare resident; 75–275 m. **ID** From White-throated by broader white supercilia, which meet on forehead and extend back to reach nape, variable blackish throat and white submoustachial stripe, and white spotting on wing-coverts. Throat spotted with variable amounts of white or grey (can sometimes appear almost white, as in White-throated). Underparts white. Juvenile similar to adult, but has browner upperparts, with rufous tips to body feathers, tertials and coverts. **Voice** Song an ascending, then descending series of clear whistles, *chee-chee-cheweechee-vi*; call a harsh *chuck*. **HH** Like Yellow-bellied, but less active and rather shy, often seeking cover when disturbed. Undergrowth and bushes near ground in more open and drier forests than White-throated.

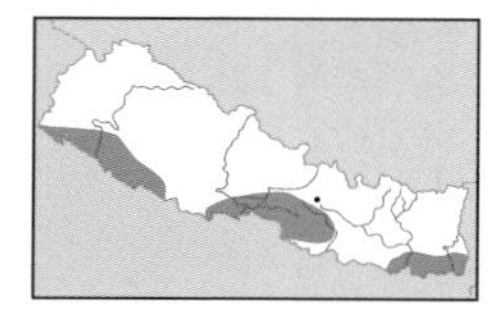

Black-naped Monarch *Hypothymis azurea* 16 cm

Local partial migrant; 100–365 m. **ID** Male almost entirely azure-blue, with black nape patch, black gorget on upper breast, beady black eye, black feathering at bill base. Female has duller blue head, lacks black nape patch or gorget, has blue-grey breast and grey-brown mantle, wings and tail. **Voice** Rasping, high-pitched *sweech-which* and a ringing *pwee-pwee-pwee-pwee*. **HH** Arboreal, often in canopy. Flits actively through foliage and branches. Captures insects during aerial sorties from a perch; also hovers in front of leaves to disturb insects. Broadleaved forests.

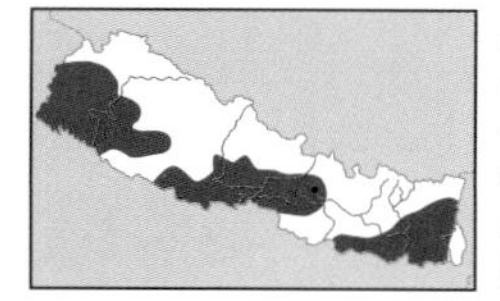

Asian Paradise-flycatcher *Terpsiphone paradisi* 20 cm

Chiefly a summer visitor, locally common; 75–1,525 m. **ID** Male has black head and crest, with white or rufous upperparts and long tail-streamers. Intermediate birds, showing rufous and white in plumage, occur. Female and immature similar to rufous male, but have shorter crest and short square-ended tail; throat and lower ear-coverts greyer. Juvenile similar to female, but has indistinct pale centres and dark fringes to breast feathers. **Voice** Song a slow warble, *peety-to-whit*, repeated quickly; calls and alarm include a nasal *chechwe* and a harsh *wee poor willie weep-poor willie*. **HH** Perches upright, often high in trees, before darting out to catch flying insects. Active and graceful with undulating flight. Open forests, well-wooded areas and groves.

♀
Common Iora
♂
Yellow-bellied
Fantail
ad
White-throated
Fantail
ad
♂
white
White-browed
Fantail
ad
♂
rufous
♂
♀
♀
Black-naped Monarch
Asian Paradise-flycatcher

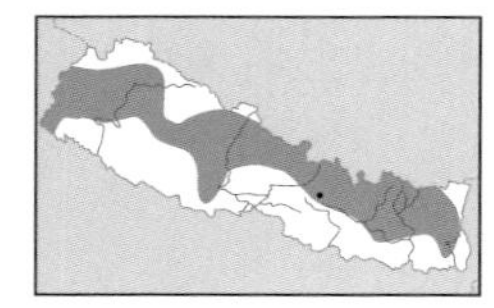

Yellow-billed Blue Magpie *Urocissa flavirostris* 61–66 cm

Common, widespread resident; mainly above 2,440 m all year, summers to 3,660 m, winters to 1,850 m (1,300 m). **ID** From Red-billed Blue by yellow or orange-yellow bill, small white crescent on nape, duller blue-grey mantle and wings. Juvenile similar to adult, but has duller head and upperparts, more extensive white on nape and dull olive-yellow bill. In west *U. f. cucullata* has white underparts and bluer mantle and wings than *U. f. flavirostris* in east, which has greyer mantle and wings and primrose-yellow wash to underparts. **Voice** A wheezy *bu-zeep-peck-peck-peck, pop-upclea, pu-pu-weer* and a high *clear-clear*. **HH** Usually in pairs or flocks of up to ten, which cross clearings in single file. Forages with agility, hopping from branch to branch, also feeds on ground, progressing in long hops. Broadleaved and coniferous forests.

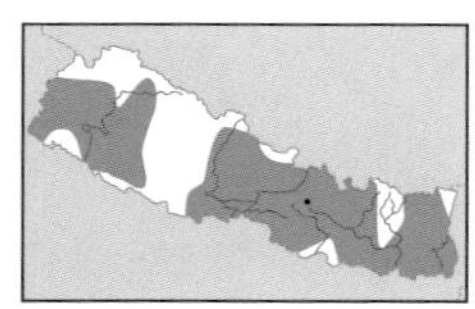

Red-billed Blue Magpie *Urocissa erythrorhyncha* 65–68 cm

Common, widespread resident; mainly 365–1,525 m, occasionally summers to 2,200 m (150–3,050 m). **ID** From Yellow-billed by red or orange-red bill, more extensive white nape (with white speckling on hindcrown) and brighter turquoise-blue mantle and wings. Juvenile has duller blue-grey upperparts, dull brownish-red bill and a more extensive white crown. **Voice** A piercing *quiv-pig-pig*, a softer *beeee-trik*, a subdued *kluk* and a sharp *chwenk-chwenk*. **HH** Similar to Yellow-billed. Broadleaved and mixed forests.

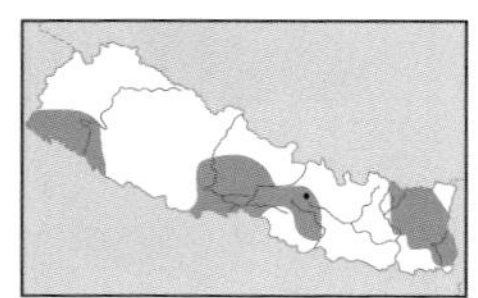

Common Green Magpie *Cissa chinensis* 37–39 cm

Locally fairly common resident; 245–1,830 m, mainly below 1,200 m (–2,300 m). **ID** Mainly lurid green, with red bill and legs, black mask, rufous-chestnut wings, black-and-white-tipped tertials and secondaries, and long, graduated black-and-white-tipped green tail. In captivity green coloration can bleach pale blue and chestnut wings can fade to olive-brown. Juvenile similar to adult, but has dull yellow bill and legs, shorter crest and paler underparts. **Voice** Very variable: harsh *chakakakakakakak* or *chkakak-wi*; high-pitched *wi-chi-chi, jao... wichitchit... wi-chi-chi, jao* with shriller *jao* notes and complex high shrill whistles combined with mimicry. **HH** In pairs or small parties, often with roaming mixed-species flocks. Forages in forest understorey or on ground under thick vegetation. Moist broadleaved forest.

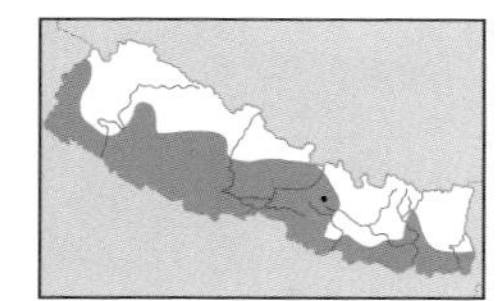

Rufous Treepie *Dendrocitta vagabunda* 46–50 cm

Widespread resident, common 75–1,050 m, uncommon to 1,370 m (–1,800 m). **ID** Adult from Grey by combination of uniform slate-grey hood (to breast), rufous-brown mantle and scapulars, pale grey wing-coverts and tertials contrasting with black of rest of wing, fulvous-buff underparts, and black-tipped silver-grey tail. In flight, pale grey wing panel, whitish subterminal tail-band, and rufous rump are useful features. Juvenile similar to adult but has browner hood (less well demarcated from mantle), buffish wash to wing-coverts, and tail feathers have pale buffish tips. **Voice** Varied harsh, metallic and mewing notes, the most distinctive a loud, flute-like *ko-ki-la*, often mixed with harsh rattles. **HH** Usually in pairs or family parties, sometimes larger groups; a frequent member of mixed-species flocks. Chiefly arboreal, keeping high in trees and climbing trunks and branches with ease. Open wooded country.

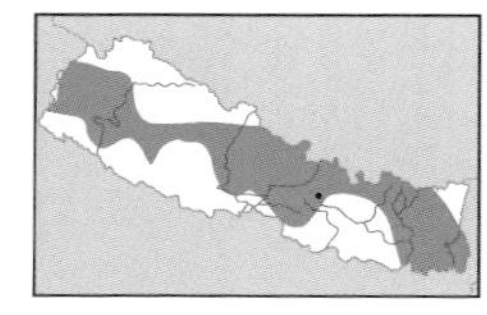

Grey Treepie *Dendrocitta formosae* 36–40 cm

Common widespread resident; breeds chiefly 1,050–2,150 m (–2,590 m), winters 915–1,525 m (–80 m). **ID** Dull-coloured treepie with blackish face contrasting with grey crown, nape and underparts, dull brown mantle, black wings with white patch at base of primaries, grey rump and rufous undertail-coverts. Juvenile similar to adult, but has narrower black forehead, dusky throat concolorous with upper breast, browner crown and nape, whitish belly, and rufous tips to wing-coverts. **Voice** Wide variety of calls, often a loud, metallic, undulating *klok-kli-klok-kli-kli*. **HH** Similar to Rufous, though typically in flocks. Broadleaved forest, second growth and well-wooded country.

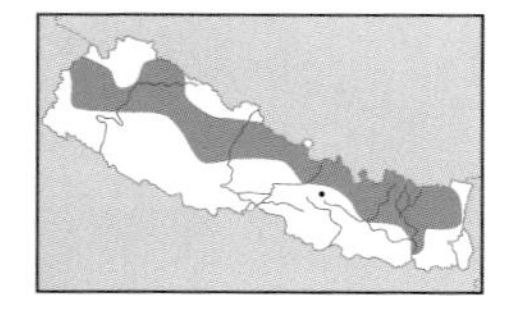

Spotted Nutcracker *Nucifraga caryocatactes* 32–35 cm

Common widespread resident; breeds mainly 2,745–3,660 m (–1,500 m), winters at least 2,135–3,050 m (–305 m). **ID** Largely brown, with boldly white-spotted head and body, white vent, and white sides and tip to tail. Juvenile duller and less cleanly spotted. **Voice** A far-carrying dry and harsh *kraaaak*; a quiet song of various piping, squeaking whistling and whining notes. **HH** Usually in pairs or family parties. Often perches on tops of tall trees, and often flicks tail, revealing white sides. Easily located by its distinctive calls. Coniferous forest.

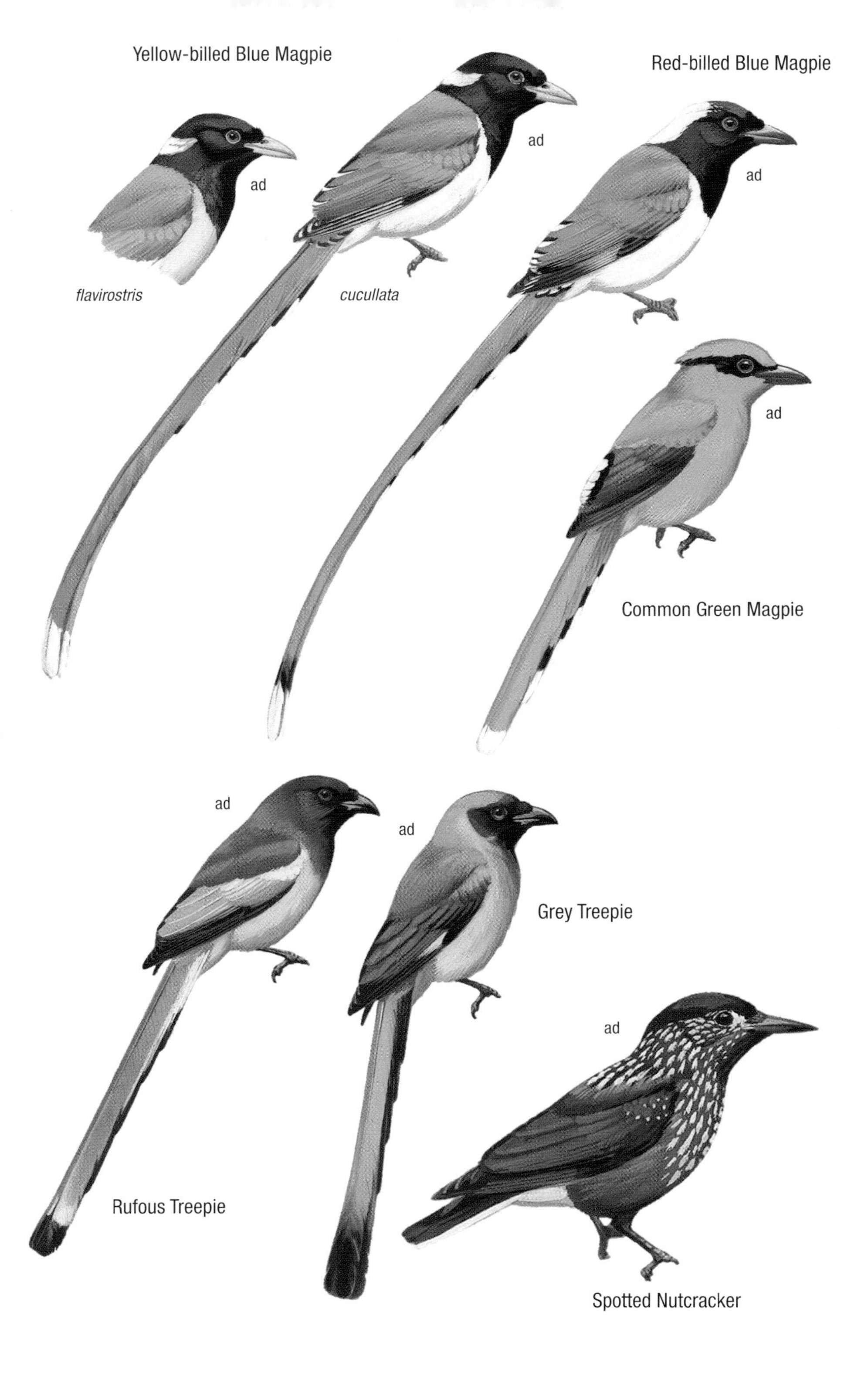
Yellow-billed Blue Magpie
ad
flavirostris
ad
cucullata
Red-billed Blue Magpie
ad
ad
Common Green Magpie
ad
ad
Grey Treepie
Rufous Treepie
ad
Spotted Nutcracker

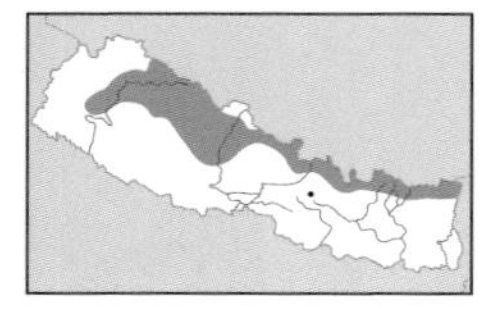

Red-billed Chough *Pyrrhocorax pyrrhocorax* 36–40 cm

Common, widespread resident; mainly above 2,400 m, summers to 5,490 m (–7,950 m), sometimes winters down to 2,135 m (–1,450 m). **ID** Longer, more downcurved, red bill is best feature from Alpine Chough. In flight, has broader wingtips with more pronounced fingers to primaries, and square-ended tail which equals width of wing base. Juvenile lacks metallic gloss of adult, has dark legs, and shorter browner bill (slimmer and more noticeably downcurved than Alpine). **Voice** Distinctive far-carrying, nasal *chaow... chaow.* **HH** Gregarious all year; often in flocks of several hundred in winter. Forages for invertebrates by probing and digging in ground and by turning over stones and dung. High mountains, alpine pastures and cultivation.

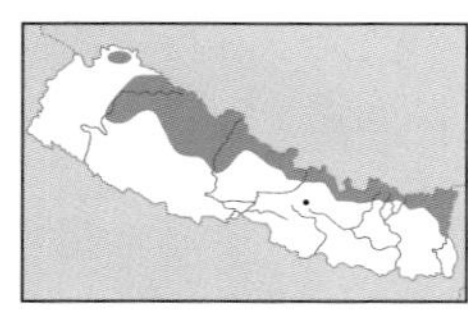

Alpine Chough *Pyrrhocorax graculus* 37–39 cm

Common, widespread resident; chiefly above 3,500 m to at least 6,250 m (2,350–8,235 m). **ID** Shorter and straighter yellow bill is best feature from Red-billed. In flight, narrower wingtips with less pronounced fingered primaries, more pronounced curve to trailing edge of wing, and longer rounded tail. Juvenile lacks metallic gloss of adult, with dark legs and duller olive-yellow bill. **Voice** A far-carrying, sweet, rippling *preeep* and a descending whistled *sweeeoo.* **HH** Similar to Red-billed, but more confiding and will scavenge around settlements. High mountains, alpine pastures and cultivation.

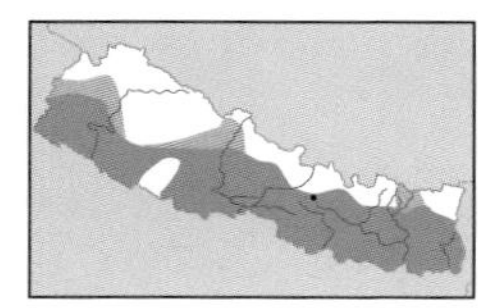

House Crow *Corvus splendens* 40 cm

Abundant widespread resident; 75–1,525 m (–2,100 m). **ID** Two-toned appearance, with paler nape, neck and breast. Adult has gloss to black of plumage, and 'collar' is well defined and paler with wear. Juvenile lacks gloss, and 'collar' is duskier and less well defined. **Voice** Main call a flat, dry *kaaa-kaaa*, weaker than Large-billed. **HH** Gregarious when feeding and roosting. Bold and cunning, omnivorous and an opportunistic feeder. Scavenges at rubbish dumps, in streets and on riverbanks. On ground moves with confident strides and occasional sideways hops. Around habitation and cultivation.

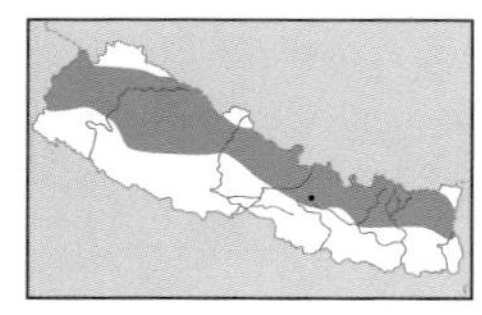

Large-billed Crow *Corvus macrorhynchos* 46–49 cm

Common and widespread resident; breeds 1,920–4,200 m, down to 1,850 m in non-breeding season. **ID** Lacks any contrast between head and neck/breast as in House Crow, and bill is stouter with more pronounced curve to culmen. Bigger with heavier bill, wedge-shaped tail, and harsher calls, compared to jungle crows. **Voice** Call of *intermedius* (W Himalayas) is very guttural *graak, graak*; E Himalayan *tibetosinensis* gives a fairly hoarse *kyarrh, kyarrh.* **HH** Usually singly, in pairs or small groups, but roosts communally in large numbers. Inquisitive, bold and omnivorous. Frequently scavenges and eats carrion. Wide habitat range.

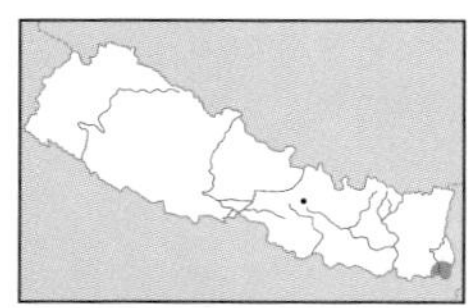

Eastern Jungle Crow *Corvus* (*macrorhynchos*) *levaillantii* 42 cm

Resident in far east; 75-1000 m. **ID** Compared with House Crow is entirely black, with heavier bill. Bill is stouter and with straighter culmen compared with Indian Jungle, but is otherwise identical except for distinctive voice. **Voice** Calls very nasal *nyark, nyark,* first rising then falling. **HH** Wooded habitats. **TN** Distinctiveness of voice needs confirmation from the region where it meets Indian Jungle Crow.

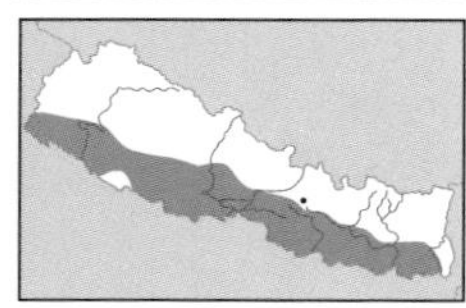

Indian Jungle Crow *Corvus* (*macrorhynchos*) *culminatus* 41 cm

Common and widespread resident; 75–2,000 m. **ID** Compared to House Crow is entirely black, with heavier bill that has more pronounced curve to culmen. Bill finer with more pronounced curve to culmen compared to Eastern Jungle. Note difference in calls. **Voice** Loud, throaty *kyearh, kyearh.* **HH** Wide range of habitats with trees. **TN** Absence of intermediates with Large-billed Crow in foothills needs confirmation.

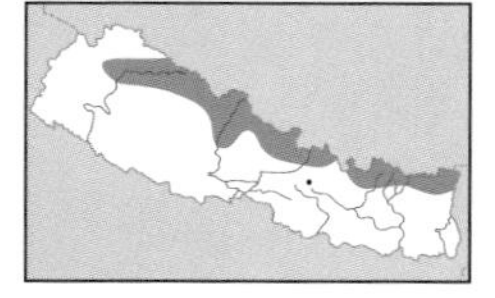

Northern Raven *Corvus corax* 58–69 cm

Fairly common resident; mainly 3,500 m to at least 5,000 m (2,500–8,235 m). **ID** From Large-billed, which can appear raven-like, by longer neck, longer and more angled wings with more pronounced fingered primaries, and more markedly wedge-shaped tail. On ground, further differences are larger size, with more rugged and shaggy look, stouter bill with straighter culmen, prominent throat hackles and flatter crown, more extensive nasal bristles (reaching to centre of bill), and wingtips fall just short of tail (noticeably short on Large-billed). **Voice** Typical call a deep, resonant croaking *wock... wock.* **HH** Powerful and agile in flight, and can soar and glide well; performs impressive aerobatics, often tumbling, diving with closed wings and rolling onto back in mid-air. Dry rocky areas above treeline and around habitation.

ad
ad
ad
Red-billed Chough
Alpine Chough
House Crow
ad
ad
Large-billed Crow
Indian Jungle Crow
ad
ad
Eastern Jungle Crow
Northern Raven

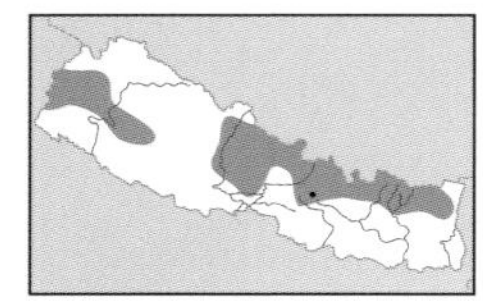

Eurasian Jay *Garrulus glandarius* 32–36 cm

Resident, widespread but local, and locally fairly common; mainly 1,800–2,440 m (900–2,750 m). **ID** Pinkish- to reddish-brown, with beady eye, stout black bill and black moustachial stripe. Wings and tail mainly black, with patches of cobalt-blue barring. In flight, prominent white rump contrasts with black tail. Flight slow and laboured, with rather jerky beats of its rounded wings. In west *G. g. bispecularis* has vinaceous-fawn body; *G. g. interstinctus* in east has darker reddish-brown body. **Voice** Loud, rasping *skaaaak-skaaaak* alarm call; also a crow-like *kraah*. **HH** Singly, in pairs or family parties in breeding season; in autumn and winter often gathers in flocks, often with blue magpies and Black-headed Jays. Moist, mainly broadleaved forest, chiefly oaks.

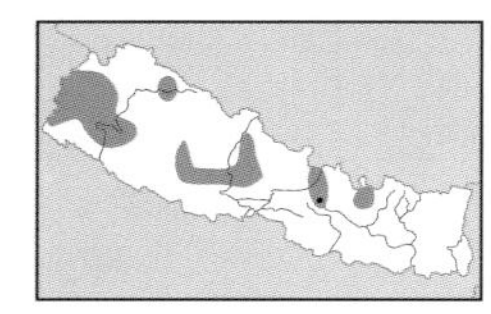

Black-headed Jay *Garrulus lanceolatus* 33 cm

Resident, common in far west, uncommon further east to Gaurishankar Conservation Area, east-central Nepal; 915–2,500 m. **ID** Black (slightly crested) head, white-streaked black throat, stout grey bill, pinkish-fawn upperparts and underparts variably washed blue-grey. Wings have patches of blue barring, similar to Eurasian Jay, but also white carpal patch and tips to tertials and flight feathers. In flight, pinkish-fawn rump (white in Eurasian), and longer mainly blue tail with black cross-bars and white tip. Juvenile similar, but duller and browner. **Voice** Alarm call similar to Eurasian Jay, but usually a single, flatter *skaaaak*, also a soft *kraah*. In autumn and winter often in parties with Eurasian Jay or Yellow-billed Blue Magpie. **HH** Mixed oak/coniferous forest.

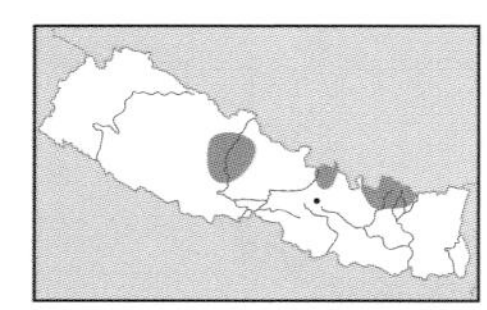

Yellow-rumped Honeyguide *Indicator xanthonotus* 15 cm

Local and very uncommon resident; 1,800–3,300 m (–610 m). **ID** Finch-like with stout bill. Bright orange-yellow forehead, narrow olive-yellow edges to feathers of mantle and wings, golden-yellow lower back and rump, and dark, square-ended tail. Inner edges of tertials white, forming parallel lines on back. Male has broad golden-yellow malar stripes, ill-defined on smaller, duller female. **HH** Closely associated with Giant Rock Bee *Apis dorsata* nests attached to vertical cliffs, and adjacent forests. Feeds on bee-comb wax, bees and other insects. Male defends bees' nests and mates with females that visit the nests to feed.

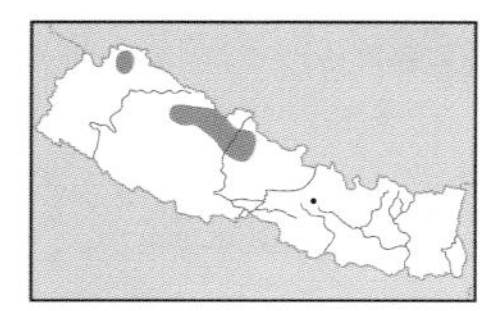

Groundpecker *Pseudopodoces humilis* 19 cm

Local and uncommon resident in Trans-Himalayas in north-west, also extreme north-east Himalayas; 3,965–5,355 m. **ID** Upright stance with downcurved black bill. Sandy-brown and buffish-white. Largely buffish-white tail with brown central feathers, black lores, whitish nape, broad white tips to alula. Juvenile has shorter bill and lacks black lores. **Voice** A munia-like *cheep* and a whistling *chip* followed by a quickly repeated *cheep-cheep-cheep-cheep*. **HH** Terrestrial. Seeks invertebrates using strong bill to peck vigorously and to probe crevices. Very active and progresses by long bounding hops. When perched bobs up and down and flicks tail. Weak, fluttering flight, prefers to escape by hopping away. High mountain steppe with scattered dwarf bushes, often on stony ground. **AN** Ground-tit.

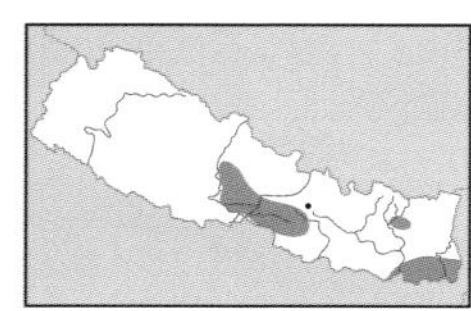

Sultan Tit *Melanochlora sultanea* 20.5 cm

Rare and very local resident, most recent records from Churia Hills, Chitwan; 250–1,370 m (–2,500 m). **ID** A huge, bulbul-like tit. Male largely glossy blue-black, with bright yellow crest and yellow underparts below black breast. Female similar, but black is duller blackish-olive (especially pronounced on throat and breast). Juvenile has shorter crest and fine yellowish-white tips to greater coverts. **Voice** Song a series of five loud, ringing *chew* notes; call a loud, rapidly uttered squeaky whistle, *tcheery-tcheery-tcheery*. **HH** Similar behaviour to other tits, but slower and more deliberate movements. Keeps to treetops alone, in pairs or in small parties. Broadleaved forest, favours evergreens.

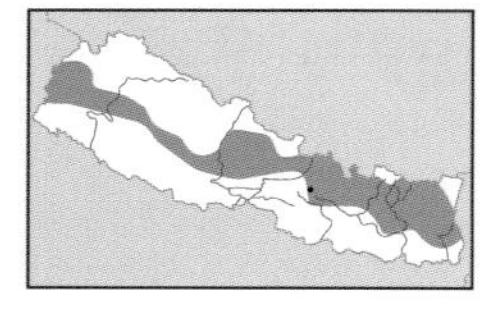

Yellow-browed Tit *Sylviparus modestus* 10 cm

Fairly common resident; breeds 2,135–3,250 m (–3,660 m), winters 1,830–2,800 m (–1,500 m). **ID** Very small, with slight crest and rather stubby bill. Olive-green upperparts, yellowish eye-ring, fine yellow supercilium and yellowish-buff underparts. From Fire-capped Tit by stouter bill, crested appearance, rather uniform wings with less distinct (olive-buff) greater coverts bar, and lack of pale rump or distinct pale tertial fringes. **Voice** Sharp, very high-pitched, well-spaced *tis* or *tis-tis-tis*; also very high-pitched, thin *tis-tis-tis-sisisisisi*. **HH** Quiet and unobtrusive. Forages mainly in canopy, also lower. Frequently hangs upside-down and constantly flicks wings. Often with mixed-species flocks of other insectivores. Mainly mixed broadleaved forests.

ad
ad
Black-headed Jay
bispecularis
Eurasian Jay
♀
♂
ad
Groundpecker
Yellow-rumped Honeyguide
♂
♀
ad
Sultan Tit
Yellow-browed Tit

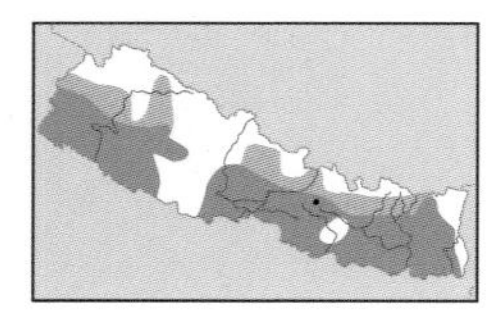

Great Tit *Parus major* 14 cm

Common, widespread resident; 100–1,600 m (–2,440 m). **ID** Black breast centre and line down belly, greyish mantle, greyish-white breast-sides and flanks, and white wing-bar. Juvenile has yellowish-white cheeks and underparts, and yellowish-olive wash to mantle. **Voice** Extremely variable. Song includes loud, clear whistling *weeter-weeter-weeter, wreet-chee-chee*; calls include *tsee tsee tsee* and harsh churrs. **HH** Sometimes joins mixed roving flocks in non-breeding season. Forages in middle and especially lower level of forests and well-wooded country. **TN** Nepal taxon often treated as part of separate species, Cinereous Tit *P. cinereus*, but this is not followed here because the situation where the two potential species meet is still not well known.

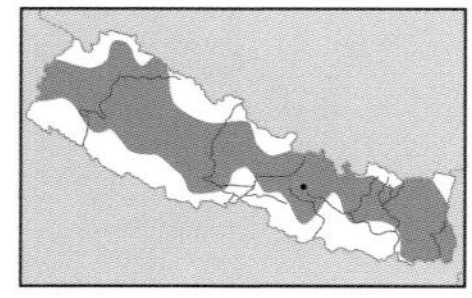

Green-backed Tit *Parus monticolus* 12.5 cm

Common, widespread resident; breeds 1,300–3,660 m, winters 1,300–2,745 m (–75 m). **ID** From Great Tit by bright green mantle and back, yellow on breast-sides and flanks, and double white wing-bars. Wings look bluish owing to blue edges to remiges. Female has duller black throat and narrower stripe on centre of belly. Juvenile duller than adult, and white cheeks and wing-bars washed with yellow. **Voice** Song includes loud, pleasant, ringing *whitee... whitee*; calls resemble Great Tit. **HH** Forages chiefly in lower, also mid-storey; occasionally on ground. Highly acrobatic, often hanging upside-down from twigs. Forests and well-wooded country.

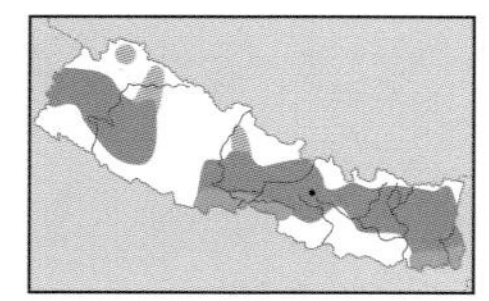

Black-lored Tit *Parus xanthogenys* 13 cm

Fairly common, widespread resident; breeds 850–2,300 m (–2,925 m), winters 915–2,135 m (–75 m). **ID** Where ranges overlap in east, best separated from Yellow-cheeked by black forehead and lores (yellow in Yellow-cheeked), black border to yellow cheeks, uniform greenish upperparts with black streaking confined to scapulars, and yellowish wing-bars. Sexes similar and juvenile only slightly duller with shorter crest. **Voice** Song includes *pui-pui-tee, pui-pui-tee* and *tsi-teuw, tsi-teuw*; calls include *tzee-tzee-wheep-wheep-wheep*, and rattled *chi-chi-chi-chi*. **HH** Habits resemble Green-backed, but mainly in upper storey, occasionally lower. Often joins mixed foraging flocks. Open forests, forest edges and clearings.

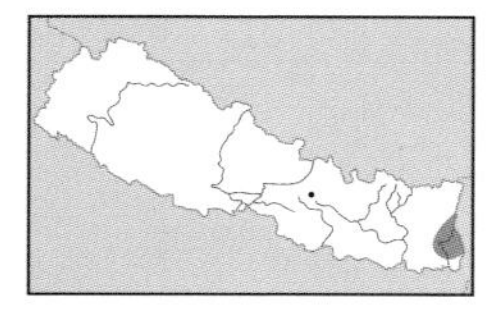

Yellow-cheeked Tit *Parus spilonotus* 14 cm

Probably former resident, previously recorded in far east; 1,980–2,400 m (–450 m). **ID** From Black-lored by yellow forehead and lores, absence of black border to yellow cheeks, black streaking on greenish mantle, and white wing-bars. Juvenile similar to adult, but duller, with shorter sooty-black crest, and has yellowish-white wing-bars. Sexes similar. **Voice** Song comprises three ringing, rapidly repeated notes, *chee-chee-pui*; calls rather like Great Tit. **HH** Typical tit, see Green-backed. Frequently in mixed-species foraging flocks. Chiefly in lower forest canopy, also small trees and bushes. Open broadleaved forest.

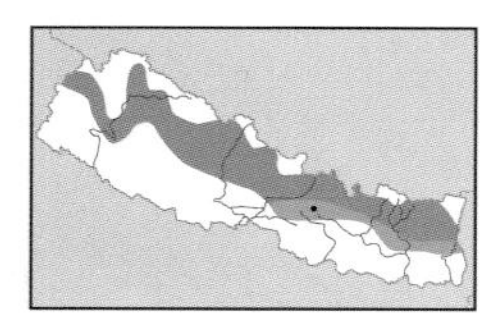

Rufous-vented Tit *Periparus rubidiventris* 12 cm

Common, widespread resident; breeds 2,550–4,000 m (–4,575 m), winters 2,400–4,270 m (–2,135 m). **ID** Where ranges overlap in west, separated from Rufous-naped Tit by smaller size, rufous belly concolorous with vent, and less extensive black bib (not reaching to belly). From 'Spot-winged' Tit (see Coal Tit) by absence of broad white tips to median and greater coverts. In east *P. r. beavani* has greyish belly. Juvenile has shorter crest and duller cap, yellowish cheeks, and bib is less clearly defined. **Voice** Wide variety of calls including *seet, piu* and *chit*; variable rattling song, e.g., *chi-chi-chi-chi*. **HH** Typical tit, see Green-backed, but feeds very actively in treetops, also on ground and in bushes. Often in mixed-species flocks. Coniferous, mixed coniferous/broadleaved and broadleaved forests. **TN** This and Rufous-naped Tit formerly placed in genus *Parus*.

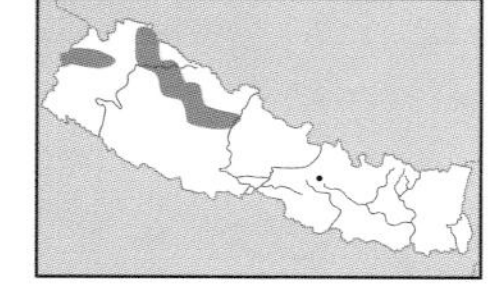

Rufous-naped Tit *Periparus rufonuchalis* 13 cm

Locally fairly common resident in north-west; 2,600–4,000 m. **ID** From Rufous-vented by larger size, more extensive black bib (extending to upper belly), grey (rather than rufous) lower belly. Also variable rufous wash to white nape and rufous patch on sides of breast. Eastern subspecies of Rufous-vented has grey belly, but otherwise above-mentioned features still hold. Juvenile has shorter crest; mantle and belly suffused brown, undertail-coverts buff, and black bib duller and less extensive. **Voice** Monotonous, repeated two-note whistling song, *whi-whee...whi-whee*. Three-noted repeated call: the first two are metallic squeaks, third a louder whistle, *tsi-tsi-peeduw*. **HH** Typical tit, see Green-backed. In canopy, bushes and on ground. Often in mixed-species flocks. Less active than other black tits. Coniferous forests.

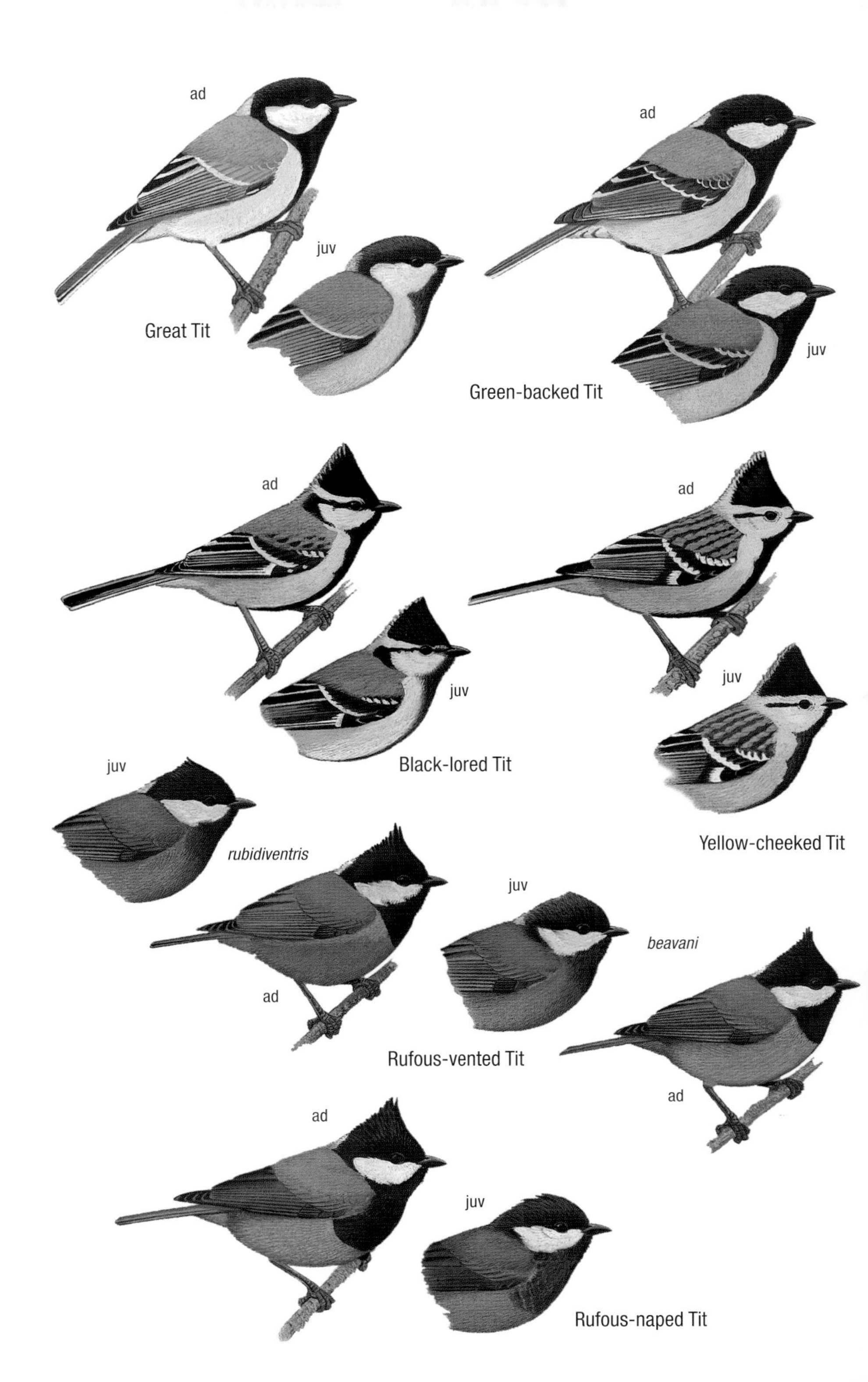
ad
juv
Great Tit
ad
juv
Green-backed Tit
ad
juv
Black-lored Tit
ad
juv
Yellow-cheeked Tit
juv
rubidiventris
ad
juv
beavani
ad
Rufous-vented Tit
ad
juv
Rufous-naped Tit

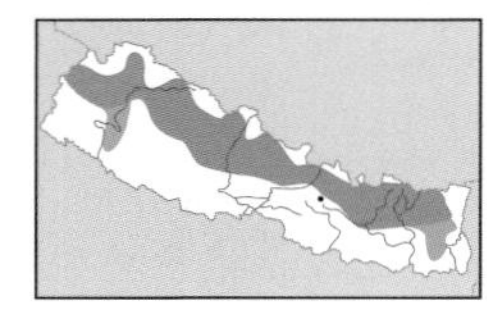

Coal Tit *Periparus ater* 11 cm

Resident, common in west, fairly common in centre and east; breeds 2,450–4,000 m (–4,250 m), winters 2,135–3,050 m. **ID** Black crest, white nape and cheeks, black throat and breast, and whitish tips to wing-coverts. Small size and whitish wing spots separate it from other black-crested tits. Juvenile lacks crest, has poorly defined bib, stronger olive cast to mantle, and yellowish wash to cheeks and breast. Eastern 'Spot-winged' Tit *P. a. melanolophus* differs from nominate race by having rufous breast-sides and flanks, dark grey belly and darker blue-grey mantle. Hybridises frequently with nominate in west-central Nepal producing a variety of intergrades. **Voice** Song a low-pitched, slow *wee-tsee... wee-tsee... wee-tsee*; calls include thin *tsi tsi, pip, pip-sziu* and plaintive *tsi-tsu-whichooh*. **HH** Typical tit behaviour. Chiefly in upper half of trees. Coniferous forests. **TN** Formerly placed in genus *Parus*. 'Spot-winged' Tit previously treated as a separate species *P. melanolophus*.

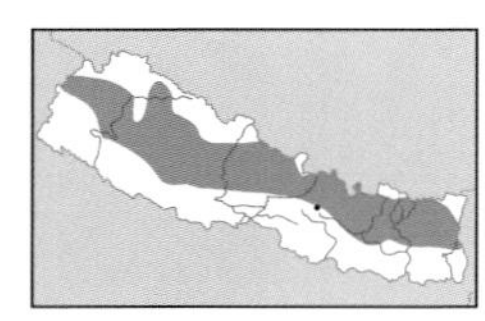

Grey-crested Tit *Lophophanes dichrous* 12 cm

Locally common resident; 2,600–4,000 m (2,450–4,270 m). **ID** Very different from other crested tits. Adult has greyish crest and upperparts, buffish-white half-collar and submoustachial stripe, greyish-buff throat, and orange-buff underparts. Juvenile has shorter crest. **Voice** Song *whee-whee-tz-tz-tz*; alarm call a rapid *cheea, cheea* and *ti-ti-ti-ti-ti*. **HH** Typical tit behaviour, but rather quiet. In middle and lower storeys, also on ground. Broadleaved, broadleaved/coniferous and coniferous forests, both open and closed forests. **TN** Formerly placed in genus *Parus*.

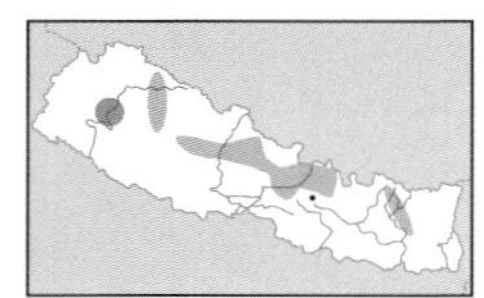

Fire-capped Tit *Cephalopyrus flammiceps* 10 cm

Very uncommon, probably resident, also erratic visitor; 2,135–3,000 m (–1,280 m). **ID** Flowerpecker-like, with greenish upperparts and yellowish to whitish underparts. Lacks crest, has sharply pointed bill. Male breeding has bright orange forecrown, chin and throat, and golden-yellow breast. Female breeding has yellowish forecrown and olive-yellow throat and breast. Adult non-breeding similar to breeding but throat whitish. Juvenile duller with pale grey underparts (no trace of yellow). In east, *C. f. olivaceus* has upperparts darker olive-green than nominate of west and centre; orange restricted to forehead and centre of throat. **Voice** Calls include high-pitched, repeated *tsit, tsit* and soft, weak *whitoo-whitoo*; song a series of high-pitched notes. **HH** Forages in treetops, also bushes. Open broadleaved forest.

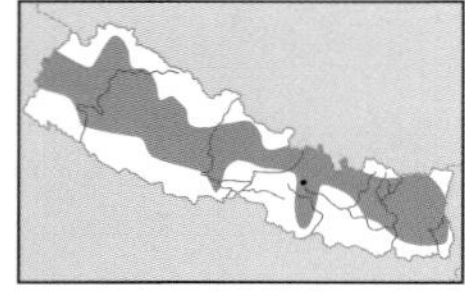

Black-throated Tit *Aegithalos concinnus* 10.5 cm

Common resident; 1,450–2,600 m (1,065–3,050 m). **ID** Adult best identified by combination of rufous crown, white chin and black throat, and grey mantle. Juvenile has white throat and indistinct black-spotted breast-band. Juvenile possibly confusable with White-throated, but lacks white forehead, has narrower and clearly defined black mask, yellow iris, and paler buff breast and belly. **Voice** Calls include a churring *trrrt trrrt* and rapid, low-pitched twittering *tir-ir-ir-ir-ir*; alarm a drier *tzit-tzit-tzit*. **HH** Typical tit. Usually in flocks. Mainly in middle and lower storeys. Open forest, edges and clearings of forest and bushes. **AN** Black-throated Bushtit.

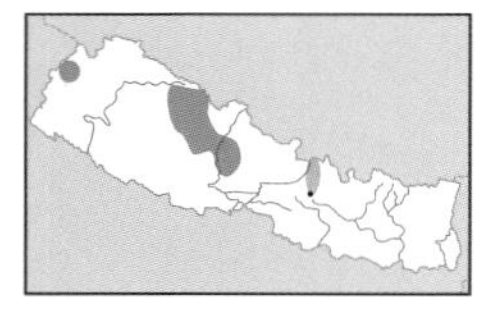

White-throated Tit *Aegithalos niveogularis* 11 cm

Local and fairly common resident in north-west; 2,800–3,965 m (–2,600 m). **ID** Adult best identified by combination of white forehead and forecrown, whitish throat and dark iris. Diffuse blackish mask and cinnamon underparts, with darker breast-band. Juvenile has dusky throat, more prominent breast-band, and paler lower breast and belly. **Voice** Song a rapid, chattering *tweet-tweet* interspersed with high-pitched *tsi-tsi* notes and short warbles; calls include a frequently uttered *t-r-r-r-r-t*. **HH** Typical tit. Bushes in broadleaved and broadleaved/coniferous forests. **AN** White-throated Bushtit.

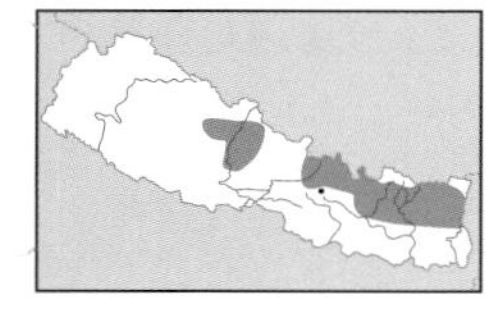

Rufous-fronted Tit *Aegithalos iouschistos* 11 cm

Frequent resident from west-central Nepal east; 2,590–3,400 m (–3,700 m). **ID** Adult from other *Aegithalos* tits by combination of broad black mask, rufous-buff forehead and centre of crown, rufous-buff cheeks and deeper cinnamon-rufous underparts, and silvery-white bib bordered by blackish chin and sides to throat. Iris yellow (brown in White-throated). Juvenile similar to adult, but crown-stripe, cheeks and underparts are paler buff. **Voice** Variety of notes similar to Black-throated, continuously uttered by flocks. **HH** Typical tit, but in non-breeding season keeps in flocks, often not with other species. Forages from treetops down to bushes. Coniferous, broadleaved/coniferous and broadleaved forests and nearby bushes **AN** Rufous-fronted Bushtit.

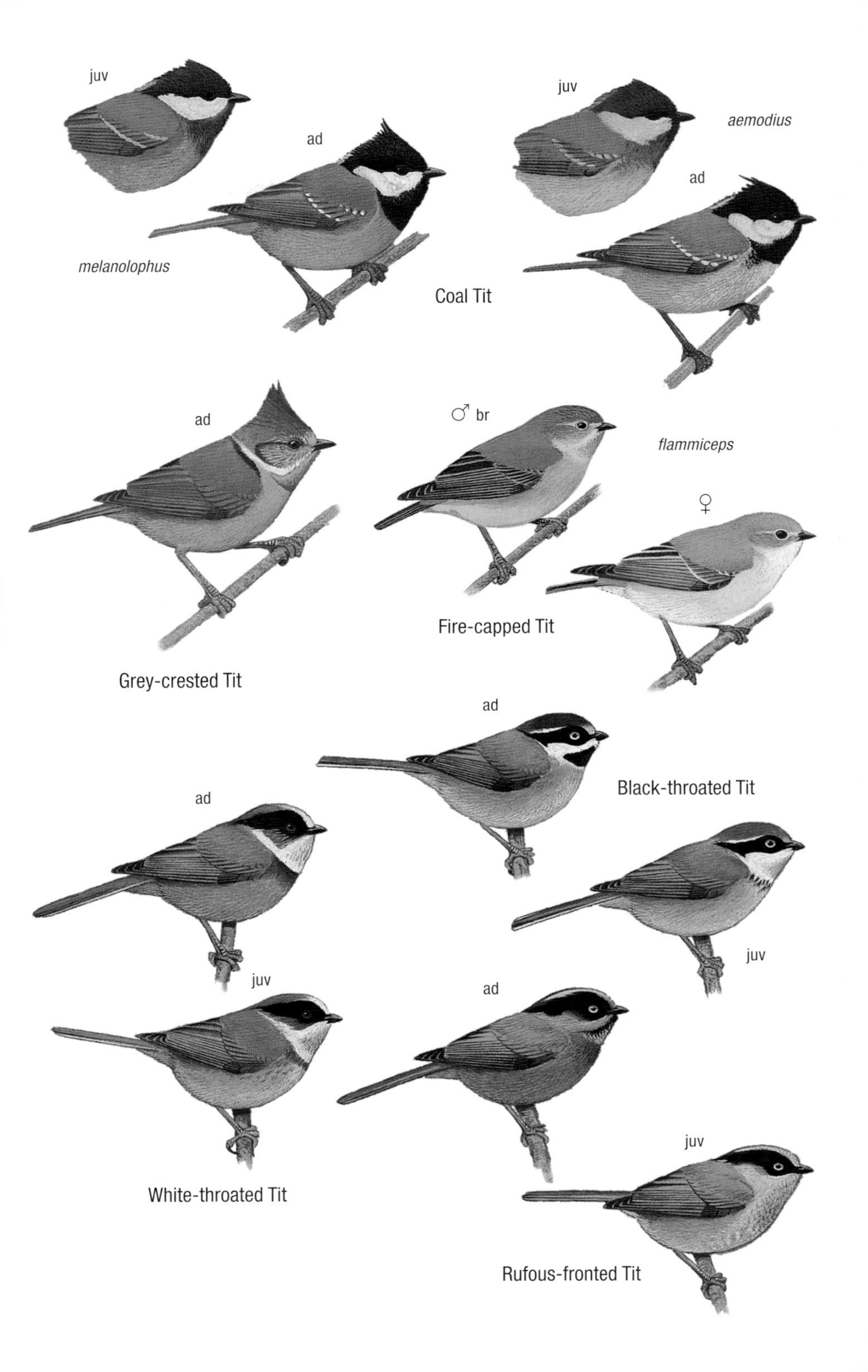
juv
ad
melanolophus
Coal Tit
juv
aemodius
ad
ad
Grey-crested Tit
♂ br
flammiceps
♀
Fire-capped Tit
ad
Black-throated Tit
ad
juv
juv
ad
White-throated Tit
juv
Rufous-fronted Tit

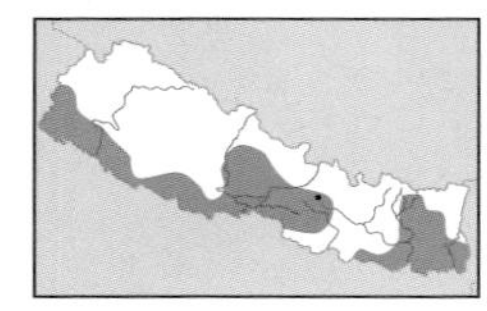

Plain Martin *Riparia paludicola* 12 cm

Locally common resident; 75–1,500 m (–2,990 m). **ID** Pale brownish-grey throat and breast, merging into dingy white rest of underparts. On some, throat paler than breast, and may show suggestion of breast-band. Underwing darker than Sand and Pale Martins, flight weaker and more fluttering, and has shallower indent to tail (but note variation in Pale). Slightly darker, richer brown upperparts than Pale, with paler rump. Juvenile has rufous fringes to upperparts (especially tertials and rump/uppertail-coverts), and throat and breast are paler and washed pinkish-buff. **Voice** Weak high-pitched twittering song; rasping *chrrr* call. **HH** Usually hawks insects over water in swift, agile sustained flight, sometimes high in air. Around rivers and lakes. **AN** Grey-throated Martin, if local form *chinensis* is split.

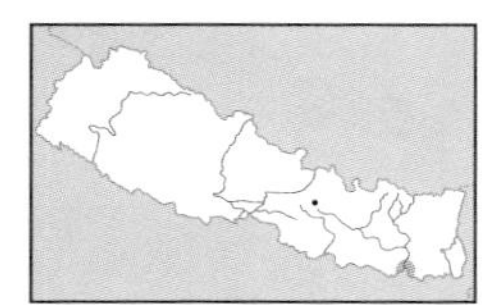

Sand Martin *Riparia riparia* 13 cm

Status uncertain because of confusion with Pale Martin; possibly rare winter visitor and passage migrant; 75–3,000 m. **ID** Adult from Plain by white throat and half-collar, and brown breast-band which is generally well defined against whitish underparts. In addition, appears stockier and more purposeful in flight, with a more prominently forked tail. Very similar to Pale, but upperparts slightly darker brown, breast-band dark and clearly defined, throat white and clearly demarcated from dark brown ear-coverts, and tail-fork deeper. Juvenile has buff fringes to upperparts and buff tinge to throat, and breast-band can be less clearly defined. **Voice** Rasping call, drier than Common House Martin. **HH** Habits like Plain. Around large waterbodies.

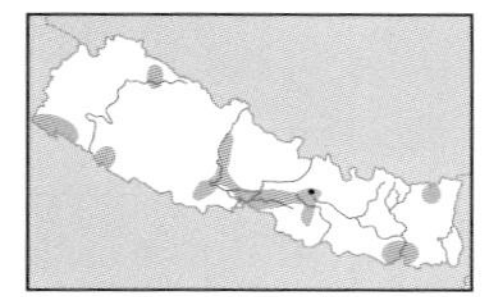

Pale Martin *Riparia diluta* 13 cm

Status uncertain because of confusion with Sand Martin; possibly scarce winter visitor and passage migrant; 75–3,000 m. Many records of Sand may refer to Pale. **ID** Very similar to Sand, but upperparts paler and greyer, breast-band pale and weakly defined, throat greyish-white and grades into pale greyish-brown ear-coverts, and tail-fork shallower. Juvenile has brownish throat recalling Plain, but stronger contrast between breast and belly, and lacks pinkish buff wash across breast of juvenile Plain. There is confusing subspecies variation; compared to nominate, *indica* is smaller and has very shallow tail-fork, and *tibetana* comparatively large and as dark as Sand. **HH** Habits like Plain. Around large waterbodies.

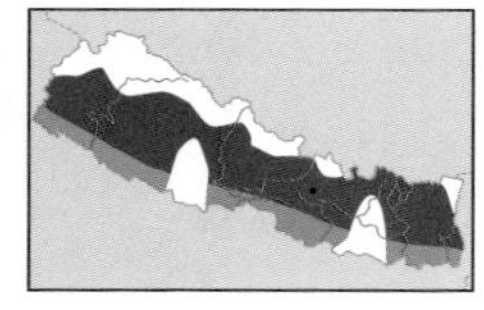

Barn Swallow *Hirundo rustica* 18 cm

Common resident and summer visitor; 75–1,830 m (–6,400 m on passage). **ID** Adult has bright red forehead and throat, blue-black breast-band and upperparts, and long tail-streamers. Underparts vary from white to rufous. Immature has duller orange forehead and throat, breast-band is browner and less well defined, upperparts duller, and has shorter tail-streamers. **Voice** Varied twittering song; call a clear *vit vit*, louder *vheet vheet* in alarm. **HH** Swift and agile flight with frequent banks and turns. Gregarious in non-breeding season, when often congregates on telegraph wires. Cultivation, towns, cities, and lakes and rivers in open country.

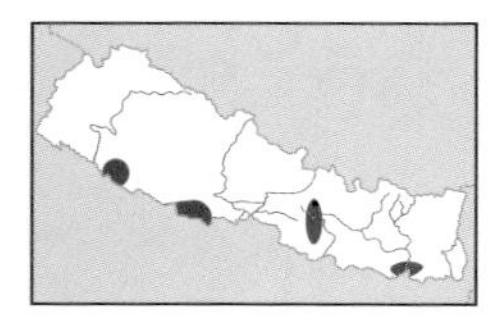

Wire-tailed Swallow *Hirundo smithii* 14 cm

Rare visitor; possibly also very local resident; 75–365 m (–1,280 m). **ID** From Barn Swallow by chestnut crown, brighter blue upperparts, glistening white underparts, and fine filamentous projections to outer tail feathers. White underwing-coverts contrast more strongly with dark underside of flight feathers. Can have pinkish or buffish wash to breast. Wire-like tail-streamers frequently broken, entirely lost, or difficult to see; tail then appears square-ended. Juvenile has brownish cast to blue upperparts and dull brownish crown. Possibly confusable with Streak-throated, but larger and proportionately longer-winged/shorter-tailed, with whiter, unstreaked underparts and underwing-coverts. White throat and breast and squarer tail help separate from juvenile Barn Swallow. **Voice** Twittering song; calls include a double *chirrik-weet*, *chit-chit* contact call and *chichip chichip* alarm call. **HH** Habits like Barn. Open country near water.

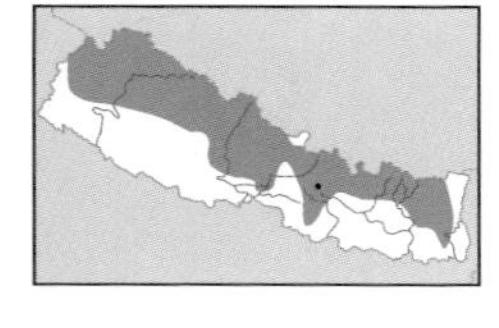

Eurasian Crag Martin *Ptyonoprogne rupestris* 15 cm

Common resident in dry areas of north-west; breeds 2,650–4,950 m, winters 915–2,975 m. **ID** Dark brown upperparts and underwing-coverts, dusky throat, dusky-brown flanks and vent, and distinct pale fringes to undertail-coverts. Juvenile has buff to rufous fringes to upperparts; compared to adult, underparts generally warmer, washed buff or rufous, and chin/throat less distinctly marked. **Voice** Song comprises series of soft twittering notes; calls include *prrrt*, *zirr* alarm and plaintive *whee*. **HH** Hawks insects close to faces of crags, rocky cliffs and gorges.

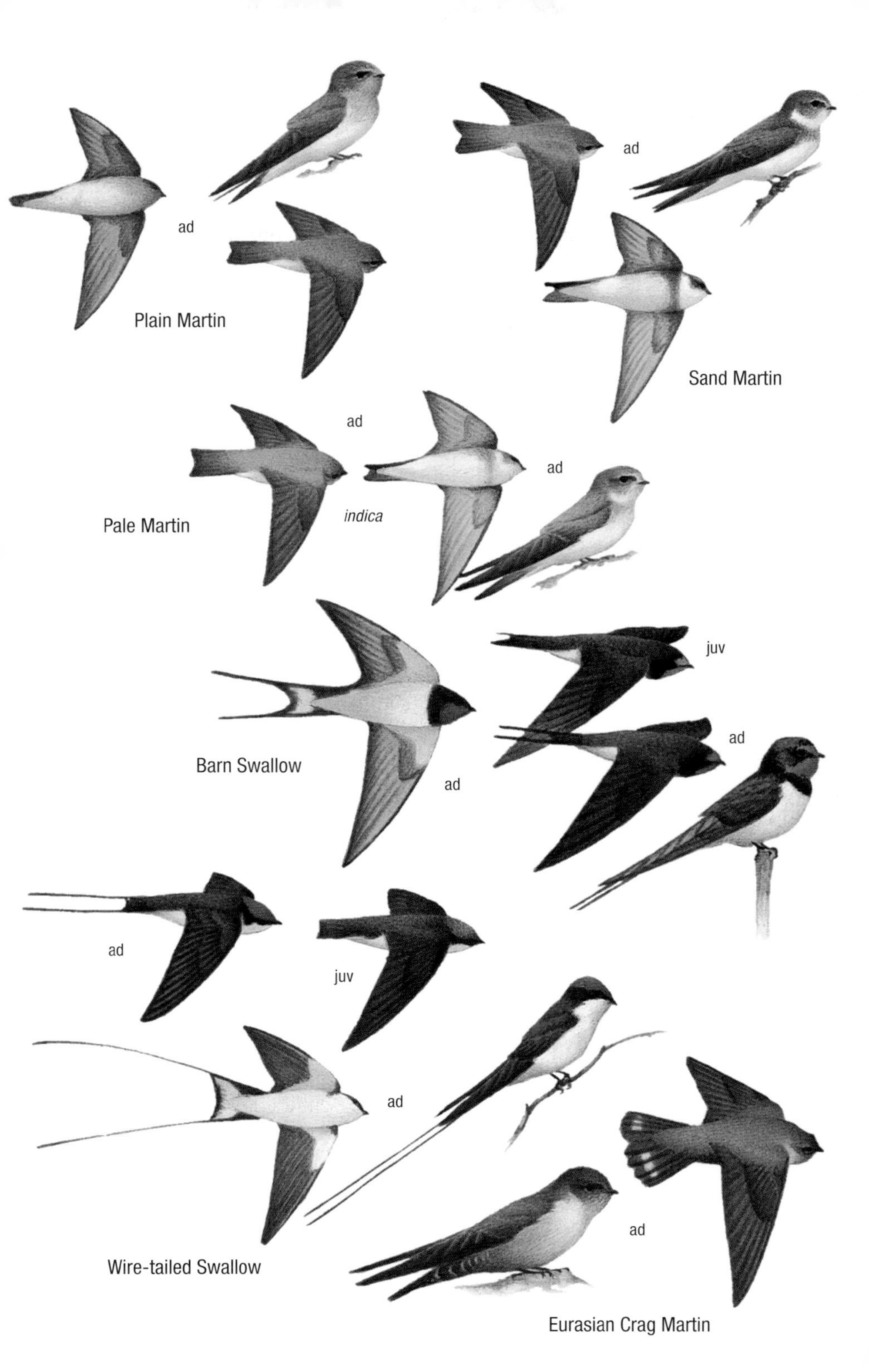

ad
Plain Martin
ad
Sand Martin
ad
ad
indica
Pale Martin
juv
ad
Barn Swallow
ad
ad
juv
ad
Wire-tailed Swallow
ad
Eurasian Crag Martin

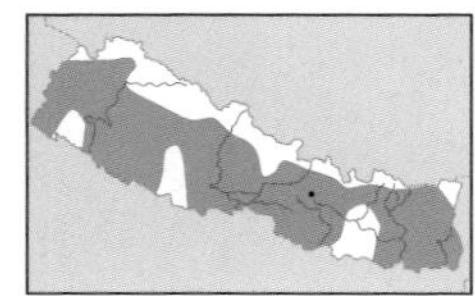

Red-rumped Swallow *Cecropis daurica* 16–17 cm

Common resident; breeds 230–2,130 m, non-breeding 75–2,745 m (–2,900 m). **ID** From Barn by rufous-orange sides of neck (creating collar and dark-capped appearance), rufous-orange rump, finely streaked buffish-white underparts, and black undertail-coverts. Compared to Barn, bulkier, has slower and more buoyant flight, gliding strongly for prolonged periods. Juvenile has duller upperparts, paler neck-sides (and collar) and rump, buff tips to tertials, and shorter tail-streamers. **Voice** Distinctive *treep* call; twittering song. **HH** Typical hirundine behaviour. Catches tiny invertebrates on wing in swift, agile, sustained flight. Perches readily on wires and other exposed perches. Often feeds in flocks with other hirundines and swift species. Breeds in habitation and forages nearby over cultivation and grassy slopes; winters in open country.

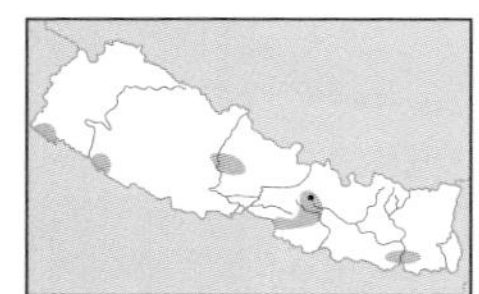

Streak-throated Swallow *Petrochelidon fluvicola* 11 cm

Rare visitor, mainly in winter, also recorded March and May; 75–1,280 m. **ID** A small, compact swallow with slight fork to long broad tail and weak and fluttering flight. Adult from other swallows by combination of lightly streaked chestnut crown and nape, dirty off-white underparts (with brown streaking on chin, throat and breast), narrow white streaks on mantle, and brownish rump. Juvenile has duller, browner crown, and brown-toned mantle and wings, with buff fringes (most obvious on scapulars and tertials). From juvenile Plain Martin by larger size, chestnut cast to crown, blue gloss to mantle, and heavily streaked throat. Streaked throat is best distinction from juvenile Wire-tailed Swallow. **Voice** A twittering *chirp* and sharp *trr trr* in flight. **HH** Typical hirundine, see Red-rumped. Usually seen over lakes and rivers. **TN** Formerly placed in genus *Hirundo*.

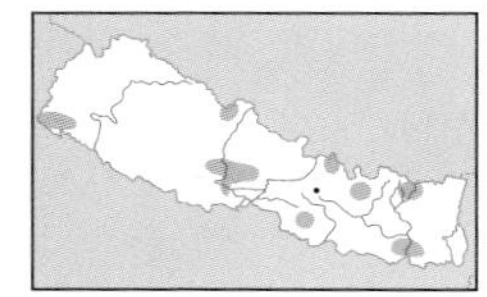

Common House Martin *Delichon urbicum* 13 cm

Very rare passage migrant; 75–3,000 m. **ID** Adult from Asian House by combination of whiter underparts, longer and more deeply forked tail, and paler underwing-coverts. Juvenile has browner (less glossy) upperparts, rather dingy below with brownish centres to some feathers of vent and undertail-coverts, whitish tips to tertials, and shallower tail-fork (thus very similar to Asian). **Voice** Soft twittering song; rolling scratchy *prrit-prrit* flight call and *jeet-jeet* in alarm. **HH** Typical hirundine, see Red-rumped Swallow. Flight is less swift with much less swooping and twisting than Barn Swallow, and it usually feeds higher than that species. Chiefly recorded over mountain valleys and over water. **AN** Northern House Martin.

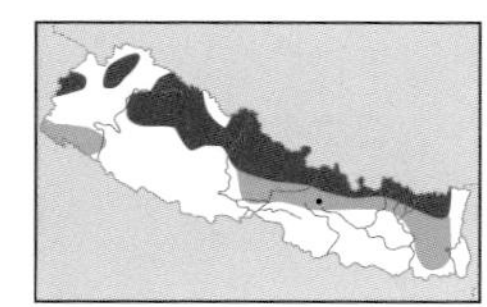

Asian House Martin *Delichon dasypus* 12 cm

Fairly common resident; breeds 3,100–4,575 m (–2,400 m), winters down to 75 m. **ID** Adult very similar to Common House and best distinguished by uniform pale dusky grey-brown wash to underparts, shallower fork to shorter tail (appearing almost square-ended when spread) and darker underwing with dark grey coverts concolorous with underside of flight feathers. Most adult Asian have variable dusky-brown centres to undertail-coverts, which are lacking on adult Common House, but present on juveniles of that species. Rump patch often looks smaller and dirty white compared to Common. Juvenile has browner upperparts, stronger dusky wash to underparts, broad white tips to tertials, and squarer-ended tail. **Voice** Like Common House. **HH** Habits like Common. Mainly seen over grassy hill slopes with cliffs and over forest.

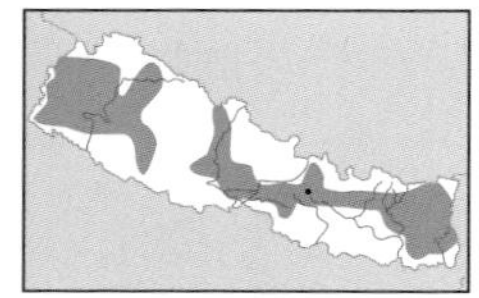

Nepal House Martin *Delichon nipalense* 13 cm

Fairly common resident; breeds 950–3,000 m (–3,865 m), winters 915–2,135 m (–75 m). **ID** Smaller and more compact than Common and Asian, with almost square-cut tail. Underwing-coverts and undertail-coverts black, contrasting sharply with white underparts. Extent of black on throat varies: only chin is black in west; in centre and east has variably mottled blackish throat, giving rise to dark-headed appearance. Juvenile duller and browner on upperparts, with buffish wash to underparts. **Voice** Call a high-pitched *chi-i*. **HH** Habits like Common House. Mainly recorded over forests, river valleys, and mountain ridges with cliffs and around villages.

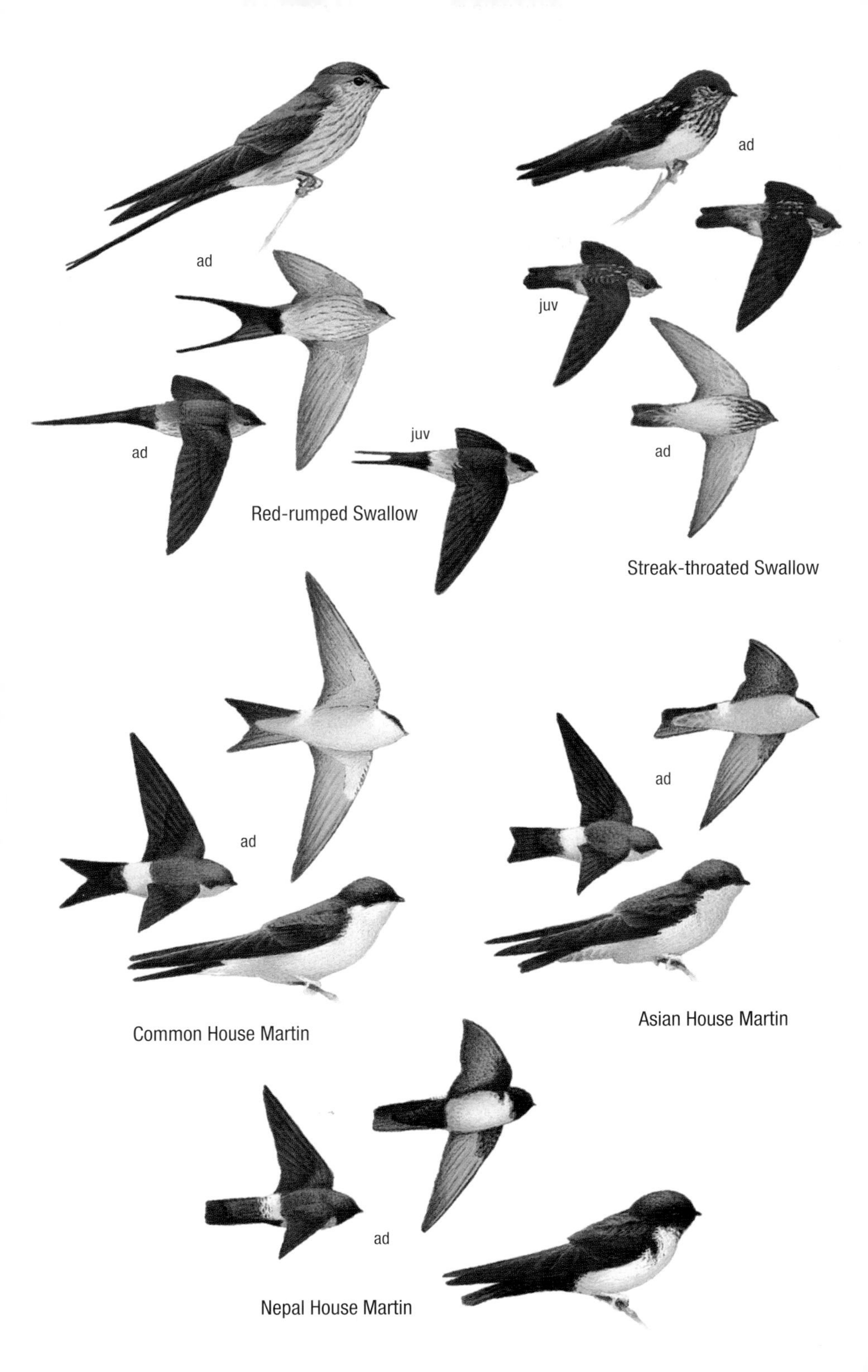
ad
ad
ad
juv
Red-rumped Swallow
ad
juv
ad
Streak-throated Swallow
ad
ad
Common House Martin
Asian House Martin
ad
Nepal House Martin

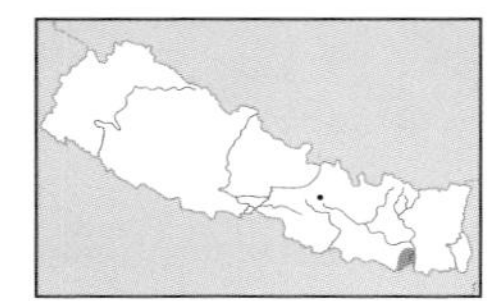

Rufous-vented Prinia *Prinia burnesii* 17 cm

Very rare and local resident; only known on Koshi River islands; 100 m. **ID** Large, long-tailed prinia with dark streaking on upperparts, whitish throat and breast, and pale rufous undertail-coverts. Lacks well-defined pale tips and dark subterminal bands visible on undertail of Striated Prinia. Recently described subspecies *nepalicola* is known only from Koshi Tappu Wildlife Reserve, Nepal. Compared to extralimital nominate, it has paler rufous undertail-coverts, thinner and fainter streaking on upperparts and sides of breast, lacks cinnamon on hind collar, and is greyer above with shorter tail. **Voice** Song a rising and falling, clear, high-pitched warbling, very different from other prinias. **HH** Skulks in dense vegetation; most easily located by its frequent calls, and when it sings in early mornings and evenings. Tall grasslands.

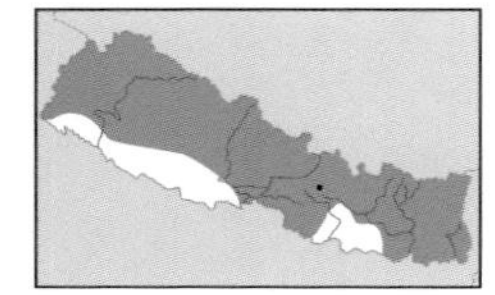

Striated Prinia *Prinia crinigera* 16 cm

Common resident; breeds 1,220–2,300 m (–3,000 m), winters 915–2,135 m (–75 m). **ID** Large, long-tailed, hill-dwelling prinia with stout bill. Adult breeding has dark grey-brown (and very worn) upperparts with ill-defined blackish streaking, dark lores, buffish-white underparts with variable dark grey mottling, narrow rufous-brown edges to remiges, and black bill. Adult non-breeding has rufous-brown (and fresh) upperparts with prominent dark brown streaking, buff lores and eye-ring, warm buff on underparts, especially flanks, broad rufous-brown fringes to remiges, and orange-tinged lower mandible; undertail has rufous-buff tips and dark subterminal bands. Female smaller and shows less marked seasonal variation than male. Juvenile has olive-brown upperparts, with indistinct streaking (mainly on crown), and warm buff or pale yellowish underparts. **Voice** Monotonous wheezy *chitzereet-chitzereet-chitzereet* song; *tchak-tchak* call. **HH** Skulking; hops and clambers actively about in bushes and low vegetation. In breeding season, males sing in open. Open hillsides with scattered low bushes and isolated trees, often on steep slopes.

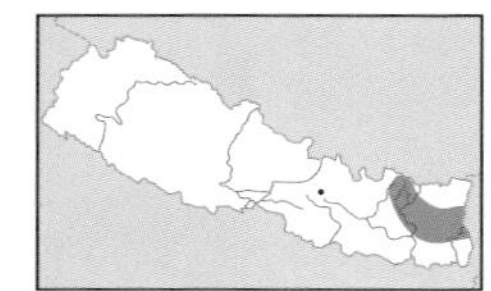

Black-throated Prinia *Prinia atrogularis* 17 cm

Fairly common resident in far east; 1,400–2,500 m. **ID** Large, long and narrow-tailed, hill-dwelling prinia. Adult breeding has black throat and breast, white moustachial stripe and breast spotting, greyish ear-coverts and breast-sides, greyish olive-brown upperparts and tail, olive-buff fringes to remiges forming panel on wing, whitish belly, and olive-buff flanks. Adult non-breeding has white supercilium, buff underparts with variable dark mottling/streaking on throat and breast, and rufous-brown edges to remiges. From Striated by white supercilium contrasting with dark lores, unstreaked upperparts (can appear diffusely streaked, especially on lower back), and longer and finer bill. Juvenile similar to non-breeding adult, but has yellowish wash on underparts and darker olive breast-band. **Voice** Loud *tulip... tulip... tulip* song; calls include soft *tp-tp-tp-tp*. **HH** Habits like Striated. Scrub and grass hillsides. **TN** Formerly known as Hill Prinia, a name now used for the extralimital potential split, *P. superciliaris*.

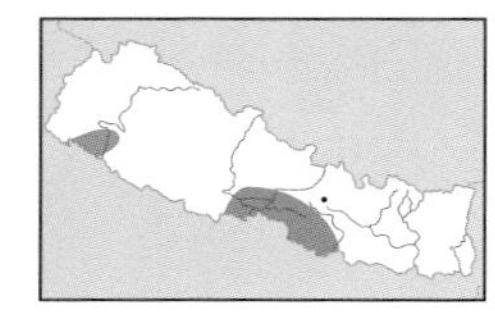

Grey-crowned Prinia *Prinia cinereocapilla* 11 cm

Local resident; 150–1,065 m. **ID** Bill fine and black in all plumages. In non-breeding plumage, supercilium orange-buff, and extends noticeably (and becomes whiter) behind eye. Noticeable contrast between purer blue-grey crown and nape and more rufescent-brown mantle and back compared to Ashy. Supercilium absent in breeding plumage. Juvenile undescribed. **Voice** *Cheeeeeeeesum-zip-zip-zip* song; rapid repeated *tzit* call. **HH** Forages low in bushes. Very active. *Themeda* grasslands close to forest. Globally threatened.

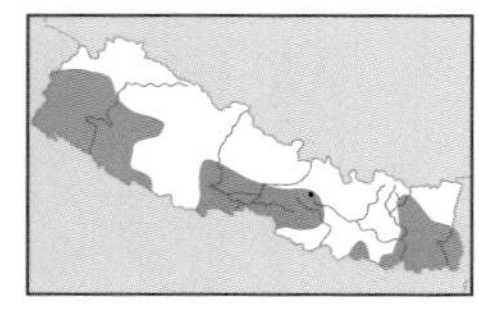

Grey-breasted Prinia *Prinia hodgsonii* 11 cm

Common resident; 75–1,200 m (–1,750 m). **ID** Adult breeding has grey 'cap' and upperparts, and variable greyish breast-band. Adult non-breeding has white supercilium and dark lores, olive-brown upperparts with rufescent cast, and white to greyish underparts. **Voice** Rhythmic undulating *tirr-irr-irr-irr* song; high-pitched *hee-hee-hee-hee* call. **HH** Gregarious in winter. Creeps about in bushes and runs mouse-like. Bushes at forest edges, scrub and second growth.

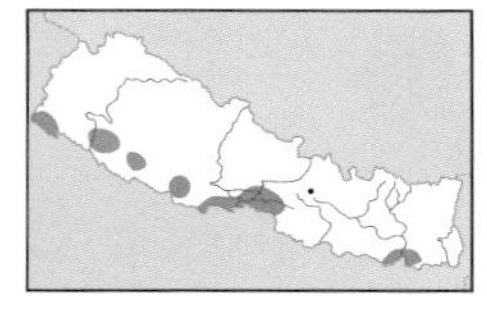

Graceful Prinia *Prinia gracilis* 11 cm

Local resident; 75–150 m (–500 m). **ID** Small prinia with streaked grey-brown upperparts, white underparts, and cross-barred tail. Much smaller than Striated Prinia. **Voice** Fast, rhythmic, wheezy warbling song, *ze-r witze-r wit*; nasal, buzzing *bzreep* call. **HH** Feeds actively in bushes, grass and on ground. Medium to short *Saccharum* grassland and bushes, especially near water.

ad
Rufous-vented Prinia
non-br
br
Striated Prinia
non-br
br
Black-throated Prinia
ad
Grey-crowned Prinia
non-br
br
Grey-breasted Prinia
ad
Graceful Prinia

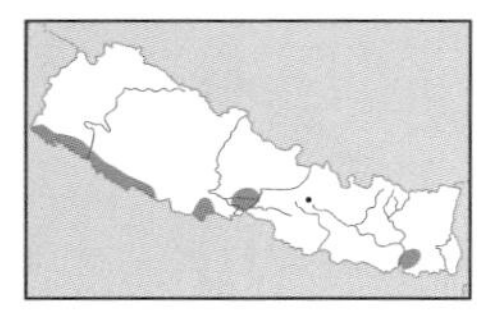

Jungle Prinia ***Prinia sylvatica*** 13 cm

Local, frequent resident in west; 75–150 m. **ID** Larger-bodied and stouter-billed than Plain Prinia, with sturdier-looking tail; supercilium usually less distinct than Plain (diagnostic when absent) Adult breeding has grey upperparts, largely white outer rectrices and black bill. Adult non-breeding has longer tail with buffish outer tips, orange-brown lower bill, and is more rufescent-brown above. Female smaller with less seasonal variation than male. Juvenile has more olive-brown upperparts, with more rufescent wings, and can have yellowish wash to underparts. **Voice** Loud, pulsing song, *zee-chu* repeated metronomically. **HH** Rather less active than small prinias. Flits about bushes and grass; occasionally popping onto top of vegetation then diving down again. Sings from an exposed perch. Scrub and tall grass in open dry areas.

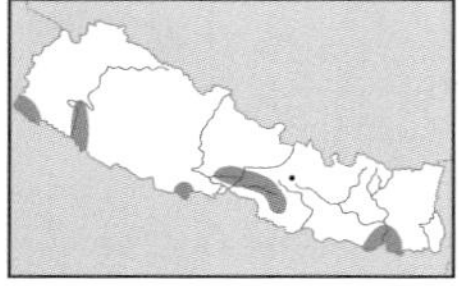

Yellow-bellied Prinia ***Prinia flaviventris*** 13 cm

Local resident; 75–300 m. **ID** Adult has fine white supercilium (sometimes lacking), white throat and breast, slate-grey forecrown and ear-coverts, dark olive-green upperparts, yellowish belly and vent. Juvenile has yellowish olive-brown upperparts (no blue-grey on head), uniform pale yellow underparts. **Voice** Sharp *chirp* followed by a five-note trilled song. **HH** Very active, foraging in tall grass and occasionally clambering to top of grass stems to look around; sometimes feeds on ground. Tall grass along riverbeds and in reedbeds

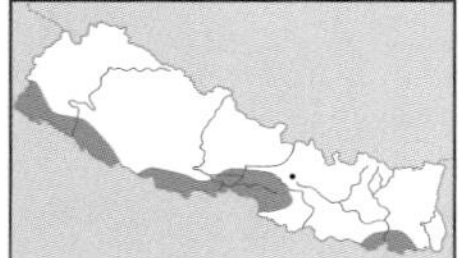

Ashy Prinia ***Prinia socialis*** 13 cm

Resident, fairly common in west, frequent in centre and uncommon and very local in east; 75–305 m (–1,280 m). **ID** White supercilium (sometimes lacking), slate-grey crown and ear-coverts, red eye, slate-grey (breeding) or rufous-brown (non-breeding) mantle, orange-buff wash on underparts, and prominent black subterminal marks (and whitish tips) to tail feathers. Juvenile has greenish upperparts and buffish-yellow underparts. **Voice** Song a wheezy *jimmy-jimmy-jimmy*. **HH** Forages low in grasses and bushes, and on ground, staying close to cover. Tall grass and scrub, reedbeds along rivers and forest edges; prefers damper habitats than Plain.

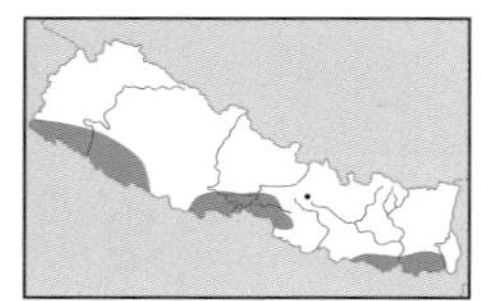

Plain Prinia ***Prinia inornata*** 13 cm

Fairly common and widespread resident; 75–305 m. **ID** Adult breeding has black bill, grey-brown upperparts and whitish underparts, with largely white outermost rectrices. Adult non-breeding has longer tail, pale base to lower mandible, warm brown upperparts, more rufescent wings and tail, buff tips and dark subterminal marks to rectrices, and warm buff wash to underparts. Juvenile more rufescent. **Voice** Song a rapid, wheezy trill, *tlick tlick tlick*. **HH** When perched, tail often held cocked and slightly fanned, and is frequently jerked and side-switched. Feeds actively in low vegetation. Normal flight is weak and jerky, and covers only short distances. Reeds, grassland, edges of cultivation, scrub and forest edges.

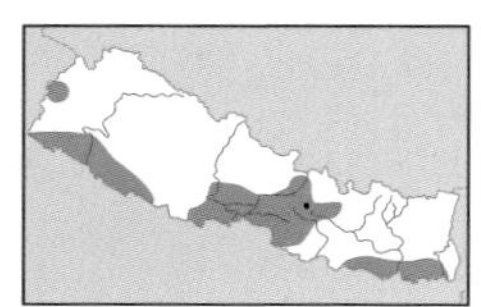

Zitting Cisticola ***Cisticola juncidis*** 10 cm

Fairly common and widespread resident; 75–1,350 m (–1,900 m). **ID** Adult breeding has diffusely streaked grey-brown crown, and often rather distinct buff or rufous rump. Adult non-breeding has longer tail, more heavily streaked upperparts and often less distinct rump; some have brighter rufous-buff upperparts, but lack the more clearly defined rufous nape of Golden-headed Cisticola. First-winter has sulphur-yellow wash to underparts and less heavily streaked upperparts. **Voice** Repeated *pip* uttered in distinctive display flight. **HH** Active and excitable, frequently flicks wings and cocks and spreads tail. Displays frequently when breeding; circles widely over its territory with bouncing flight, beating wings as it rises, drops a little then rises again, while singing. Paddyfields and other crops, and dry and marshy grasslands.

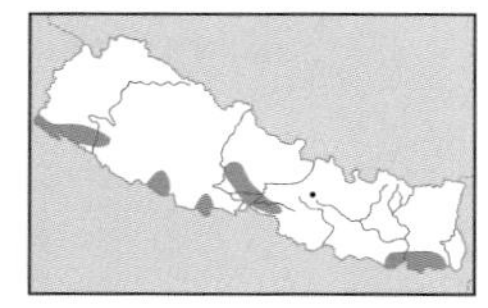

Golden-headed Cisticola ***Cisticola exilis*** 10 cm

Local resident; 75–150 m (–915 m). **ID** In all plumages from Zitting by blacker tail with narrow buffish or greyish tips (broader white tips in Zitting), unstreaked rufous nape and sides of neck, and rufous (rather than whitish) supercilium. Breeding male has unstreaked creamy-white crown and underparts, rufous-brown wash to nape and sides of neck, and unstreaked olive-grey rump. Female and non-breeding male have heavily streaked crown and mantle, and more closely resemble Zitting. Tail much longer in non-breeding plumage. First-winter has yellow wash to underparts. **Voice** Song comprises 1–2 jolly doubled notes introduced by a buzzy wheeze: *bzzeeee... joo-ee*, the wheeze often repeated separately. **HH** Habits similar to Zitting; display flight faster, less jerky and ends in nose-dive at high speed. Open large expanses of medium to short, dense grassland. **AN** Bright-headed Cisticola.

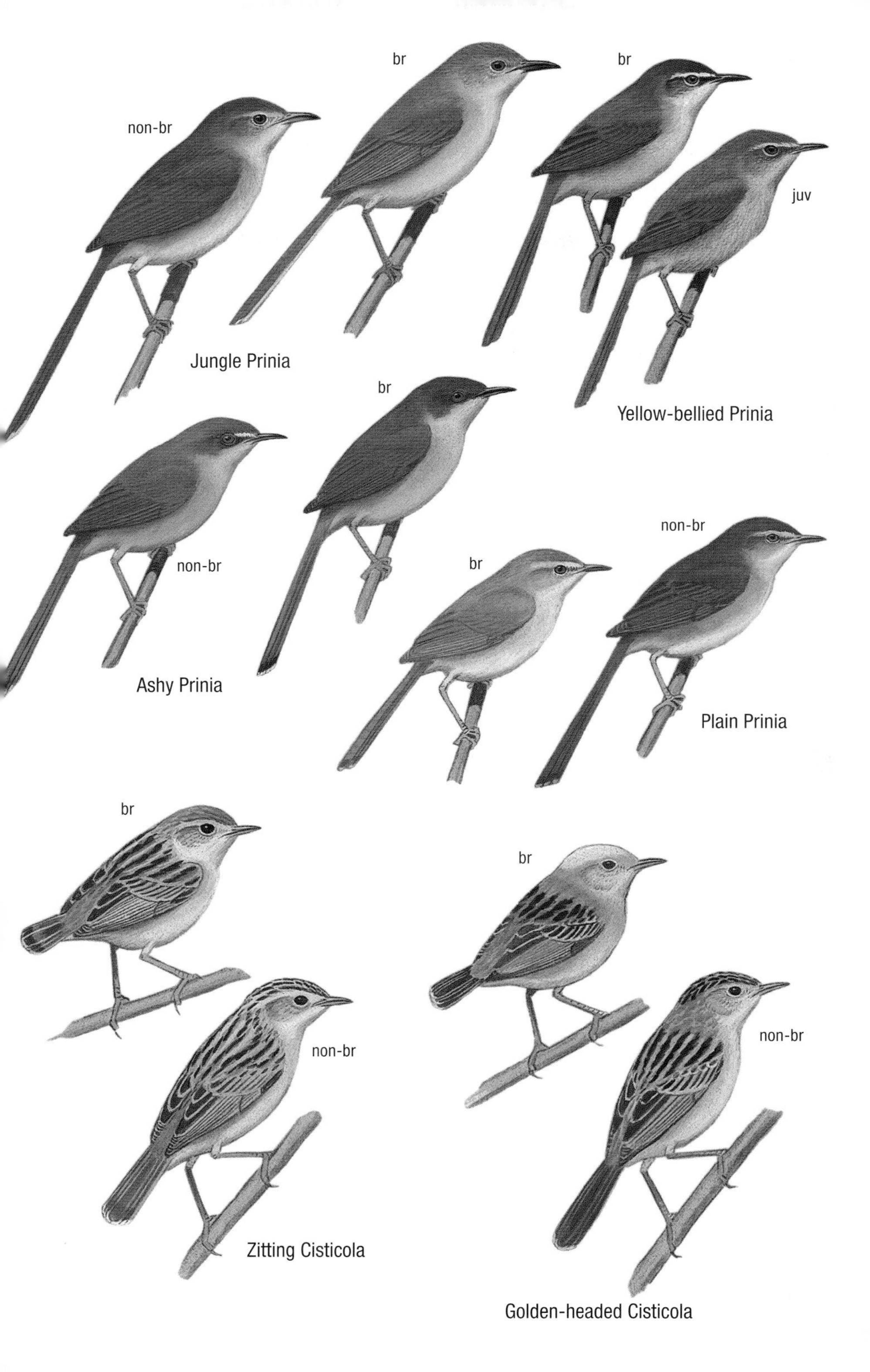

non-br
br
br
juv
Jungle Prinia
Yellow-bellied Prinia
br
non-br
Ashy Prinia
br
non-br
Plain Prinia
br
non-br
Zitting Cisticola
br
non-br
Golden-headed Cisticola

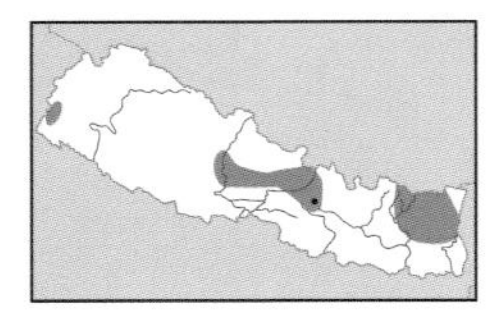

Striated Bulbul *Pycnonotus striatus* 23 cm

Local and frequent resident; 1,525–2,690 m. **ID** A crested, green bulbul with bold yellowish-white streaking on underparts, and fine white streaking on crest, ear-coverts and upperparts. Bright yellow lores, eye-ring and throat, and yellow undertail-coverts. In flight, olive-green tail has pale yellow tips to outer feathers. Juvenile duller and less heavily streaked, with shorter crest. **Voice** A distinctive, repeated disyllabic *chi-chirp*. **HH** Forms small flocks in non-breeding season; less noisy than other bulbuls. Feeds on berries, also insects taken in short aerial sallies. Strong direct flight. Mainly broadleaved evergreen forest.

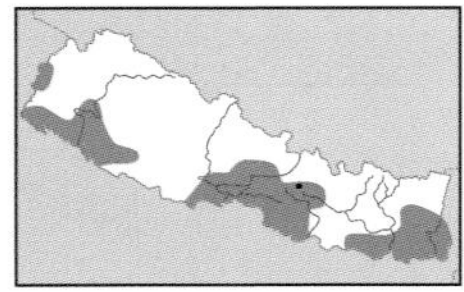

Black-crested Bulbul *Pycnonotus* (*melanicterus*) *flaviventris* 22 cm

Widespread and locally frequent resident; 75–915 m (–1,380 m). **ID** Black-headed bulbul with olive-green upperparts and yellow underparts. Erect black crest, uniform olive-green wings, uniform olive-brown tail and yellow iris. Juvenile has dull black head, shorter crest and paler yellow underparts. **Voice** Song is sweet, musical and lacks harsh notes. **HH** Rather quiet and retiring for a bulbul. Singly, in pairs or sometimes in groups in lower forest storey and bushes. Broadleaved forest with dense undergrowth and thick secondary forest.

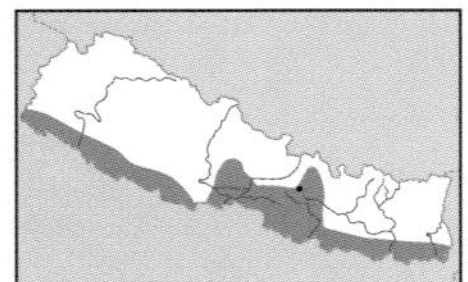

Red-whiskered Bulbul *Pycnonotus jocosus* 20 cm

Locally common and widespread resident; 75–455 m (–1,500 m). **ID** Striking with glossy black crown and erect crest, red patch behind eye, white patch on lower ear-coverts bordered by black moustachial stripe, white underparts with complete or broken breast-band, and red vent. Lacks white rump (see Red-vented Bulbul). Juvenile lacks red 'whiskers'; crest and nape are paler brown compared to adult, and has paler, rufous-orange vent. **Voice** Calls include a lively *pettigrew* or *kick-pettigrew*. **HH** Active, noisy and confiding. Often perches conspicuously atop small bushes. Sometimes in flocks with other bulbuls outside breeding season. Open forest and second growth.

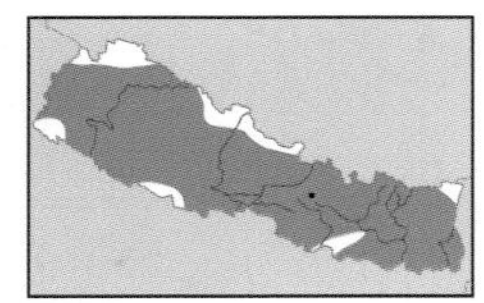

Himalayan Bulbul *Pycnonotus leucogenys* 20 cm

Common and widespread resident; mainly 350–2,400 m (250–3,050 m). **ID** White-cheeked bulbul, with brownish-grey upperparts and yellow vent. Prominent forward-pointing brown crest, black crescent at rear of white cheek-patch, and variable narrow white supercilium. Hybridises with Red-vented. **Voice** Song a variable combination of melodious phrases, *we-did-de-dear-up*, *whet-what*, and *who-lik-lik-leer*; also a *plee-plee-plee* flight call and a *wik-wik-wik-wiker* alarm. **HH** Bold and confiding, very lively and conspicuous. Often perches in open, flicking wings; constantly moving from bush to bush. Flight strong and undulating. Dry habitats: open dry scrub, bushes around towns, villages and cultivation, and second growth.

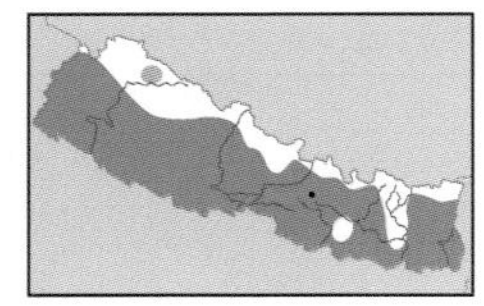

Red-vented Bulbul *Pycnonotus cafer* 20 cm

Common and widespread resident; 75–1,500 m (–2,135 m). **ID** Short-crested black head, white rump, white-tipped black tail, and red vent. White rump helps separate it from other bulbuls in flight. Juvenile has browner head, rufous edges to flight feathers, buffish cast to white rump, and lacks white tips to tail. Hybridises with Himalayan. **Voice** Cheery *be-care-ful* or *be quick-quick*; alarm call a sharp repetitive *peep*. **HH** Bold, tame and quarrelsome. In pairs or small loose flocks according to season. Feeds mainly on fruits and berries, also insects taken in aerial sallies. Open deciduous forest, second growth and trees around habitation.

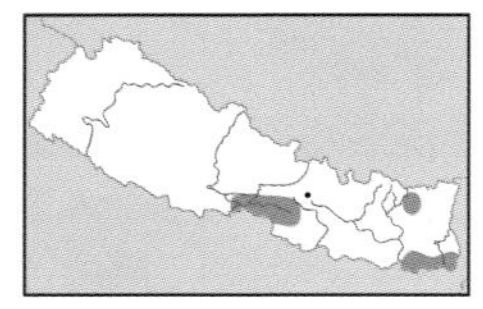

White-throated Bulbul *Alophoixus flaveolus* 22 cm

Local resident, locally frequent in Chitwan National Park, rare elsewhere; 150–455 m (–1,800 m). **ID** Adult striking with prominent brownish crest and puffed-out white throat. Also whitish lores and grey ear-coverts (streaked white), yellow breast and belly, and olive-green upperparts with rufous-brown cast to wings and tail. Juvenile has browner upperparts, more rufescent wings and tail, and brownish wash to underparts. **Voice** A chacking *chi-chack chi-chack chi-chack* and nasal *cheer*. **HH** Noisy and heard more often than seen. Creeps and clambers about bushes and lower forest storey in flocks; sometimes ascends to canopy. Often puffs out throat feathers. Undergrowth in dense broadleaved evergreen forest and second growth.

ad
Striated Bulbul
ad
juv
Black-crested Bulbul
ad
Red-whiskered Bulbul
ad
Himalayan Bulbul
juv
ad
White-throated Bulbul
ad
Red-vented Bulbul

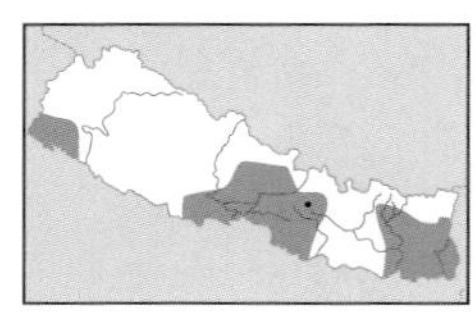

Ashy Bulbul *Hemixos flavala* 20 cm

Widespread, generally uncommon resident; 305–1,525 m. **ID** Distinctive, crested bulbul, with black mask, tawny ear-coverts, grey upperparts and greyish-brown tail, olive-yellow wing panel, white throat, and pale grey breast merging into whitish belly. Juvenile similar to adult, but has shorter crest, browner upperparts and duller wing panel. **Voice** A loud ringing call of 4–5 notes, the second or third highest and last two descending. **HH** In pairs in breeding season, otherwise in noisy parties. Arboreal; forages in middle and upper storeys on berries, nectar and insects, the last sometimes caught in aerial sallies. Broadleaved forest; also forest edges in winter.

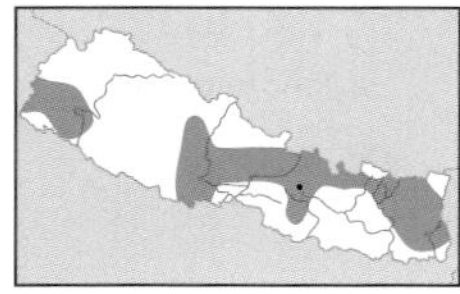

Mountain Bulbul *Ixos mcclellandii* 24 cm

Fairly common and quite widespread resident; mainly 1,830–2,135 m (915–2,285 m). **ID** Distinctive bulbul, with white streaking on shaggy brown crest, white-streaked greyish throat, and cinnamon-brown breast with buff streaking. Greenish mantle, wings and tail, and yellowish vent. Juvenile has shorter crest, uniform grey throat and browner upperparts. **Voice** A metallic ringing or squawking. **HH** In pairs or small parties, usually in forest canopy, also bushes. Rather shy. Often puffs out throat feathers. Feeds chiefly on berries and fruits. Forest and second growth. **TN** Formerly placed in genus *Hypsipetes*.

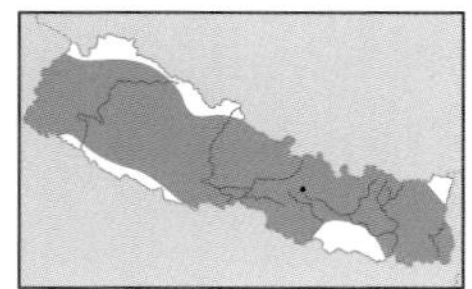

Black Bulbul *Hypsipetes leucocephalus* 25 cm

Common and widespread resident; 75–3,050 m, most frequent 1,830–2,135 m. **ID** Slate-grey bulbul, with slight crest. Bright red bill, legs and feet, pale fringes to undertail-coverts, and shallow fork to tail (at times recalling a drongo). Juvenile lacks crest, has whitish throat, grey breast-band, brownish cast to upperparts, and brownish bill, legs and feet. **Voice** Song a monotonously repeated series of 3–4 rising and falling notes, also a variety of loud, screeching and mewing notes. Large flocks produce a continuous, shrill, nasal, chattering babble. **HH** Gregarious when not breeding, in flocks of up to 100. Very noisy. Forages chiefly in treetops, moving restlessly. Feeds mainly on berries, also insects. Mainly broadleaved forest.

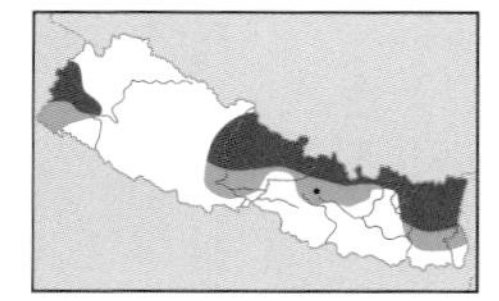

Chestnut-headed Tesia *Oligura castaneocoronata* 8 cm

Fairly common and widespread resident; 2,135–4,000 m summer; 800–1,830 m winter (–250 m). **ID** Adult has bright chestnut 'hood', prominent white crescent behind eye, dark olive-green mantle and wings, and bright yellow underparts with olive-green sides of breast and flanks. Juvenile has dark olive upperparts with brownish cast (lacking chestnut head), and dark rufous underparts. **Voice** Song an explosive *cheep-cheeu-chewit*; sharp, explosive *whit* call. **HH** Typically skulks in forest undergrowth, usually near ground. Singing males sometimes briefly emerge from cover. Active and inquisitive. Flight is weak and species is loath to fly. Thick undergrowth in moist forest. **TN** Formerly placed in genus *Tesia*.

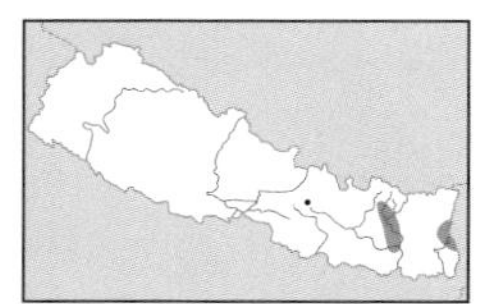

Slaty-bellied Tesia *Tesia olivea* 9 cm

Rare and local resident in east; 1,000–2,200 m. **ID** Best separated from Grey-bellied by uniform dark slate-grey underparts, and yellowish-green crown distinctly brighter than mantle (although some have duller and less contrasting crown, and may show hint of brighter supercilium). Additional features are less prominent black stripe behind eye, and finer bill with brighter orange or orange-red lower mandible without dark tip. Juvenile reportedly similar to juvenile Grey-bellied, but has darker olive-green underparts. **Voice** Song comprises 4–6 measured whistles followed by an explosive tuneless jumble of notes; calls include a sharp *tchirik*. **HH** Habits like Grey-bellied. Dense low undergrowth with ferns in moist forest, especially evergreen.

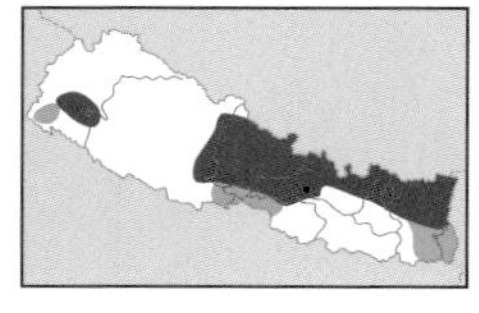

Grey-bellied Tesia *Tesia cyaniventer* 9 cm

Frequent resident; summers 1,525–2,440 m; winters 800–1,830 m (-75 m). **ID** From Slaty-bellied by paler grey underparts, becoming almost whitish on throat and centre of belly, and by concolorous olive-green crown and mantle, with brighter lime-green supercilium. Also more prominent black stripe behind eye, and stouter bill with dark tip and yellow basal two-thirds to lower mandible. Juvenile has dark olive-brown upperparts and olive cast to grey underparts; like adult, has brighter green supercilium and dark eye-stripe. **Voice** Much slower song than Slaty-bellied and lacks its explosive jumble of notes; loud and rattling *trrrrrk* call. **HH** Skulking, very active and always on the move. Tangled undergrowth and ferns in thick forest, usually near small streams; ravines favoured when breeding, shady broadleaved forest in winter.

ad
ad
Mountain Bulbul
juv
Ashy Bulbul
ad
Black Bulbul
juv
juv
ad
ad
ad
Chestnut-headed Tesia
Slaty-bellied Tesia
Grey-bellied Tesia

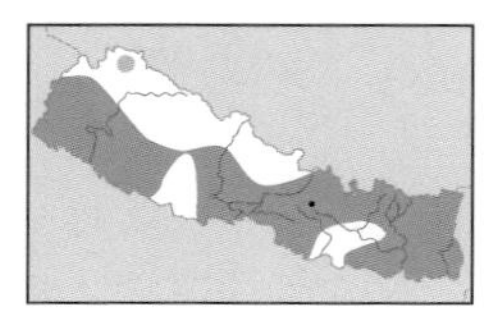

Common Tailorbird *Orthotomus sutorius* 13 cm

Widespread resident; 75–1,830 m. **ID** Long (slightly downcurved) pale bill, rufous forehead and forecrown, greenish upperparts, and dull whitish or buffish underparts. Breeding male has elongated central tail feathers. Juvenile lacks rufous on forecrown. See Appendix 1 for differences from Mountain Tailorbird. **Voice** Song a loud *pitchik-pitchik-pitchik*. **HH** Confiding. Hunts actively, usually low down in vegetation and on ground; occasionally climbs quite high in trees. Tail constantly cocked over back. Flies weakly and over short distances. Bushes in gardens, cultivation edges and forest edges.

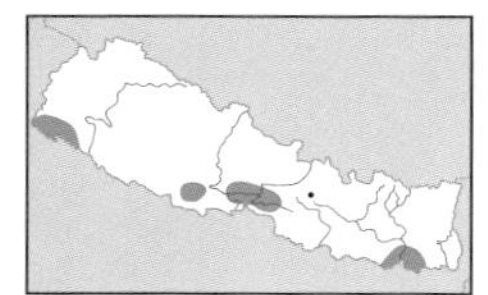

Striated Grassbird *Megalurus palustris* 25 cm

Rare and local resident; 75–250 m. **ID** A large, babbler-like warbler with brownish-buff upperparts prominently streaked black, whitish supercilium, rufous crown, whitish underparts with fine brown streaking on breast and flanks, and long, graduated tail. In fresh plumage, supercilium and underparts strongly washed pale yellow and streaking on underparts partially obscured. **Voice** Song a clear, drawn-out whistle, ending in a short explosive *wheeechoo*. **HH** Often perches conspicuously and has habit of jerking its tail and flicking one wing half-open, then the other. Feeds in dense vegetation and can climb rapidly up and down stems. In breeding season males sing from exposed perches and in short gliding flight. Tall wet grassland and reedbeds.

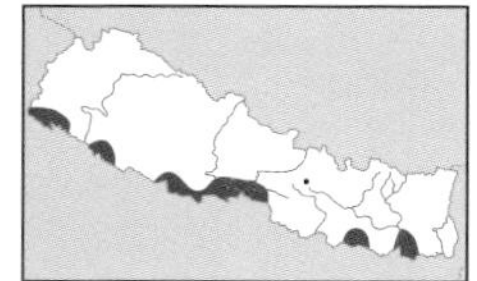

Bristled Grassbird *Chaetornis striata* 20 cm

Very local, mainly summer visitor; 75–250 m. **ID** From Striated by smaller size, stout mostly dark bill, less distinct supercilium (barely apparent behind eye), and shorter (broader) tail with pale shafts, buffish-white tips and more prominent dark cross-barring (undertail appears blackish with broad whitish tips, although these can be lost due to wear). In addition, breast and flanks lack the streaking apparent on worn Striated. In fresh plumage, underparts washed buff and feathers of upperparts and wings have broad buff fringes. Upperparts become greyer-brown and underparts whiter when worn. **Voice** Song a disyllabic *trew-treuw* repeated at 2–3-second intervals. **HH** When breeding males perform song flight, rising *c.*20 m or more, circling widely for *c.*10 minutes or longer before descending; also sings from exposed perch. Favours tall *Saccharum spontaneum* grassland with bushes. Globally threatened.

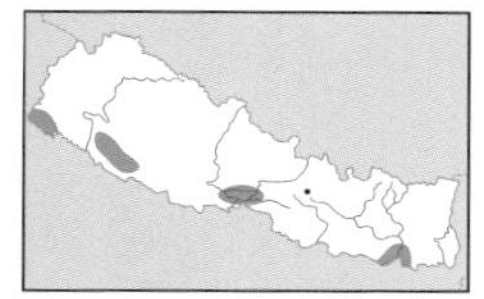

Rufous-rumped Grassbird *Graminicola bengalensis* 18 cm

Very local resident; 150–270 m. **ID** Dark upperparts with rufous streaking on crown and lower mantle, and white or buff streaking on nape and upper mantle. Largely rufous rump and wings. Tail blackish broadly tipped white. Underparts white with rufous-buff breast-sides and flanks. **Voice** Song a subdued high *er-wi-wi-wi-, you-wuoo, yu-wuoo*, followed by a series of harsh notes and ending in wheezy sounds. **HH** Outside breeding season skulks in grass. When breeding males are very noisy, singing from reed-tops and soaring into air. Relatively undisturbed, dense, tall grasslands.

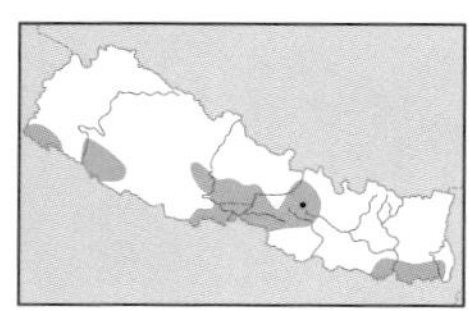

Thick-billed Warbler *Phragamaticola aedon* 19 cm

Widespread winter visitor; fairly common at Koshi, uncommon elsewhere; 75–1,500 m. **ID** From large *Acrocephalus* by short, stout bill (lacking dark tip), rounded head (with crown feathers often raised), and 'plain-faced' appearance (lacks prominent supercilium and any hint of darker eye-stripe). Tail appears long and graduated, and wings short. In fresh plumage, upperparts have rufous suffusion, especially to fringes of remiges, rump and uppertail-coverts (not apparent in Clamorous Reed Warbler), and is warmer buff on breast and flanks. **Voice** Gives a hard *shak*, tongue-clicking *tuc* and loud chattering calls. **HH** Skulks in dense bushes, usually moving with clumsy, heavy hops. Tall grass, scrub, reeds and bushes at edges of forest and cultivation, generally drier habitats than Clamorous. **TN** Formerly placed in *Acrocephalus*.

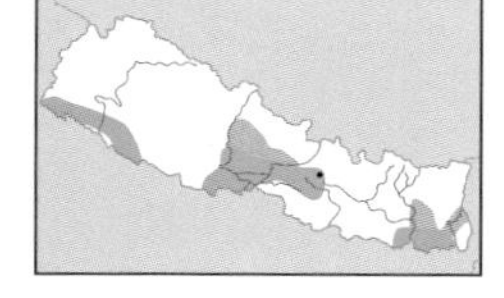

Lesser Whitethroat *Sylvia curruca* 13 cm

Widespread, very uncommon winter visitor and passage migrant; 75–1,500 m (–2,750 m). **ID** Brownish-grey upperparts, dull whitish underparts (can have pinkish flush to breast in fresh plumage), slate-grey crown (greyer and slightly darker than mantle), and darker lores and ear-coverts (forming diffuse mask). Bill blackish and legs and feet grey. Can show hint of paler supercilium and pale buffish fringes to tertials and secondaries in fresh plumage. **Voice** A soft, low-pitched, rather scratchy warbling song and dry rattle, the two often run together. **HH** Chiefly keeps low in vegetation, often venturing into open. If flushed flies low and jerkily into cover. Typically has horizontal carriage. Mainly scrub and acacias.

♂
♀
Common Tailorbird
ad worn
ad fresh
Striated Grassbird
ad
Bristled Grassbird
ad
Rufous-rumped Grassbird
ad
Thick-billed Warbler
ad
Lesser Whitethroat

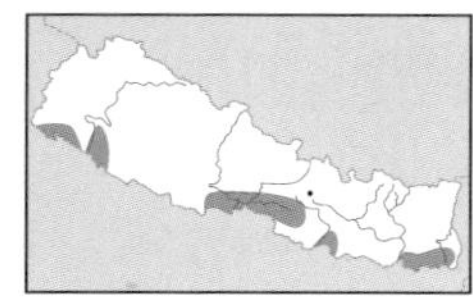

Pale-footed Bush Warbler *Cettia pallidipes* 11 cm

Very local resident; 75–250 m (–915 m). **ID** From Brown-flanked by whiter underparts (contrasting with brownish-olive breast-sides and flanks), more rufescent upperparts, shorter (square-ended) tail, pale pinkish legs and feet, and different vocalisations. **Voice** Song is loud, explosive *zip... zip-tschuk-o-tschuk*; *chik-chik* call. **HH** Typical bush warbler habits; very skulking in non-breeding season, less so when breeding. Calls frequently. Forages by flitting and hopping about in vegetation close to ground. Reluctant to fly, and usually only covers short distances at low level before dropping into dense cover again. Tall grasses and bushes at forest edges and clearings.

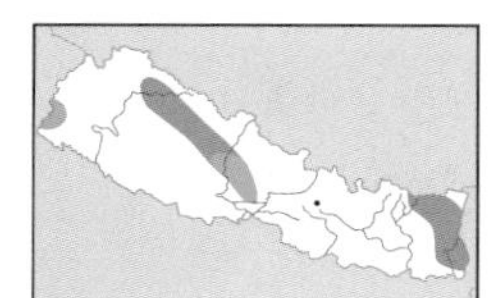

Brown-flanked Bush Warbler *Cettia fortipes* 12 cm

Resident, locally fairly common in far east and uncommon to rare in west; summers 1,800–3,200 m, winters 1,400–2,135 m (–250 m). **ID** From Pale-footed by duskier underparts, mainly pale buffish-grey with brownish-olive flanks. Also longer (rounded) tail, brownish legs and feet, more olive coloration to upperparts, less prominent supercilium and eye-stripe, and different vocalisations. Nominate in east has warmer, rufous-brown upperparts compared to *C. f. pallidus* in west, brownish-buff (rather than buffish-grey) throat and breast, and buffish (rather than greyish-white) supercilium. Juvenile has yellow underparts; confusable with Aberrant Bush Warbler, but upperparts browner. **Voice** Song a loud whistle, *weeee*, followed by explosive *chiwiyou*; *pallidus* gives harsh *tchuk*, *fortipes* a *chuk* and *tyit-tyu-tyu*. **HH** Typical bush warbler behaviour, see Pale-footed. Open areas with small groups of trees and bushes; forest edges and small ravines.

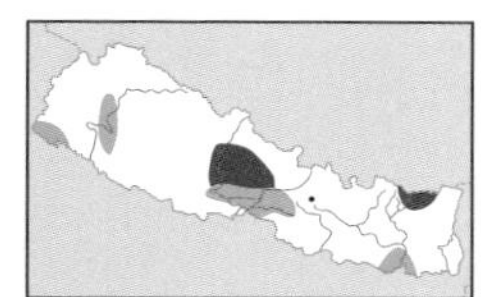

Chestnut-crowned Bush Warbler *Cettia major* 13 cm

Uncommon resident; breeds 3,550–3,680 m, winters 75–250 m. **ID** From Grey-sided by larger size and more robust appearance, larger bill, longer supercilium (indistinct and rufous-buff in front of eye), and whiter underparts (especially throat and centre of breast). Juvenile lacks chestnut on crown, and is more olive on upperparts and underparts, with greyish-buff supercilium behind eye; whiter throat and belly separate from juvenile Grey-sided. **Voice** Song comprises an introductory note followed by 3–4-note explosive warble. Call very similar to Grey-sided. **HH** Typical bush warbler behaviour, see Pale-footed. Breeds in rhododendron shrubberies and bushes in forest; winters in reedbeds.

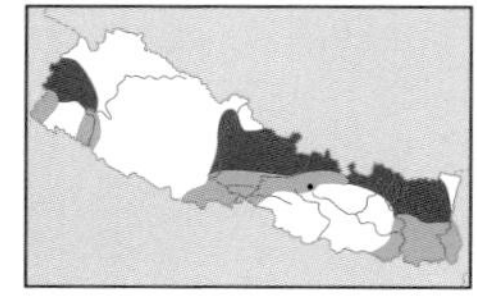

Aberrant Bush Warbler *Cettia flavolivacea* 12 cm

Fairly common resident; breeds 2,440–3,950 m, winters 75–1,830 m. **ID** From other bush warblers by yellowish-green cast to olive upperparts, yellowish supercilium, and buffish-yellow to olive-yellow underparts. Confusable with Tickell's Leaf Warbler, but has longer, 'loosely attached' and rounded tail, usually held slightly cocked, more rounded wings, and grating call accompanied by much wing-flicking. **Voice** Song a short warble, followed by a long inflected whistle, *dir dir-tee teee-weee*. Call a *brrrt-brrrt*, different from any *Phylloscopus*. **HH** Typical bush warbler. Breeds in bushes at forest edges, in clearings and shrubberies; winters in grassland and scrub.

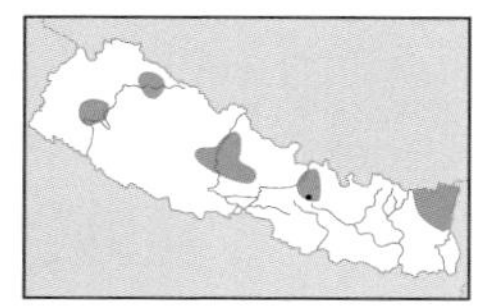

Hume's Bush Warbler *Cettia brunnescens* 9.5 cm

Local resident; 1,900–3,600 m. **ID** From Brown-flanked by smaller and finer bill, paler rufous-brown upperparts with strong olive cast (especially to lower back and rump), noticeable rufous fringes to tertials, and paler underparts with yellowish belly and flanks (yellow sometimes barely apparent). **Voice** Song a thin, high-pitched *see-saw see-saw see-saw* etc.; *chrrt chrrt chrrt* call. **HH** Typical bush warbler. Bamboo stands.

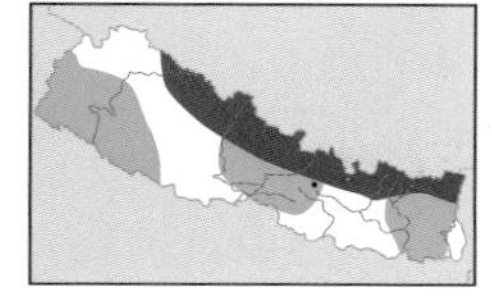

Grey-sided Bush Warbler *Cettia brunnifrons* 10 cm

Fairly common and widespread resident; breeds 2,745–4,000 m, winters 75–2,135 m. **ID** From Chestnut-crowned by smaller size, smaller bill, shorter supercilium (whitish-buff and well defined in front of eye), and greyer underparts. Juvenile lacks chestnut crown, has rufous-brown upperparts and brownish-olive underparts. **Voice** Song a loud wheezing *sip ti ti sip*, repeated continually. Call a bunting-like *pseek*. **HH** Typical bush warbler, but less skulking in breeding season. Breeds in high-altitude shrubberies and bushes at forest edges; winters in scrub and forest undergrowth in damp areas.

ad
Pale-footed Bush Warbler
ad
fortipes
juv
ad
pallidus
Brown-flanked Bush Warbler
juv
ad
ad
Aberrant Bush Warbler
Chestnut-crowned Bush Warbler
ad
juv
ad
Hume's Bush Warbler
Grey-sided Bush Warbler

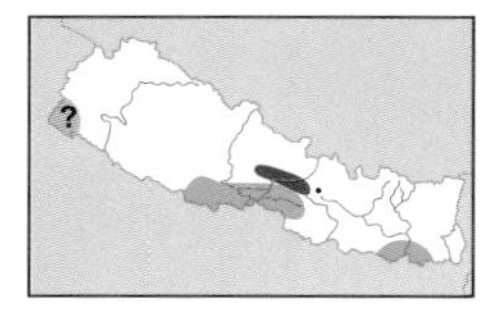

Spotted Bush Warbler *Bradypterus thoracicus* 13 cm

Rare on breeding grounds, local and frequent in winter; breeds 3,350–3,850 m, winters 75–250 m. **ID** From other bush warblers by spotting on throat and breast (sometimes indistinct), greyish supercilium, dark olive-brown upperparts with rufescent cast, grey ear-coverts and breast, olive-brown flanks, and boldly patterned undertail-coverts. Juvenile lacks grey on underparts, has cloudy olive-brown spotting and faint olive-yellow wash to underparts. See Appendix 1 for differences from Baikal Bush Warbler. **Voice** Repeated, rhythmic *trick-i-di* song; calls include long, drawn-out, harsh *tzee-eenk*. **HH** Habits similar to other bush warblers, see Pale-footed. Breeds in high-altitude shrubberies, grass and bushes; winters in reedbeds and tall grass near water.

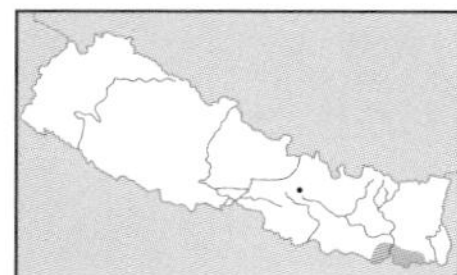

Chinese Bush Warbler *Bradypterus tacsanowskius* 14 cm

Very rare winter visitor and passage migrant, mainly at Koshi. **ID** Very similar to Brown Bush Warbler (see Appendix 1), but has greyer-brown upperparts with olive cast, olive grey-brown ear-coverts and breast (latter often with yellowish wash), olive-buff flanks, and pale tips to undertail-coverts (lacking or only just apparent on Brown). Variable (sometimes indistinct) whitish supercilium. Some have fine spotting on lower throat. Juvenile has strong yellowish wash to breast and flanks, yellowish-buff undertail-coverts, and stronger olive wash to upperparts. **Voice** *Chirr chirr* call. **HH** Typical bush warbler. Reedbeds, grass and bushes, and paddyfields.

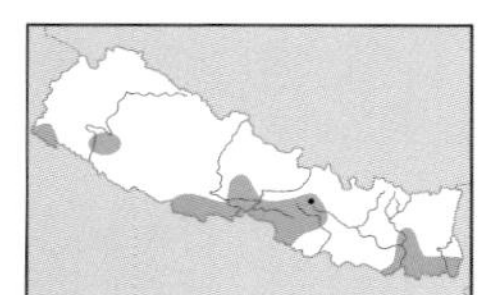

Clamorous Reed Warbler *Acrocephalus stentoreus* 19 cm

Uncommon winter visitor and passage migrant; 75–915 m (–1,340 m). **ID** Large *Acrocephalus* with long bill, short primary projection, whitish supercilium, and unstreaked underparts. Lacks white at tip of tail, which also helps to distinguish from Oriental Reed Warbler (see Appendix 1). In fresh plumage, olive cast to upperparts and variable buff wash to breast and flanks. **Voice** Loud repeated *karra-karra-karet-karet* song; loud and deep, hard *tak* or soft *karrk* call. **HH** Often forages in bushes. Movements slower and less agile than smaller *Acrocephalus*. Reedbeds and bushes around wetlands. **TN** The local subspecies *brunnescens* is sometimes treated as a separate species, Indian Reed Warbler.

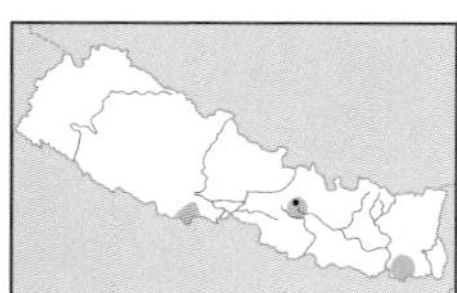

Blunt-winged Warbler *Acrocephalus concinens* 13 cm

Very rare and local visitor and passage migrant 75–1,525 m. **ID** From Paddyfield by shorter and less distinct supercilium (typically barely extends beyond eye), absence of dark border above supercilium (sometimes not apparent on Paddyfield), lack of dark stripe behind eye, longer and stouter bill (lacking dark tip) and longer tail (accentuated by shorter primary projection). Confusable with Blyth's Reed; when fresh, has more rufous tones, with warm buff breast and flanks; when worn retains rufous on rump. Also appears shorter-billed than Blyth's Reed, with longer and more rounded tail, more prominent fringes to tertials, and shorter and more rounded primary projection. **Voice** Calls include quiet *tcheck* and soft drawn-out *churr*. **HH** Habits resemble Paddyfield. Tall grasses, reedbeds and bushes along riverbanks.

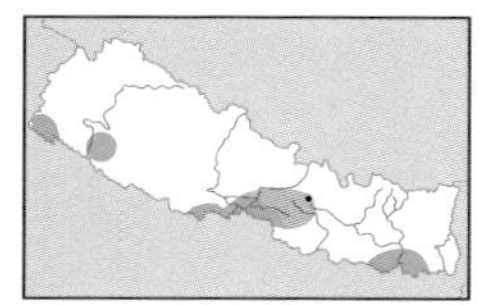

Paddyfield Warbler *Acrocephalus agricola* 13 cm

Local visitor, mainly to lowlands, chiefly at Koshi; 75–150 m (–1,340 m). **ID** Compared to Blyth's Reed has more prominent white supercilium behind eye (often with diffuse dark upper edge), more pronounced dark eye-stripe, shorter bill usually with well-defined dark tip, and typically shows dark centres and paler edges to tertials (uniform on Blyth's). Rufous cast to mantle and rump in fresh plumage, when Blyth's is more olive on upperparts, and flanks are more strongly washed buff. Worn upperparts greyer or sandier, but retain rufous cast to rump (absent on Blyth's). Legs and feet yellowish-brown to pinkish-brown (dark grey in Blyth's). **Voice** Soft *dzak* call. **HH** Skulking and lively, adept at climbing plant stems. Frequently flicks and cocks tail and raises crown feathers. Usually only flies short distances low over reeds with tail slightly spread and depressed. Reedbeds and paddyfields.

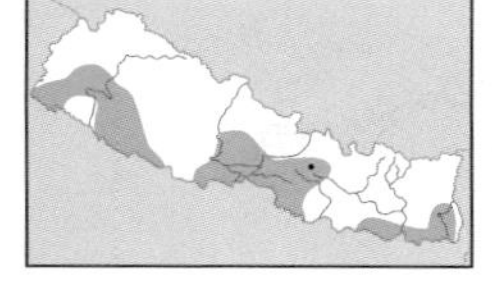

Blyth's Reed Warbler *Acrocephalus dumetorum* 14 cm

Fairly common and quite widespread winter visitor and passage migrant; 75–1,525 m (–2,900 m). **ID** Compared to Paddyfield has longer bill (usually lacking well-defined dark tip), olive-brown to olive-grey upperparts, and uniform wings. Supercilium comparatively indistinct and barely apparent behind eye. Noticeable olive cast to upperparts and edges of remiges in fresh plumage, but in first-winter upperparts can have slight rufous cast. **Voice** Fairly soft *chek* or grating *chek-tchr* call. **HH** Typically hops and creeps about in bushes, also trees and ground cover. Frequently flicks, raises and fans tail. Bushes and trees at edges of forest, cultivation, grassland and gardens

ad

ad

Spotted Bush Warbler

1st-win

ad fresh

ad

brunnescens

Clamorous Reed Warbler

Chinese Bush Warbler

ad fresh

Blunt-winged Warbler

ad worn

ad fresh

Paddyfield Warbler

ad worn

ad fresh

Blyth's Reed Warbler

ad worn

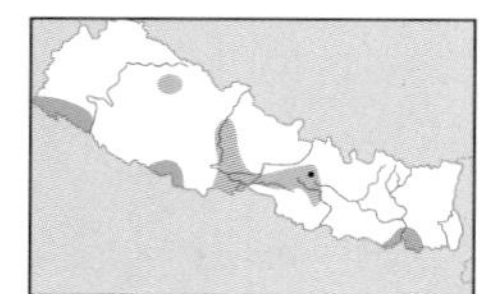

Booted Warbler *Iduna caligata* 12 cm

Very uncommon passage migrant and winter visitor; 75–2,810 m. **ID** Small, with square-ended tail, and short undertail-coverts. Upperparts brownish, with buff on sides of breast and flanks in fresh plumage; greyish-brown on upperparts, and whiter on underparts in worn plumage. Often shows faint whitish edges and tip to tail and fringes to tertials. Supercilium usually reasonably distinct and square-ended behind eye, and can be bordered above by diffuse dark line. Possibly confusable with Common Chiffchaff, but lacks any greenish or olive tones, and has pale base to lower mandible and pale legs. **Voice** Hard *chur chur* call. **HH** Forages at all levels from canopy to bushes, undergrowth and sometimes on ground. Moves restlessly in foliage and on branches capturing insects mostly by gleaning, and also in air by fluttering from end of a branch. Scrub and bushes at cultivation edges in dry habitats, favours acacias. **TN** Formerly in genus *Hippolais*.

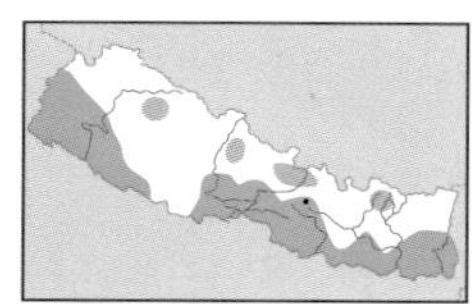

Common Chiffchaff *Phylloscopus collybita* 11 cm

Fairly common and quite widespread winter visitor; 75–1,370 m (–2,800 m). **ID** Whitish or buffish supercilium, and greyish to brownish upperparts with olive-green cast to rump, wings and tail. Underparts whitish, with buffish or greyish on sides of breast and flanks. Blackish bill and legs, less prominent supercilium (with prominent whitish crescent below eye) and absence of wing-bar separate from Greenish Warbler. **Voice** Calls include a plaintive *peu* or *hweet*. **HH** Forages at all levels, from tall treetops to bushes, undergrowth and sometimes on ground. Feeds chiefly by gleaning, but also hovers and makes short aerial sallies. Light forest, second growth and bushes. **TN** The subspecies in Nepal, *tristis*, is often treated as a separate species, Siberian Chiffchaff.

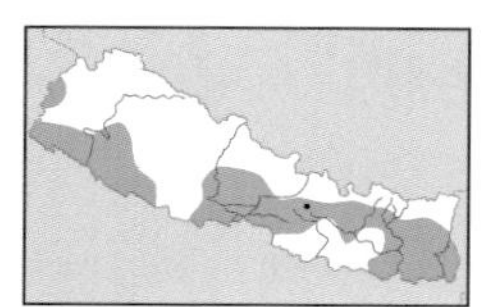

Dusky Warbler *Phylloscopus fuscatus* 11 cm

Frequent winter visitor; 75–1,600 m. **ID** Whitish underparts, often with buff on sides of breast and flanks, and dark brown to paler greyish-brown upperparts. Appears stockier than Common Chiffchaff, with more prominent supercilium and stronger dark eye-stripe, lacks olive-green edges to wing feathers and yellow at bend of wing, with paler legs and pale base to lower mandible. Two races occur: nominate (described above) and *P. f. weigoldi* (darker above, duskier below, with hint of yellow on underparts in fresh plumage). **Voice** Hard *chack chack* call diagnostic. **HH** Secretive, skulking in dense low cover, often on or close to ground, and sometimes in lower branches. Forest edges and bushes, especially near water.

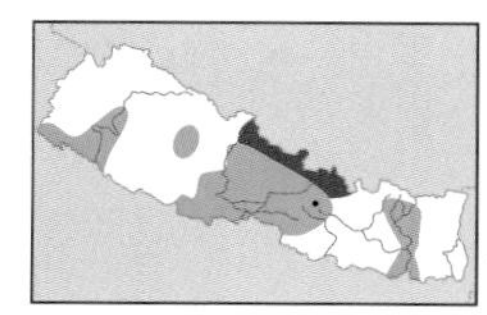

Smoky Warbler *Phylloscopus fuligiventer* 10 cm

Resident, rare on breeding grounds where probably under-recorded, but locally fairly common in wintering areas; breeds 3,900–5,000 m, winters 75–915 m. **ID** From Dusky and Sulphur-bellied by smaller size, short-looking tail, darker sooty-olive upperparts (with greenish tinge in fresh plumage), short, indistinct supercilium (with bold white crescent below eye), and mainly dusky-olive underparts (with oily yellow centre). **Voice** Monotonous *tsli-tsli-tsli-tsli-tsli* song; throaty *thrup thrup* call. **HH** When breeding creeps about in low scrub and clambers among boulders; in winter forages close to water's edge, sometimes darting out and hovering briefly over water. Breeds in high-altitude shrubberies; winters in dense undergrowth near water.

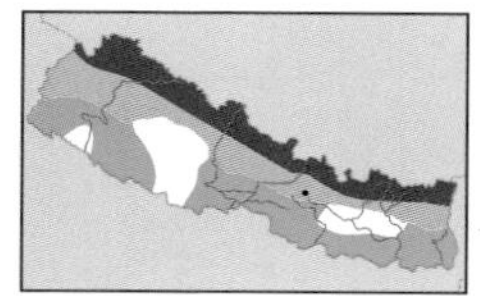

Tickell's Leaf Warbler *Phylloscopus affinis* 11 cm

Common and widespread resident; breeds 2,550–4,880 m, winters 75–1,190 m. **ID** From Sulphur-bellied by greenish-brown upperparts, greenish edges to wing feathers, and brighter lemon-yellow underparts. Supercilium similar in coloration to throat, and eye-stripe more clearly defined than Sulphur-bellied, contrasting with yellowish ear-coverts. Worn birds may lack greenish cast to upperparts, and have paler yellow supercilium and underparts. **Voice** Short *chip...whi-whi-whi-whi*; *chit* call not as hard as Dusky. **HH** When breeding flits actively among low bushes, rocks and on ground; in winter usually in bushes, also trees. Breeds in open country with bushes; winters in bushes at edges of forest and cultivation.

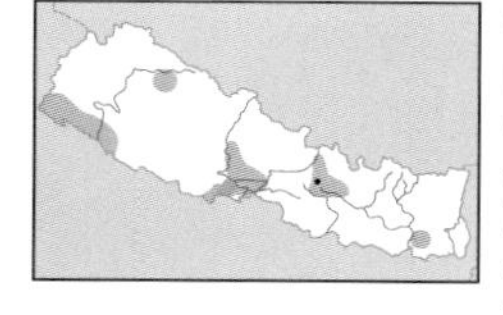

Sulphur-bellied Warbler *Phylloscopus griseolus* 11 cm

Rare passage migrant, mainly in April; very rare winter visitor; 75–2,700 m. **ID** From Tickell's by colder brownish-grey upperparts lacking any greenish tones, greyish-white edges to remiges, and duller buffish underparts (with yellow purest on belly). Supercilium bright sulphur-yellow, brighter than throat, with grey-brown ear-coverts and sides of breast. **Voice** Soft *quip* call, distinct from Tickell's. **HH** Distinctive habit of climbing trees and walls/rock faces. Winters in rocky areas and around old buildings.

ad worn
tristis
ad
ad fresh
Booted Warbler
Common Chiffchaff
ad
Dusky Warbler
ad
Smoky Warbler
ad
Tickell's Leaf Warbler
ad
Sulphur-bellied Warbler

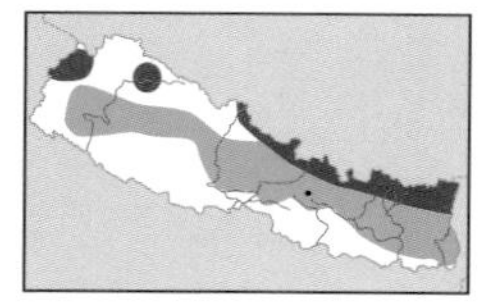

Buff-barred Warbler *Phylloscopus pulcher* 10 cm

Common and widespread resident; breeds 3,250–4,300 m, winters 915–3,050 m (–75 m). **ID** From Hume's by buffish-orange wing-bars, white in tail, yellowish supercilium, and small yellowish rump patch. Poorly defined dull yellowish centre to rear crown. Two intergrading races occur: *P. p. kangrae* in west is brighter than nominate race in centre and east, with more yellowish-olive upperparts and cleaner yellow underparts. **Voice** High-pitched twittering song; short, sharp *swit* call. **HH** Usually in mixed flocks of other insectivores in non-breeding season. Forages actively among foliage and twigs. Breeds mainly in subalpine forest; winters in broadleaved forest.

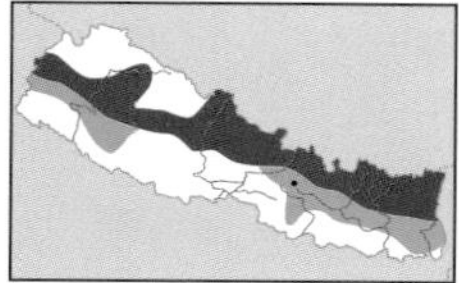

Ashy-throated Warbler *Phylloscopus maculipennis* 9 cm

Fairly common and widespread resident; breeds 1,800–3,500 m, winters 1,525–2,900 m (–500 m). **ID** Small with yellow rump and double yellowish wing-bars. From Lemon-rumped by greyish-white supercilium and crown-stripe, greyish throat and breast, yellow belly, and white on tail. **Voice** Song *ti-ti-whee-tew*; short *swit* call. **HH** Singly or in foraging parties, often with other insectivores. Mainly in trees, also bushes. Very restless, flutters and hovers among foliage and twigs. Broadleaved and broadleaved/coniferous forests; also secondary forest in winter.

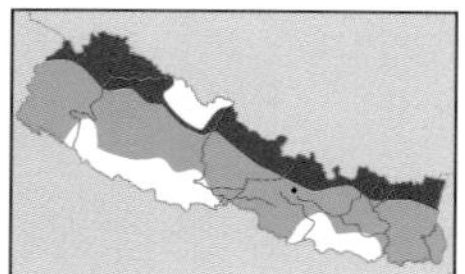

Lemon-rumped Warbler *Phylloscopus chloronotus* 9 cm

Common and widespread resident; 2,750–4,000 m, winters 275–2,750 m (–75 m). **ID** From similar species by combination of broad yellowish-white supercilium and crown-stripe (contrasting with dark olive sides of crown), double yellowish-white wing-bars, well-defined yellowish (sometimes almost whitish) rump, and whitish underparts. Lacks white in tail. Two intergrading races occur: *P. c. simlaensis* in west has brighter yellowish-green upperparts, and brighter yellow supercilium and crown-stripe, compared to nominate in centre and east. **Voice** Two songs: one a drawn-out thin rattle followed by a series of stammering notes; the other a stuttering series of notes with alternating pitch; high-pitched *uist* call. **HH** With other insectivores in non-breeding season. Extremely active, frequently hovering in and around foliage. Breeds in broadleaved, coniferous and mixed broadleaved/coniferous forests, also shrubbery above treeline; winters in forest and second growth.

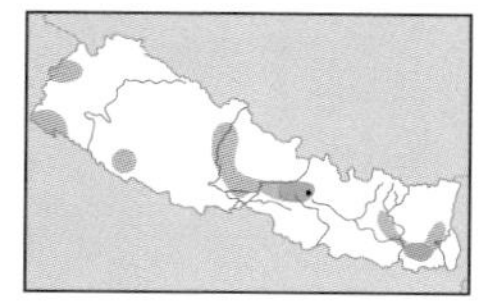

Yellow-browed Warbler *Phylloscopus inornatus* 10–11 cm

Uncommon winter visitor and passage migrant; 75–2,590 m. **ID** In fresh plumage, brighter greenish-olive upperparts compared to Hume's, with yellowish-white supercilium, ear-coverts and wing-bars; further, median coverts wing-bar is typically well defined, and underparts very white with variable amounts of yellow. In worn plumage, close in coloration to Hume's, but has pale base to lower mandible, and legs are paler flesh-brown or greyish-brown (bill and legs darker in Hume's). There are important vocal differences. **Voice** Loud rising *che-wiest* call. **HH** Hunts actively in trees and bushes, favours sunny edges of forest clearings. Groves and open forest.

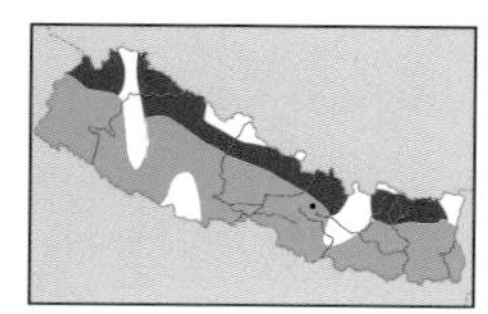

Hume's Leaf Warbler *Phylloscopus humei* 10–11 cm

Common and widespread resident and passage migrant; breeds 3,280–3,980 m, winters 75–2,135 m (–2,560 m). **ID** Compared to Yellow-browed has greyish-olive upperparts, with variable yellowish-green on mantle and back, and browner crown, while supercilium, ear-coverts and greater coverts wing-bar are buffish-white. Median coverts bar poorly defined, but double wing-bar can be apparent. Bill all dark, legs blackish-brown. Supercilium, wing-bars and underparts white when worn, upperparts much greyer. **Voice** Song a repeated *wesoo*, often followed by descending high-pitched *zweeeeeeeoooo*; disyllabic *whit-hoo* and sparrow-like *chwee* calls. **HH** Forages in trees and bushes, occasionally descending to ground and favours sunny edges of forest clearings. Frequent member of mixed foraging flocks in winter. Breeds in coniferous forest; winters in forest and second growth.

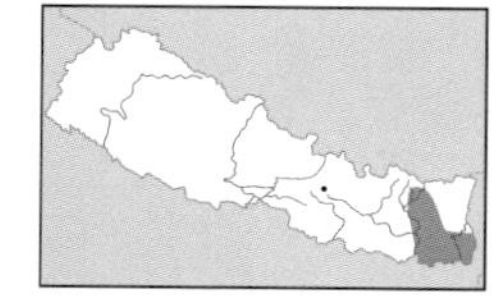

Yellow-vented Warbler *Phylloscopus cantator* 10 cm

Rare and local resident in east; 75–1,525 m. **ID** From Blyth's Leaf by yellow throat, upper breast and undertail-coverts contrasting with white lower breast and belly. In addition, smaller in size, has brighter yellow supercilium and crown-stripe contrasting with darker lateral crown-stripes, and has brighter yellowish-green upperparts. **Voice** Song consists of several single notes on same pitch ending in two slurred notes: *seep, seep, seep to-you* with accent on *you*; double call is not softer than other *Phylloscopus* with accent on second note. **HH** Often in mixed itinerant flocks in winter. Forages in lower storey. Has habit of spreading tail and flicking it upwards when calling. Breeds in dense, moist evergreen broadleaved forest; winters in more open forest.

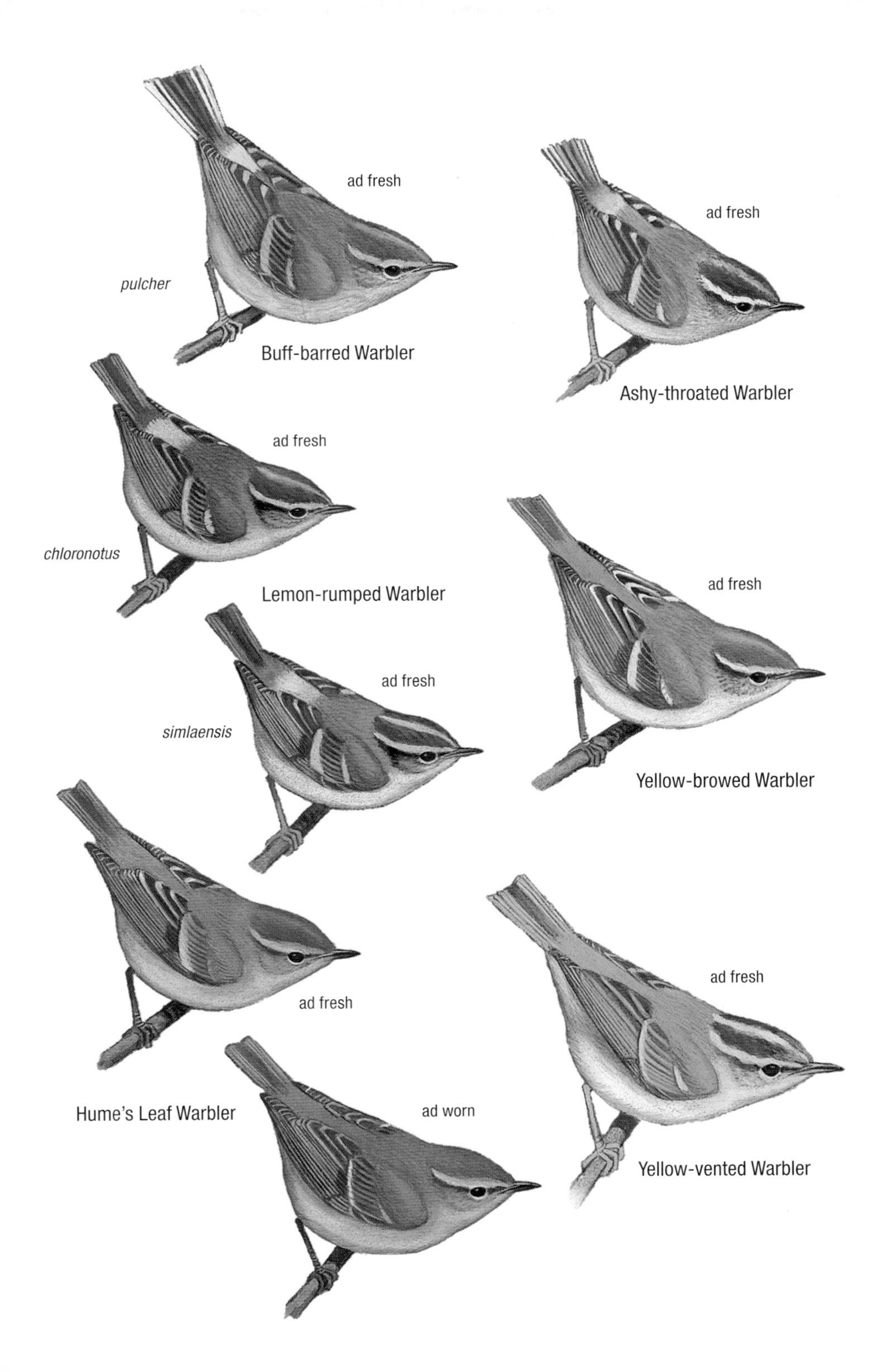

ad fresh
pulcher
Buff-barred Warbler
ad fresh
Ashy-throated Warbler
ad fresh
chloronotus
Lemon-rumped Warbler
ad fresh
simlaensis
ad fresh
Yellow-browed Warbler
ad fresh
Hume's Leaf Warbler
ad worn
ad fresh
Yellow-vented Warbler

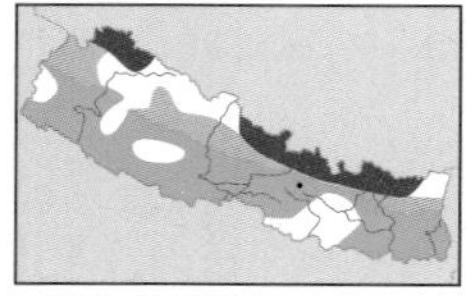
including Green Warbler

Greenish Warbler *Phylloscopus trochiloides* 9.5–10.5 cm

Common summer visitor, winter visitor and passage migrant; breeds 3,000–4,270 m, winters 75–1,830 m. **ID** Variable *Phylloscopus* with uniform crown, fine wing-bars and distinctive call. In fresh plumage *P. t. viridanus* (winter visitor) has olive-green upperparts, single white wing-bar, yellowish-white supercilium and whitish underparts with faint yellowish suffusion. When worn, upperparts duller and greyer and underparts whiter. Nominate (breeding visitor with some overwintering) has darker oily green upperparts (with darker crown), mottled ear-coverts, dusky underparts with diffuse oily yellow wash, and darker bill (orange at base of lower mandible); often shows trace of second (median coverts) wing-bar. Nominate can appear very similar to Large-billed Leaf, and best identified by call. **Voice** *P. t. viridanus* has loud, slurred and abrupt *chi-weee* call; loud, repeated *chi-chi-chi-chiwee-chiweee* song. *P. t. trochiloides* has *chis-weet* call and *chis-weet, chis-weet* song. **HH** Forages from canopy to low bushes. Breeds in broadleaved and coniferous forests and in subalpine shrubbery; winters in well-wooded areas.

Green Warbler *Phylloscopus (trochiloides) nitidus* 10–11 cm

Uncommon passage migrant. **ID** In fresh plumage, upperparts brighter and purer green than Greenish, and has one, sometimes two, slightly broader and yellower wing-bars, while supercilium and cheeks are noticeably yellow and underparts have much stronger yellow suffusion. When worn, upperparts duller, although still brighter than Greenish, and supercilium and underparts retain yellowish wash. **Voice** Trisyllabic *chis-ru-weet* call. **HH** Open forest and wooded areas.

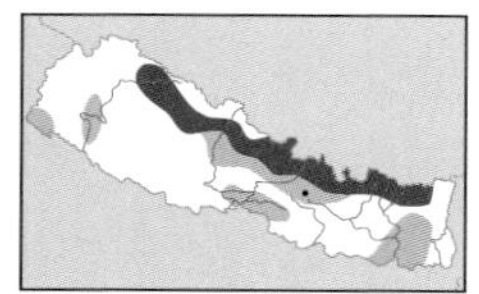

Large-billed Leaf Warbler *Phylloscopus magnirostris* 13 cm

Fairly common summer visitor, rare in winter; breeds 2,440–3,800 m, 250 m (winter). **ID** From *viridanus* Greenish by larger size, larger and mainly dark bill, darker oily green upperparts (with darker crown), more striking yellowish-white supercilium, and broader dark eye-stripe with greyish mottling on ear-coverts. Underparts tend to look rather dirty, often with diffuse streaking and oily yellow wash (but can be whiter and much like Greenish). Has yellowish-white greater coverts wing-bar, often with trace of second (median coverts) bar. Very similar to nominate Greenish and best separated by distinctive vocalisations. Also larger, has larger bill with more pronounced hooked tip, more prominent supercilium and broader dark eye-stripe. **Voice** Clear, whistled, upward-inflected *der-tee* call; loud song, five syllables in three descending notes. **HH** Mainly in trees. Breeds in forest along streams and rivers.

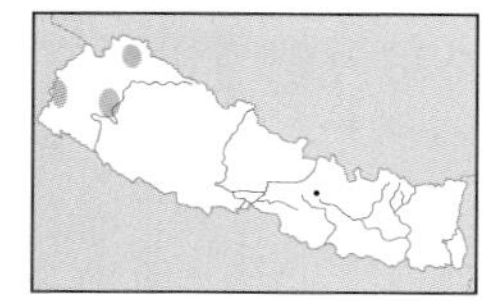

Tytler's Leaf Warbler *Phylloscopus tytleri* 11 cm

Rare and local passage migrant in west; 2,135–3,050 m. **ID** From Greenish by long, slender, mainly dark bill, shorter tail, lack of wing-bar and different call. Supercilium tends to look finer, and is offset by rather broad and well-defined dark olive lores and eye-stripe. In fresh plumage, dark greenish cast to upperparts, supercilium yellowish-white, and has variable yellowish wash to ear-coverts and underparts. In worn plumage, supercilium and underparts whitish, and upperparts greyer. **Voice** Double *y-it* call; distinctive, repeated *pi-tsi-pi-tsu* song. **HH** Forages mainly by gleaning, also makes aerial sallies. Bushes at forest edges.

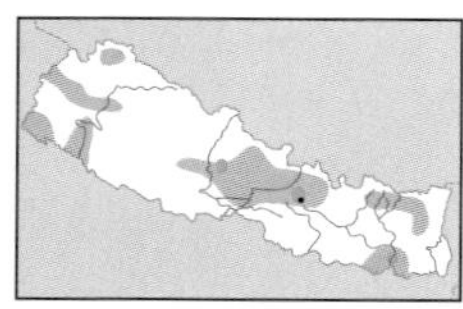

Western Crowned Warbler *Phylloscopus occipitalis* 11 cm

Uncommon to rare winter visitor and passage migrant, mainly in spring; 75–2,900 m. **ID** Very similar to Blyth's Leaf but appears larger and more elongated, with larger and longer-looking bill. Upperparts generally duller greyish-green, and whitish underparts strongly suffused grey. Median and greater coverts wing-bars are less prominent and head pattern tends to be less striking (supercilium and crown stripe duller and contrast less with dusky-olive sides to crown, which may be darker towards nape). **Voice** Constantly repeated *chit-weei* call. **HH** Forages at all levels. Broadleaved forests.

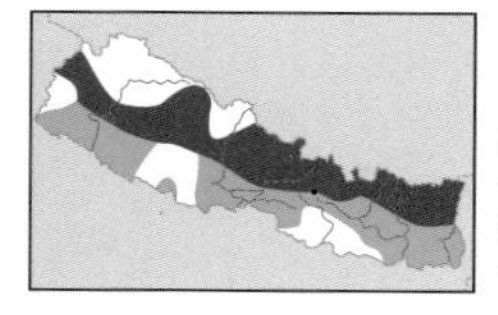

Blyth's Leaf Warbler *Phylloscopus reguloides* 11 cm

Common and widespread resident; breeds 1,750–3,800 m, winters 75–1,500 m (–2,700 m). **ID** Very similar to Western Crowned. Head pattern tends to be more striking, with yellower supercilium and crown-stripe, and darker lateral crown-stripes (can be almost black). Underparts generally have distinct yellowish wash, and upperparts are darker and purer green. Wing-bars more prominent (being broader, often divided by dark panel across greater coverts). **Voice** Constantly repeated *kee-kew-i* call; *ch-ti-ch-ti-chi-ti-ch-ti-chee* trilled song. **HH** Unlike other leaf warblers clings upside-down to trunks like a nuthatch. Breeds in broadleaved and coniferous forests, and winters in forests.

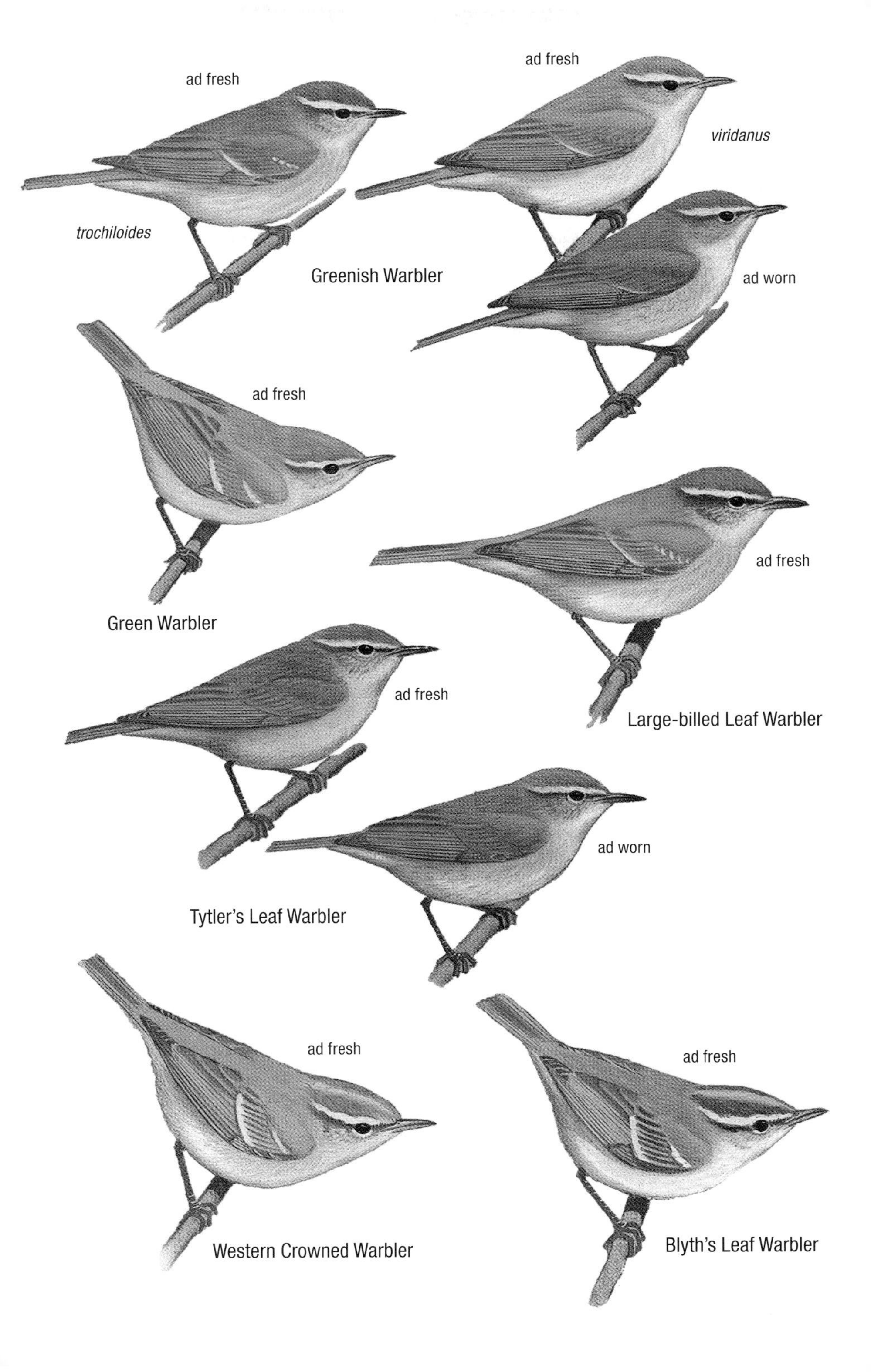
ad fresh
ad fresh
viridanus
trochiloides
Greenish Warbler
ad worn
ad fresh
ad fresh
Green Warbler
ad fresh
Large-billed Leaf Warbler
ad worn
Tytler's Leaf Warbler
ad fresh
ad fresh
Western Crowned Warbler
Blyth's Leaf Warbler

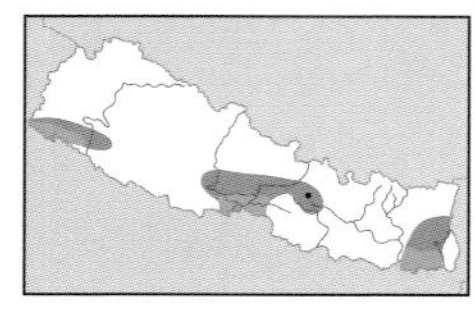

Green-crowned Warbler *Seicercus burkii* 11 cm

Fairly common, widespread resident; breeds 1,550–2,050 m, winters 250–915 m (–75 m), possibly higher. **ID** Has yellow eye-ring, yellowish-green face and green crown. Compared to Whistler's, sides of crown are blacker, has narrower eye-ring (broken at rear) and usually lacks wing-bar. See Whistler's for further details. **Voice** Song comprises varied, quite rich phrases, e.g. *weet-weeta-weeta-weet* interspersed by short trills. **HH** Feeds in similar manner to *Phylloscopus* warblers by gleaning from foliage and twigs, and by making frequent aerial sallies. In pairs or mixed feeding flocks of warblers, tits and babblers. Forages in forest understorey and second growth. **AN** Golden-spectacled Warbler.

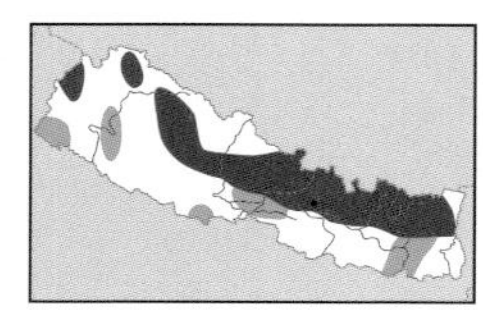

Whistler's Warbler *Seicercus whistleri* 11–12 cm

Common and widespread resident; breeds 2,130–3,800 m, winters 75–2,135 m. **ID** Very similar to Green-crowned; dark sides of crown not as black and diffuse on forehead, and yellow eye-ring broader at rear. Generally, upperparts duller greyish-green, underparts duller yellow, and wing-bar usually more distinct. Shows more white in outer tail feathers; much white on basal half of outer web of outermost tail feathers (generally lacking white in this area in Green-crowned). **Voice** Song a simple *witchu-witchu* is best means of separation from Green-crowned. **HH** Behaviour very similar to Green-crowned. Forest understorey, second growth and high-altitude shrubberies. **TN** Formerly treated as conspecific with Green-crowned Warbler.

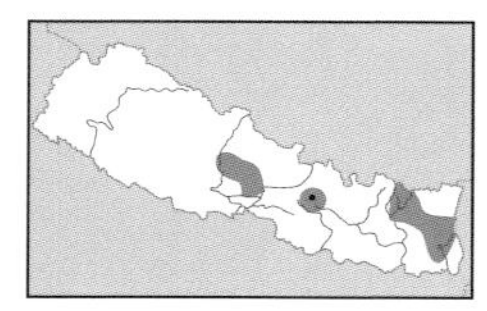

Grey-cheeked Warbler *Seicercus poliogenys* 10 cm

Rare resident; 2,440–3,200 m (–250 m). **ID** From other *Seicercus* by combination of prominent white eye-ring, whitish chin and upper throat, dark grey ear-coverts, and more uniform dark grey head, with poorly defined crown-stripe and diffuse dark sides to crown. **Voice** Song a combination of slurred notes, rather like a *Phylloscopus*, phrases include *tisi-titsi-chi*; *chi-chi-chi-chi-chi*; *twechi-chewchi-chew*. **HH** Flits actively through undergrowth and bushes of dense forest; frequently makes aerial sallies, and flutters and hovers among foliage and twigs. Often joins mixed foraging flocks. Bamboo and dense undergrowth in broadleaved evergreen forest.

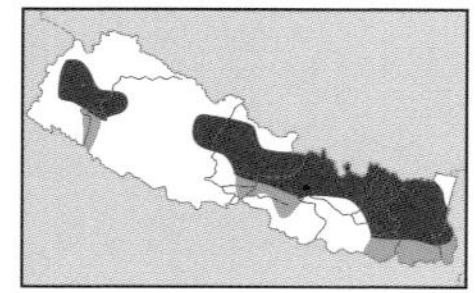

Chestnut-crowned Warbler *Seicercus castaniceps* 9.5 cm

Frequent resident; breeds 1,800–2,750 m, winters 1,000–2,285 m (–250 m). **ID** From other *Seicercus* by combination of small size, chestnut crown with diffuse dark brown lateral crown-stripes, bright lemon-yellow rump, grey sides of head with white eye-ring, grey throat and upper breast contrasting with white lower breast and belly, and bright yellow flanks. Grey on sides of head, double yellow wing-bars, prominent yellow rump and lack of white supercilium are best features for separation from Broad-billed Warbler. **Voice** Song of 5–7 notes extremely high-pitched, sibilant and slightly undulating. **HH** Singly, in pairs or with mixed foraging flocks of warblers and tits. Flits restlessly through middle and upper forest storeys. Gleans from foliage and twigs, frequently making aerial sallies, hovering and fluttering. Broadleaved forest, mainly of oaks.

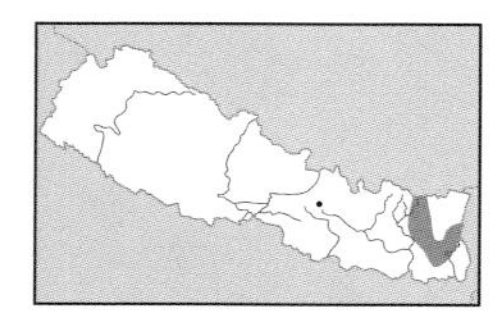

Broad-billed Warbler *Tickellia hodgsoni* 10 cm

Rare and local in east, probably resident; 2,195–2,300 m. **ID** Distinctive warbler with chestnut forehead and crown, greyish-white supercilium and dark grey eye-stripe, greyish ear-coverts, throat and breast, and yellow belly, flanks and vent. Slightly yellower rump. Chestnut crown, grey throat and breast, and white on tail are best features from Yellow-bellied. **Voice** Song a series of very thin, high-pitched whistles, repeated at intervals; also rapid, rather metallic jumbled notes. **HH** Often in company with itinerant mixed flocks of small birds. Frequents forest understorey: bamboo and other undergrowth in dense evergreen broadleaved forest.

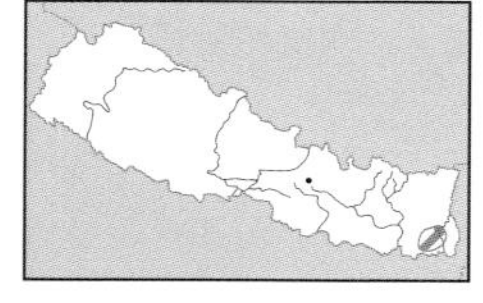

Rufous-faced Warbler *Abroscopus albogularis* 8 cm

Probably a former resident; 305–1,200 m; no recent records. **ID** Readily identified by pale rufous face (lacking prominent eye-ring), buff crown (with blackish lateral crown-stripes), blackish mottling on throat, yellow band across breast, whitish rump, uniform wings (no wing-bars) and lack of white in tail. **Voice** Sibilant, high-pitched repetitive whistles *titiriiiii, titiriiiii titiriiiii* etc. **HH** Very active. Often with itinerant mixed flocks of small birds. Usually keeps to undergrowth and lower branches, in bamboo clumps at edges, and more open areas of moist broadleaved and evergreen forest.

ad

Green-crowned Warbler

ad

Whistler's Warbler

ad

Grey-cheeked Warbler

ad

Chestnut-crowned Warbler

ad

Broad-billed Warbler

ad

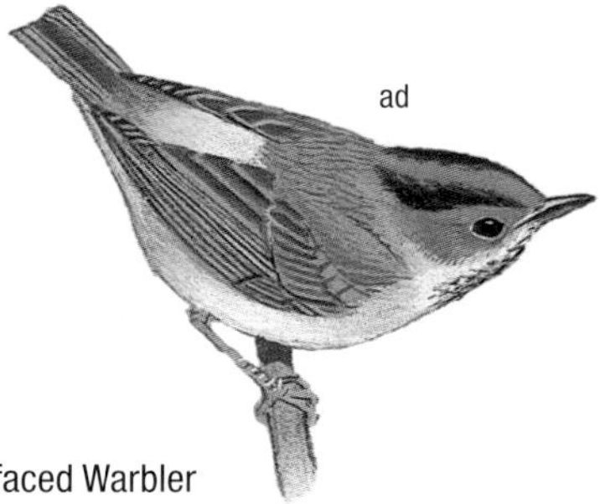

Rufous-faced Warbler

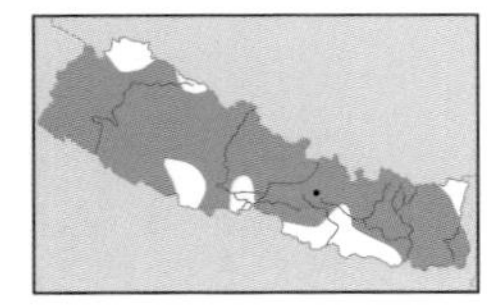

Grey-hooded Warbler *Phylloscopus xanthoschistos* 10 cm

Common and widespread resident; breeds 1,000–2,750 m (–4,250 m); mainly winters 750–2,200 m (75–2,750 m). **ID** From *Seicercus* by combination of greyish-white supercilium, and grey crown and mantle. Has diffuse pale grey central crown-stripe, and darker lateral crown-stripes. Two intergrading races occur: *P. x. albosuperciliaris* in west has paler grey lateral crown-stripes and mantle (with brownish cast) compared to nominate in east, which is purer and darker grey in these areas, with noticeable contrast between grey mantle and green back. **Voice** High-pitched *psit-psit*, plaintive *tyee-tyee* call. Song a brief, incessantly repeated high-pitched warble, *ti-tsi-ti-wee-tee*. **HH** Active warbler, hunts restlessly and feeds by gleaning, making short aerial sallies and hovering in front of sprigs. Often with other small insectivores when not breeding. Lower canopy and bushes in forest and second growth; prefers edges and open forest. **TN** Formerly in genus *Seicercus*.

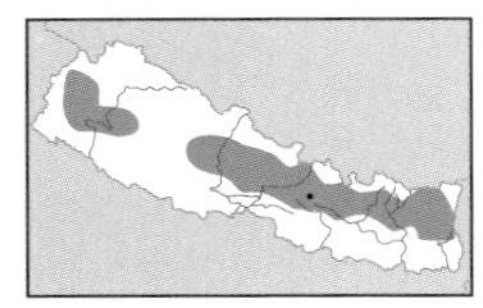

Black-faced Warbler *Abroscopus schisticeps* 9 cm

Local resident; 1,525–2,700 m. **ID** Striking yellow supercilium and throat, and black mask. Grey crown and nape, and uniform olive-green upperparts lacking wing-bars. Superficially resembles Yellow-bellied Fantail but has much shorter tail. **Voice** Rapid, high-pitched *tz-tz-tz-tz-tz* alarm call. **HH** Usually in fast-moving parties of 10–15 birds, often with other warblers and babblers in non-breeding season. Forages actively in mid-storey and in tall bushes in moist oak and mixed broadleaved forests, especially in moss-covered trees, bamboo and thick undergrowth.

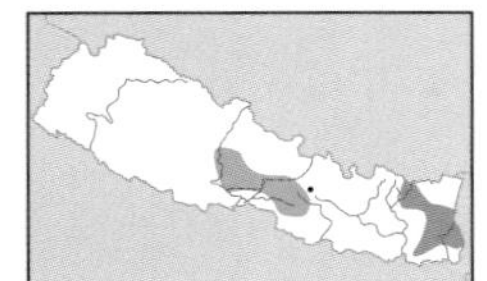

Yellow-bellied Warbler *Abroscopus superciliaris* 9 cm

Uncommon winter visitor to Chitwan, very local and very uncommon elsewhere, probably resident; 245–1,525 m (–2,300 m). **ID** Dull-coloured compared to *Seicercus* and other *Abroscopus* warblers. White supercilium, dark crown and eyestripe, yellowish-olive upperparts, white throat and yellow rest of underparts. Confusable only with Grey-hooded, and best distinguished by white throat, yellowish-olive (rather than grey) mantle, and lack of white on tail. Separated from all *Phylloscopus* warblers by combination of rather long bill, brownish-grey crown, white throat and yellow rest of underparts, and by fairly narrow tail, lacking prominent undertail-coverts, giving distinctive profile. Tail appears pale brownish-buff from below. **Voice** Song a halting ditty of six notes, ascending at end; *chrrt chrrt chrrt* call. **HH** Singly, in pairs or in mixed, roving flocks of insectivores. Very active, feeds chiefly by making short aerial sallies. Forest edges, wooded ravines, often near forest streams, and usually near bamboo in moist broadleaved forest.

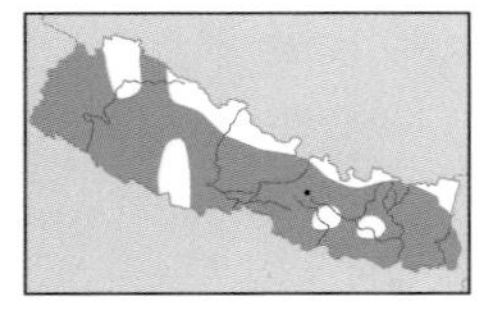

Oriental White-eye *Zosterops palpebrosus* 10 cm

Common and widespread resident; breeds up to 2,440 m, winters 75–1,370 m. **ID** Distinctive, with prominent white eye-ring, black bill and lores, green to yellowish-green upperparts, bright yellow throat and vent, and whitish rest of underparts with variable greyish wash. **Voice** Plaintive *cheer* or *prree-u* call; tinkling jingle song. **HH** Outside breeding season in flocks of up to 50, which continually utter plaintive contact calls. Favours flowering shrubs and trees. Forages actively and often hangs upside-down. Open broadleaved forest and wooded areas.

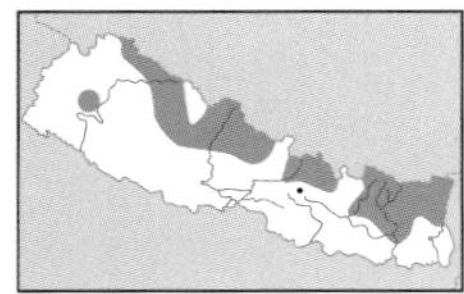

Goldcrest *Regulus regulus* 9 cm

Frequent resident; breeds up to 4,000 m, winters 2,200–3,050 m. **ID** Small and plump, with greenish upperparts, recalling *Phylloscopus* warbler. Best distinguished by very plain face, without supercilium and eye-stripe, by pale ring around large dark eye, and by brilliant yellow to orange crown bordered by black. Juvenile lacks striking crown pattern. **Voice** Trisyllabic, high-pitched *see-see-see* call; similarly high-pitched *see-seesisyu-seesisyu-seesisyu-sweet* song. **HH** Arboreal, often in mixed flocks with tits, treecreepers and leaf warblers. Forages amongst foliage, usually in tree canopy where constantly flits from twig to twig. Mainly coniferous forest.

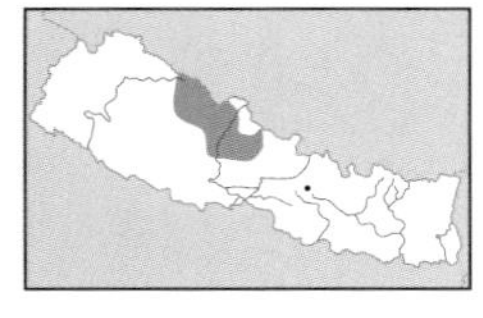

White-browed Tit Warbler *Leptopoecile sophiae* 10 cm

Frequent resident in Trans-Himalayas; 2,700–4,575 m all year. **ID** White supercilium, rufous crown, greyish upperparts, white outer tail feathers, and bluish flanks and rump. Male has lilac wash to crown, and violet-blue sides to head, throat and breast. Female has pale grey-brown head-sides and greyish-white underparts (with lilac wash to flanks). **Voice** Slow, subdued *teet* call; song a sweet loud chirping. **HH** Often in small parties. Forages very actively in bushes and undergrowth where it flits and hops among branches and foliage; can hang upside-down like a tit. Occasionally perches on top of a bush. Frequently cocks tail. Dwarf scrub in semi-desert.

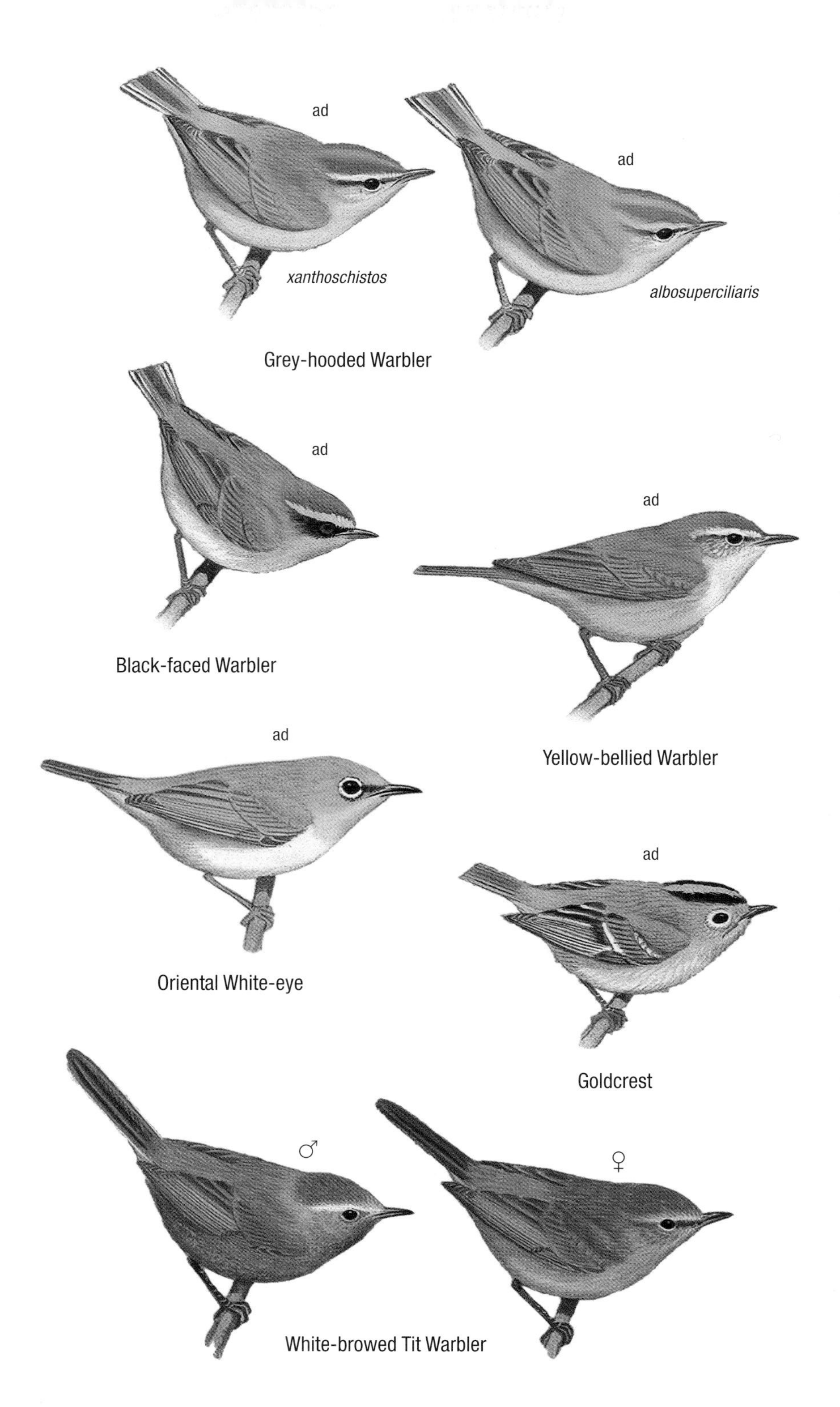
ad
ad
xanthoschistos
albosuperciliaris
Grey-hooded Warbler
ad
ad
Black-faced Warbler
ad
Yellow-bellied Warbler
ad
Oriental White-eye
Goldcrest
♂
♀
White-browed Tit Warbler

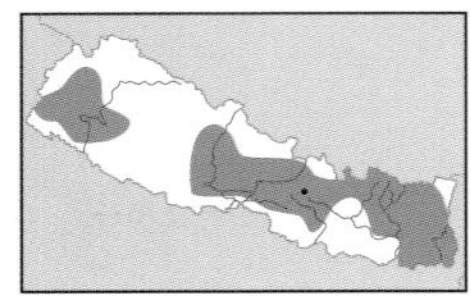

Rusty-cheeked Scimitar Babbler *Pomatorhinus erythrogenys* 25 cm

Locally common resident; 305–2,440 m. **ID** Rufous lores and ear-coverts. Unmistakable throughout west and central Himalayas: lacks white supercilium, and has rufous forehead, ear-coverts, sides of breast, flanks and vent. *P. e. haringtoni* in east has greyer or noticeably grey-streaked/spotted throat and upper breast. **Voice** Loud, far-carrying three-noted *quoit* or *khwoi* calls followed immediately by staccato *quit* from female; also a ringing *khwir jot-khwir jot* often followed by *khuwiyiu* and quickly answered by female. **HH** In pairs in breeding season, small parties at other times. Moves rapidly through undergrowth keeping well concealed. Rummages and probes among dead leaves for insects. Short, floppy flight and bouncing hops on ground. Thick scrub and undergrowth at forest edges, secondary jungle, bushy hillsides, and bushes at field edges.

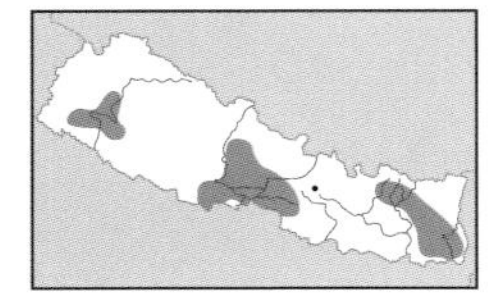

White-browed Scimitar Babbler *Pomatorhinus schisticeps* 22 cm

Local resident; 245–1,695 m. **ID** Striking white supercilium contrasting with black ear-coverts, and downcurved yellow bill. From similar Streak-breasted by clean white centre to breast and belly, chestnut sides to breast with variable white streaking, larger size and larger, more curved bill. Juvenile smaller, with smaller bill and olive-brown crown, more closely resembles Streak-breasted, but centre of breast white and sides of breast chestnut. **Voice** Single note followed by a trilled hoot or evenly spaced three-noted call that varies from a hoot to a whistle. **HH** Often in parties with other babblers. Forages chiefly in thick forest undergrowth and dense scrub, favouring bamboo thickets.

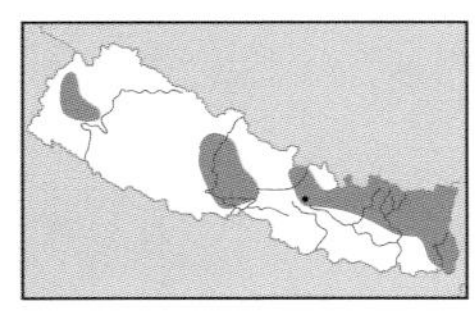

Streak-breasted Scimitar Babbler *Pomatorhinus ruficollis* 19 cm

Fairly common and widespread resident; 1,300–2,700 m. **ID** Striking white supercilium contrasting with dark ear-coverts, and downcurved yellow bill. From similar White-browed by diffuse olive-brown streaking on breast and belly, merging into olive-brown on flanks, distinct rufous patch on sides of neck extending across nape as diffuse band, smaller size, and smaller and less downcurved bill. Eye dark (yellow in White-browed). Juvenile has brighter rufous upperparts, rufous breast with some white streaking, and smaller, almost straight bill. **Voice** Call a soft, musical *of-an-on*. Song varied, typically including rising *pouki-wurki pouki-wurki* or falling *prrurti-witeu-witeu*. **HH** Habits similar to Rusty-cheeked. Sometimes with other babblers. Dense undergrowth in open forest, thick broadleaved forest, and dense scrub on hillsides.

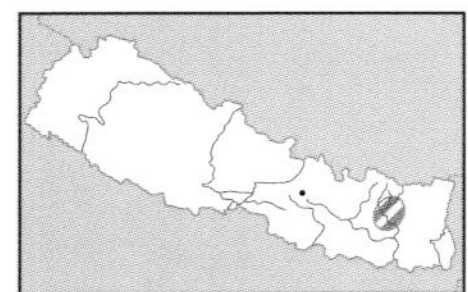

Coral-billed Scimitar Babbler *Pomatorhinus ferruginosus* 22 cm

Probably former resident; 1973 records from Arun Valley in far east; 2,745–3,660 m. **ID** From White-browed and Streak-breasted, by blood-red bill, deep rufous breast and belly, and bright rufous feathering on forehead and lores. **Voice** Soft whistles, oriole-like mewing and short squeaks combined with harsh scolding. **HH** Skulking and secretive. Bamboo thickets and dense undergrowth in broadleaved evergreen forest.

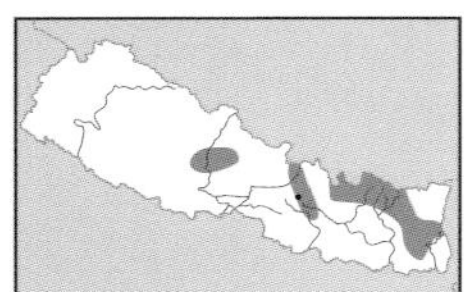

Slender-billed Scimitar Babbler *Xiphirhynchus superciliaris* 20 cm

Uncommon and local resident from west-central Nepal eastwards; 1,500–3,000 m. **ID** Long, slender, downcurved black bill, fine and feathery white supercilium contrasting with slate-grey crown and ear-coverts, grey-streaked white throat, and deep rufous underparts. **Voice** Repeated powerful tremulous *pwoorr* or slower *toop-toop-toop-toop* with female interjecting a two- to four-noted whistle, *bu-tu-wheip-wheip*. **HH** Skulking and shy. Forages in similar fashion to other scimitar babblers by probing leaf litter, moss and bark crevices; sometimes also nectar. Bamboo thickets and other dense undergrowth in moist, broadleaved forests.

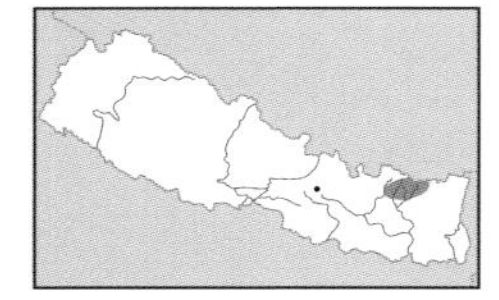

Long-billed Wren Babbler *Rimator malacoptilus* 12 cm

Very rare and local resident, only recorded in Makalu Barun National Park and its buffer zone; 1,770–3,260 m. **ID** Large, short-tailed wren babbler with long, downcurved bill. Buffish-white throat and buff belly, broad buff streaking on brown breast and flanks, and brown upperparts with fine buff shaft-streaking. Also narrow dark moustachial stripe, less distinct malar stripe, and rufous-brown vent and undertail-coverts. **Voice** Series of short, clear bell-like whistles, *pee*, repeated every 3–4 seconds; churring *prurr prurr prrit* in alarm. **HH** Very skulking; keeps mainly to ground, hopping within low vegetation and poking among leaf litter. Dense bamboo undergrowth, ferns and mossy boulders in thick mixed rhododendron forest and dense tangled thickets in subtropical broadleaved forest.

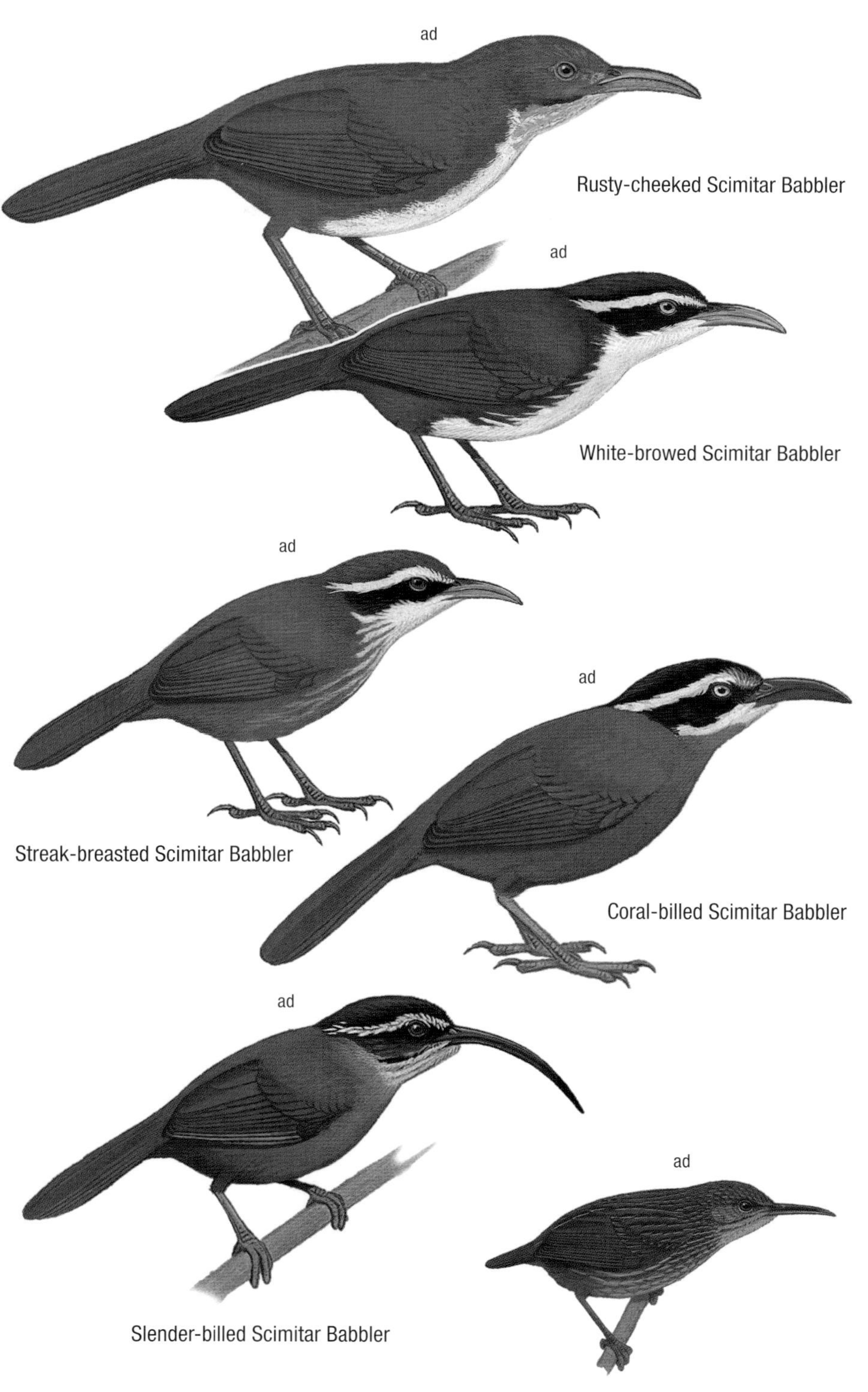
ad
Rusty-cheeked Scimitar Babbler
ad
White-browed Scimitar Babbler
ad
Streak-breasted Scimitar Babbler
ad
Coral-billed Scimitar Babbler
ad
Slender-billed Scimitar Babbler
ad
Long-billed Wren Babbler

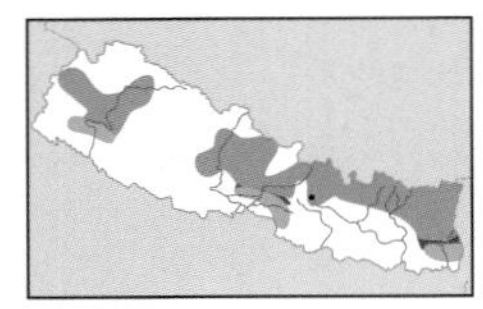

Scaly-breasted Wren Babbler *Pnoepyga albiventer* 10 cm

Fairly common resident; breeds 2,440–4,000 m, winters 300–2,285 m. **ID** From Pygmy Wren Babbler by larger size. Usually shows well-defined buff spotting on sides of crown and sides of neck (occasionally over entire crown and mantle), which is lacking in Pygmy. Juvenile considerably smaller with uniform dark chestnut-brown upperparts, while underparts are sooty-grey to brown and lack bold scaling. Two intergrading races occur; *P. a. pallidior* in west and centre has paler olive-brown upperparts than nominate in east. **Voice** Strong warbling song, *tzee-tze-zit-tzu-stu-tzit*, rising then ending abruptly; high-pitched *tzit* call. **HH** Males sing from rock or tree root; very skulking in non-breeding season. Keeps on or very close to ground. Dense undergrowth in broadleaved forests, particularly near streams.

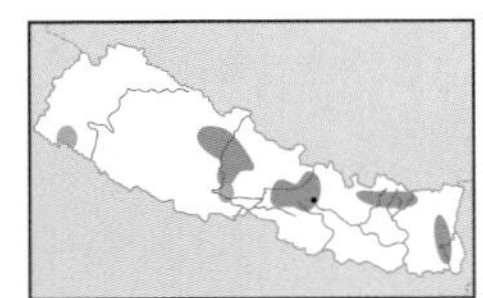

Nepal Wren Babbler *Pnoepyga immaculata* 10 cm

Locally common resident; 275 m (winter) to 2,700 m (breeding). **ID** Similar to Scaly-breasted, with longer and heavier bill; best separated by song. Feathers of underparts more elongated in shape, resulting in narrower arrowhead-shaped black centres and pale fringes to feathers (giving rise to streaked appearance). Upperparts lack buff spotting on crown, mantle and wing-coverts (particularly characteristic are unmarked areas above and behind eye and on neck-sides). White morphs tinged ochre, appearing distinctly dirty compared to most Scaly-breasted; fulvous morph is deep rusty-yellow on underparts (deeper rust-coloured in Scaly-bellied). Upperparts paler olive-brown than nominate Scaly-breasted. **Voice** High-pitched piercing notes, speeding up towards end and repeated, *si-su-si-si-swi-si-si-si*. **HH** Habits similar to Scaly-breasted. Tall herbage at forest edge, in second growth, and amongst boulders and scrub.

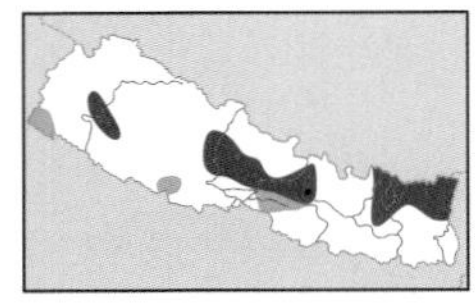

Pygmy Wren Babbler *Pnoepyga pusilla* 9 cm

Frequent resident; breeds 1,500–2,590 m, winters 915–1,770 m (–275 m). **ID** From Scaly-breasted by smaller size (appears similar to Eurasian Wren or Grey-bellied Tesia) and distinctive song. Spotting on upperparts confined to lower back and wing-coverts (lacking well-defined buff spotting on crown and neck). Juvenile has uniform dark brown upperparts, and underparts are sooty-grey without bold scaling. **Voice** Song is loud, slowly drawn-out *see-saw*, repeated monotonously; *tzit* and *tzook* calls. **HH** Habits and habitat similar to Scaly-breasted.

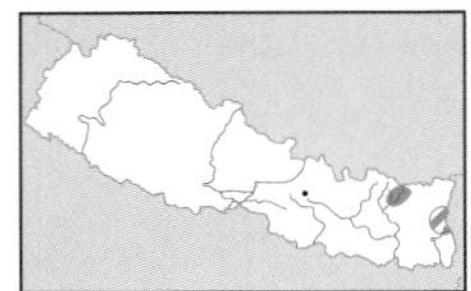

Rufous-throated Wren Babbler *Spelaeornis caudatus* 9 cm

Very rare, very local in far east; 2,135–2,440 m. **ID** From Pygmy by grey ear-coverts, whitish chin, rufous-orange throat and breast (white-barred or -spotted rather than scaled appearance to breast and belly, no buff spotting on wing-coverts, and short, square-ended tail. Some are paler orange-buff on throat and breast. Juvenile resembles adult, but is smaller, lacks dark scaling on upperparts, and lacks white spotting and black barring on underparts. **Voice** Sudden burst of 3–5 rapidly repeated *swediddy* notes; warble of 2–4 notes, *swichu-wichu-wichu*; *dzik* call similar to Scaly-breasted. **HH** Habits similar to Scaly-breasted. Fallen trees, mossy rocks and ferns in moist broadleaved evergreen forest.

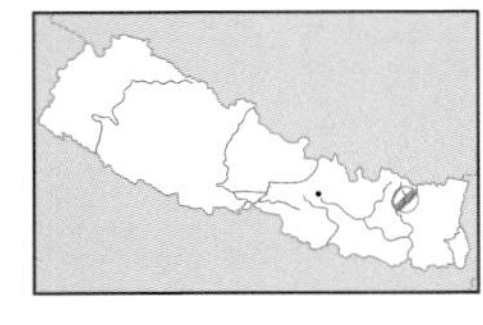

Spotted Elachura *Elachura formosa* 10 cm

Very rare, very local resident in far east; no recent records; 1,785 m. **ID** Dark brown with a noticeable tail. From other wren babblers by broad dark brown barring on rufous-brown wings and tail. Irregular white flecking on grey-brown sides of head and upperparts (especially prominent on nape, sometimes forming collar, and wing-coverts), and white and dark brown mottling, and dark brown vermiculations on buffish underparts. Underparts can appear dark, with pattern difficult to see in field. Lacks intense barring on mantle, back, wing-coverts and underparts of Eurasian Wren. **Voice** High-pitched, faltering whistle song, *did-did-did-dit, did-di-di-did* constantly repeated; spluttering *put-put-put...* trill. **HH** Habits similar to Scaly-breasted. Undergrowth in dense, moist forest. **TN** Formerly placed in genus *Spelaeornis* and called Spotted Wren Babbler, but recently placed in monospecific family.

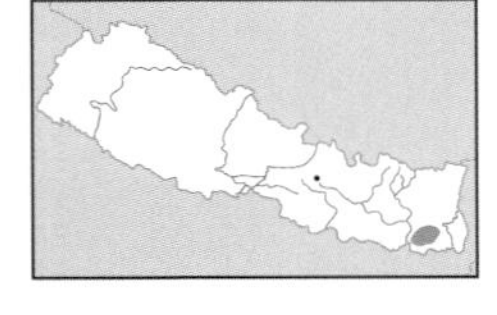

Himalayan Wedge-billed Babbler *Sphenocichla humei* 18 cm

Very rare, local resident in far east; 500 m. **ID** An extraordinary-looking babbler, with wedge-shaped bill, broad but diffuse grey supercilium and spotting on sides of neck, boldly marked upperparts, diffuse brown barring on wings and tail, and sturdy blackish legs and feet. **Voice** Song includes loud, melodious whistles, often in duet. **HH** Climbs tree trunks; also forages in undergrowth like other babblers. Vegetation by small forest streams. **AN** Blackish-breasted Babbler.

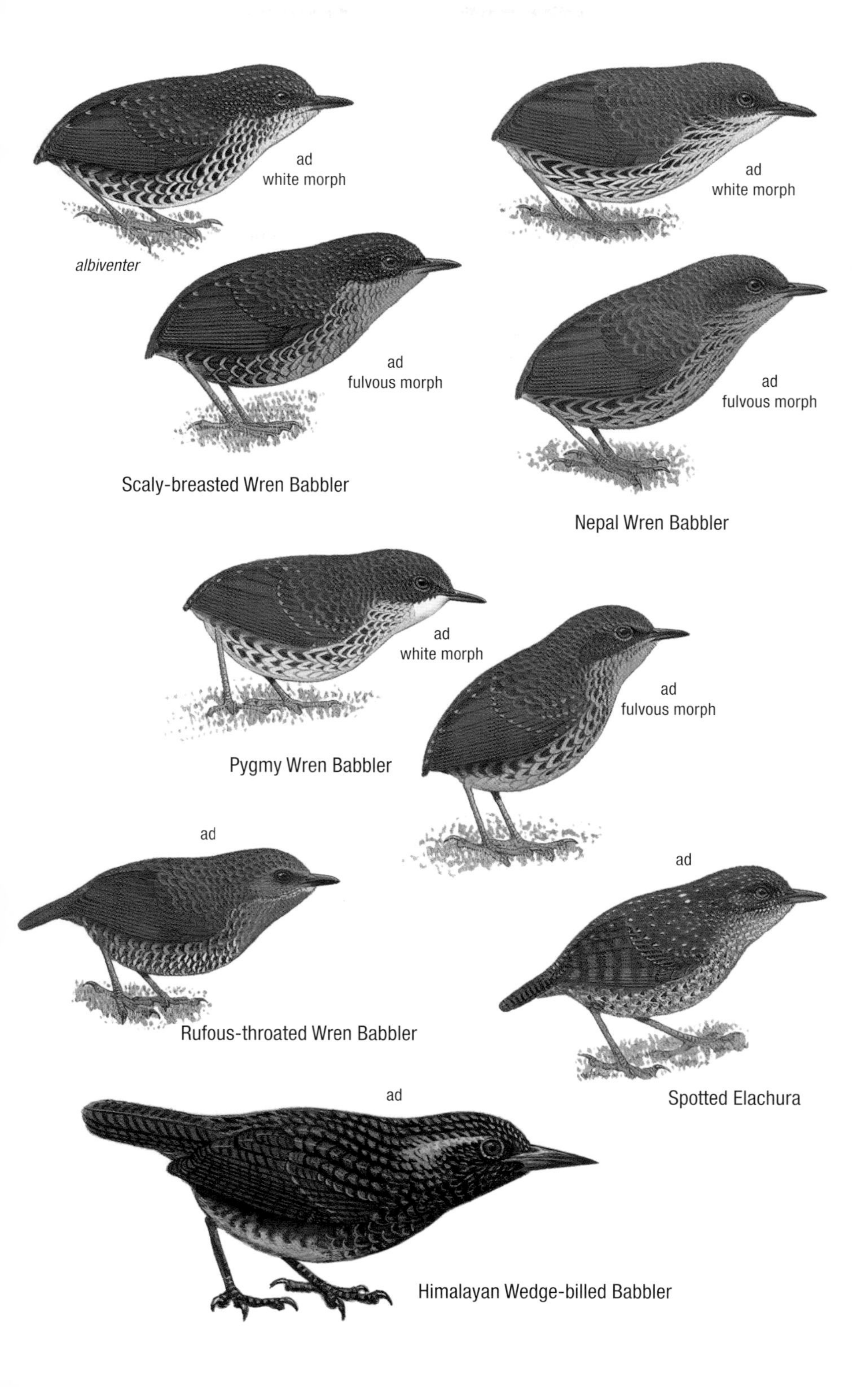
ad
white morph
albiventer
ad
fulvous morph
Scaly-breasted Wren Babbler
ad
white morph
ad
fulvous morph
Nepal Wren Babbler
ad
white morph
Pygmy Wren Babbler
ad
fulvous morph
ad
Rufous-throated Wren Babbler
ad
Spotted Elachura
ad
Himalayan Wedge-billed Babbler

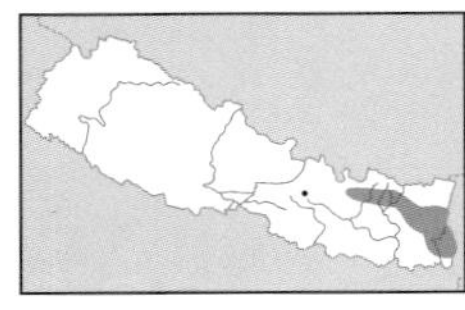

Rufous-capped Babbler *Stachyridopsis ruficeps* 12 cm

Fairly common resident in east; 1,220–2,745 m. **ID** Well-defined rufous cap. In addition, has bright olive upperparts, pale yellow throat (faintly streaked black), and yellowish-buff face and underparts. Juvenile as adult, but duller, with paler underparts and paler crown. **Voice** Song *pee-pi-pi-pi-pi-pi pi*; four-noted whistling call, *whi-whi-whi-whi*. **HH** In pairs or fast-moving parties with other species according to season. Dense scrub, undergrowth and bamboo thickets in humid broadleaved forest. **TN** Formerly placed in genus *Stachyris*.

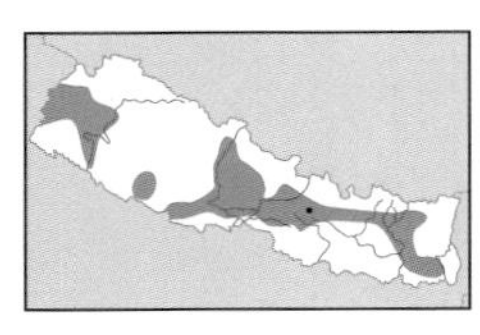

Black-chinned Babbler *Stachyridopsis pyrrhops* 10 cm

Fairly common and widespread resident; 245–2,440 m. **ID** Small, rather bull-headed babbler, with black lores, chin and centre of throat. Also olive-buff crown with dark streaking, buffish ear-coverts, orange-buff underparts and olive-brown upperparts (lacking any rufous). Iris red. **Voice** Pleasant bell-like piping, *whit-whit-whit-whit*, repeated at irregular intervals; soft melodious churring contact call, *t-r-r-r-r tr-eee tu-r-r-r*. **HH** Depending on season, in pairs or loose parties up to eight, often with other species. Active and restless. Undergrowth in open forest, second growth and forest edges. **TN** Formerly placed in genus *Stachyris*.

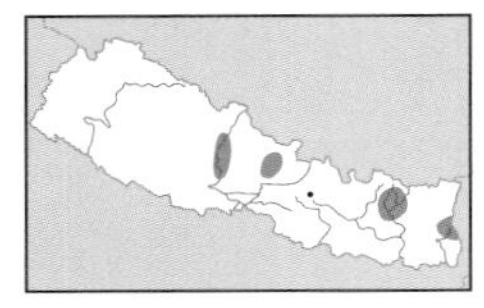

Golden Babbler *Stachyridopsis chrysaea* 10 cm

Very local resident, mainly in Annapurna Conservation Area; 1,800–2,440 m. **ID** Distinctive babbler with yellow forehead and black-streaked yellow crown, black lores and moustachial stripe, olive-green upperparts and yellow underparts. **Voice** Song like slower variation of Rufous-capped, but pauses after first note. **HH** Usually in mixed-species flocks outside breeding season. Continually on the move. Bamboo and dense undergrowth in humid, broadleaved, evergreen forests. **TN** Formerly placed in genus *Stachyris*.

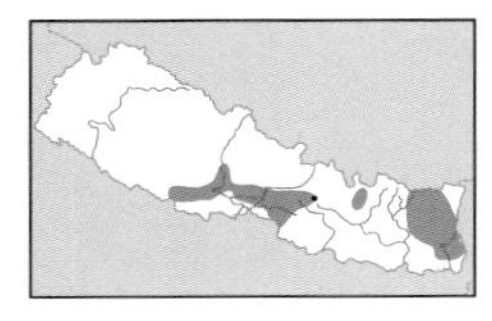

Grey-throated Babbler *Stachyris nigriceps* 12 cm

Frequent resident from west-central Nepal east; 245–2,000 m. **ID** From *Stachyridopsis* babblers by combination of blackish crown with white streaking, black lateral crown-stripes, greyish supercilium, grey chin and throat, and buff underparts. **Voice** High-pitched, tinkling trill, prefaced by single note and a brief pause. **HH** Usually keeps well inside cover. Often in itinerant, mixed foraging parties in non-breeding season. Undergrowth and bamboo thickets in moist broadleaved forests and second growth.

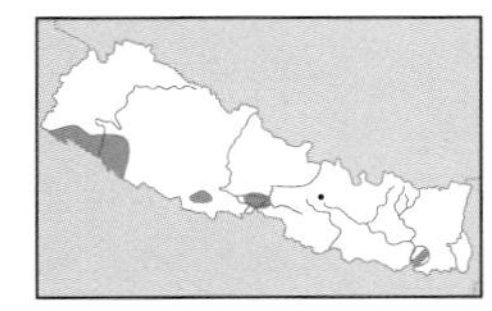

Tawny-bellied Babbler *Dumetia hyperythra* 13 cm

Rare and local resident in western lowlands; 150–305 m. **ID** Fairly long-tailed, rufous-coloured babbler. Forehead and forecrown rufous-buff, throat and underparts orange-buff, mantle olive-brown, and wings and tail brown, the tail with faint barring. Eye pale and bill pale pinkish-brown. Juvenile has brighter, more rufous-brown upperparts, lacking rufous-brown forehead and duller underparts. **Voice** Whistling seven-note song; soft *tack-tack* call and flocks utter *sweech-sweech*. **HH** In loose flocks of *c.*10 birds which restlessly follow each other. Skulking and if alarmed, flock disperses and dives into thick cover. Tall grass and scrub.

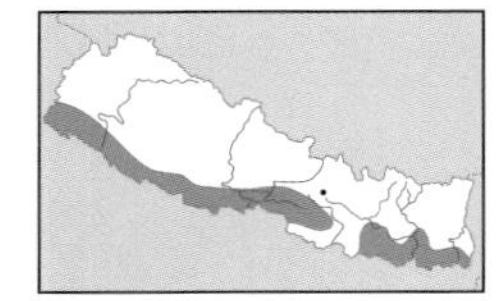

Pin-striped Tit Babbler *Macronus gularis* 11 cm

Resident, common in Chitwan National Park, local and generally uncommon elsewhere; 75–760 m. **ID** Rather scruffy-looking babbler with rufous-brown cap, yellow eyes, pale yellow lores and supercilium, olive mantle, brown wings and tail, and pale yellow underparts with fine black streaking on throat and breast. **Voice** Loud, monotonous *chuk-chuk-chuk* repeated up to 15 times; *bizz-chir-chur* alarm. **HH** Particularly noisy babbler readily located by distinctive call. Creeps and clambers about unobtrusively, searching for insects. In pairs or parties, often with other species, depending on season. Bushes, creepers, tangles and undergrowth in broadleaved forest. **AN** Striped Tit Babbler.

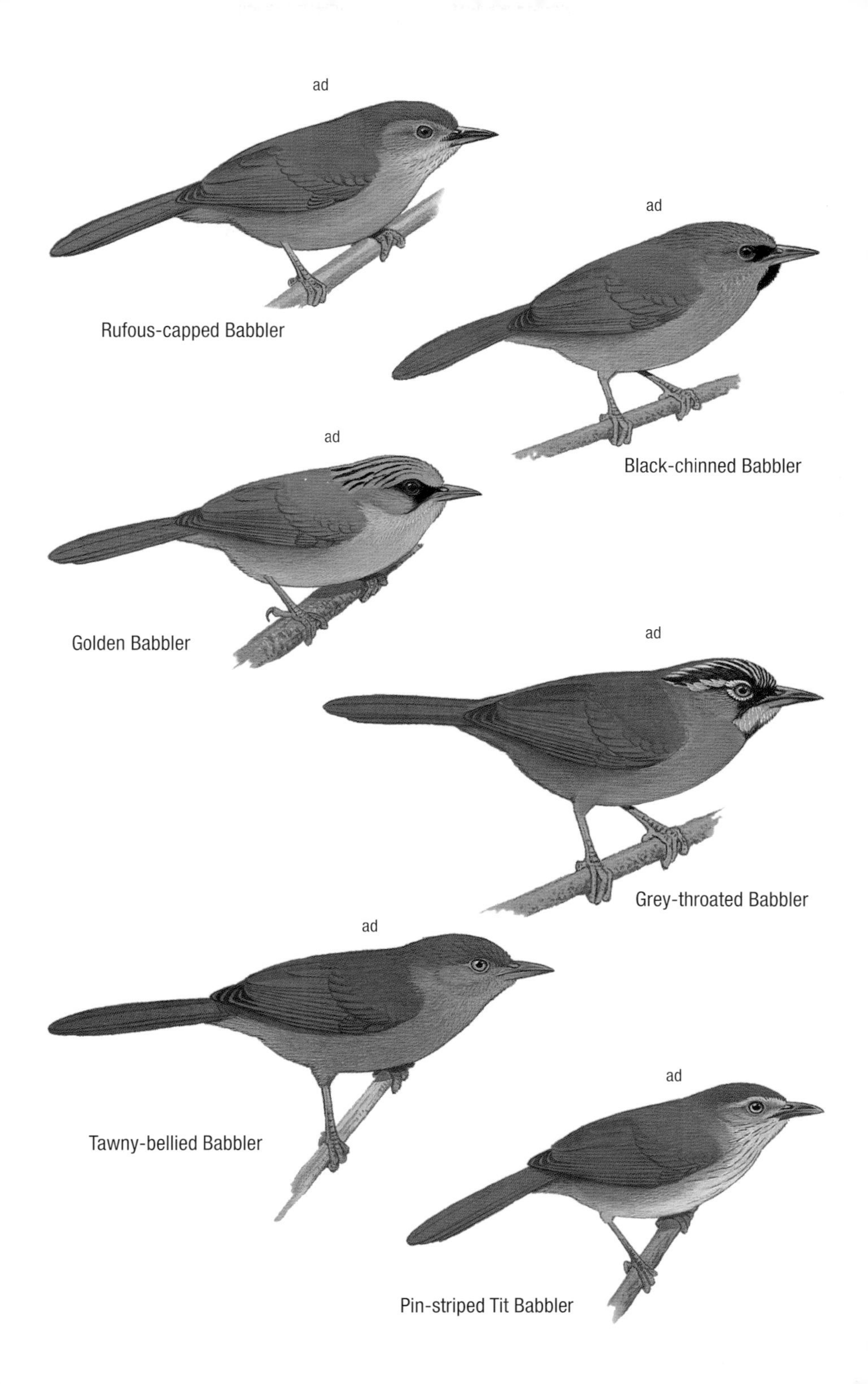
ad
Rufous-capped Babbler
ad
Black-chinned Babbler
ad
Golden Babbler
ad
Grey-throated Babbler
ad
Tawny-bellied Babbler
ad
Pin-striped Tit Babbler

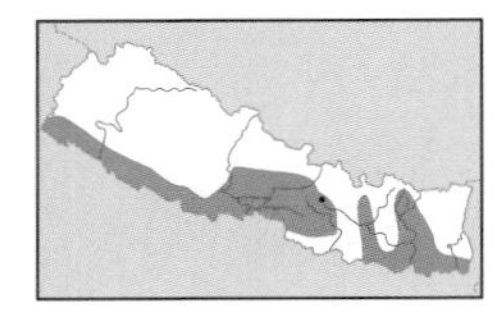

Puff-throated Babbler *Pellorneum ruficeps* 15 cm

Locally fairly common resident; 75–915 m (–1,675 m). **ID** Comparatively long-tailed. Rufous crown, prominent buff supercilium, white throat (often puffed out) and bold brown spotting/streaking on breast and sides of neck. **Voice** Halting, impulsive song, *sweee ti-ti-hwee hwee hwee ti swee-u*, rambles up and down scale; calls include a plaintive whistled *ne-menue* and throaty churrs. **HH** Very skulking, staying on or near ground. Runs over forest floor or makes long hops when foraging. In pairs or family parties. Bushy undergrowth in broadleaved forests, dense scrub and second growth.

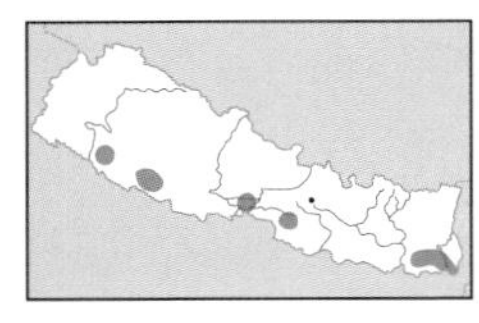

Abbott's Babbler *Malacocincla abbotti* 17 cm

Rare and local resident, mainly in east; 75–275 m. **ID** Top-heavy, with large bill and short tail. Unspotted white throat and breast, grey lores and supercilium, rufous uppertail-coverts and tail, and rufous-buff flanks and vent. **Voice** Song of 3–4 whistled notes with last note highest; sometimes duets, with mate giving 1–2 *peep* notes. **HH** Singly or in pairs. Skulking and mainly terrestrial. Reluctant to fly, if disturbed seeks cover. Dense tangles and thickets in moist forest especially at edges along stream banks, favours broadleaved evergreen forest.

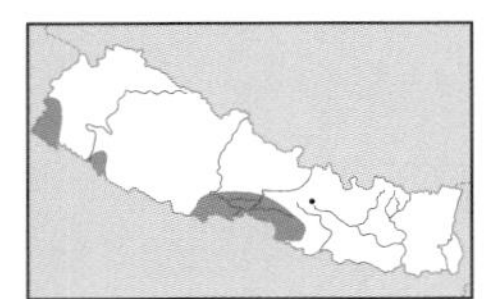

Chestnut-capped Babbler *Timalia pileata* 17 cm

Local resident; 150–300 m. **ID** Stocky, thick-necked babbler with bright chestnut cap, reddish iris, white forehead and supercilium, thick black bill and black lores (resulting in masked appearance). White throat and breast (streaked finely with black), bordered slate-grey on sides of neck, buffish-olive flanks and vent, and buff belly. Tail faintly barred. **Voice** Fast, high-pitched, descending trill, varying in length; *tzt* contact note and *pic-pic-pic* alarm. **HH** In small parties in non-breeding season. Usually keeps within thick cover although may sun itself for a few seconds or sing in open. Typically clambers up and down grass stems, systematically searching for insects. Tall, moist lowland grasslands.

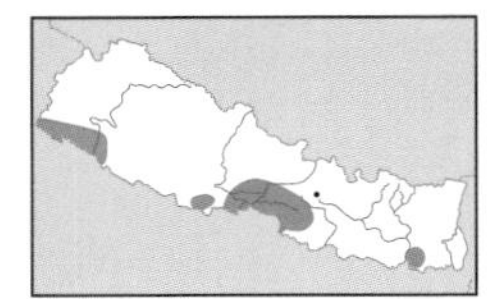

Yellow-eyed Babbler *Chrysomma sinense* 18 cm

Local resident. **ID** Long-tailed babbler with rounded head and stout, dark bill (recalling a giant prinia in shape). Most distinctive features are yellow iris and broad orange orbital ring, white lores and supercilium, chestnut upperparts, striking white throat and breast merging into richer buff lower belly and flanks, and yellow legs and feet. Juvenile has browner bill, dark eye and duller eye-ring. **Voice** Song a variable, rapid twittering trill *tri-rit-ri-ri-ri-ri* ending in a two-noted *toway-twoh*; call a loud, plaintive *teeuw-teeuw-teeuw*, repeated 2–4 times. **HH** In parties of 12 or more outside breeding season. Skulking, spending most time concealed in tall grass and bushes, frequently foraging near ground. Often calls briefly from a conspicuous perch before diving into cover. Laboured, jerky flight. Tall moist grassland; thickets, bamboos and reeds, often near water.

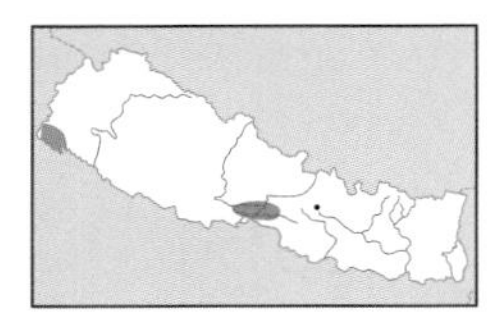

Jerdon's Babbler *Chrysomma altirostre* 17 cm

Very rare and local resident; only Nepal localities are Sukla Phanta Wildlife Reserve and Chitwan National Park; 150–250 m. **ID** From Yellow-eyed by paler yellowish-brown bill, greyish lores and supercilium, grey throat and breast, brown iris, dull yellowish-green orbital ring, and fleshy-brown legs and feet. Upperparts rich chestnut-brown with rich buff on belly, flanks and vent. **Voice** Song comprises series of undulating two-toned whistles, *twee-too whit-tooh*, lacking rapid trills of Yellow-eyed. **HH** Habits similar to Yellow-eyed. Reedbeds and tall grassland. Globally threatened.

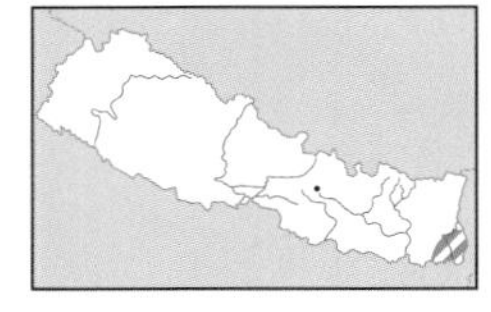

White-hooded Babbler *Gampsorhynchus rufulus* 23 cm

Probably former resident or winter visitor to far east; no recent records; 600–1,400 m. **ID** Bull-headed, long-tailed babbler. Adult has white head and underparts (with buff wash on flanks), contrasting with rufous-brown upperparts and tail; tail tipped pale buff. Iris and bill strikingly pale. Variable white on wing-coverts is usually obscured by feathers of mantle. Juvenile has rufous-orange crown and ear-coverts, rufous-brown upperparts, and whitish throat, becoming buff on underparts. **Voice** Usual call a harsh, stuttering rattle or cackle; contact calls soft and quiet *wit*, *wet* and *wyee* notes. **HH** Gregarious and often with other species. Forages in dense undergrowth and bamboo in moist broadleaved evergreen forest, also in second growth in evergreen biotope.

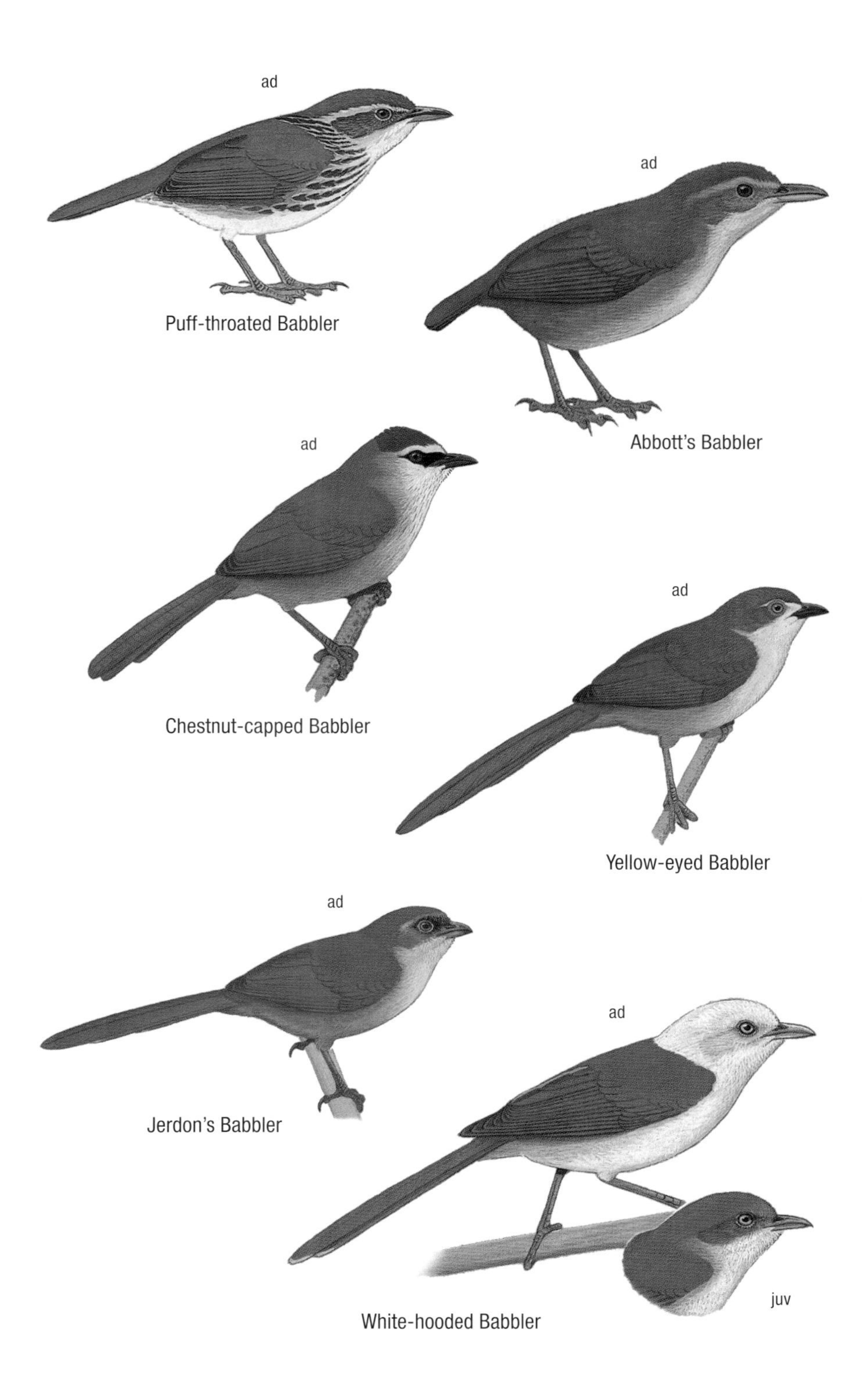
ad
Puff-throated Babbler
ad
Abbott's Babbler
ad
Chestnut-capped Babbler
ad
Yellow-eyed Babbler
ad
Jerdon's Babbler
ad
juv
White-hooded Babbler

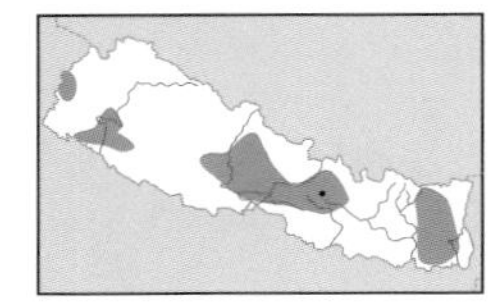

Spiny Babbler *Turdoides nipalensis* 25 cm

Nepal's only endemic bird. Local and frequent resident; breeds 1,500–2,135 m, winters 500–1,830 m. **ID** Downcurved black bill, white on face, dark brown upperparts without prominent streaking, white iris, and strong, fine black streaking on throat and breast. Noticeable variation in pattern of head and colour of underparts: some show largely white face and throat, and white centre of breast and belly with dark shaft-streaking restricted to lower throat and breast; others have white restricted to lores and malar, with buff lower throat and breast, which are more extensively covered in dark shaft-streaking. **Voice** Distinctive, harsh ringing whistling song starts with a few whistles, then ascends scale. Call a clear *el-el-el-el-el* and low *churr* alarm. Female has loud *wick-er-wick-er-wick-er*; pair sometimes burst into wild crescendo of screaming calls. **HH** Very skulking except early in breeding season when males often sing in open. Dense scrub on hillsides; favours thicker areas.

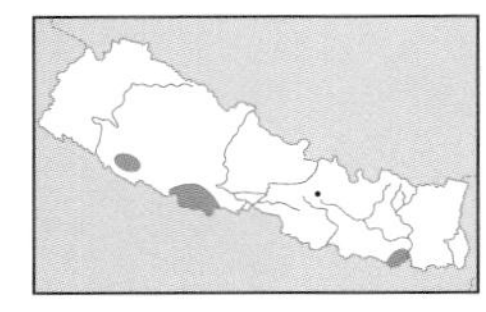

Common Babbler *Turdoides caudata* 23 cm

Local resident; 100–300 m. **ID** From Striated by unstreaked whitish throat and unstreaked centre to breast (with streaking on underparts confined to breast-sides). Also slightly smaller, with colder whitish or greyish-buff underparts, yellowish legs and feet, and darker, more orange-brown iris. **Voice** A series of pleasant, rapidly repeated fluty whistles, and a louder, more drawn-out *pieuu-u-u pie-u-u pi-e-u-u*; also a high-pitched *qwe-e-e qwe-e-e* alarm. **HH** Flock members maintain constant conversation of whistles, trills and squeaks as they move about. Feeds chiefly on ground, moving with springing hops; sometimes in bushes. Dry cultivation and scrub.

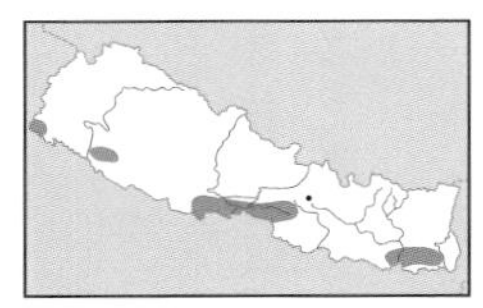

Striated Babbler *Turdoides earlei* 21 cm

Resident, locally common in protected areas, rare elsewhere; 75–305 m. **ID** From Common by streaked or mottled appearance to fulvous throat and breast (unstreaked and white in Common). Additional features are slightly larger size, deeper buff breast and belly, blue-grey legs and feet (varying to olive-brown), and golden-yellow iris. **Voice** Song is a loud repeated series of *tiew-tiew-tiew-tiew* notes, interspersed by *quip-quip-quip* calls from other group members. **HH** Habits similar to Common. In noisy flocks, individuals follow each other. Clambers about vegetation when foraging. Tall grass and reedbeds in damper habitats.

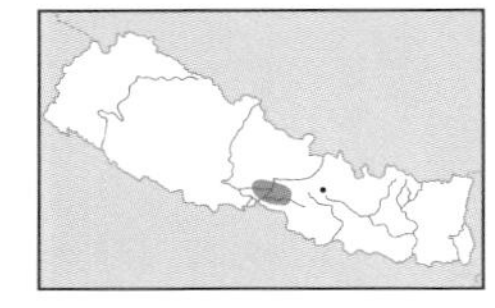

Slender-billed Babbler *Turdoides longirostris* 23 cm

Very local resident, only recorded in Chitwan National Park; 150–250 m. **ID** From other babblers by fine, downcurved blackish bill, whitish lores and buff ear-coverts, unstreaked dark rufous-brown upperparts, whitish throat, and unstreaked deep buff underparts. **Voice** Vocalisations include a shrill, rather high-pitched, strident series of notes, a rather clear, high-pitched *wii-wii-jiu-di*, even, fairly strident 4–6-noted *chiu-chiu-chiu-chiu* and a discordant high-pitched, repeated *tiu-tiu-tiu*. **HH** In noisy flocks outside breeding season; also singly and in pairs. Skulks in dense cover and usually remains hidden except in breeding season. Tall grass, especially near water. Globally threatened.

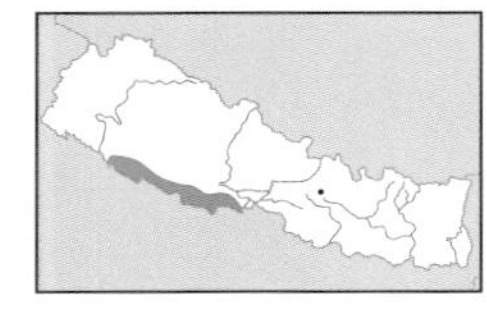

Large Grey Babbler *Turdoides malcolmi* 28 cm

Locally distributed and uncommon resident in western lowlands, near Indian border; 75–105 m. **ID** Large, pale grey babbler with darker grey mottling on upperparts and prominent white sides to tail. Additional features include greyish-pink throat and breast (without dark mottling), pale grey forehead, dark grey lores, pale yellow iris, dull-coloured bill, and brownish-grey legs and feet. Juvenile smaller, with paler grey upperparts which lack dark mottling. **Voice** Monotonous, plaintive, drawling *kay-kay-kay-kay* (flatter and less squeaky than Jungle) and a noisy chattering in alarm. **HH** Habits similar to Jungle, but less skulking. Open dry scrub, cultivation and gardens.

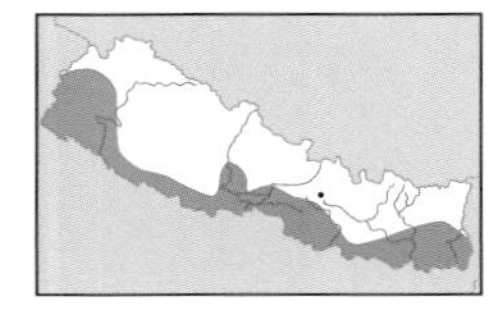

Jungle Babbler *Turdoides striata* 25 cm

Common and widespread resident; 75–1,220 m (–2,300 m). **ID** Greyish, often rather scruffy-looking babbler with pale streaking on mantle and underparts. From Large Grey by smaller and stockier appearance, shorter, broader-looking tail lacking prominent white sides, mottling on throat and breast, and yellowish bill (can be horn-brown in winter). **Voice** A harsh *ke-ke-ke* which frequently becomes a chorus of excited, discordant squeaking and chattering. **HH** Gregarious, noisy and excitable, birds calling to each other almost continuously. Mainly feeds on ground, hopping about, busily turning over leaves. Unlike other *Turdoides*, has characteristic habit of fluffing out its rump feathers and drooping wings and tail. Secondary scrub, bushes in cultivation and gardens.

ad
Spiny Babbler
ad
ad
Common Babbler
ad
Striated Babbler
ad
Slender-billed Babbler
ad
Large Grey Babbler
ad
Jungle Babbler

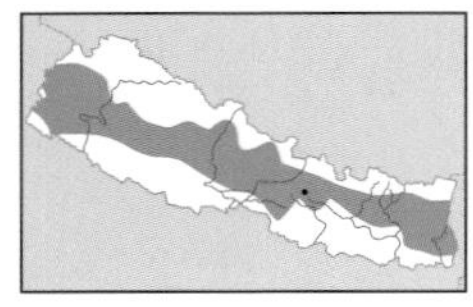

White-throated Laughingthrush *Garrulax albogularis* 28 cm

Common, widespread resident; 300–3,500 m. **ID** Large, with striking white throat and upper breast, brownish-olive breast-band and pale rufous-orange underparts. Additional features include black lores, white eyes, rufous-orange forehead, and indistinct grey panel on wing. In flight, broad white tips to tail feathers. Juvenile similar to adult, but has less distinct breast-band and paler rufous-buff belly and flanks. **Voice** Wheezy call and variety of squeals, hisses and shrill, squeaky laughter. **HH** Like all laughingthrushes characteristically breaks into concert of loud hissing, chattering and squealing at least sign of danger. Often feeds on ground, progressing via long springy hops, rummaging among leaf litter, flicking leaves aside and into air, and digging for food with its strong bill. Gregarious. Forest undergrowth and second growth.

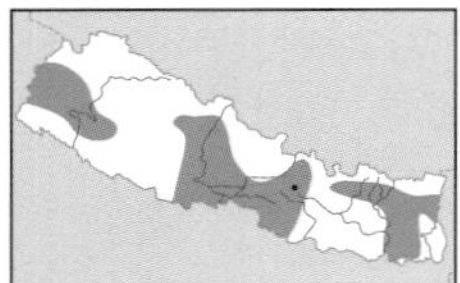

White-crested Laughingthrush *Garrulax leucolophus* 28 cm

Common, widespread resident; 305–2,135 m. **ID** Large, with white crest and black mask. Also, contrasting white throat and upper breast, grey nape, chestnut mantle and band on lower breast, and dark olive-brown wings and tail. **Voice** Frequent bursts of cackling laughter. **HH** Habits similar to White-throated. Broadleaved forest and second growth.

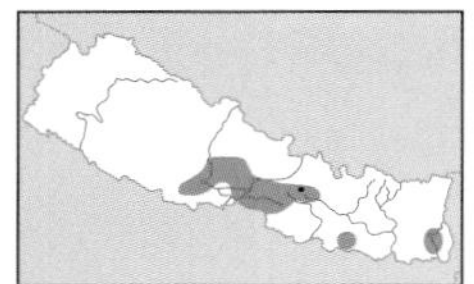

Lesser Necklaced Laughingthrush *Garrulax monileger* 27 cm

Local resident, frequent at Chitwan but rare elsewhere; 150–2,000 m. **ID** Very similar to Greater, with white supercilium, black necklace, rufous-orange flanks, greyish-white panel on wing, and bold white tips to largely blackish outer tail feathers. Smaller than Greater with finer, dark bill, yellow eye with dark eye-ring, dark lores, and brownish (rather than slate-grey) legs and feet. Lower black border of ear-coverts does not extend to bill, and white throat is bordered by rufous-orange adjacent to necklace. Has narrower necklace often almost obscured by rufous at centre (although can be uniformly broad on some birds), and white of breast-sides extends as crescent beneath black necklace. Also olive-brown (not dark grey) primary-coverts concolorous with rest of coverts. Juvenile similar, but has dusky necklace. **Voice** Loud, mellow *tee-too-ka-kew-kew-kew* song. **HH** Habits similar to White-throated. Often with Greater Necklaced. Breeds in moist, dense broadleaved hill forests; riverine and mixed forests in winter.

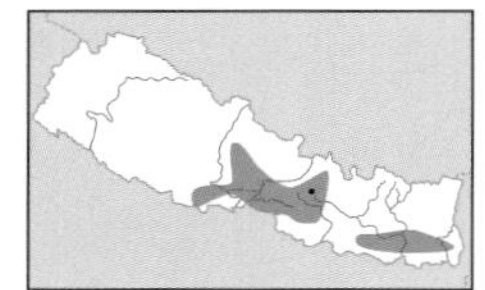

Greater Necklaced Laughingthrush *Garrulax pectoralis* 29 cm

Local and uncommon resident; mainly at Chitwan, also in far east; 100–1,400 m. **ID** Larger than Lesser Necklaced, with stouter (paler-based) bill, dark eye and yellow eye-ring, complete black moustachial stripe (bordering either black or white, or streaked black-and-white ear-coverts), uniform buff or white throat without two-toned appearance, blackish primary-coverts contrasting with mantle and wings, broader necklace clearly defined in centre, and slate-grey (rather than brownish) legs and feet. As Lesser Necklaced, broad white tip to blackish outer tail in flight. **Voice** Loud *what-what-who-who* song. **HH** Habits similar to White-throated. Breeds in moist dense broadleaved hill and sal forests; riverine and mixed forests in winter.

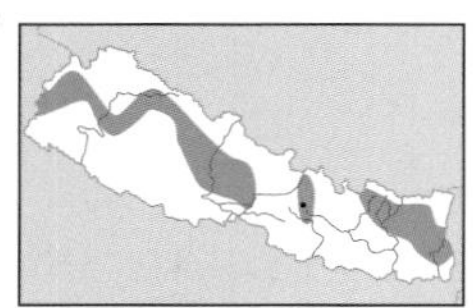

Striated Laughingthrush *Garrulax striatus* 28 cm

Common, widespread resident; 1,200–2,850 m. **ID** Large with stout black bill and floppy crest, affording dome-headed appearance. Chestnut crown, russet upperparts and brownish underparts profusely covered with white to buffish-white streaking. Wings more rufous-brown, with greyish-white panel. Lack of dark barring on wings and tail helps to distinguish from barwings. **Voice** Varied calls: whistling *hoo-wee...chew-chew* and *hooooooo-weeeeeee-chew*, gurgling *which-we-we-heet-chuuu* and soft *poor-poor*. **HH** Habits similar to White-throated, but more arboreal. Dense broadleaved forest.

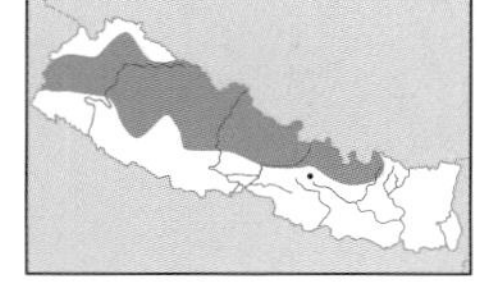

Variegated Laughingthrush *Garrulax variegatus* 24 cm

Fairly common resident; 2,100–4,100 m. **ID** A complex-patterned, mainly grey and olive laughingthrush without spotting or scaling. Rufous-buff forehead and malar stripe, black mask and centre of throat, rufous greater coverts, black primary-coverts and patch on secondaries, golden-olive wing panel, rufous vent, and grey subterminal band and white tip to tail. **Voice** Loud, penetrating whistle *pit-we-weer*; alarm calls include rapid, squeaking *queek-queek-queek-queek*. **HH** Habits similar to White-throated. Thick undergrowth in open forests; bushes at forest edges and rhododendron shrubberies.

ad
White-throated Laughingthrush
ad
White-crested Laughingthrush
ad
Lesser Necklaced Laughingthrush
ad
Greater Necklaced Laughingthrush
ad
Striated Laughingthrush
ad
Variegated Laughingthrush

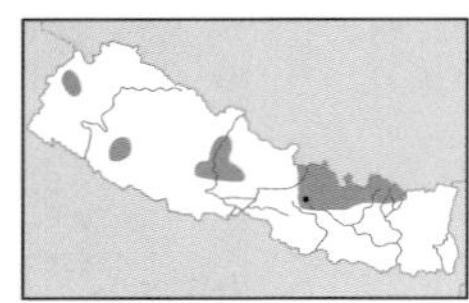

Rufous-chinned Laughingthrush *Garrulax rufogularis* 22 cm

Local and very uncommon resident; 915–2,135 m. **ID** Irregular black spotting and barring on upperparts and underparts. Rufous chin and upper throat, blackish cap, diffuse black moustachial stripe, buff lores, irregular black and grey banding on wings, and black subterminal band and rufous tip to tail. Two intergrading races occur: *G. r. occidentalis* in west and centre has olive-brown upperparts, rufous ear-coverts and buffish-olive supercilium; nominate in east has rufous olive-brown upperparts and tail, grey ear-coverts and dull rufous supercilium. **Voice** Squeals, chuckles and chatters; pleasant whistling song, *swee-tu...tu-tu-wee-u*. **HH** Less gregarious and less noisy than most laughingthrushes. Secretive and usually keeps out of sight. Dense undergrowth in broadleaved forest and scrub.

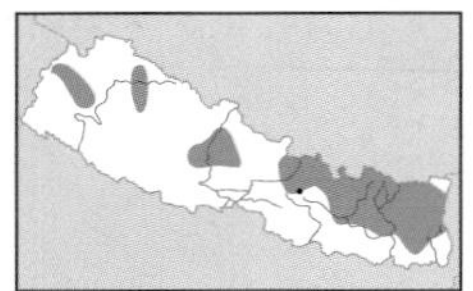

Spotted Laughingthrush *Garrulax ocellatus* 32 cm

Locally fairly common resident; 2,135–3,660 m. **ID** Large, chestnut laughingthrush, with profuse black-based white spotting on chestnut upperparts. Also blackish cap, rufous supercilium, lores and chin, deep chestnut ear-coverts, blackish throat grading into black barring on breast, buff lower breast and belly, chestnut-and-black wings with grey panel and white tips to primaries, and white tips to chestnut, grey and black tail. **Voice** Whistled *tu wee tu wee tu witty-o* song with human-like quality; also a subdued *pie pie pie pie*. **HH** Often associates with Black-faced. Inquisitive and easily observed. Undergrowth and bamboo in broadleaved and coniferous forest and rhododendron shrubberies.

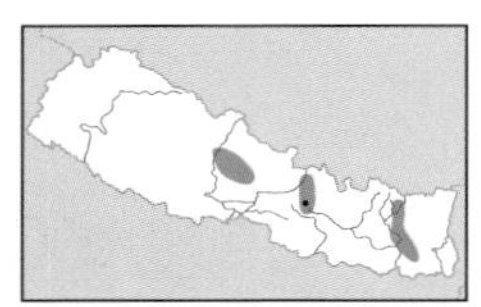

Grey-sided Laughingthrush *Garrulax caerulatus* 25 cm

Rare, local resident; 1,370–2,745 m. **ID** Rufous brown and white, with grey breast-sides and flanks, bluish-slate eye-patch, black face and white cheeks, black scaling on crown, and more rufescent edges to flight feathers and tail. **Voice** Squealing *klee-loo*, jovial *ovik-chor-r-r*, loud *joy-to-weep, poo-ka-ree, new-jeriko*; *chik-chik-chik* alarm call. **HH** Undergrowth in dense, moist broadleaved forests and bamboo thickets.

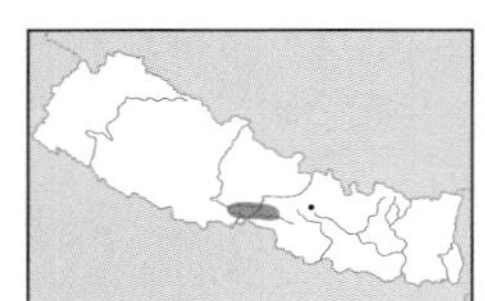

Rufous-necked Laughingthrush *Garrulax ruficollis* 23 cm

Very local resident in Chitwan National Park and its western buffer zone; 150–275 m. **ID** Small, mainly olive-brown laughingthrush, with prominent rufous patch on sides of neck, and black face and throat. Also has grey crown and nape, and rufous vent and centre of lower belly. Tail uniform brownish-black. Juvenile duller, with browner crown. **Voice** An incredibly varied and vocal songster. Vocalisations include shrill whistles that run up the scale, scolding whistles, descending trills, chittering babbles and hoarse squawks. **HH** Skulks on ground and in undergrowth or low bushes. Tall grassland at forest edges, riverine forests with young *Acacia* trees densely mixed with shrubs and grasses.

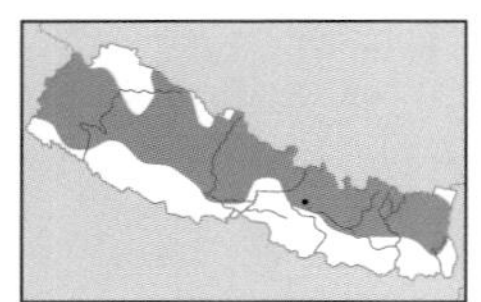

Streaked Laughingthrush *Garrulax lineatus* 20 cm

Fairly common resident; 1,065–3,905 m. **ID** Small, finely streaked, rufous-coloured laughingthrush. Fine dark streaking on crown and nape, brighter rufous ear-coverts, fine white streaking on mantle and underparts, and grey-tipped olive-brown tail with diffuse black subterminal band on outer feathers. **Voice** Song a short rapid trill followed by a loud ringing whistle higher in pitch than Variegated; also a plaintive whistle, *peah-tee-teeh-teeh*. **HH** Shuffles on ground like a rodent, and creeps systematically through grass clumps and low bushes, often flicking wings and tail. Hill scrub, forest edges, second growth and bushes at cultivation edges.

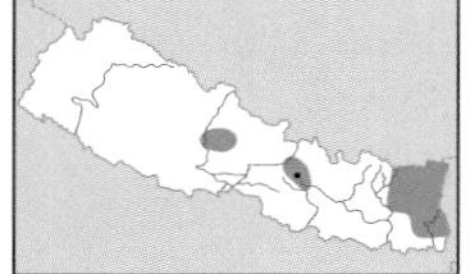

Blue-winged Laughingthrush *Garrulax squamatus* 25 cm

Local and very uncommon resident; 1,220–2,700 m. **ID** From Scaly Laughingthrush by black supercilium, silvery-blue outer webs to primaries forming prominent wing panel (brighter and more extensive than on Scaly), largely rufous wings with darker outer edge, chestnut-brown flanks and vent, rufous uppertail-coverts, and rufous-tipped dark tail. Striking white eye (brown in juvenile). Male has grey cast to crown and blacker tail. Female has browner crown and dark olive tail. **Voice** Song comprises *cur-white-to-go*; scratchy *seek* alarm. **HH** Very secretive; if alarmed dives into thick cover. Dense undergrowth in moist broadleaved evergreen forest and bamboo thickets.

ad
Rufous-chinned Laughingthrush
occidentalis
ad
rufogularis
ad
Spotted Laughingthrush
ad
Grey-sided Laughingthrush
ad
Rufous-necked Laughingthrush
ad
Streaked Laughingthrush
ad
Blue-winged Laughingthrush

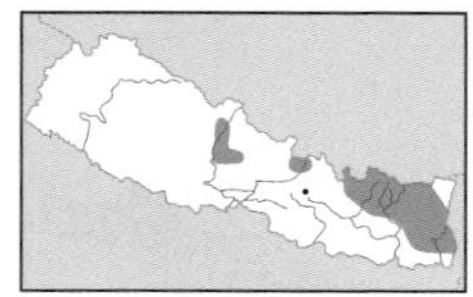

Scaly Laughingthrush *Garrulax subunicolor* 23 cm

Local, uncommon resident; 2,750–4,600 m. **ID** From Blue-winged by more uniform head (lacking black supercilium), yellowish-olive wing patch, less extensive and paler blue-grey panel on primaries, olive belly and vent, olive uppertail-coverts and central tail feathers concolorous with rest of upperparts, and white tips to outer tail feathers. Eye can be strikingly yellow or dark. **Voice** Usual song a two- or three-part wolf whistle (descends on last note) that is buzzy and more slurred than several similar laughingthrush songs. **HH** Habits similar to other laughingthrushes, see White-throated. Thick undergrowth in moist broadleaved forests and rhododendron shrubberies.

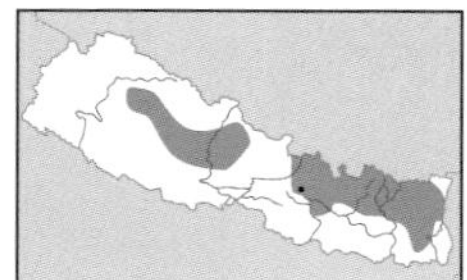

Black-faced Laughingthrush *Garrulax affinis* 25 cm

Locally common resident; 2,750–4,600 m. **ID** Distinctive, mainly rufous-brown laughingthrush with black supercilium and ear-coverts, and white malar stripe and patches on sides of neck. Black chin and centre of throat, greyish-white scaling on breast, variable grey mottling on upperparts, olive-yellow panel on mainly blue-grey wings, black primary-coverts patch, and olive-yellow tail with broad grey tip. Two races occur: *G. a. bethelae* in east has deeper rufous-brown underparts with indistinct greyish fringes to breast compared to nominate in the west and centre. **Voice** Song a loud *tew-wee-to-whee-to-whee-you-whee*; alarm a repeated, rapid *dze*. **HH** Habits similar to other laughingthrushes, see White-throated, but less shy and more conspicuous than most. Broadleaved, coniferous and mixed forest; also shrubberies above treeline.

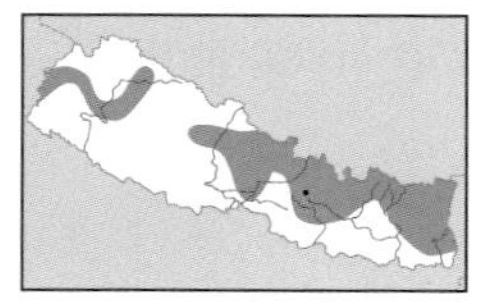

Chestnut-crowned Laughingthrush *Garrulax erythrocephalus* 28 cm

Common, widespread resident; breeds 1,800–3,200 m, winters 900–2,750 m. **ID** In east (*nigrimetus*) crown is grey streaked black, with some chestnut on forehead and nape. Birds from central and west Nepal (*kali*) have chestnut crown. Other features include dark scaling/spotting on mantle and breast, olive-yellow wings, and olive-yellow tail-sides. **Voice** Song comprises repeated *to-ree-rear*; *m-u-r-r* alarm. **HH** Habits similar to other laughingthrushes, see White-throated. Dense undergrowth in broadleaved forests.

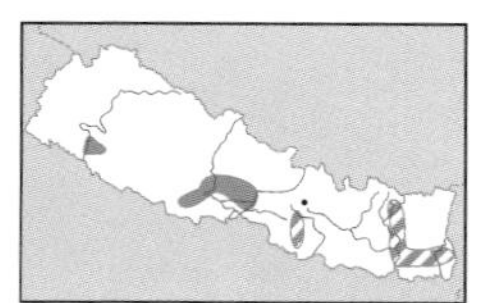

Silver-eared Mesia *Mesia argentauris* 15 cm

Very rare and local resident; 305–1,830 m. **ID** Striking orange-yellow bill, black cap and moustachial stripe, silver-grey ear patch, orange-yellow throat and breast, crimson and yellow wing-panels, and yellow sides to black tail. Male has crimson uppertail- and undertail-coverts. Female has olive-yellow uppertail-coverts and orange-buff undertail-coverts. Juvenile has brown crown. **Voice** Cheerful descending whistled song, *che tchu-tchu che-rit*; calls include a flat piping *pe-pe-pe-pe-pe* and chattering notes. **HH** Constantly on move. In pairs or foraging flocks according to season. Seeks insects and fruit among foliage. Inhabits thickets and bushes in forest edges, clearings and ravines; favours broadleaved evergreen biotope.

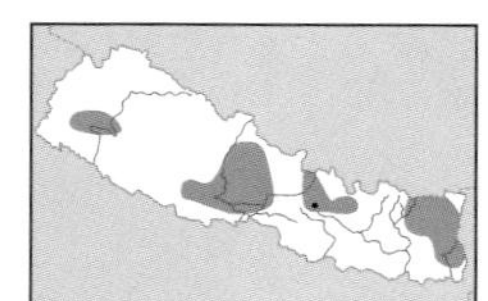

Red-billed Leiothrix *Leiothrix lutea* 13 cm

Resident, fairly common in east, frequent or uncommon in central and west-central areas, rare in west; 915–2,745 m (–250 m). **ID** Stocky with domed head and forked tail. Red bill, creamy-white face, dark moustachial stripe, and yellow throat merging into orange breast. Forked black tail partly overlain by long, white-tipped uppertail-coverts. Crimson, orange and yellow edges to wing feathers (crimson edges lacking in female). Female duller than male, with paler throat and breast. Juvenile has buff throat and grey breast and flanks (no yellow on underparts). **Voice** Rapid, thrush-like song of up to 15 fluted notes; harsh scolding rattles, a clear piping *pe-pe-pe-pa* and rapid *pu-pu-pu-pu-pu*. **HH** Lively babbler, forages in thick undergrowth and wooded ravines in moist broadleaved forests.

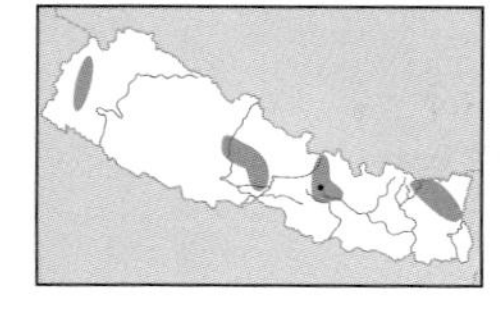

Himalayan Cutia *Cutia nipalensis* 20 cm

Very local resident; 1,095–2,700 m. **ID** A stocky, slow-moving, nuthatch-like babbler. Male has blue-grey crown and blue-black mask, rufous mantle, white underparts with bold blackish barring on buffier flanks, and blue-grey wing-panel. Female has duller grey crown, brown mask and black-streaked olive-brown mantle. Both sexes have black tail just showing beyond long uppertail-coverts, and bright orange legs and feet. Juvenile duller than respective adult, with black barring restricted to lower flanks. **Voice** A loud ringing series of 10–15 sharp *toot* notes, falling off at end; pair members converse with short, buzzy squawks and single or double toots as they travel; other calls include an upturned *hweet* and scolding *trt*. **HH** Arboreal, usually in canopy. Forages among foliage and mossy branches and trunks. Mossy humid broadleaved evergreen forests.

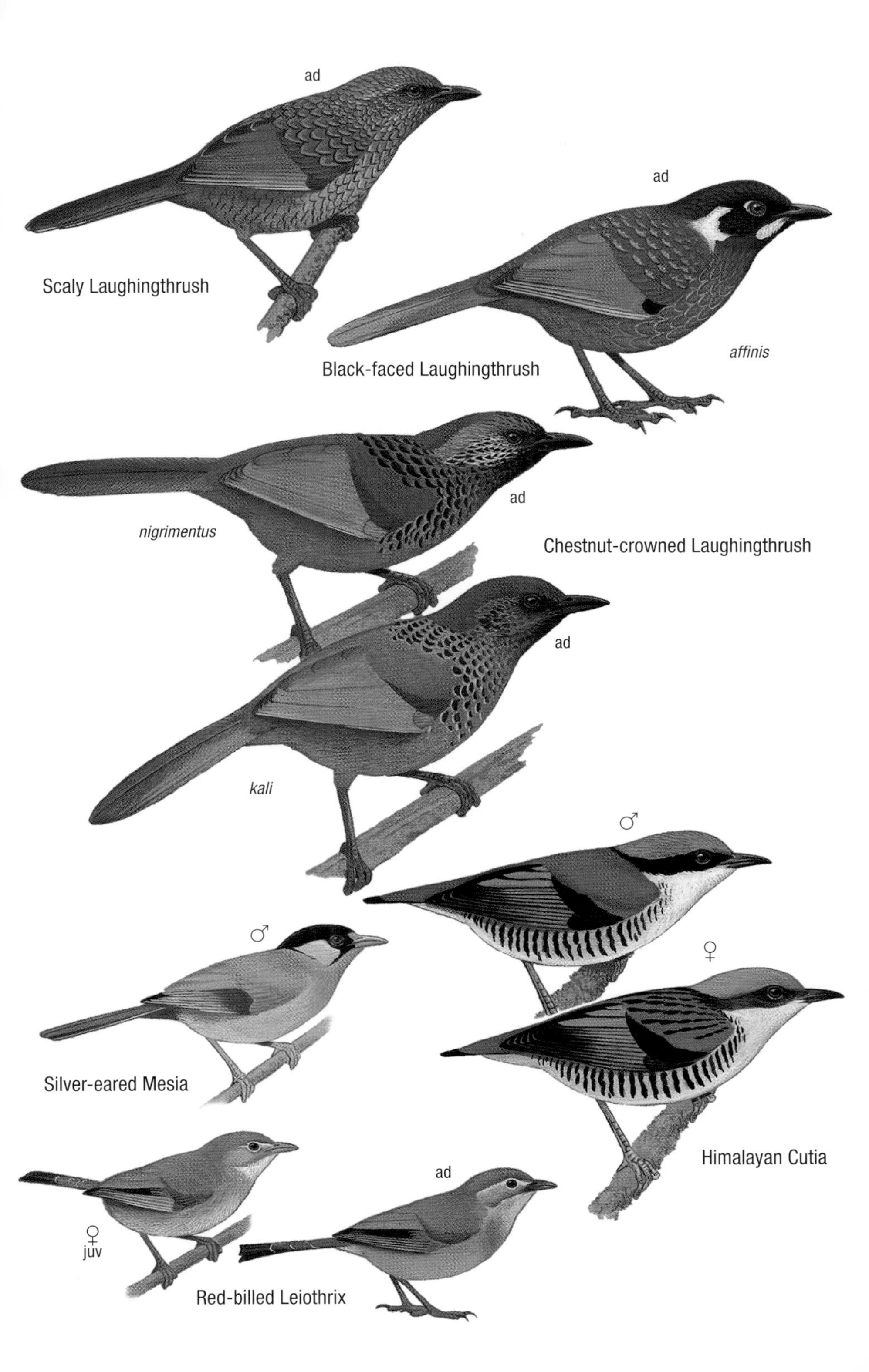
ad
Scaly Laughingthrush
ad
Black-faced Laughingthrush
affinis
nigrimentus
ad
Chestnut-crowned Laughingthrush
ad
kali
♂
♂
Silver-eared Mesia
♀
Himalayan Cutia
ad
♀
juv
Red-billed Leiothrix

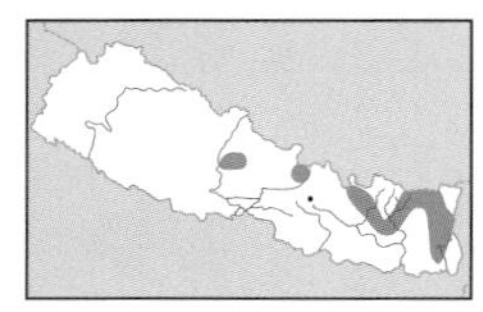

Black-headed Shrike Babbler *Pteruthius rufiventer* 17 cm

Rare, local resident; 1,700–2,500 m (–3,230 m). **ID** Male has black hood, grey throat and breast, rufous-brown mantle, olive-yellow patch on sides of breast, pinkish-buff underparts, and rufous tips to black tertials and tail. Female duller with grey hood, olive-green mantle (variably marked black), and olive wings and tail. Rufous uppertail-coverts, grey throat and breast, brownish-pink underparts, and rufous tip to tail separate female from female White-browed Shrike Babbler. **Voice** Bright, repeated *pew-pew-peee-tu* song. **HH** Found in bushes near ground but also up to forest canopy. Lethargic. Dense, moist, mossy, evergreen broadleaved forests.

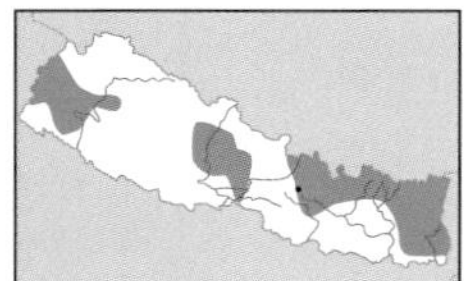

White-browed Shrike Babbler *Pteruthius flaviscapis* 16 cm

Frequent resident; breeds 1,800–2,550 m, winters 1,500–2,135 m (–305 m). **ID** Male has black cap with white supercilium, grey upperparts, white tips to black wings, rufous tertials, and whitish underparts with pinkish flanks. Female has grey cap, olive mantle and largely yellowish-olive wings and tail. Larger size and rufous tertials separate female from Green. Unmarked mantle, rufous tertial patch, uniform buffish-white underparts, white primary tips, and yellowish tip to tail separate female from female Black-headed. **Voice** Song comprises rhythmic three- or six-noted *yip-yip-yip* or *yip-dip-dip* with stress on first or last note and sometimes extending to six notes; short *pink* and grating *churr* in alarm. **HH** Arboreal; feeds mainly in canopy. Rather slow-moving. Broadleaved forests, favours oaks. **AN** Blyth's Shrike Babbler *P. aeralatus*; if eastern form *validirostris* is split; Himalayan Shrike Babbler, if western form *ripleyi* is split.

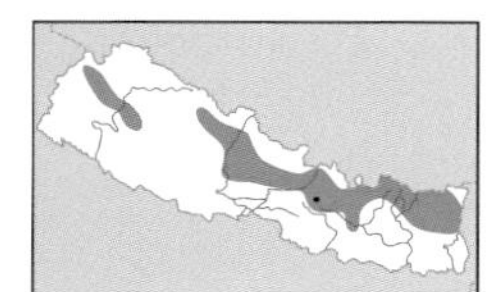

Green Shrike Babbler *Pteruthius xanthochlorus* 13 cm

Frequent, widespread resident; 1,980–3,355 m. **ID** Small and stocky with large-headed appearance. Grey cap, stubby blackish bill, olive-green upperparts, white or yellowish-white greater coverts wing-bar, blackish primary-coverts, greyish-white throat and breast, and pale yellow belly. Sexes similar, but female has paler olive-grey crown. Juvenile has olive-brown crown and mantle, paler underparts and yellowish wing-bar. *P. x. occidentalis* (in west) has blue-grey crown, which is dark grey on male *P. x. xanthochlorus* (in east, although female is much as male *xanthochlorus*). **Voice** Rapid, tit-like *whee-tee whee-tee whee-tee* song; grating *chaa* call. **HH** Forages unobtrusively in trees. Rather sluggish. Broadleaved, coniferous and mixed forests.

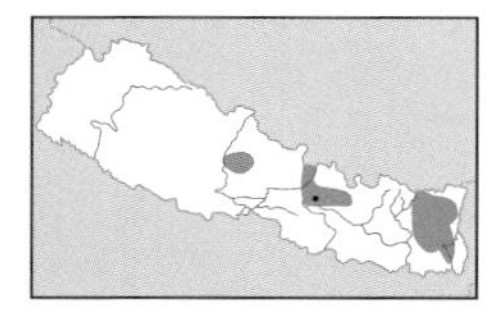

Black-eared Shrike Babbler *Pteruthius melanotis* 11 cm

Local resident; breeds 1,800–2,440 m, winters 1,500–2,000 m (–305 m). **ID** A small, strikingly patterned shrike babbler. Male has chestnut throat and breast, and white wing-bars. Female has chestnut reduced to malar region and buff wing-bars. Both sexes have greenish crown and mantle, and yellow cheeks bordered black. Juvenile similar to female, but has olive-brown upperparts (lacking grey patch on nape), and underparts are paler lacking any chestnut. **Voice** Bright *tew wee, tew we tew-wee* song; *dz-wee-tik* call and rasping alarm. **HH** Arboreal and rather lethargic. Humid broadleaved evergreen forests.

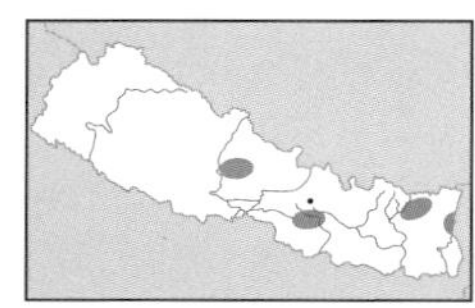

Rusty-fronted Barwing *Actinodura egertoni* 23 cm

Rare, local resident; 1,500–2,400 m. **ID** Slimmer and longer-tailed than Hoary-throated Barwing, and best separated by combination of rufous 'front' to head, stout yellowish bill, uniform grey crown and nape contrasting with mantle (without prominent streaking), greyish-buff (rather than rufous-brown) barring on wings, and rufous-brown and more diffusely barred tail (lacking blackish terminal band). Juvenile has crown and nape rufous-brown (concolorous with mantle), thus rufous forehead less prominent. **Voice** Three-note whistle, *ti-ti-ta*, the first note accentuated, the last lower; also feeble *cheep*. **HH** Forages by clambering among bushes and undergrowth; sometimes in canopy. Dense thickets in humid, broadleaved evergreen forest.

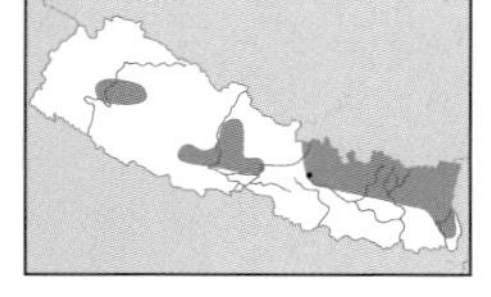

Hoary-throated Barwing *Actinodura nipalensis* 20 cm

Locally fairly common resident; 1,980–3,000 m (1,500–3,500 m). **ID** Lacks rufous 'front' of Rusty-fronted. Other differences are unstreaked greyish throat and breast, prominent buffish-white shaft-streaking on dark brown crest, grey ear-coverts contrasting with dark moustachial stripe, diffuse white streaking on mantle and scapulars, and shorter, more strongly barred tail. **Voice** One or two whistles followed by a soft trill, with whistles sometimes added to end of verse; a series of long, whistling notes and jay-like alarm. **HH** Usually frequents upper half of forest trees. Forages among ferns and mosses growing on trees, often clinging to trunk or hanging upside-down from branches. Mossy oak–rhododendron forest.

♂
♀
Black-headed Shrike Babbler
♂
♀
White-browed Shrike Babbler
♂
occidentalis
Green Shrike Babbler
♂
♀
Black-eared Shrike Babbler
ad
Rusty-fronted Barwing
ad
Hoary-throated Barwing

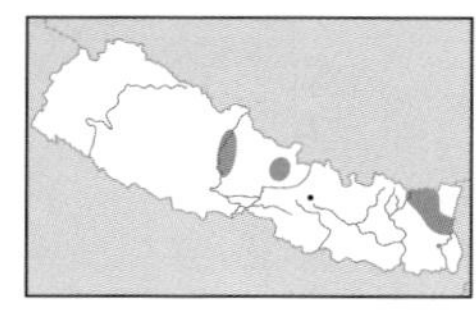

Golden-breasted Fulvetta *Lioparus chrysotis* 11 cm

Local resident with fragmented distribution; 2,200–3,050 m. **ID** Dark grey head with silver-grey ear-coverts, grey upperparts, golden-yellow underparts, and orange-yellow wing-panel and sides to tail. **Voice** Thin high-pitched series of five notes, slightly descending; alarm a high-pitched buzz. **HH** In non-breeding season, in flocks of up to 50, often with parrotbills and White-browed Fulvettas. Tame and active. Often hangs upside-down like a tit when foraging. At low and medium heights in bamboo or undergrowth. Bamboo stands and bamboo understorey in broadleaved forests. **TN** Formerly placed in genus *Alcippe*.

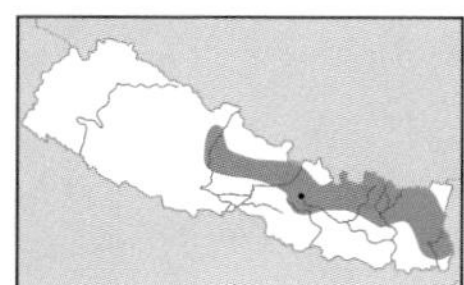

Rufous-winged Fulvetta *Pseudominla castaneceps* 10 cm

Resident. Locally common resident; breeds 1,825–2,745 m (–3,505 m), winters 1,525–2,745 m. **ID** From White-browed Fulvetta by black band across wing-coverts, and whitish pattern on cheeks (with short black moustachial stripe). Also has chestnut crown and nape (streaked buffish-white), white supercilium contrasting with black eye-stripe, and rufous wing panel. **Voice** Usual call a wheezy, descending trill *tsi-tsi-tsi-tsi-tsirr*; also a quiet *chip*; song a rich, undulating and descending warble, *tju-tji-tju-tji-tju-tji-tju*. **HH** In flocks of up to 30, often in mixed feeding flocks with other small babblers, tits and warblers. Constantly on the move at low and medium heights in forest, sometimes in foliage of bushes and undergrowth, also climbs up mossy trunks like a nuthatch. Moist forest and undergrowth in broadleaved forests, also second growth. **TN** Formerly placed in genus *Alcippe*.

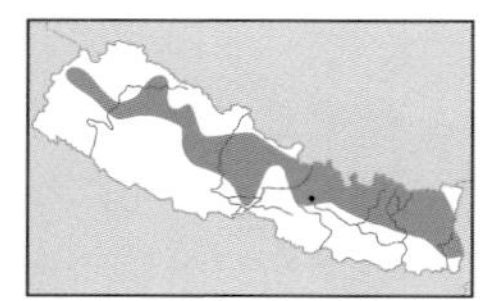

White-browed Fulvetta *Fulvetta vinipectus* 11 cm

Common, widespread resident; breeds 2,400–4,200 m, winters 2,135–3,000 m (–1,525 m). **ID** Broad white supercilium, contrasting with dark crown and ear-coverts, and black and grey panels in wing. Two races occur. In west and centre nominate race has unstreaked white throat. *F. v. chumbiensis* in east has brown streaking on throat. **Voice** Song a rapid *chit-it-it-it-or-key* and a *tew-tu-tu-wheeee*; *vek vek vek* and *czzzzzz* alarm calls. **HH** When not breeding, in flocks, sometimes with other babblers and tits. Inquisitive and tame. Forages restlessly low down in bushes and undergrowth. Tit-like actions, but slower moving. Bushes in broadleaved and coniferous forests, and high-altitude shrubberies. **TN** Formerly placed in genus *Alcippe*.

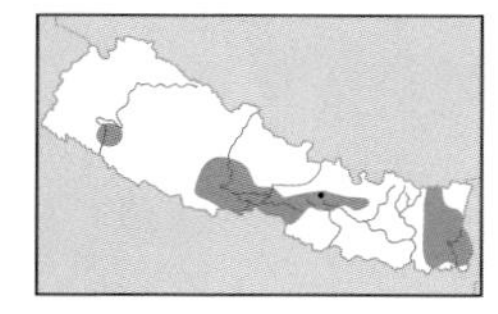

Nepal Fulvetta *Alcippe nipalensis* 12 cm

Resident; common on hills around Kathmandu Valley, uncommon elsewhere; winters 245–1,830 m, breeds to 2,285 m. **ID** Distinctive fulvetta with grey head and blackish lateral crown-stripes, prominent white eye-ring, olive-brown upperparts, and whitish underparts with buff flanks. **Voice** Short buzzes and metallic *chit* notes; also a short fast trill of varying speed. **HH** Found in energetic parties, continually calling as they move. Forages mainly in bushes, undergrowth and small trees, sometimes on ground. Usually quite shy, coming into open infrequently. Dense undergrowth in moist forests and second growth.

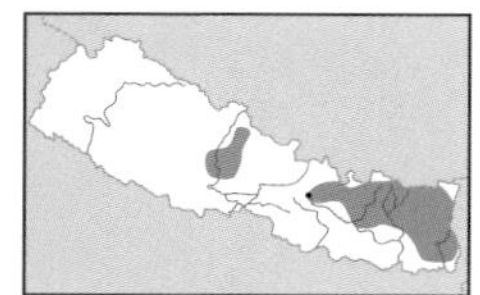

Red-tailed Minla *Minla ignotincta* 14 cm

Resident; locally common in east, uncommon in central and west-central areas; breeds 1,830–3,400 m, winters 760–2,285 m. **ID** A slim, long-tailed babbler. Black crown and ear-coverts, broad white supercilium and striking pale eye. Also prominent white tips to black tertials and yellowish underparts. Male differs from female in having deep maroon-brown (rather than olive-brown) upperparts, red (rather than orange-yellow) wing panel, and red (rather than pinkish) sides and tip to tail. **Voice** Varied calls including a high-pitched *wi-wi-wi*, a loud repeated *chik*, high-pitched *tsi*, high-pitched *chititit* and tit-like *whi-whi-te-sik-sik*. **HH** In roving parties with other species. Forages in trees; often searches mossy trunks and branches like a treecreeper, or clings upside-down to small twigs like a tit, but more slow-moving. Mainly moist, dense broadleaved or mixed forest.

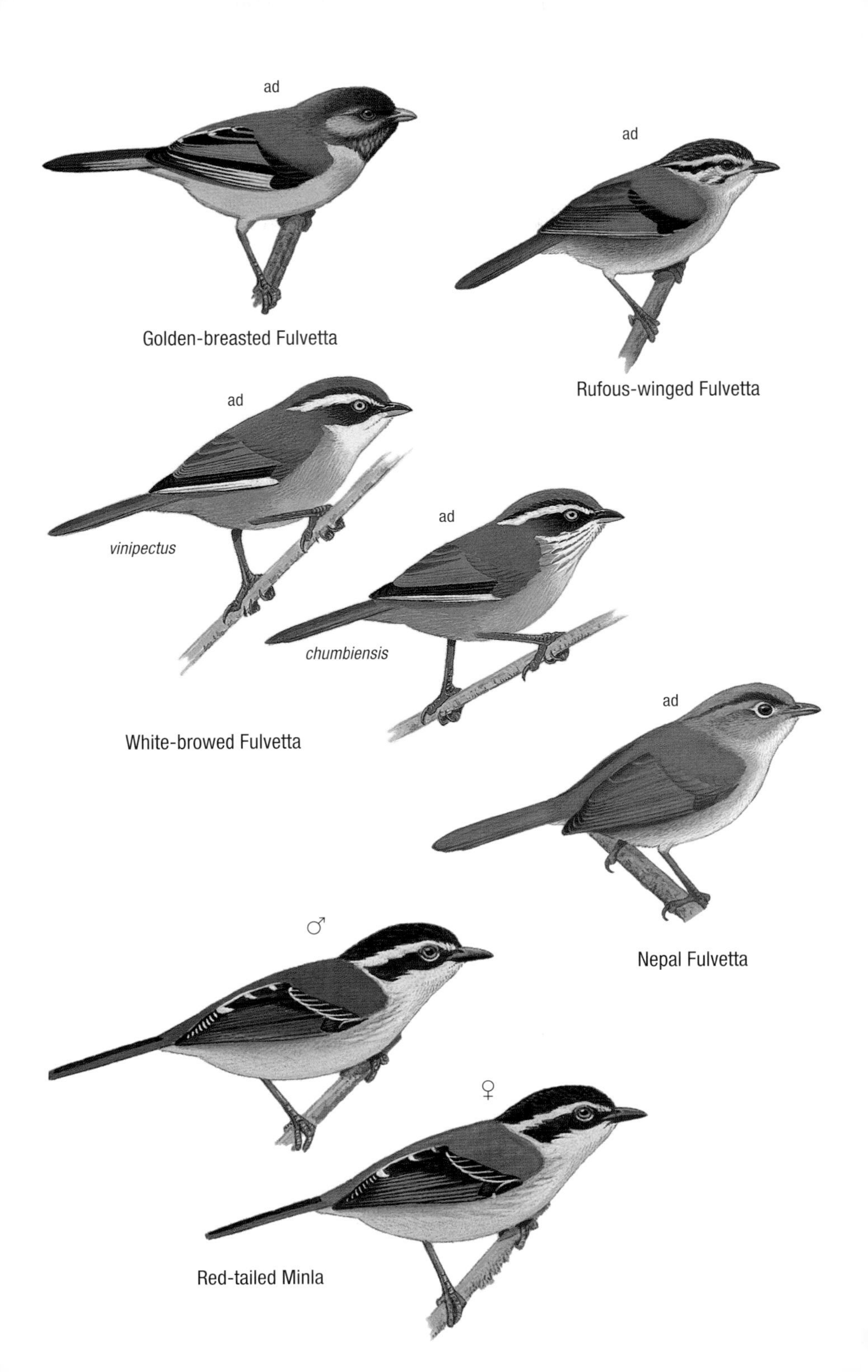

ad
Golden-breasted Fulvetta
ad
Rufous-winged Fulvetta
ad
vinipectus
ad
chumbiensis
White-browed Fulvetta
ad
Nepal Fulvetta
♂
♀
Red-tailed Minla

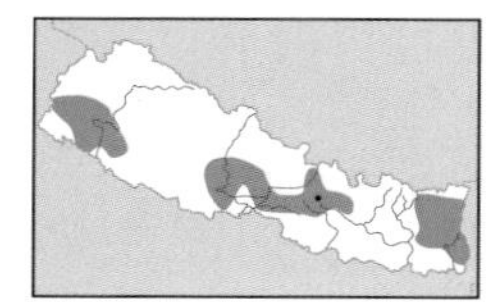

Blue-winged Siva *Siva cyanouroptera* 15 cm

Resident, fairly common in east, frequent or uncommon in west-central areas and west; breeds to 2,440 m (–2,745 m), winters 1,000–1,830 m (–2,285 m). **ID** Slim babbler with long square-ended tail. Although colourful, can look drab at long range, with dark-capped, pale-faced appearance, fulvous-brown mantle, vinous-grey underparts, and blue panels in wings and tail. At close range has pale violet-grey crown and nape, with violet-blue streaking on forehead and sides of crown. White underside to dark-bordered tail striking from below. Juvenile has brownish-grey crown and nape, and buffish underparts. **Voice** A clear, whistled *pi-piu* with emphasis on first note, and loud *swit*. **HH** Usually in small parties with other species, often babblers, warblers and flycatchers. Typically feeds in bushes, sometimes at middle height in trees. Bushes and undergrowth in dense forests and well-wooded country. **TN** Formerly placed in genus *Minla*. **AN** Blue-winged Minla.

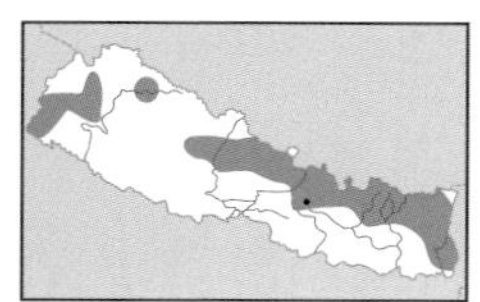

Bar-throated Siva *Siva strigula* 14 cm

Common resident; breeds 2,440–3,750 m, winters 1,400–2,745 m (–1,000 m). **ID** Orange crown and nape, grey ear-coverts with black moustachial stripe, and black-and-white barring on throat. Upperparts olive-brown and underparts yellowish. Black primary-coverts patch, grey fringes to black-centred tertials, and orange-yellow wing-panel. Shows orange-yellow at sides and tip of tail when seen from below. Juvenile has less distinct throat barring and duller crown and nape. **Voice** A slurred whistle *jo-ey, joey dii*, the last note highest; also a ringing metallic *chew*. **HH** Gregarious; in small noisy flocks, usually with other species, and continually on the move. Frequents bushes and lower canopy. Broadleaved forests especially of oak. **TN** Formerly placed in genus *Minla*. **AN** Chestnut-tailed Minla.

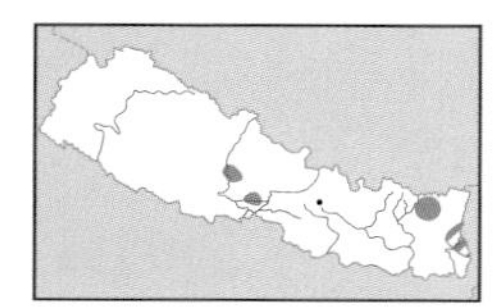

Rufous-backed Sibia *Leioptila annectans* 18 cm

Very rare and local, probably resident; 1,450–2,650 m. **ID** Small, short-tailed sibia. Black cap, black-and-white-streaked upper mantle merging into rufous back and rump, and white underparts with deep buff flanks and vent. Black wings have white fringes to tertials and grey edges to primaries and secondaries. Black tail tipped white (showing as large white spots at tip of undertail). Yellow base to lower mandible and yellow legs and feet. **Voice** Sings with repeated loud, strident jolly phrases. **HH** In small parties. Arboreal, spends most time in canopy of tall trees. Creeps along branches and flits from branch to branch, sometimes climbs like a nuthatch; tends to keep close to trunk. Dense, moist broadleaved evergreen forests. **TN** Formerly placed in genus *Heterophasia*.

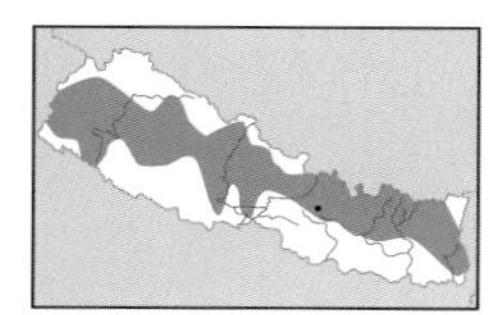

Rufous Sibia *Malacias capistratus* 21 cm

Common, widespread resident; breeds 1,980–3,000 m, winters 1,050–2,750 m (–800 m). **ID** Black cap, rufous or cinnamon-buff nape and underparts, grey tip and black subterminal band to rufous tail, and grey panelling on mainly black wings. Three races occur. Nominate (in west) has paler cinnamon-buff nape and underparts, and grey-brown mantle and back. *M. c. nigriceps* and *M. c. bayleyi* (centre and east) have deeper rufous nape and underparts; *bayleyi* has a brighter rufous-brown mantle almost concolorous with nape. **Voice** Song a clear, flute-like *tee-dee-dee-dee-dee-o-lu*, the first five notes on same pitch, the sixth lowest and last in between; call a rapid *chi-chi* and alarm a harsh *chrai-chrai-chrai*. **HH** In pairs or small parties, depending on season. Conspicuous, noisy and arboreal. Forages among foliage or on moss-covered branches and trunks in middle levels and canopy, sometimes lower; hops rapidly from branch to branch; can cling to slender stems while probing flowers, and to tree trunks while searching bark crevices. Sometimes sallies to catch flying insects. Broadleaved forest, especially of oak–rhododendron. **TN** Formerly placed in genus *Heterophasia*.

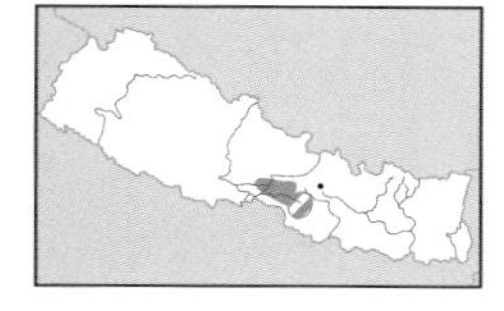

Long-tailed Sibia *Heterophasia picaoides* 30 cm

Very rare and local, probably resident; only recent records from Chitwan district; 305–900 m. **ID** From other sibias by long tail with greyish-white tips, grey head and upperparts (with slightly darker lores), paler grey underparts, and dark grey wings with white patch on secondaries. **Voice** Calls include thin, metallic, high-pitched *tsittsit* and *tsic* notes, interspersed by dry, even-pitched rattling. **HH** Arboreal, frequents mainly middle levels and canopy. Often feeds on nectar of flowering trees; also on flower buds, insects and seeds. Broadleaved evergreen forest.

ad
Blue-winged Siva
ad
Bar-throated Siva
ad
Rufous-backed Sibia
ad
nigriceps
ad
capistratus
Rufous Sibia
ad
Long-tailed Sibia

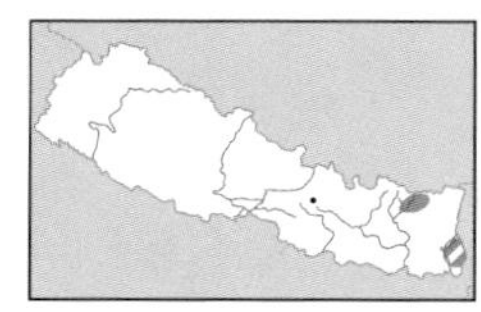

White-naped Yuhina *Yuhina bakeri* 13 cm

Rare, very local resident, only recent records from Makalu Barun National Park buffer zone; 915–2,200 m. **ID** From other yuhinas by combination of stout bill, rufous crest, white nape (especially prominent when crest raised), blackish lores, white streaking on rufous ear-coverts, white shaft-streaking on mantle, and fine brown streaking on pinkish-buff breast. **Voice** Ringing *zee-zee or zeuu-zuee*; high-pitched alarm notes. **HH** In pairs or small parties, depending on season; often with other small babblers, tits and warblers. Forages actively, mainly in middle level of forest and bushes. Broadleaved, evergreen forest.

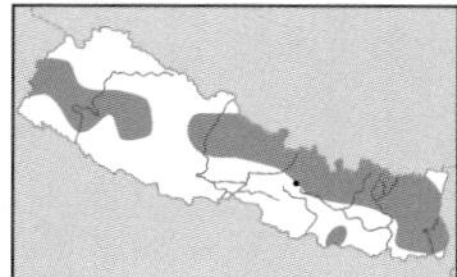

Whiskered Yuhina *Yuhina flavicollis* 13 cm

Common, widespread resident; breeds 1,830–2,745 m, winters 800–2,745 m. **ID** Separated from other yuhinas by combination of grey-brown crest, prominent black moustachial stripe, yellowish to rufous hind collar, and prominent white eye-ring. Sides of breast and flanks olive-brown, streaked white. Compared to nominate in east, *Y. f. albicollis* (in west) has paler hind collar, which is yellowish-buff in centre. **Voice** A thin, rather squeaky *swii swii-swii*, a clear, metallic ringing note and loud ringing notes issued in chorus from flocks. **HH** In non-breeding season, in mixed feeding parties with other small insectivores. Not shy. Hunts energetically in bushes and middle level of forest; flits from branch to branch, sometimes makes aerial sallies after insects. Bushes and lower storey of broadleaved forests, wooded area and second growth.

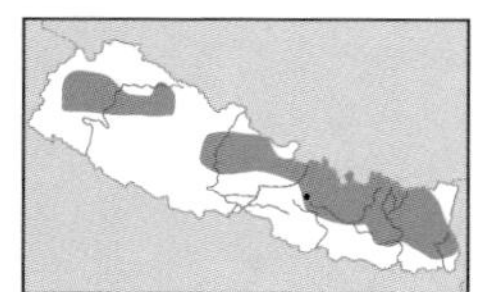

Stripe-throated Yuhina *Yuhina gularis* 14 cm

Common, widespread resident; breeds 2,435–3,700 m, winters 1,700–3,050 m (–1,400 m). **ID** Largest yuhina, best distinguished by combination of olive-brown crest, black streaking on pale vinaceous throat, rufous-orange wing panel contrasting with black primaries, and brownish-orange belly and vent. **Voice** Nasal, descending *queee* call. **HH** In non-breeding season, typically in parties, often with other small babblers, tits and warblers. Usually frequents taller bushes and lower branches in forest. Often feeds on nectar of flowering rhododendrons. Broadleaved and broadleaved/coniferous forest, favouring oak and rhododendron forest.

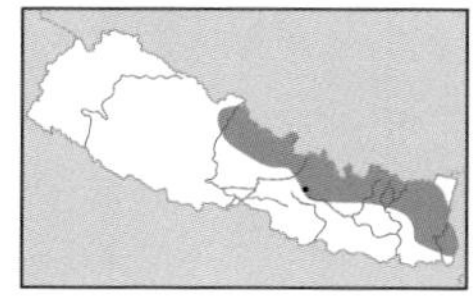

Rufous-vented Yuhina *Yuhina occipitalis* 13 cm

Common, widespread resident; breeds 2,400–3,600 m, winters 1,830 m to at least 2,745 m. **ID** Rufous lores, nape patch, belly and vent, grey crest, ear-coverts and hindneck, fine black moustachial stripe and vinaceous underparts. Reddish bill and legs. **Voice** A constant chittering; high-pitched strong *zee-zu-drrrrr, tsip-ch-e-e-e-e* song; *z-e-e... zit* alarm call. **HH** Habits similar to Stripe-throated, but forages more in treetops. Broadleaved forest, especially oak and rhododendron.

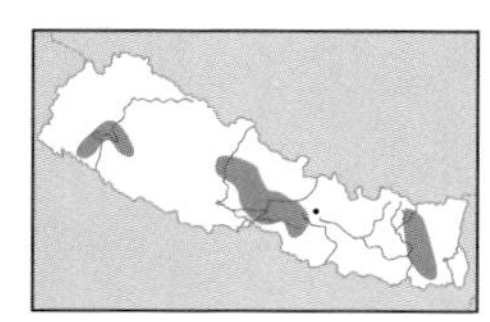

Black-chinned Yuhina *Yuhina nigrimenta* 11 cm

Uncommon resident; 250–1,500 m. **ID** Small, short-tailed, mainly olive-brown yuhina with grey-streaked black crest on greyish head, black chin and lores, and white to pale grey throat and upper breast contrasting with buffish rest of underparts. Red lower mandible with dark tip, and legs and feet orange-yellow. **Voice** Flocks utter constant twittering and buzzing; calls include a high *de-de-de-de* and *zee-zoe-zen*; soft *whee-to-whee-de-der-n-whee-yer* song. **HH** When not breeding, in constantly twittering flocks, often in mixed parties with other small insectivores. Very restless, forages busily, mainly in low bushes, sometimes up into canopy; hangs upside-down on twigs in tit-like fashion. Broadleaved evergreen forest and second growth.

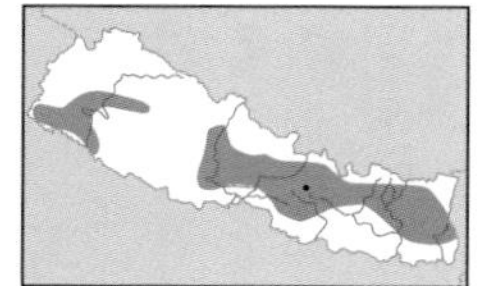

White-bellied Erpornis *Erpornis zantholeuca* 11 cm

Local resident; frequent from west-central Nepal east, rare further west; 150–2,285 m. **ID** Crested, olive-yellow and white, with beady black eye. Undertail-coverts bright yellow. Pinkish bill and legs. Juvenile duller, with brownish cast to upperparts. **Voice** Song a short, high-pitched, descending trill, *si-i-i-i-i*. Calls include a subdued, metallic *chit* and *cheaan* alarm. **HH** Often found singly, occasionally in small parties, sometimes with other small birds. Although lively, it is quiet and unobtrusive. Forages chiefly at middle levels, also in undergrowth and canopy. Broadleaved forests and second growth, especially edges and clearings. **TN** Formerly placed in genus *Yuhina* but not related to other yuhinas.

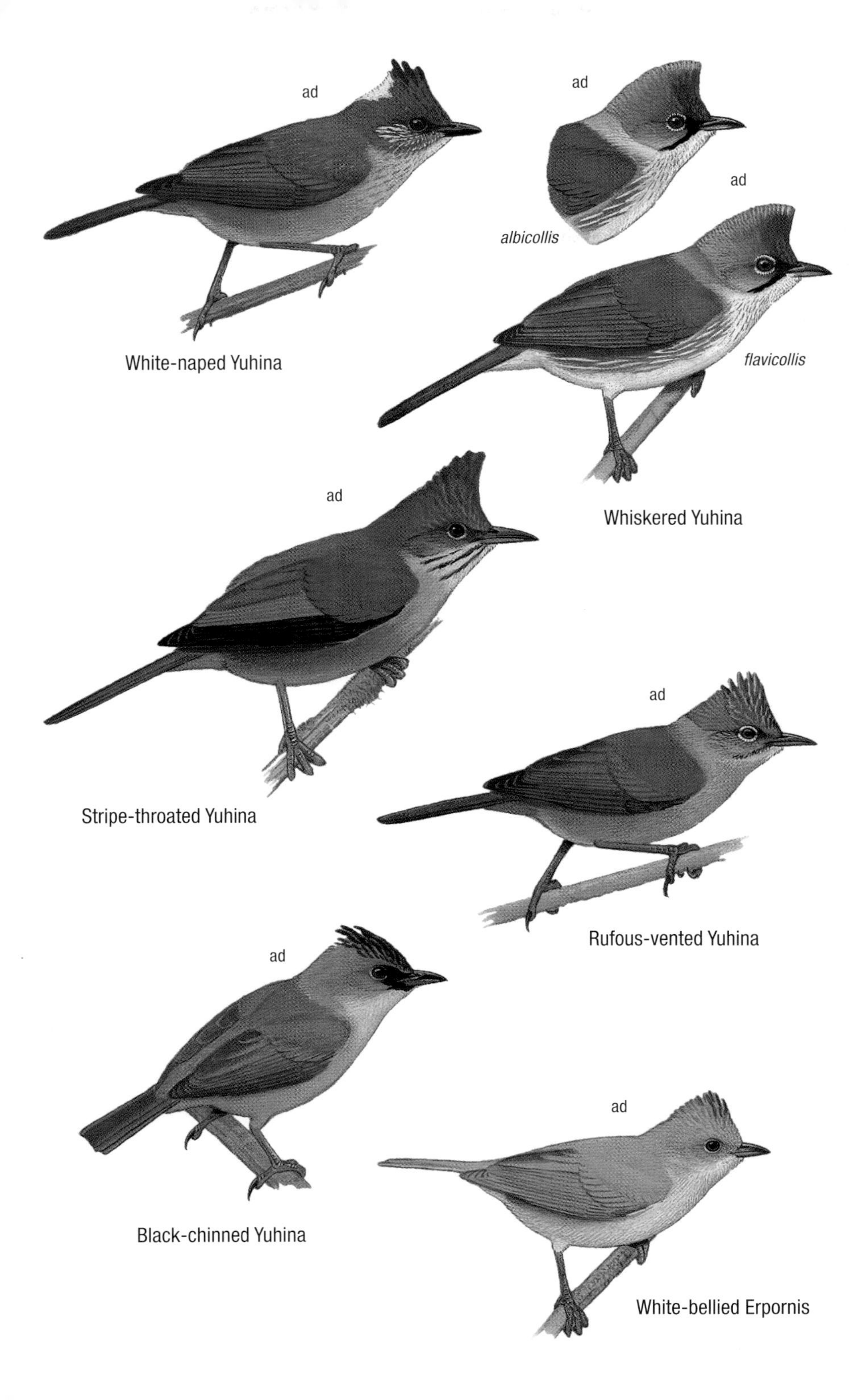
ad
ad
albicollis
ad
White-naped Yuhina
flavicollis
ad
Whiskered Yuhina
ad
Stripe-throated Yuhina
Rufous-vented Yuhina
ad
ad
Black-chinned Yuhina
White-bellied Erpornis

PLATE 104: PARROTBILLS, MYZORNIS AND FAIRY BLUEBIRD

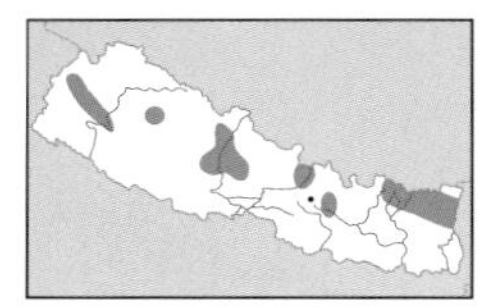

Great Parrotbill *Conostoma aemodium* 28 cm

Local, uncommon resident; 2,700–3,660 m. **ID** Large 'parrotbill', rather laughingthrush-like in shape and behaviour. From Brown by larger size and paler grey-brown coloration, larger cone-shaped bill, buffish-white forehead contrasting with brown lores, paler and more uniform grey-brown sides of head, throat and breast, and grey sides to tail. Lacks dark lateral crown-stripes of Brown. **Voice** Halting, loud song of 2–4 clear whistled notes; also more plaintive nasal wheeze, squeals, cackles and chirrs; alarm a *churrrrrr*. **HH** Singly, in pairs or small parties. Tends to keep hidden, though not shy. Slow-moving, hops and clambers like a laughingthrush. Bamboo stands in forest, usually oak–rhododendron or fir–rhododendron.

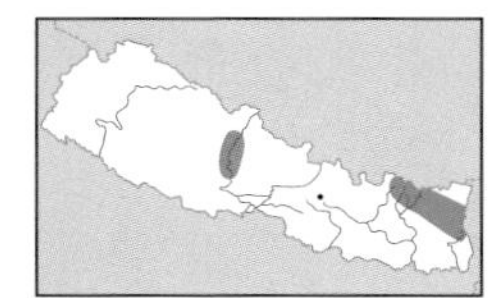

Brown Parrotbill *Cholornis unicolor* 21 cm

Very local, uncommon resident; 2,450–3,400 m. **ID** From Great by smaller size and stocky shape, smaller and much stouter bill, diffuse blackish lateral crown-stripes broadening on nape, diffuse grey supercilium behind eye and grey eye-ring, diffuse grey mottling on dull brown ear-coverts, dusky-grey and rather untidy-looking throat and breast merging into olive-brown flanks and belly, and browner tail. Rufous-brown panel on wings, as Great. **Voice** Call *chirrup* and loud *churrr-churrr* in alarm. **HH** Usually in small parties that skulk in undergrowth and frequently call to each other. Sometimes hangs upside-down when feeding. Dense stands of bamboo or dwarf rhododendron. **TN** Formerly placed in genus *Paradoxornis*.

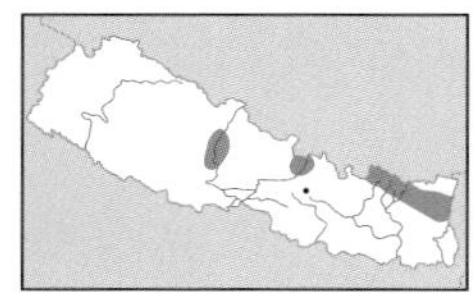

Fulvous Parrotbill *Suthora fulvifrons* 12 cm

Local resident; 2,700–3,400 m. **ID** Small, comparatively long-tailed parrotbill, lacking black, white and grey pattern of Black-throated Parrotbill. Main differences from latter are fulvous crown and supercilium, olive-brown (rather than black) lateral crown-stripes, lack of striking white malar patch, and fulvous face and underparts. **Voice** Probable song thin, high-pitched *si-si ssuuu-juuu* with harsher final note; also *si-si-sissu-suu-u* and *si-si-sissu-suue* with thin last notes. Short contact calls. **HH** Usually in flocks that move rapidly through bamboo and maintain contact with continual twitter. Climbs with agility on bamboo stems, sometimes hanging upside-down. Dense bamboo stands. **TN** Formerly placed in genus *Paradoxornis*.

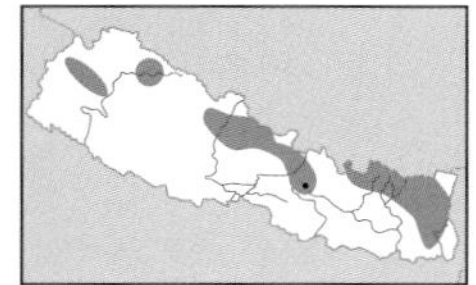

Black-throated Parrotbill *Suthora nipalensis* 10 cm

Local resident; 2,000–3,000 m (1,040–3,500 m). **ID** Small parrotbill. Broad blackish lateral crown-stripes and black throat, white eye-patch and malar patch, and black primary-coverts patch. *S. n. nipalensis* in west has grey crown and ear-coverts; *S. n. humii* in east has brownish-orange crown and ear-coverts and brighter orange-brown mantle. **Voice** Flocks give general hubbub of dry, chattering trills. **HH** Habits similar to Fulvous, but more fast-moving and often in mixed feeding flocks with other babblers and tits. Bamboo and thick undergrowth in broadleaved forests. **TN** Formerly placed in genus *Paradoxornis*.

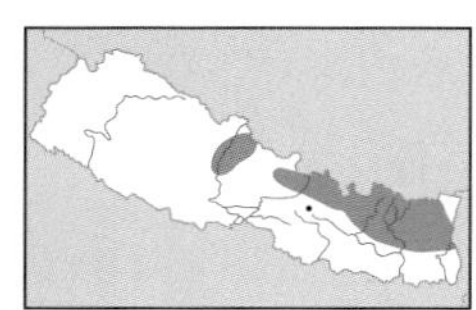

Fire-tailed Myzornis *Myzornis pyrrhoura* 12 cm

Local and uncommon resident; 2,000–3,950 m. **ID** Mainly brilliant emerald-green, with fine black bill, black eye-stripe and scaling on crown, and red-and-orange panels on black wings. Sexes similar but female has orange-buff throat (throat/upper breast red in male), greyish belly and flanks (more orange in male), less extensive white tips to secondaries and primaries, and duller red sides to tail. **Voice** A *trrrr-trrrr-trrr* preceded by a high-pitched squeak; repeated *tzip* in alarm; very high-pitched *tsi-tsi* contact notes. **HH** Usually forages in bushes. Often probes blossoms, especially rhododendrons, for nectar; also seeks invertebrates in foliage, hovering in front of flowers, creeping up mossy trunks and rocks, and making short flights. Rhododendron and juniper shrubberies; mossy oak–rhododendron forest and bamboo stands.

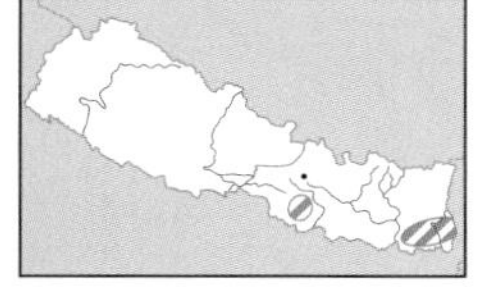

Asian Fairy Bluebird *Irena puella* 25 cm

Very rare, probably former resident; 75–365 m. **ID** Male has glistening violet-blue upperparts and black underparts. Female and first-year male entirely dull blue-green, with dusky lores and blackish flight feathers. Both sexes have striking red eye. Juvenile entirely dull brown. **Voice** Liquid *tulip wae-waet-oo* and shorter liquid notes. **HH** Forages for fruit and nectar, usually in canopy. Broadleaved evergreen and moist deciduous forest.

ad
Great Parrotbill
ad
Brown Parrotbill
ad
Fulvous Parrotbill
ad
nipalensis
ad
Black-throated Parrotbill
humii
♂
♀
Fire-tailed Myzornis
♀
♂
Asian Fairy Bluebird

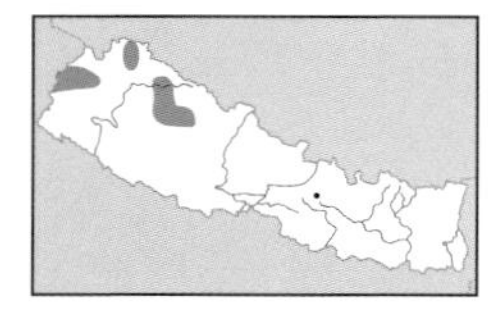

Kashmir Nuthatch *Sitta cashmirensis* 12 cm

Locally fairly common resident in north-west; 2,400–3,505 m. **ID** Compared to Chestnut-bellied, has uniform undertail-coverts and lacks clearly defined white cheeks. Larger and longer billed than White-tailed, with more pronounced white cheeks, no white at base of tail, and diagnostic loud, rasping, jay-like *jhreee* calls. Female similar to male, but underparts paler, with flanks and undertail-coverts more deep cinnamon, and whitish cheek patch less clearly defined. Female therefore very similar to White-tailed, but ear-coverts whiter and underparts more pinkish-cinnamon. **Voice** Song is series of rapidly repeated high-pitched whistles. **HH** Habits like White-tailed, but often feeds in understorey, occasionally on ground. Forests in dry areas.

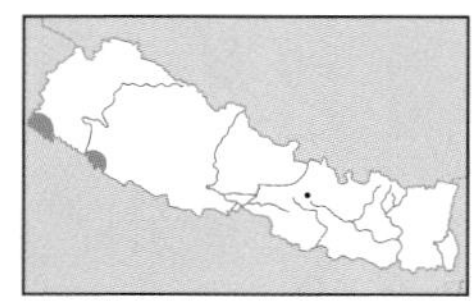

Indian Nuthatch *Sitta castanea* 12 cm

Status uncertain, probably resident; 150 m. **ID** Compared to Chestnut-bellied is smaller, with shorter, slender bill. Scalloping on undertail-coverts is grey (same colour as mantle), and crown/nape are distinctly paler than mantle, while underparts of male are darker reddish-chestnut. **Voice** Song comprises a rapid trill. **HH** Habits like White-tailed; usually forages in upper half of trees, occasionally on ground. Broadleaved forests and groves. **TN** Considered here to be conspecific with next species as Chestnut-bellied Nuthatch *S. castanea*.

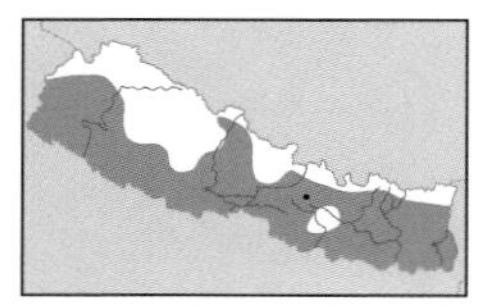

Chestnut-bellied Nuthatch *Sitta* (*castanea*) *cinnamoventris* 12.5 cm

Common, widespread resident; 150–1,540 m. **ID** From Kashmir and White-tailed Nuthatches by whitish scalloping on undertail-coverts. Male always has striking white cheek patch contrasting with rather uniform orange-brown underparts. Female similar, but underparts paler, and is more similar to Kashmir, but underparts are richer, darker and more uniform cinnamon-brown, and has more clearly defined white chin and cheeks. Lacks white at base of central tail feathers of White-tailed, also underparts darker and more uniform, and cheek patch more prominent. **Voice** Song a single pure whistle every few seconds; calls include high mouse-like *seet*, lower and squeakier *vit* and full *chup*. **HH** Habits like White-tailed. Broadleaved forests and groves.

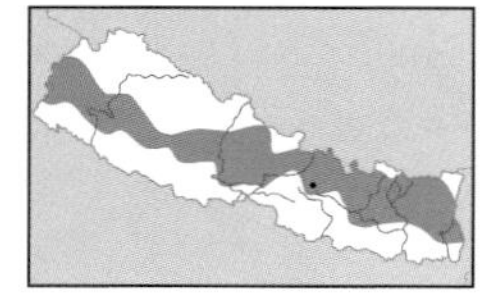

White-tailed Nuthatch *Sitta himalayensis* 12 cm

Common, widespread resident; breeds 1,800–3,140 m (–3400 m), winters to 915 m. **ID** From Kashmir by white at base of central tail feathers (although this can be very difficult to see in field), and different call. Also smaller, with relatively shorter bill. Sexes similar. White-tailed has less distinct cheek patch compared to Kashmir (cheeks off-white or buff, and ear-coverts more cinnamon-orange). From female Chestnut-bellied by uniform undertail-coverts, and underparts paler and less uniform. **Voice** Song a series of clear whistles; calls include squeaky *nit* and soft *chak*, harder *chak'kak*, which may be rapidly repeated as monotonous rattle, and shrill *shree*. **HH** Like all nuthatches an agile climber; moves with ease upwards, downwards, sideways and upside-down over trunks and branches, progressing via series of jerky hops. Unlike woodpeckers and treecreepers, usually begins near treetop and works down. Frequent member of feeding flocks of tits, warblers etc. Inhabits upper half of trees, occasionally low bushes. Broadleaved forests.

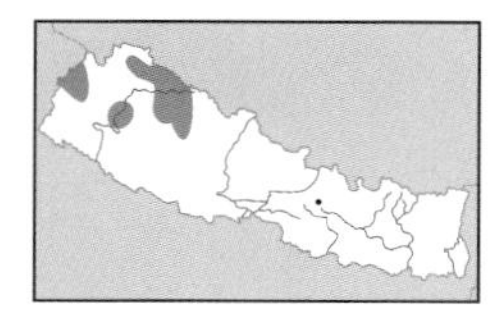

White-cheeked Nuthatch *Sitta leucopsis* 12 cm

Locally frequent resident in north-west; 2,745–3,900 m. **ID** Very distinctive, with black crown and nape, white face and throat with beady eye, and whitish underparts with buffish wash, becoming rufous on rear flanks and undertail-coverts. Juvenile has faint barring on underparts. **Voice** Call likened to young goat bleating; song comprises continuous, rapid series of low-pitched squeaks; call similar, but slower and in shorter bursts; in alarm a series of single notes. **HH** Habits similar to White-tailed but feeds in upper canopy, most easily located by distinctive call. In breeding season, males sometimes perch conspicuously on tall treetops, calling and flicking wings. Coniferous forests.

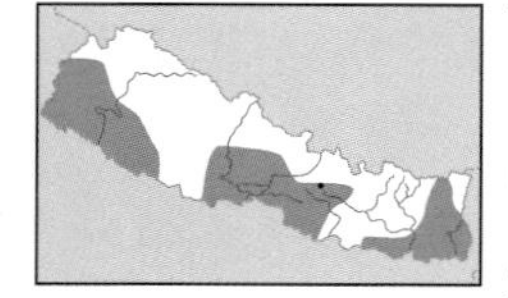

Velvet-fronted Nuthatch *Sitta frontalis* 10 cm

Locally fairly common resident; 75–2,015 m. **ID** Striking violet-blue upperparts, black forehead, black-tipped red bill, startling yellow iris and eye-ring, and lilac suffusion to ear-coverts and underparts. Male has black eye-stripe extending behind eye (lacking in female) with stronger lilac suffusion on underparts (more cinnamon, less lilac in female). Juvenile has blackish bill and duller and greyer upperparts; underparts lack lilac suffusion and are washed orange-buff. **Voice** Song is fast, hard rattle. **HH** Habits similar to White-tailed but more active than other Nepal nuthatches. Forages from canopy down to undergrowth, but not on ground. Open broadleaved forest and well-wooded areas.

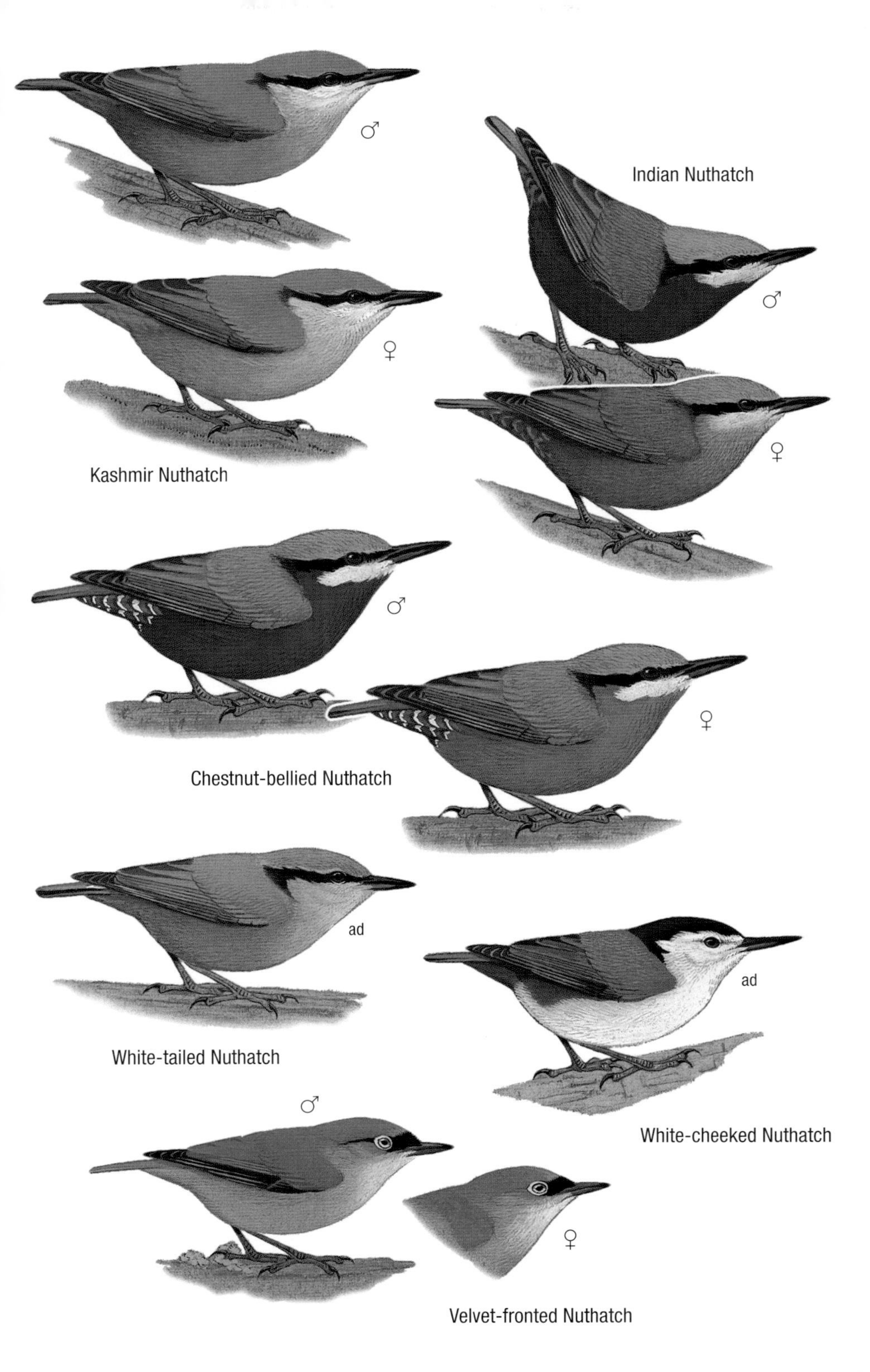
♂
Indian Nuthatch
♀
♂
♀
Kashmir Nuthatch
♂
♀
Chestnut-bellied Nuthatch
ad
ad
White-tailed Nuthatch
♂
White-cheeked Nuthatch
♀
Velvet-fronted Nuthatch

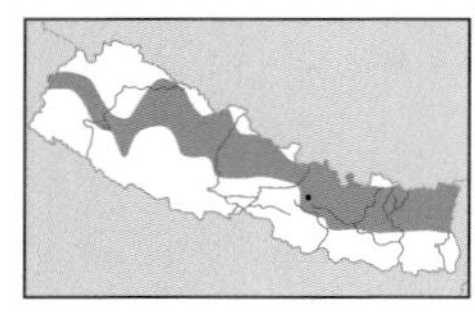

Eurasian Wren *Troglodytes troglodytes* 9 cm

Widespread, locally common resident; mainly 2,500–4,725 m (2,135–5,300 m). **ID** Small and squat, with stubby tail, and rather long, pointed bill. Sooty-brown all over, with indistinct buffish supercilium, barred wings and tail, and barring on belly and flanks. **Voice** Powerful, rapidly delivered warbling and trilling song; calls include hard *chek* and rattling *churr*. **HH** Breeds on rocky slopes; winters around villages and fallen forest trunks. **AN** Winter Wren.

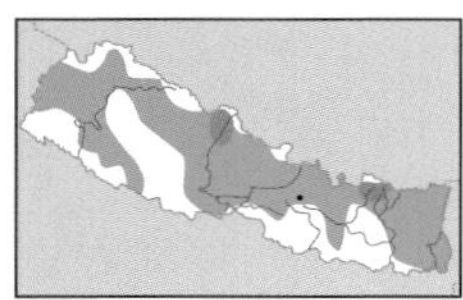

Wallcreeper *Tichodroma muraria* 16 cm

Locally fairly common resident and winter visitor, mainly recorded in non-breeding season; mainly 245–5,000 m (80–5,730 m). **ID** Long, downcurved, black bill. In flight, wings rounded and reveal largely crimson wing-coverts and bases to black flight feathers, with two rows of white spots on primaries. Also white corners to tail. Adult male breeding has black throat and upper breast. Adult female breeding has whitish chin and variable blackish patch on lower throat and upper breast. Adult non-breeding has white throat and upper breast, and brown cast to crown. Juvenile has straighter bill, and underparts are paler and more uniform grey. **Voice** Song a repeated variable sequence of high whistles, increasing in strength and speed; thin piping and whistling calls. **HH** Rocky cliffs and gorges, gravel beds and rock debris; also stony riverbeds in winter.

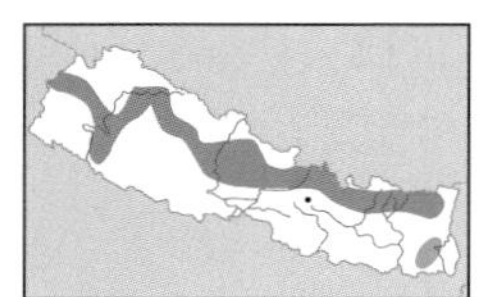

Hodgson's Treecreeper *Certhia hodgsoni* 12 cm

Locally fairly common resident; breeds 3,000–4,100 m, winters 2,000–3,655 m. **ID** From Bar-tailed Treecreeper by combination of (generally) shorter, less downcurved bill, uniform (unbarred) tail, and more prominent buff banding on wing. From Brown-throated by whitish throat and breast, dull brown tail, and more pronounced white supercilium. From Rusty-flanked by lack of prominent white border to rear of ear-coverts and duller buffish flanks. **Voice** Song a high-pitched, rising and then descending *tzee-tzee-tzizizi*; thin, piercing *tsee-tsee-tsee* call. **HH** Mainly coniferous forest sometimes mixed with birch or rhododendron. **TN** Formerly treated as conspecific with Eurasian Treecreeper *C. familiaris*.

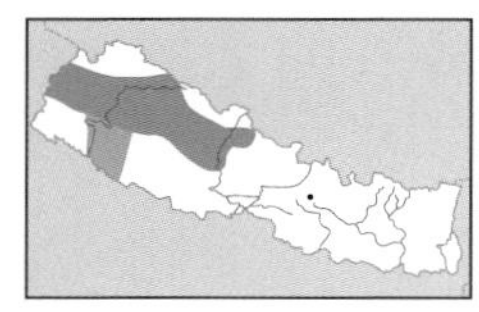

Bar-tailed Treecreeper *Certhia himalayana* 12 cm

Fairly common resident in west; breeds 2,200–3,660 m, winters down to 1,800 m (–305 m). **ID** From other treecreepers by combination of (generally) longer, more downcurved bill and dark cross-bars on tail. Supercilium and pale banding on wings less distinct than Hodgson's. White throat and dull whitish or dirty greyish-buff underparts are further differences from Brown-throated and Rusty-flanked. **Voice** Song a high-pitched trill, *chi-chi-chi-chiu-chiu-chiu-chu*; weak high-pitched thin *tsi-tsi* call. **HH** Breeds mainly in coniferous forest, prefers open forests and edges.

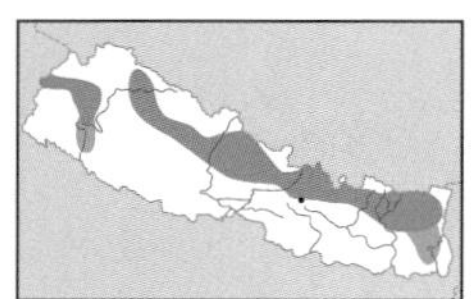

Rusty-flanked Treecreeper *Certhia nipalensis* 12 cm

Locally fairly common resident; breeds 2,550–3,660 m, winters 1,830–3,505 m. **ID** From other treecreepers by combination of shorter, straighter bill, well-defined buffish supercilium, which continues around (and contrasts with) dark ear-coverts, and creamy-buff breast and belly with warm rufous flanks. Supercilium of Hodgson's may extend, rather indistinctly, behind ear-coverts, but ear-coverts patch is paler and smaller. Also has unbarred tail (compare Bar-tailed) and white throat (compare Brown-throated). **Voice** Distinctive song: two slow notes followed by rapid, penetrating, accelerating trill, ending abruptly; high, thin *sit* call often doubled, and a penetrating *zip*. **HH** Moist broadleaved and coniferous forests.

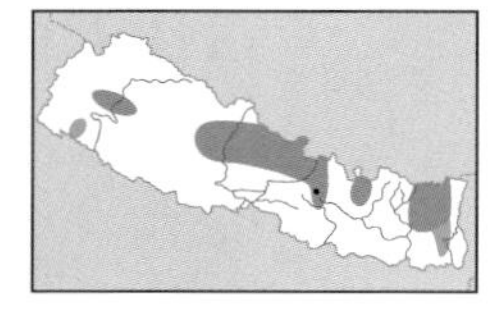

Brown-throated Treecreeper *Certhia discolor* 12 cm

Locally fairly common resident; breeds 2,000–2,750 m (–3050 m), winters down to 1,800 m (–305 m). **ID** From other treecreepers, where ranges overlap, by brownish-buff throat and breast, becoming paler on belly and flanks. Also tail is unbarred, and supercilium less distinct compared to Rusty-flanked. **Voice** Song a long monotonous rattle; calls include explosive *chit* and higher, thinner *seep*. **HH** Broadleaved forests, prefers edges and clearings.

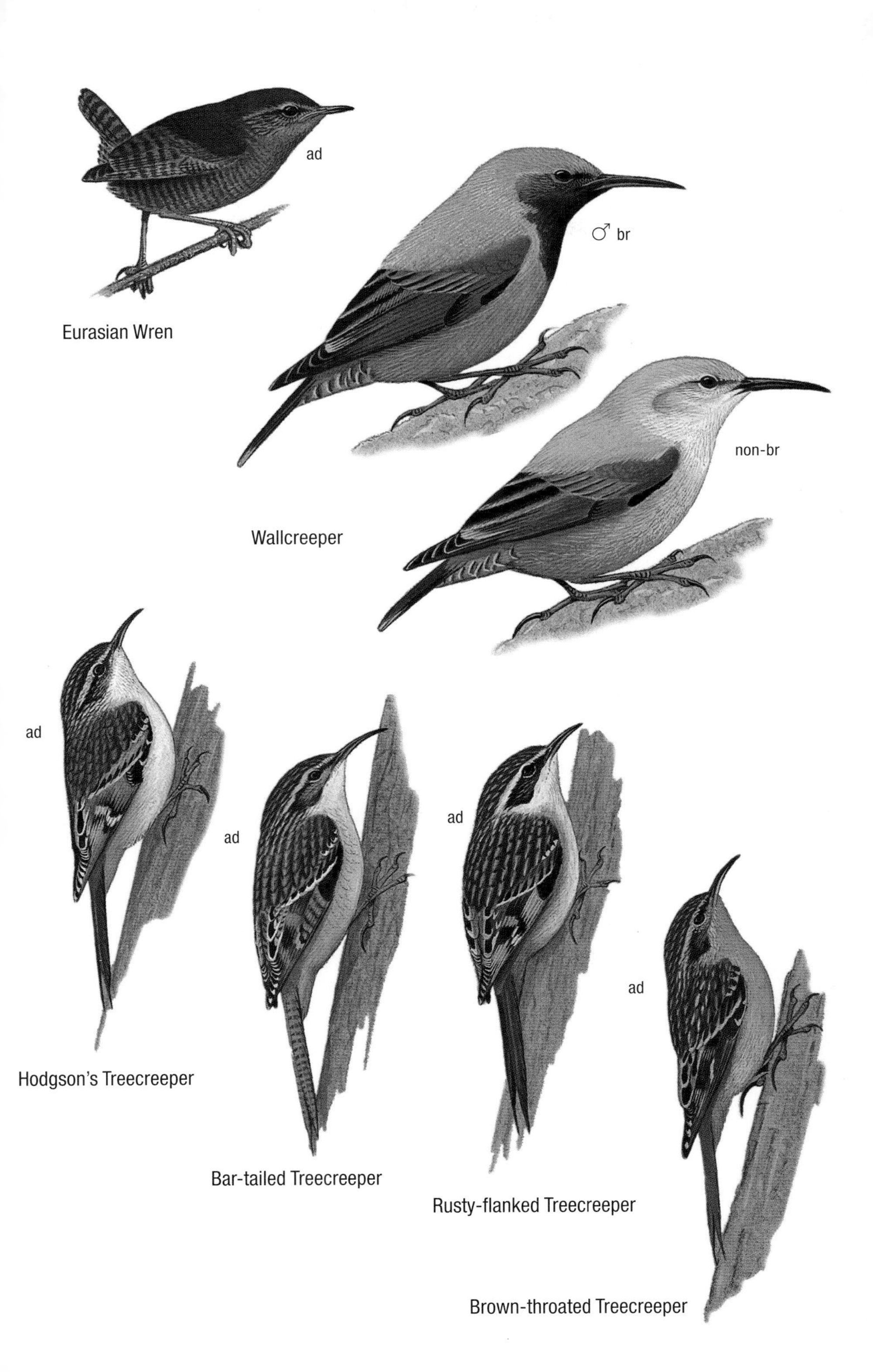

ad
♂ br
Eurasian Wren
non-br
Wallcreeper
ad
ad
ad
ad
Hodgson's Treecreeper
Bar-tailed Treecreeper
Rusty-flanked Treecreeper
Brown-throated Treecreeper

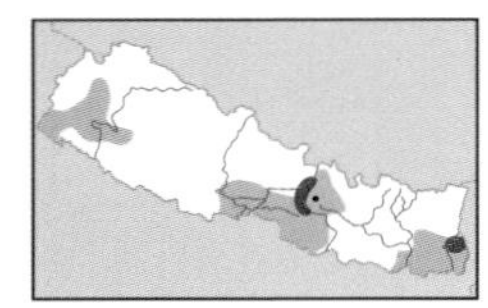

Spot-winged Starling *Saroglossa spiloptera* 19 cm

Breeding migrant recorded all year; movements poorly understood; frequent below 915 m, uncommon to 1,830 m. **ID** White wing patch and whitish iris. Male has blackish mask, reddish-chestnut throat, pale rusty-orange breast, dark-scalloped greyish upperparts and rufous tail. Female has browner upperparts and whitish underparts with greyish-brown markings on throat and breast. Juvenile similar to female, but has buff wing-bar, more uniform upperparts and dark eye. **Voice** Continuous harsh, unmusical jumble of discordant notes; calls include explosive scolding *kwerrh* and grating nasal *schaik*. **HH** Prefers to feed on nectar. In noisy flocks, often with mynas and drongos in flowering or fruiting trees. Open broadleaved forest, well-wooded areas; favours flowering trees.

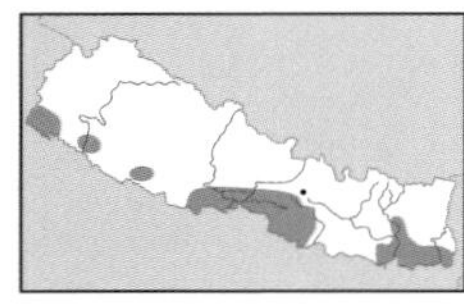

Common Hill Myna *Gracula religiosa* 29 cm

Resident, frequent in east, uncommon in central Nepal, rare in west; 75–455 m. **ID** Large myna with yellow wattles and large orange to yellow bill. Plumage entirely 'black' except prominent white wing patches. Adult has purple-and-green gloss to plumage and bright orange bill. Juvenile has duller yellowish-orange bill, paler yellow wattles and less gloss to plumage, with non-glossy brownish-black underparts. **Voice** Extremely varied, loud piercing whistles, screeches, croaks and wheezes. **HH** Active and noisy. Feeds in trees, mainly on fruits and berries, also flower buds, nectar and insects, sometimes in fruiting bushes. Moist broadleaved forest.

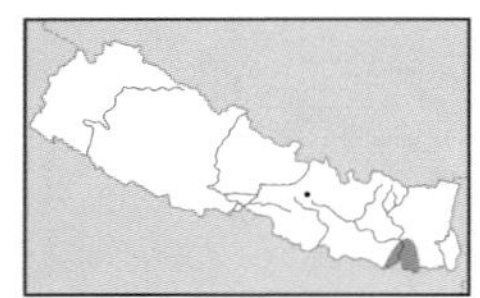

Great Myna *Acridotheres grandis* 25 cm

Rare, very local resident; has bred in Koshi area in east, recently colonised Nepal and may spread; 100 m. **ID** Similar to Jungle Myna, but has uniform blackish-grey upperparts (little contrast between crown and mantle and rump and tail), and uniform dark grey underparts (including belly and flanks), strongly contrasting with white undertail-coverts. Further, has more prominent frontal crest, all-yellow bill, and reddish to orange-brown iris. Juvenile browner and lacks prominent frontal crest; throat diffusely mottled with white on some. Brown belly with diffuse brownish-white fringes and broad whitish tips to brown undertail-coverts are best distinctions from juvenile Jungle. **Voice** Song very similar to Common. **HH** Habits very similar to Jungle. Cultivation, grassland. **TN** Formerly treated as conspecific with White-vented Myna *A. cinereus*.

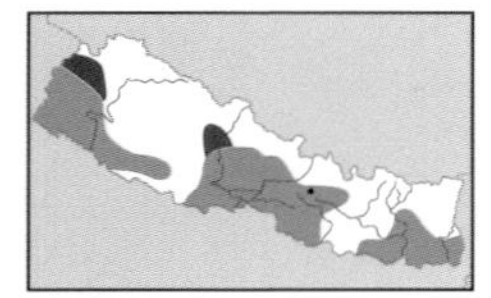

Jungle Myna *Acridotheres fuscus* 23 cm

Common, widespread resident; 75–1,525 m (–2,200 m). **ID** Adult resembles Bank Myna, but has more prominent frontal crest, white patch at base of primaries, white tip to tail and lacks bare orbital skin. Eyes pale. Black of crown and ear-coverts merges into grey or grey-brown of upperparts (less distinct 'cap' than Bank). Bill orange, with dark blue base to lower mandible. Juvenile browner, with darker brown head; pale shafts on ear-coverts, pale mottling on throat, all-yellow bill, and frontal crest much reduced. **Voice** Song similar to Common. In pairs, family parties or flocks according to season. Less bold than Common, not as commensal and less of a scavenger. Strides jauntily on ground when foraging, also feeds in trees and bushes. Cultivation near well-wooded areas, edges of habitation.

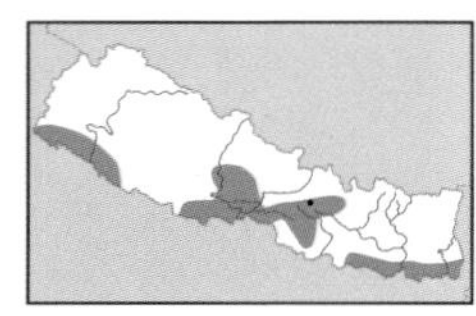

Bank Myna *Acridotheres ginginianus* 23 cm

Local resident; 75–305 m (–1,370 m). **ID** From Common by smaller size, bluish-grey coloration, small frontal crest, orange-red orbital patch, orange-yellow bill, red eye, orange-buff patch at base of primaries and on underwing-coverts, and orange-buff tip to tail. Has capped rather than hooded appearance of Common. Juvenile duller and browner than adult, with buffish-white wing patch and rufous-buff tips to tail. **Voice** Similar to Common, but not as loud and strident. **HH** Habits similar to Common, but more gregarious, Feeds chiefly on ground; also regularly rides on backs of grazing animals. Cultivation, moist grassland near habitation.

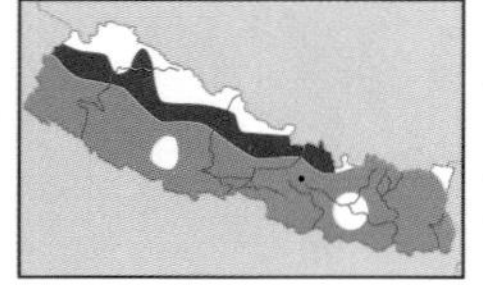

Common Myna *Acridotheres tristis* 25 cm

Abundant, widespread resident; 75–1,830 m all year; occasionally to 3,050 m in summer and 2,135 m in winter. **ID** Brownish myna with yellow orbital skin, white wing patch and white tail tip. Adult has glossy black on head and breast merging into maroon-brown of rest of body. Juvenile duller, with brownish-black head and paler brown throat and breast. **Voice** Song disjointed, noisy and tuneless, with gurgling and whistling, and much repetition; distinctive alarm call, a harsh *chake-chake*. **HH** Usually in pairs or small parties. Bold, tame and pugnacious. Scavenges in built-up areas; flocks follow grazing cattle or feed in cultivation. Omnivorous diet. Forms large noisy communal roosts. Habitation, cultivation.

♀
♂
Spot-winged Starling
ad
Common Hill Myna
ad
ad
juv
Great Myna
juv
Jungle Myna
ad
ad
juv
Bank Myna
ad
ad
Common Myna

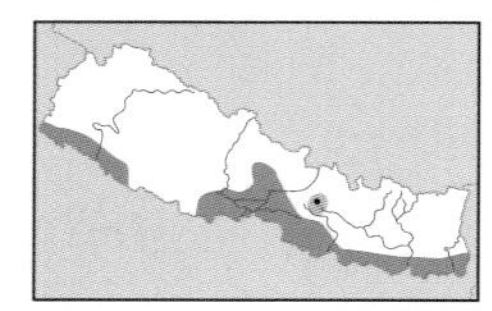

Asian Pied Starling *Gracupica contra* 23 cm

Common and widespread resident; 75–305 m (–1,370 m). **ID** Black and white, with white cheek patch and scapular line. Orange orbital skin and base to large, pointed yellowish bill. In flight, white uppertail-coverts contrast with black tail. Juvenile has black plumage replaced by brown; white cheeks washed brown and less distinct and breast-band is not clearly defined. **Voice** Assorted high-pitched musical, liquid notes. **HH** Mainly terrestrial. Seeks invertebrates by digging in damp ground, also eats cereal grain, fruit and flower nectar. Usually in small parties in non-breeding season and forms noisy communal roosts. Cultivation, damp grassland and habitation. **TN** Formerly placed in genus *Sturnus*. **AN** Pied Myna.

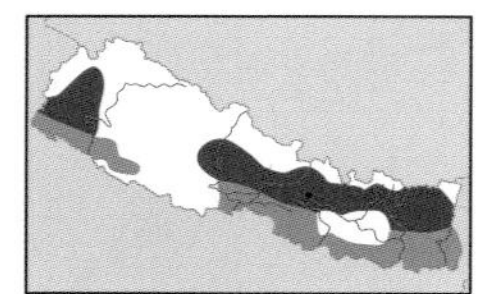

Chestnut-tailed Starling *Sturnia malabarica* 20 cm

Common and widespread resident; 75–1,370 m. **ID** Adult has grey head and upperparts, with whitish forehead and throat, and whitish lanceolate feathers across crown, nape and sides of neck and breast. Underparts rufous (variable in extent) and tail mainly chestnut with grey central feathers. Bill yellow with bluish base, eye whitish. Female more uniformly pale grey, with underparts paler rufous-buff. Juvenile has pale sandy-grey upperparts and greyish-white underparts. **Voice** Sharp disyllabic metallic note and a mild tremulous whistle. **HH** Chiefly arboreal, also in flowering bushes and sometimes on ground. Forages by hopping about for nectar, berries, figs and insects. Usually in flocks, often with other starlings and mynas, constantly chattering and squabbling. Open wooded areas, groves, villages and towns. **TN** This and the next two species were formerly placed in the genus *Sturnus*.

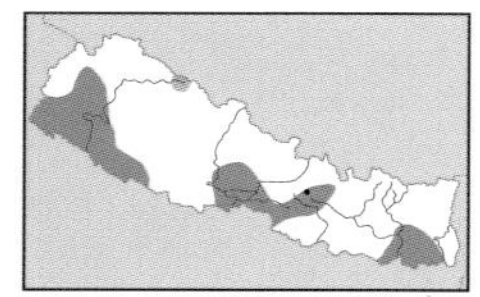

Brahminy Starling *Sturnia pagodarum* 21 cm

Resident; locally frequent in far west, very uncommon elsewhere; 75–915 m (–3,050 m). **ID** Myna-like profile. In flight, shows white sides and tip to dark tail, and uniform wings without white patch. Adult has black crest, and rufous-orange sides of head and underparts. Yellowish bill with blue base, and blue or yellow skin behind eye. Juvenile lacks crest, but has grey-brown cap, paler orange-buff underparts, duller bill and eye-patch. **Voice** Song is a short, gurgling drawn-out cry followed by a bubbling yodel. **HH** Forages for invertebrates on ground in sprightly manner, also takes fruits, berries and nectar in trees. Less gregarious than most starlings. Dry, well-wooded areas, dry cultivation with groves and thorn scrub.

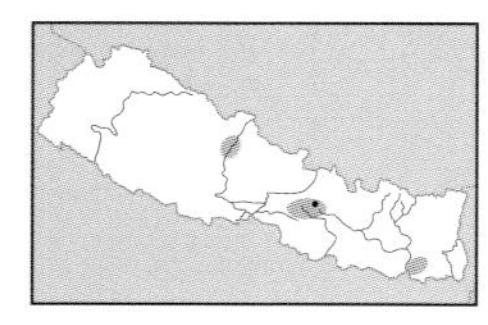

Rosy Starling *Pastor roseus* 21 cm

Uncommon passage migrant; 75–250 m. **ID** Adult has blackish head with shaggy crest, pinkish mantle and underparts, and blue-green gloss to wings. In non-breeding and first-winter plumage much duller; pink of plumage partly obscured by buff fringes; black by greyish fringes. Juvenile mainly sandy-brown, with stout yellowish bill and broad pale fringes to wing feathers. **Voice** Flight call a loud clear *ki-ki-ki*; also a *shrr* like Common Starling, and rattling *chik-ik-ik-ik* when feeding. **HH** Habits similar to Common. Cultivation and damp grassland.

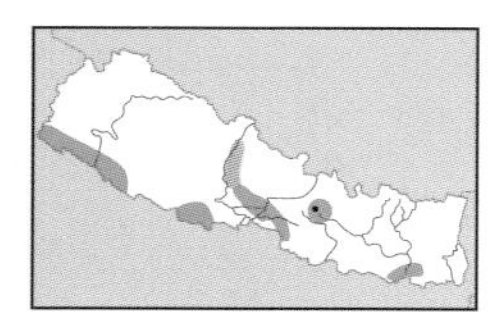

Common Starling *Sturnus vulgaris* 21 cm

Uncommon and widespread, mainly a passage migrant, also winter visitor; 75–1,500 m (–2,805 m). **ID** Adult breeding is metallic green and purple with yellow bill. Adult non-breeding has dark bill; upperparts heavily spangled buff, wing feathers have broad buff fringes, and underparts are boldly spotted white. Juvenile entirely dusky brown, with whiter throat, and buff fringes to wing-coverts and flight feathers. **Voice** Song a varied combination of chirps, twitters, clicks and drawn-out whistles, with much mimicry; flocking birds give slurred *scree-scree*. **HH** Forms flocks in which birds are energetic, restless, quarrelsome and constantly alert; often whole flock takes off and wheels round together. Forages mainly on ground with confident, waddling walk; often stalks around grazing livestock. Feeds chiefly on invertebrates, also fruits and berries. Cultivation and damp grassland.

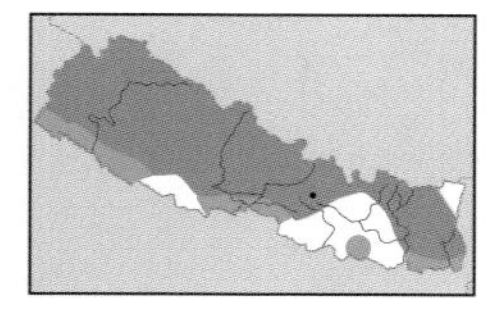

Blue Whistling Thrush *Myophonus caeruleus* 33 cm

Common, widespread resident; breeds 1,500–4,800 m (–470 m), winters 75–2,745 m. **ID** Adult dark blue-black, with head and body spangled with glistening silvery-blue. Forehead, shoulders and fringes to wings and tail brighter blue. Stout yellow bill. Juvenile browner, without blue spangling. Wings and tail duller blue than adult's. **Voice** Melodic, rambling, whistling song; calls include shrill rasping *tzet* and shrill *kree*. **HH** Streams and rivers in forests and wooded areas.

juv
ad
Asian Pied Starling
juv
♂
Chestnut-tailed Starling
juv
ad
Brahminy Starling
ad
juv
Rosy Starling
non-br
br
juv
Common Starling
ad
Blue Whistling Thrush

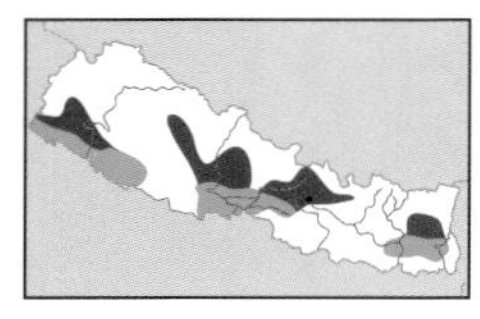

Orange-headed Thrush *Zoothera citrina* 21 cm

Fairly common partial migrant, mainly summer visitor; breeds 75–1,830 m, winters 75–250 m (–915 m). **ID** Adult has orange head and underparts, and white shoulder patch; male has blue-grey mantle, female an olive-brown wash to mantle. Juvenile has buffish-orange streaking on upperparts and mottled breast. White bands on underwing in flight. May show diffuse dark vertical bar at rear of ear-coverts. **Voice** Rich, sweet, variable song. **HH** Often on forest paths at dusk. Moist, shady places in forests and woodlands.

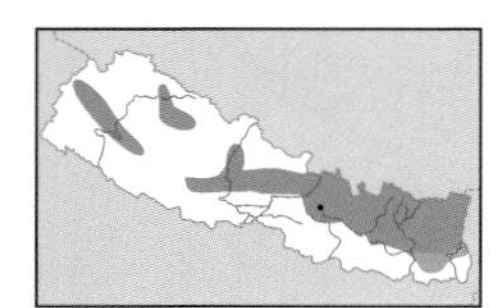

Plain-backed Thrush *Zoothera mollissima* 27 cm

Widespread, locally fairly common resident; breeds 3,000–4,000 m, winters 1,500–2,400 m (915–2700 m). **ID** From very similar Long-tailed Thrush by indistinct (or lack of) wing-bars (can show narrow buff tips to median and greater coverts). In addition, belly and flanks generally more clearly scaled black, more rufescent upperparts, especially uppertail-coverts/rump and tail, less pronounced pale wing panel, and shorter tail. In flight, broad whitish banding across underwing (as does Long-tailed). **Voice** Song similar to Scaly Thrush, a series of chirps, trills and squeaks; begins with *plee-too, plee-chu* or *ti-ti-ti* and usually ends with a descending *chup-ple-oop*. **HH** Secretive and shy. Male often sings from treetops hidden in foliage. Breeds in fir forests and open grassy slopes with scattered bushes near treeline; winters in forest and open country with bushes. **TN** The newly described Himalayan Forest Thrush *Z. salimalii* may occur in far eastern Nepal; it has a larger, all-dark bill, more rufous-toned upperparts, dark loral stripe, and usually lacks dark patch at rear of ear-coverts. Its song is more musical and thrush-like. **AN** Alpine Thrush.

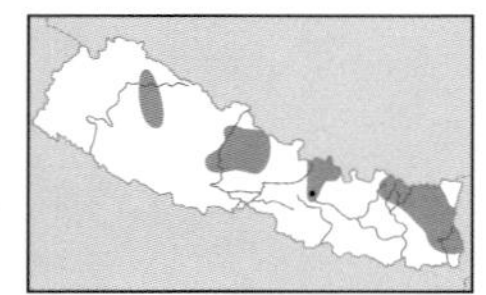

Long-tailed Thrush *Zoothera dixoni* 27 cm

Locally frequent resident; breeds 2,100–4,520 m, winters 1,500–2,700 m. **ID** From Plain-backed by comparatively broad and prominent wing-bars (buff tips to median coverts form distinct spotting); belly and flanks more sparsely marked with black, and flank markings appear more bar-like (on some, almost spotted). Further subtle differences include greyer upperparts, more boldly marked face (with more clearly defined dark malar and ear-coverts spot), more pronounced pale panel on wing, and longer tail. **Voice** Song comprises long-sustained, rambling series of mainly harsh notes; poorly known call. **HH** Habits like Plain-backed. Breeds in forest undergrowth; in winter, also second growth and open country with bushes.

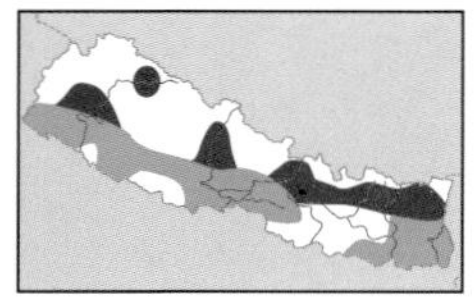

Scaly Thrush *Zoothera dauma* 26–27 cm

Locally fairly common partial migrant; breeds 2,320–3,300 m (–3,540 m), winters 75–1,500 m. **ID** Boldly scaled with pale face, large black eye and dark patch on ear-coverts. From Plain-backed and Long-tailed by bold black scaling on golden-olive upperparts, and golden-olive panels on wing, with dark bar at tip of primary-coverts. Juvenile has more barred than scaled upperparts, and breast is distinctly spotted. **Voice** Song a slow broken *chirrup... chwee... chwee... weep... chirrol... chup*; grating *tsshhh* call. **HH** Shy and retiring. Forages among leaf litter. Thick forest with dense undergrowth, often near water; also well-wooded areas in winter.

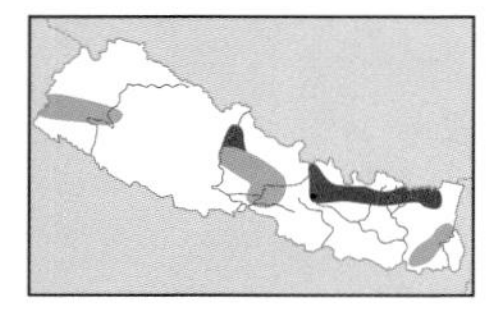

Long-billed Thrush *Zoothera monticola* 28 cm

Uncommon resident; breeds 2,285–3,850 m, winters 915–2,500 m (–75 m). **ID** From Dark-sided by larger size and bill, more uniform head-sides (dark lores, diffuse dark malar stripe and narrow white throat patch), dark slaty-olive upperparts, darker and more uniform breast and flanks (both with diffuse dark spotting), and dark spotting on whitish belly. Juvenile has pale shaft-streaks on upperparts, buff tips to wing-coverts and buffish underparts with bold dark spotting. **Voice** Song comprises loud, slow plaintive whistle of 2–3 notes; alarm a loud *zaaaaaaaa*. **HH** Crepuscular, skulks on forest floor; prefers to escape by hopping into dense undergrowth. Dense moist forests, usually near streams.

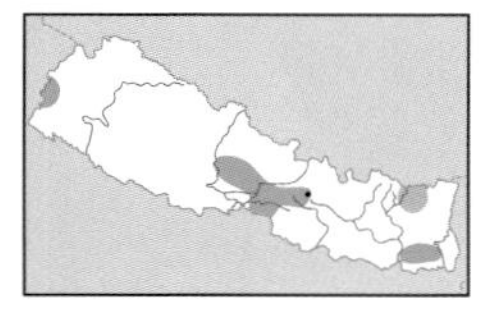

Dark-sided Thrush *Zoothera marginata* 25 cm

Rare, probably a winter visitor and passage migrant; 150–2,500 m. **ID** From Long-billed by smaller size, smaller bill, rufous-brown upperparts and wing panel, and paler underparts with prominent scaling on breast and flanks; also more strongly patterned sides of head (variable, but usually has paler lores, more distinct dark and pale patches on ear-coverts, pale crescent behind). Juvenile has pale shaft-streaks on upperparts, prominent buff tips to wing-coverts. **Voice** Song a thin whistle; soft, deep, guttural *tchuck* call. **HH** Habits like Long-billed. Dense forest near streams.

♂
♀
Orange-headed Thrush
ad
Plain-backed Thrush
ad
Long-tailed Thrush
ad
Scaly Thrush
ad
Long-billed Thrush
ad
Dark-sided Thrush

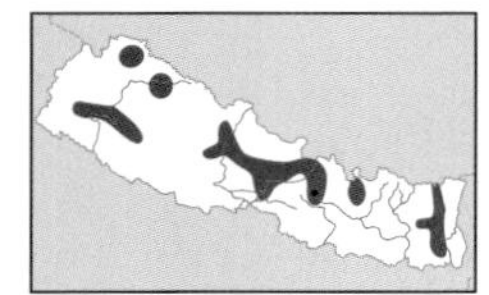

Pied Thrush *Zoothera wardii* 22 cm

Very uncommon, chiefly a summer visitor (very rare in winter); 1,500–2,400 m (–3,100 m). **ID** Adult male has white supercilium, white wing-bars and tips to tertials/secondaries, white barring on rump, white-and-black-barred flanks, and yellowish bill and legs. Female has buff supercilium, olive-brown upperparts, buff wing-bars and tips to tertials, buff spotting on olive-brown breast, and white belly and flanks (with prominent dark scaling). First-winter male has throat and breast mottled buff, and supercilium and greater coverts wing-bars are buffish-white. Juvenile has buff streaking on mantle and breast. **Voice** Song comprises 2–4 sweet high-pitched notes, the last a rattle; spitting *ptz-ptz-ptz-ptz* alarm call. **HH** Open broadleaved forest and second growth with scattered trees.

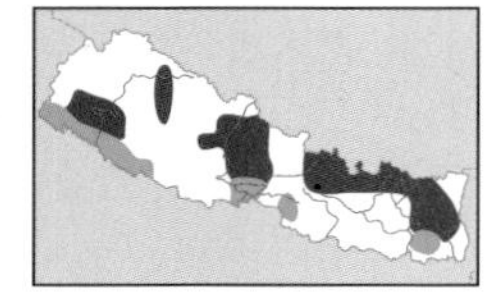

Tickell's Thrush *Turdus unicolor* 21 cm

Widespread, locally frequent resident; breeds 1,500–2,450 m (–2,745 m), winters 75–250 m. **ID** Small, compact thrush with rather plain face, small yellowish or pale brown bill, and pale legs. Male pale bluish-grey, with whitish belly and vent. Yellow bill and fine yellow eye-ring. First-winter male similar but has pale throat and submoustachial stripe, dark malar stripe, and often spotting on breast. Juvenile has dark olive-brown upperparts with fine orange-buff shaft-streaks, orange-buff tips to coverts, and heavily barred and spotted breast and flanks. **Voice** Song of weak monotonous repeated disyllabic or trisyllabic phrases, e.g. *chilliyah-chilliyah*, *tirlee-tirleechelia-chelia*; soft *juk-juk* call. **HH** Breeds in open broadleaved forests and scrubby forests; also winters in well-wooded areas.

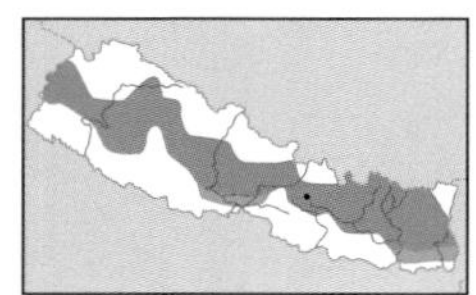

White-collared Blackbird *Turdus albocinctus* 27 cm

Fairly common resident; breeds 2,400–3,445 m (–3,750 m), winters 1,525–3,050 m (–80 m). **ID** Adult male mainly black, with white throat and broad white collar; bill and legs yellow. Female has variable pale greyish-white to buffish collar, and rest of plumage rufous-brown with pale feather fringes on underparts. Juvenile lacks collar; has orange-buff streaking on upperparts, orange-buff tips to coverts (forming double wing-bar), and orange-buff underparts with dark brown spotting and barring. **Voice** Song comprises melancholy series of soft descending whistles, *hoo-ee, hoo-ou, hoo-uu*; coarse chuckling chatter call. **HH** Forests, especially clearings and edges.

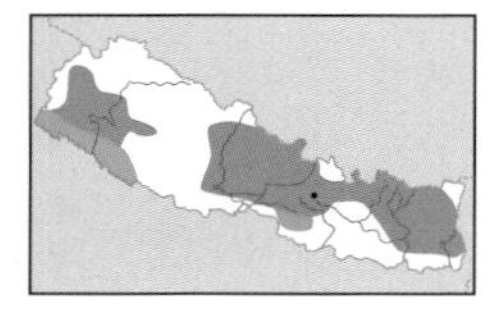

Grey-winged Blackbird *Turdus boulboul* 28 cm

Fairly common resident; mainly breeds 2,100–2,745 m (1,850–3,300 m), winters 1,400–1,980 m (–75 m). **ID** Adult male black, with pale grey panel on wing. In fresh plumage, prominent whitish fringes to belly and vent. Bill orange and legs yellowish. Female olive-brown; has paler rufous-brown panel on wing (with greater coverts paler buffish or greyish towards tips, contrasting with dark brown primary-coverts). Juvenile has orange-buff streaking to upperparts, orange-buff tips to median coverts, and brown barring on orange-buff underparts; wing panel similar to adult. **Voice** Rich melodious song, with repeated two-note whistles; *chook-chook-chook* call. **HH** Moist broadleaved forest, prefers thicker habitat to White-collared, winters in open forest, clearings and edges.

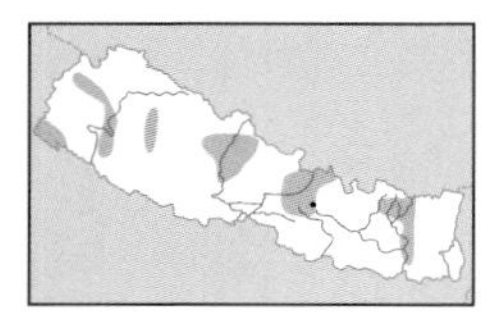

Tibetan Blackbird *Turdus* (*merula*) *maximus* 26–29 cm

Erratic visitor, mainly in winter and spring; 3,305–4,800 m (–75 m). **ID** Male black and lacks yellow orbital ring. Female uniform dark brown, without paler throat. Juvenile has rufous-buff underparts with diffuse dark brown barring and spotting, back and rump variably spotted and barred rufous-buff, and crown and mantle tend to be rather uniform dark brown (male) or pale brown (female). **Voice** Rattling *chak-chak-chak* call. **HH** Winters in juniper stands or shrubberies and grassy slopes. **TN** Often now regarded as a separate species from both Common Blackbird *T. merula* and Indian Blackbird *T. simillimus*.

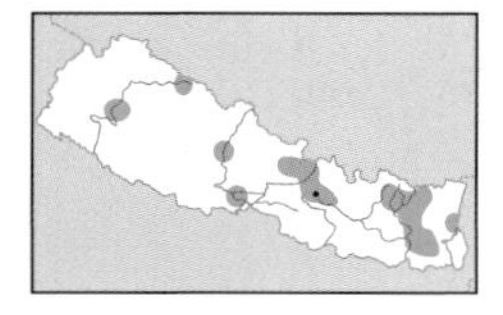

Chestnut Thrush *Turdus rubrocanus* 27 cm

Very uncommon winter visitor and passage migrant; 915–2,745 m (–3,100 m). **ID** Male has grey head with buffish-grey collar; rest of body mainly chestnut, with blackish wings and tail. Female very similar, but duller; head and hindneck pale brownish-grey (lacking distinct collar), and wings and tail brown. Juvenile has buff shaft-streaking on upperparts and dark spotting and barring on underparts; back, rump and uppertail-coverts have distinct chestnut cast. Compared to nominate race, vagrant male *T. r. gouldii* has darker slate-grey head and neck (lacking collared effect); female has darker brownish-grey head. **Voice** A *kwik* in alarm. **HH** Open wooded areas and orchards.

♂
♀
Pied Thrush
♂
1st-win
♂
♀
Tickell's Thrush
♂
♀
White-collared Blackbird
♂
♀
Grey-winged Blackbird
♂
♀
Tibetan Blackbird
♂
♂
rubrocanus
gouldii
♀
Chestnut Thrush

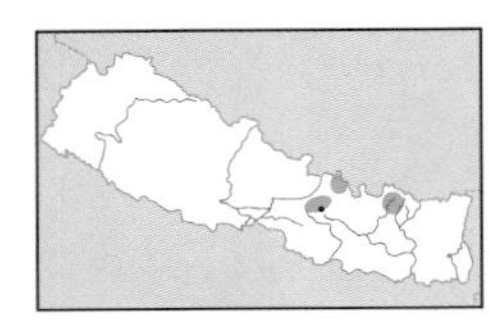

Kessler's Thrush *Turdus kessleri* 27 cm

Erratic and local winter visitor; 3,440–4,700 m (–2,500 m). **ID** Male from Chestnut Thrush by black head, neck and upper breast, and creamy-white mantle and lower breast. Female mirrors pattern of male: head, neck and breast greyish-brown, mantle variably pale grey-brown to greyish-cream but always contrasts with hindneck, and shows variable, diffuse buffish division between brown upper breast and ginger-brown rest of underparts; rump and uppertail-coverts have distinct ginger cast. **Voice** Soft *dug dug* call. **HH** Junipers, *Berberis* and fields.

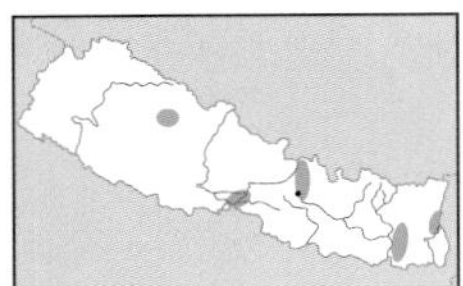

Eyebrowed Thrush *Turdus obscurus* 23 cm

Very rare winter visitor; 1,500–2,800 m (–250 m). **ID** Striking features are white supercilium and white crescent below eye, contrasting with dark lores. Peachy-orange flanks contrasting with white belly. Adult male has blue-grey head, including throat, with just a small area of white on chin. Female has olive-brown crown and nape, browner ear-coverts, white throat and submoustachial stripe, dark malar stripe, narrow grey gorget on upper breast, and duller orange breast and flanks. First-winter similar to female, but has fine greater coverts wing-bar; first-winter males brighter, with more grey on ear-coverts and upper breast. See Appendix 1 for differences from Grey-sided Thrush **Voice** Thin drawn-out *tseep* call. **HH** Open forest.

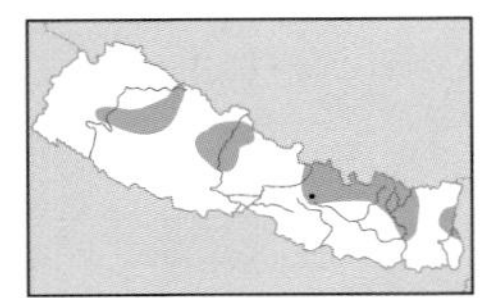

Red-throated Thrush *Turdus ruficollis* 25 cm

Uncommon winter visitor; mainly 2,400–4,100 m. **ID** Uniform grey upperparts and wings. Always shows reddish-orange at sides of tail, which can be very prominent in flight (undertail can appear entirely orange). Adult male has red supercilium, throat and breast (with narrow white fringes in fresh plumage), grey upperparts, and whitish underparts. Female similar to male, but typically has white or buffish throat, black-streaked malar stripe, and red of breast is gorget of spotting. First-winter has white tips to greater coverts and pale-fringed tertials. First-winter male resembles adult female. First-winter female less heavily marked, with finely streaked breast and flanks; usually shows rufous wash to supercilium, throat and/or breast. **Voice** Calls include a soft *jak, tack-tack*, varied harsh notes and shriller *pee-wit chip-chi-chip*. **HH** Forest edges, pastures with scattered trees.

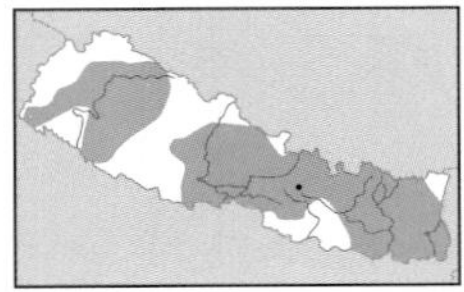

Black-throated Thrush *Turdus atrogularis* 26.5 cm

Fairly common winter visitor; 75–4,200 m. **ID** Adult male has black supercilium, throat and breast (with narrow white fringes in fresh plumage), grey upperparts and whitish underparts. Female similar to male, but typically has white or buffish throat, black-streaked malar stripe, and black gorget of spots across breast. First-winter has fine white supercilium, white tips to greater coverts and pale-fringed tertials. First-winter male resembles adult female. First-winter female less heavily marked, with finely streaked breast and flanks. **Voice** Calls as Red-throated. **HH** Highly gregarious in winter. Habitat as Red-throated. **TN** Formerly considered conspecific with Red-throated Thrush *T. ruficollis*, under the name Dark-throated Thrush.

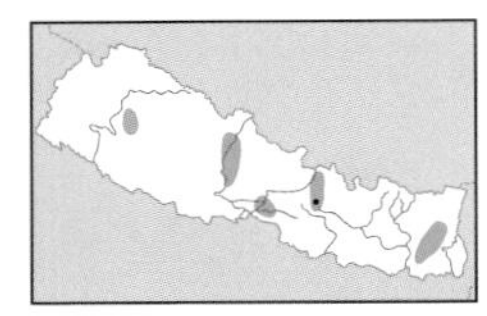

Dusky Thrush *Turdus eunomus* 24 cm

Irregular and rare winter visitor; 75–3,175 m. **ID** Adult male has broad white supercilium and throat that contrasts with dark crown and ear-coverts, chestnut wing panel, rufous-brown mantle with dark feather centres, double gorget of blackish spotting on breast, and bold spotting on flanks contrasting with white of underparts. Female similar, but usually duller and less strikingly patterned, and usually has more distinct black-streaked malar stripe. First-winter variable, but duller than adult: crown and ear-coverts greyer, supercilium less pronounced, double gorget of spotting less distinct, upperparts greyer, and has browner (and less distinct) wing panel. **Voice** Calls include a shrill *shrree* and rather harsh *chack-chack*. **HH** Cultivation, pastures with scattered trees, forest edges. **TN** Formerly treated as conspecific with Naumann's Thrush *T. naumanni*.

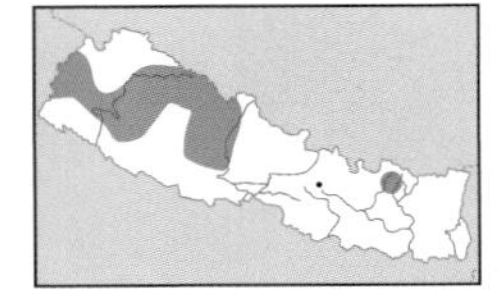

Mistle Thrush *Turdus viscivorus* 27 cm

Locally common resident; breeds 2,400–3,800 m, winters 2,135–3,050 m (1,525–3,660 m). **ID** Large size, pale grey-brown upperparts, whitish edges to wing feathers and spotted breast. Juvenile has buffish-white spotting to upperparts; lacks golden-buff bands on wing of Scaly Thrush. **Voice** Song is loud, ringing and rather melancholy, with short repeated phrases and long pauses, *tree-twee wooh......tree-twee-wooh ....tree-tili-terooh tree-tili-terooh*, calls include a characteristic loud rattle, and staccato *tuck-tuck-tuck*. **HH** Breeds in open forest and juniper shrubberies; open grassy slopes and forest edges in winter.

♂
♀
Kessler's Thrush
♂
♀
1st-win
Eyebrowed Thrush
♂
♀
♀
1st-win
Red-throated Thrush
♀
1st-win
♂
♀
Black-throated Thrush
♂
1st-win
Dusky Thrush
ad
Mistle Thrush

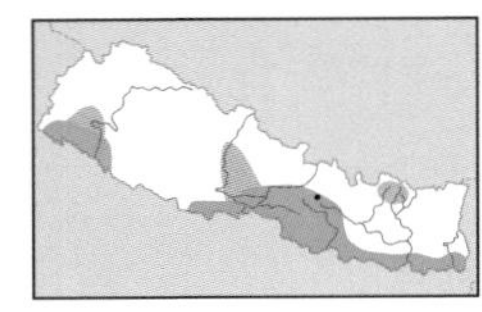

Bluethroat *Luscinia svecica* 15 cm

Widespread, locally common winter visitor and passage migrant; 75–1,370 m (–3,455 m on passage). **ID** Prominent white supercilium and rufous tail-sides in all plumages. Male has variable blue, black and rufous pattern to throat and breast (obscured by whitish fringes in fresh plumage). Female has black submoustachial stripe and band of black spots across breast; older females can have breast-bands of blue and rufous. **Voice** Deep *chack* or *chack-chack* calls. **HH** Secretive and terrestrial. Often cocks and fans tail. Damp ravines, scrub, reeds, tall grass and cultivation; often near water.

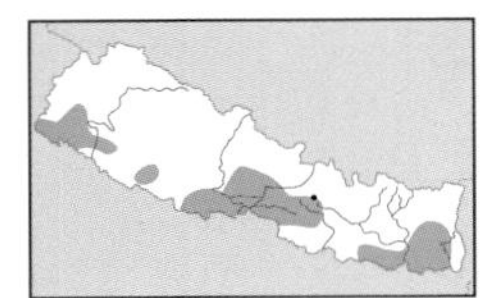

Siberian Rubythroat *Luscinia calliope* 14 cm

Frequent winter visitor and passage migrant; 75–1,370 m. **ID** Male lacks black breast-band and white sides and tip to tail of White-tailed, and has olive-brown upperparts. Female from female White-tailed by olive-brown upperparts, olive-buff wash to breast and flanks, lack of white tip to tail, and has pale brown or pinkish legs. First-winter as adult (i.e. male has red throat), with retained juvenile buff tips to greater coverts and tertials. **Voice** Calls include a loud, clear double whistle *ee-uh* and hard *schak*. Long, pleasant scratchy warbling song, often heard on spring passage. **HH** Shy and skulks in bushes and thick undergrowth.

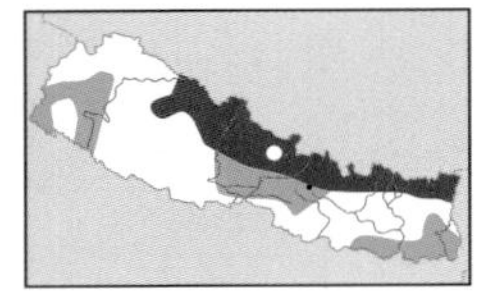

White-tailed Rubythroat *Luscinia pectoralis* 14 cm

Widespread, frequent resident; breeds 3,300–4,800 m, winters 75–1,340 m. **ID** Male from Siberian by black breast (fringed white when fresh), greyer upperparts, and blackish tail with white sides and tip. Female from Siberian by grey-brown upperparts, grey breast and flanks contrasting with belly, white tail tip, black legs. First-winter as adult female, but buff tips to greater coverts and tertials. Vagrant male *L. p. tschebaiewi* has white moustachial stripe. **Voice** Song a long series of rising and falling warbling trills and twitters, but can be uttered in short bursts; call a harsh *ke*. **HH** Habits similar to Siberian, but when breeding male sings from top of large boulder or bush. Breeds in subalpine shrubberies; winters in marshy grassland and dense scrub.

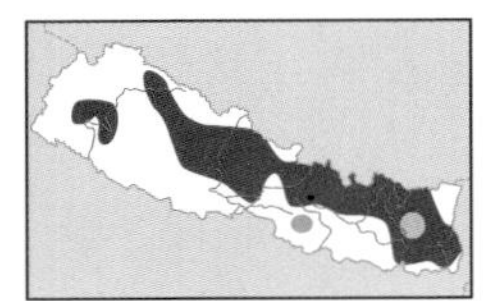

Indian Blue Robin *Luscinia brunnea* 15 cm

Fairly common summer visitor, rare in winter; breeds 2,440–3,355 m, winters 245–1,750 m (–75 m). **ID** From White-browed Bush Robin by horizontal stance, short tail (frequently bobbed and fanned), and long pale legs and large feet. Male also by shorter and broader white supercilium, (usually) black ear-coverts, and whitish centre to belly and vent. Female has olive-brown upperparts, and orange-buff to brownish-buff underparts with striking white throat, belly and vent; lacks prominent supercilium. First-year male variable; some have buffish supercilium, dull blue upperparts, and dull orange breast and flanks. **Voice** Song comprises 3–4 piercing whistles followed by rapid, tumbling notes, *tit-tit-titwit-tichu-tichu-chuchu-cheeeh*; hard *tek-tek-tek* call. **HH** Secretive; even in breeding season male usually stays out of sight. Breeds in dense undergrowth in moist forests; winters in forest and scrub.

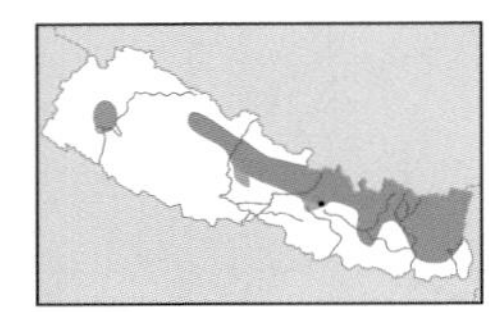

White-browed Bush Robin *Tarsiger indicus* 15 cm

Frequent resident, mainly from west-central Nepal east; breeds 3,000–4,000 m, winters 2,100–3,050 m (–75 m). **ID** Upright stance, long tail (frequently cocked) and dark legs separate from Indian Blue Robin. Male also has longer and finer supercilium, greyer upperparts, and entirely rufous-orange underparts. Female has long (sometimes part-concealed) buffish-white supercilium, which curves down behind eye, and orange-buff throat concolorous with underparts. First-summer male can breed in female-like plumage. **Voice** Call a repeated *trrrr*; song a bubbling, double-phrased *shri-de-de-dew....shri-de-de-dew*. **HH** Feeds mainly on ground under dense bushes, by searching foliage, and by sallying for insects. Breeds in dense forest undergrowth and bushes at forest edges; winters in damp places in dense forest understorey.

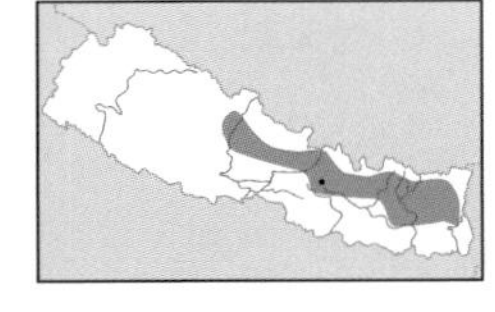

Rufous-breasted Bush Robin *Tarsiger hyperythrus* 15 cm

Locally frequent resident; breeds 3,200–4,200 m, winters 2,135–3,050 m. **ID** Carriage and profile as Himalayan Bluetail. Long legs distinguish from blue flycatchers. Male has dark blue upperparts, blackish ear-coverts, glistening blue supercilium and shoulders, and rufous-orange underparts. Female has blue tail; compared to female Himalayan Bluetail has orange-buff throat, and browner breast and flanks. **Voice** Alarm a *duk-duk-duk-squeak*; lisping warbling song *zeew.. zee..zee..zee*. **HH** Mostly frequents lower forest storey. Not shy; sometimes perches quietly in open for short periods. Often near streams; breeds in forest edges and clearings; winters in moist forest undergrowth.

♀ 1st-win
♂ non-br
♂ br
Bluethroat
♂
♀
Siberian Rubythroat
♂
tschebaiewi
♂
♀
pectoralis
White-tailed Rubythroat
♂
♀
Indian Blue Robin
♂
♀
White-browed Bush Robin
♂
♀
Rufous-breasted Bush Robin

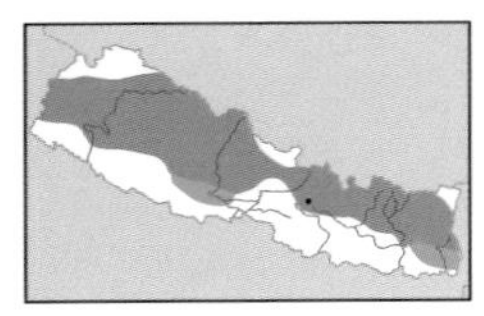

Himalayan Bluetail *Tarsiger (cyanurus) rufilatus* 15 cm

Common, widespread resident; breeds 3,000–4,000 m, winters 1,370–2,745 m. **ID** White throat, orange flanks, blue tail, and redstart-like stance. Male has blue upperparts and breast-sides. Female has olive-brown upperparts and breast-sides. **Voice** Call a deep croaking *tock-tock*; song a rather soft and weak *churrh-cheee* or *dirrh-tu-tu-dirrh*. **HH** Mainly frequents bushes. Like other bush robins continually flicks wings and fans tail. Forest understorey and dense bushes in forest clearings and at forest edges. **TN** Treated here as conspecific with Red-flanked Bluetail *T. cyanurus* (see Appendix 1).

Golden Bush Robin *Tarsiger chrysaeus* 15 cm

Locally fairly common, widespread resident; breeds 3,500–4,200 m, winters 1,700–2,800 m. **ID** Male has blackish mask, orange supercilium, orange scapular line, orange underparts, and orange rump and sides to black tail. Female has duller (and less distinct) buffish-orange supercilium and underparts, uniform golden-olive upperparts, and pale orange uppertail-coverts and sides to olive-brown tail. Both sexes have long pale legs and pale lower mandible. Juvenile has tail similar to adult. **Voice** Song a high wispy *tze-du-tee-tse* ending in a low *chur-r-r-r*; purring croak *trr-trr* and harder *tcheck-tcheck* calls. **HH** Very skulking in winter, easier to see in breeding season. Keeps in low bushes or on ground. Breeds in subalpine shrubberies and forest undergrowth; winters in forest undergrowth and dense scrub.

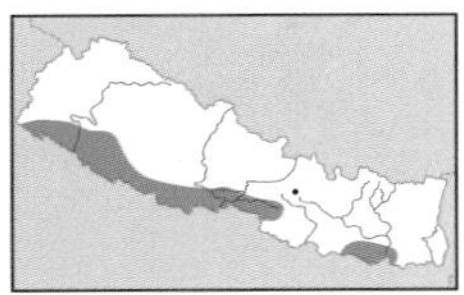

Indian Robin *Saxicoloides fulicatus* 19 cm

Local resident, frequent in west, rare in centre and east; 75–760 m. **ID** Reddish vent and black tail in all plumages. Male has white shoulders and black underparts. Female has greyish underparts. Juvenile darker brown than female; lacks spotting and scaling typical of juvenile chats, but throat lightly mottled. **Voice** Very short, high-pitched warbling song; alarm is clear two-tone whistle, *pi-peear*. **HH** Bold, sprightly and terrestrial. Frequently flips tail, has habit of holding it vertically over back or further forward almost touching its head. Sparse scrub in dry stony areas and at cultivation edges.

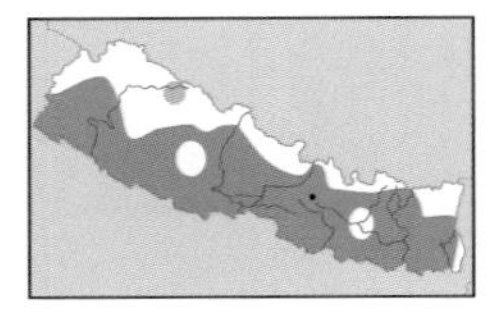

Oriental Magpie Robin *Copsychus saularis* 20 cm

Widespread resident, common 75–1,525 m, frequent up to 2,000 m (–3,050 m). **ID** In all plumages, white wing patch and white sides to long tail. Male has glossy blue-black head, upperparts and breast. Female has bluish-grey head, upperparts and breast. Juvenile has indistinct orange-buff spotting on upperparts, and orange-buff wash and diffuse dark scaling on throat and breast. **Voice** Spirited, clear and varied whistling song; plaintive *swee-ee* or *swee-swee*, alarm a harsh *chr-r*. **HH** Confiding and conspicuous. Partly crepuscular. Tail usually held cocked and frequently lowered and fanned, then closed and jerked up while wings are dropped and flicked. Gardens, groves and open dry broadleaved forest.

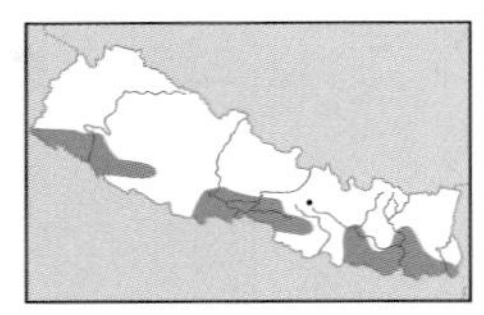

White-rumped Shama *Copsychus malabaricus* 25 cm

Locally fairly common resident; 75–365 m. **ID** Long, graduated dark tail with white sides and rump. Male has glossy blue-black upperparts and breast, rufous-orange underparts. Female duller, with brownish-grey upperparts; tail shorter and squarer. Juvenile has orange-buff spotting on upperparts, and orange-buff throat and breast with fine scaling. **Voice** Song comprises rich melodious phrases; call a musical *chir-chur* and *chur-chi-churr*, alarm a harsh scolding. **HH** Close to ground in undergrowth and low trees in broadleaved forest.

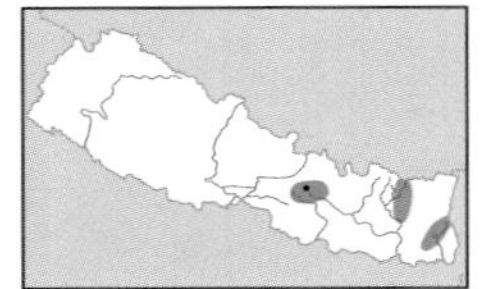

Purple Cochoa *Cochoa purpurea* 30 cm

Rare and local, probably resident; 915–2,255 m. **ID** Adult male dull purplish-grey with lilac-blue crown, black mask, lilac panelling on wing, and lilac tail with black tip. Adult female recalls male (with similar pattern to wings and tail) but has rusty-brown upperparts and brownish-orange underparts. Juvenile has black scaling to crown, indistinct buff streaking and spotting on upperparts, orange-buff underparts with bold black barring, and buff tips to wing-coverts; wings and tail as adult. See Appendix 1 for Green Cochoa. **Voice** Song a flute-like *peeeee*; also *peeee-you-peeee* like a shepherd's bamboo flute; low chuckling call. **HH** Quiet, unobtrusive and rather lethargic. Usually keeps among foliage; seen most often in spring when males sing from tall treetops. Mainly dense, moist broadleaved evergreen forest.

Himalayan Bluetail
♂
♀
♂
♀
Golden Bush Robin
♀
♂
♂
Indian Robin
♀
Oriental Magpie Robin
♂
♀
White-rumped Shama
♂
♀
Purple Cochoa

PLATE 114: REDSTARTS

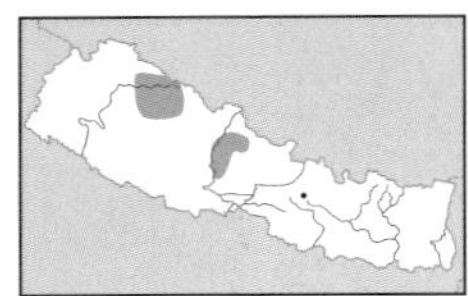

Eversmann's Redstart *Phoenicurus erythronotus* 16 cm

Erratic and very uncommon winter visitor; 2,300–3,300 m. **ID** Large size. Can appear rather shrike-like. Male has black mask and scapular line, grey crown, rufous throat and mantle, and white on wing; colours heavily obscured by pale fringes in non-breeding and first-winter plumages. Female has double buffish wing-bars and broad buff fringes to tertials. **Voice** Soft croaking *gre-er* call. **HH** Tail can be held slightly cocked; flicks it vigorously upwards but does not vibrate it like most redstarts. Drops wings in flycatcher-like manner. Dry habitats amongst scrub, on stone walls and fields. **AN** Rufous-backed Redstart.

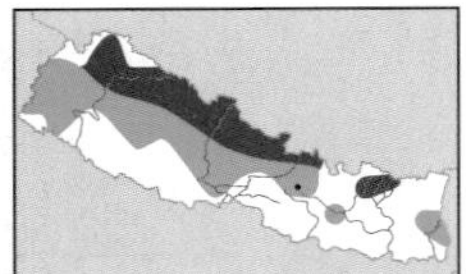

Blue-capped Redstart *Phoenicurus coeruleocephala* 15 cm

Resident, locally common in north-west and central areas; frequent in central Nepal, rare further east; breeds 2,900–3,700 m (–4,270 m), winters 1,370–2,900 m (–150 m). **ID** Male has blue-grey cap, black tail, and white on wing; coloration heavily obscured by brown fringes in non-breeding and first-winter plumages (when crown appears browner). Female has grey underparts, prominent double wing-bar, blackish tail, and chestnut rump. Juvenile has dark brown barring on upperparts and underparts; juvenile male has broad white edges to tertials. **Voice** Pleasant repetitive warbling song; call a rapid *tit-tit-tit tik-tik*. **HH** Frequents bushes and lower tree branches. Vibrates tail, when on ground sometimes raises and lowers it slowly. Breeds on rocky slopes with open forest; winters in open forest and second growth.

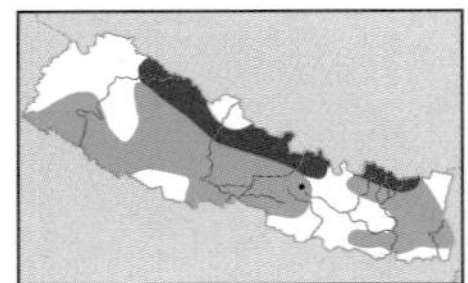

Black Redstart *Phoenicurus ochruros* 15 cm

Widespread resident; common in breeding areas, 2,560–5,200 m (–5,700 m), locally fairly common in winter, 75–700 m. **ID** Male has blackish upperparts, black breast and rufous underparts. Female and first-year male almost entirely dusky-brown with rufous-orange wash on lower flanks and belly. Juvenile has diffuse dark scaling on upperparts and underparts, and fine buff greater coverts bar. **Voice** Song a scratchy trill, followed by a short wheezy jingle; calls include a short *tsip*, scolding *tucc-tucc* and rapid rattle. **HH** Vibrates tail more frequently than other redstarts. Perches on rocks, bushes or other low vantage points. Breeds in Tibetan steppe habitat; winters in cultivation and riverine forest.

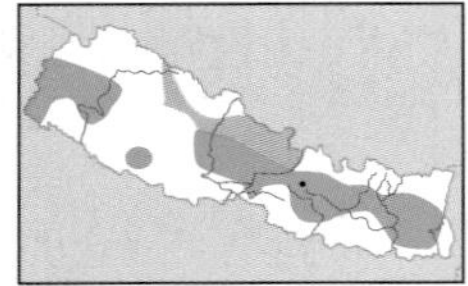

Hodgson's Redstart *Phoenicurus hodgsoni* 15 cm

Widespread, frequent winter visitor; 760–2,800 m, 75–5,350 m on passage. **ID** Male (both breeding and non-breeding) has grey upperparts, white wing patch, and black throat and upper breast. Female has dusky-brown upperparts and grey underparts; very similar to Black but has whitish area on belly (and lacks rufous-orange wash to lower flanks and belly of latter). First-winter male as female. **Voice** Rattling calls including *prit* and *trr*. **HH** Open grassy areas with bushes, stony riverbeds with trees, bushes in cultivation, and open forest.

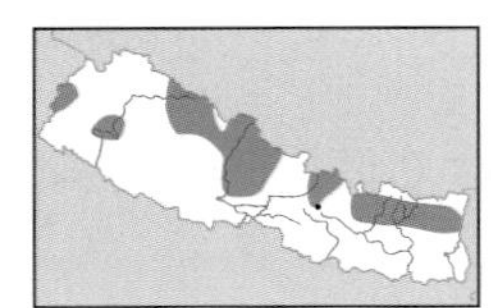

White-throated Redstart *Phoenicurus schisticeps* 15 cm

Locally common resident and winter visitor; breeds 3,050–4,400 m, winters 2,500–3,400 m (2,200–3,965 m). **ID** In all plumages has white throat, white wing patch, rufous rump and dark tail. Adult male has blue crown and nape (with brighter blue forehead), black ear-coverts and mantle, and rufous on scapulars and underparts. In fresh plumage, has rufous fringes to head and upperparts, and buff fringes to underparts. Adult female has grey-brown head and mantle and paler grey-brown breast, becoming more buffish on belly. **Voice** Calls include drawn-out *zieh* followed by a rattle; song a quiet series of short trills. **HH** Fairly shy, but often perches conspicuously on posts and bushes. Dry open habitats: edges and clearings of open coniferous forests; open barren slopes, and dry bushy vegetation on rocky slopes.

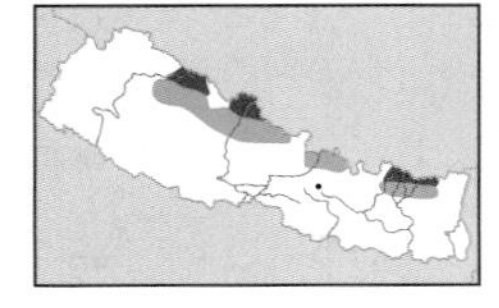

Güldenstädt's Redstart *Phoenicurus erythrogastrus* 18 cm

Resident and winter visitor; uncommon in breeding season, 2,900–3,700 m (–4,270 m), locally fairly common in some winters, 1,370–2,900 m (–150 m). **ID** Large size and stocky appearance. Male has white cap, black upperparts and large white patch on wing. Slightly duller in fresh plumage due to indistinct grey feathers fringes. Female has buff-brown upperparts and buffish underparts. First-winter as adult. **Voice** Song comprises short, pleasant whistling phrase, *teet-teet-teet* or *tit-tit-titer*; calls include weak *lik* and harder *tek*. **HH** Frequently perches on boulders, walls and low bushes. Breeds in dry rocky alpine meadows; winters in stony pastures and scrub patches. **AN** White-winged Redstart.

♂ br
♂
1st-win
♀
Eversmann's Redstart
♂ br
♂
1st-win
♀
Blue-capped Redstart
♂
♀
Black Redstart
♂ br
♀
Hodgson's Redstart
♂
♀
White-throated Redstart
♂
♀
Güldenstädt's Redstart

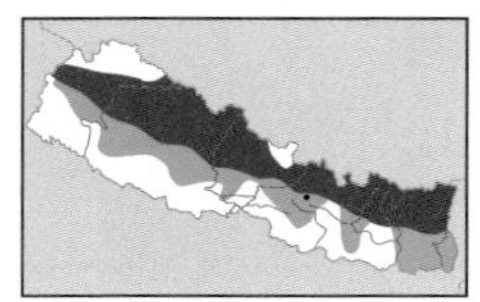

Blue-fronted Redstart *Phoenicurus frontalis* 15 cm

Common, widespread resident; breeds 3,350–4,900 m, winters 1,000–3,050 m (–455 m). **ID** Orange rump and tail-sides, with black centre and tip to tail in all plumages. Male has blue head and upperparts, and chestnut-orange underparts; heavily obscured by rufous-brown fringes in non-breeding and first-winter plumage. Female has dark brown upperparts and underparts, with orange wash to belly; tail pattern best feature from other female redstarts. **Voice** Song comprises 1–2 harsh trilling warbles then short whistling phrases, repeated with minor variations; calls include *ee-tit*. **HH** Unlike most redstarts, flicks tail up and down, and does not vibrate it. Breeds in subalpine shrubberies; winters in bushes and open forest.

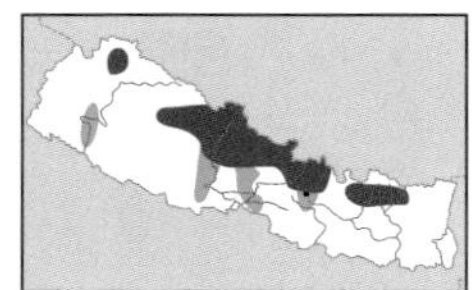

White-bellied Redstart *Hodgsonius phaenicuroides* 15 cm

Frequent summer visitor, 2,900–4,270 m (–2,450 m); very rare in winter, 250 m. **ID** Long, graduated tail. Male almost entirely dark slaty-blue with white belly and rufous tail-sides, and white spots on alula. Female has olive-brown upperparts, white throat and belly, and chestnut on tail. First-year male resembles female, but is darker brown; head, mantle and breast have some blue, and tail has blue cast. Juvenile has pale chestnut spotting on upperparts and dark scaling on underparts. **Voice** Song comprises 3–4 whistling notes, the second a drawn-out whistle that rises then falls, followed by a briefer, lower-pitched note; grating croak call, *chack-chack-chack*. **HH** Shy and skulking. Often cocks and fans tail. Breeds in subalpine shrubberies; thick undergrowth and forest edges in winter. **TN** Sometimes placed in genus *Luscinia*.

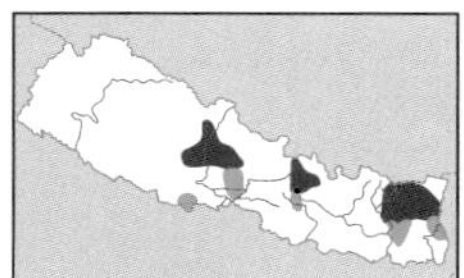

White-tailed Robin *Myiomela leucura* 18 cm

Locally fairly common resident; breeds 1,900–2,745 m, winters 75–915 m. **ID** White patches on tail in all plumages (visible as tail is slowly dipped and spread). Male blue-black, with glistening blue forehead and shoulders; concealed white patch on side of neck. Female olive-brown, with whitish lower throat. First-year male similar to female but blue on uppertail-coverts. **Voice** Song a loud, clear, rather hurried jangling of 7–8 notes; calls include a thin one- or two-note whistle and a low *tuc*. **HH** Usually keeps in cover. Undergrowth in dense, moist broadleaved forest.

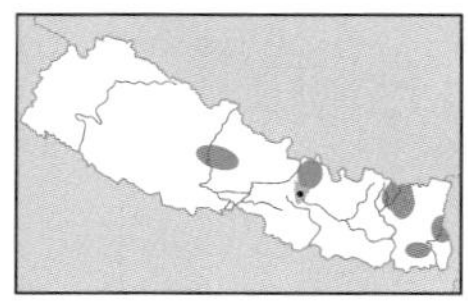

Gould's Shortwing *Heteroxenicus stellatus* 13 cm

Very rare and local, probably resident; breeds 3,200–4,200 m; to 450 m on passage. **ID** Adult has chestnut upperparts, slate-grey underparts (blacker around face) with white star-shaped spotting on belly and flanks, and fine greyish-white supercilium extending to eye. Juvenile has rufous streaking on head, mantle and breast, and greyish-black belly and flanks with broad whitish V-shaped spots. **Voice** Song begins with series of very high-pitched notes that gradually become louder and accelerate; *tik-tik* alarm. **HH** Hops among fallen branches and roots on ground like a mouse. Breeds in dense rhododendron and bamboo growth, juniper shrubberies, and thick undergrowth in fir and rhododendron forest. **TN** Formerly placed in genus *Brachypteryx*.

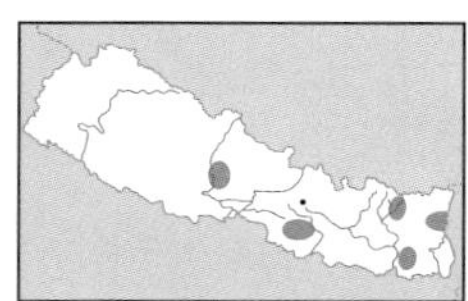

Lesser Shortwing *Brachypteryx leucophris* 13 cm

Very rare, presumably resident; 250–2,135 m. **ID** Smaller and shorter tailed than White-browed, with long pinkish legs. Male pale slaty-blue with white throat and belly. Female has rufous-brown upperparts, white throat and belly, rufous-brown wash and diffuse scaling on breast and flanks. Both sexes have fine white supercilium (often obscured). First-year male similar to female but greyer above and on breast. **Voice** Brief melodious warbling song, accelerating into rapid jumble; calls include hard *tock-tock* and plaintive whistle. **HH** Very skulking. Thick undergrowth in moist broadleaved forest.

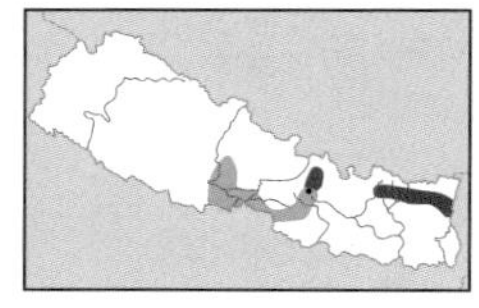

White-browed Shortwing *Brachypteryx montana* 15 cm

Very uncommon resident; breeds 2,650–3,660 m, winters 245–2,375 m. **ID** Larger than Lesser with longer tail and dark legs. Male dark slaty-blue, with black lores and fine white supercilium. Female has brown upperparts with more rufescent wings, and brownish underparts with paler belly. Rufous-orange lores and more uniform brownish underparts (lacking striking white throat and belly) separate from female Lesser. Immature male similar to female, but has fine white supercilium. Juvenile has orange-buff spotting on underparts. **Voice** Song comprises high-pitched penetrating whistles, often introduced by slower disyllabic notes; calls include a scolding rattle and penetrating whistled *hweeep*. **HH** Very secretive. Undergrowth in moist, dense evergreen forest and thickets in damp ravines.

♂
♀
Blue-fronted Redstart
♂
♀
White-bellied Redstart
♂
♀
White-tailed Robin
ad
Gould's Shortwing
'blue' ♂
♀
Lesser Shortwing
♂
♀
♂
imm
White-browed Shortwing

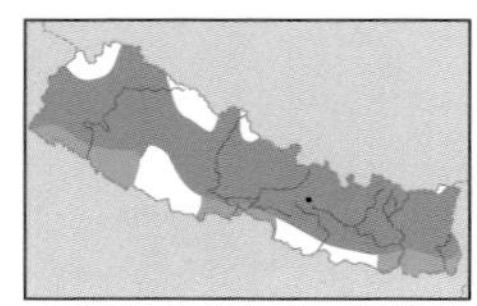

Plumbeous Water Redstart *Rhyacornis fuliginosa* 12 cm

Common, widespread resident; mainly breeds 1,525–3,750 m (600–4,420 m), winters 75–2,560 m. **ID** Stocky and short-tailed; constantly flicks open tail while moving it up and down. Male slaty-blue, with rufous-chestnut tail. Female and first-year male have black-and-white tail and white-spotted grey underparts. Juvenile resembles female (same tail pattern), but has buff spotting on browner upperparts and black scaling on buffish underparts. **Voice** Song a rapidly repeated, insect-like *streeee-treee-tree-treeeh*; strident *peet-peet* alarm. **HH** Flits restlessly between rocks in streams, frequently making flycatcher-like aerial sallies after insects. Fast-flowing mountain streams and rivers.

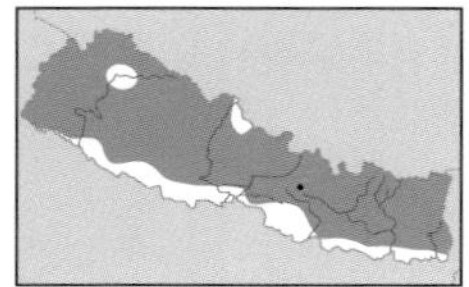

White-capped Redstart *Chaimarrornis leucocephalus* 19 cm

Common, widespread resident; breeds 1,830–5,000 m (–5,350 m), mainly winters 915–1,525 m (75–3,500 m). **ID** Adult has white cap and rufous tail with broad black terminal band. Juvenile has black fringes to white crown and blackish underparts with rufous fringes. **Voice** Song a weak, drawn-out, undulating whistle, *tieu-yieu-yieu-yieu*; far-carrying, upward-inflected *tseeit tseeit* call. **HH** Flies actively from stone to stone in streams. Distinctive habit of pumping and fanning tail, sometimes tilting it right over back, often accompanied by a deep curtsy. In summer, will forage some distance from water in alpine meadows. Mainly mountain streams and rivers. **AN** White-capped Water Redstart.

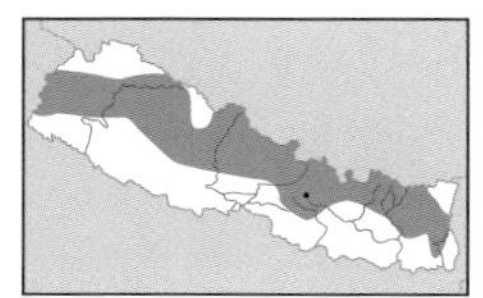

Little Forktail *Enicurus scouleri* 12 cm

Fairly common, widespread resident; mainly breeds 1,830–4,000 m (1,150–4,240 m), winters 900–1,830 m (–75 m). **ID** Small with short tail. Black tail with prominent white sides, black band on white rump, and prominent white forehead. Juvenile lacks white forehead, has brownish-black upperparts, and white underparts with dark scaling on throat and breast. **Voice** Generally silent. **HH** Always close to water; unlike other forktails is not dependent on tree cover. Continually fans and closes tail while moving it slowly up and down. Often forages by standing on partially submerged rocks. Flight low over water. Rocky mountain streams, often near waterfalls; in winter also larger, slower-moving rivers.

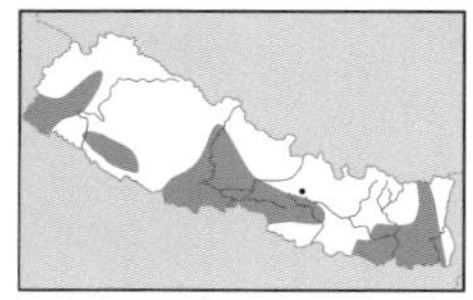

Black-backed Forktail *Enicurus immaculatus* 23 cm

Resident; frequent from west-central Nepal east, uncommon in west; 75–1,370 m. **ID** From Slaty-backed by black (rather than slate-grey) crown and mantle, generally smaller bill, and more white on forehead. Juvenile has shorter tail, lacks white forehead and supercilium, has brownish-black upperparts and dark scaling on white breast. **Voice** Hollow *huu* call, like Grey Bushchat, followed by a shrill *zeee*. **HH** Habits similar to Spotted. Fast-flowing streams in moist broadleaved forest.

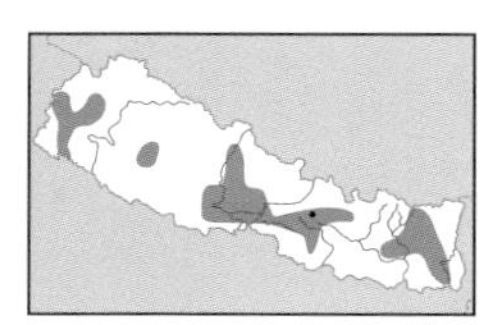

Slaty-backed Forktail *Enicurus schistaceus* 25 cm

Locally fairly common resident; 900–1,675 m (–450 m). **ID** From Black-backed by slate-grey (rather than black) crown and mantle, contrasting with black throat and wing-coverts. Also bill generally larger, and has less white on forehead. Juvenile has shorter tail, lacks white forehead and supercilium, has brown upperparts, and dark scaling on white breast. **Voice** Mellow *cheet* or metallic *teenk* calls. **HH** Habits similar to Spotted. Prefers marginal habitats, creeping amongst boulders, or wading at water's edge. Wooded lake margins and large fast-flowing streams in forest.

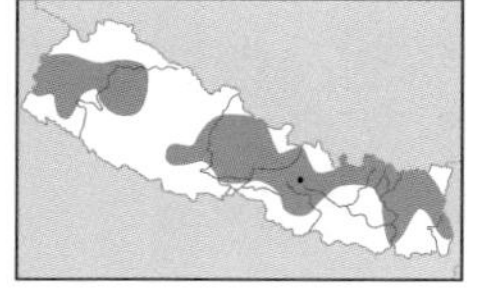

Spotted Forktail *Enicurus maculatus* 27–28 cm

Widespread resident, frequent in west and central Nepal, very uncommon in east; breeds 1,370–3,100 m, winters 290–2,745 m. **ID** From other forktails by combination of large size and very long tail, white spotting on black mantle (forming white collar towards nape), prominent white forehead, and black of throat extending to breast. Juvenile lacks white forehead, has brownish-black upperparts, paler brownish-black throat and breast with white streaking, brown mottling on upper belly, and brown flanks. *E. m. guttatus* in east differs from nominate (in west and centre) in being smaller, lacking white scaling on black breast, and having fewer and smaller white spots on mantle. **Voice** Shrill, rasping *kreee* or *tseek* call; also a creaky *cheek-chik-chi-chi-chik-chik*. **HH** Always close to water. Characteristic habit of constantly swaying its tail slowly up and down. Very restless, and frequently turns from side to side. Walks daintily over stones at water's edge or hops from stone to stone. Graceful, undulating flight low over water. Rocky streams in forest and shaded wooded ravines; avoids rivers, lakes and open country.

Plumbeous Water Redstart

White-capped Redstart

Little Forktail

Black-backed Forktail

Slaty-backed Forktail

Spotted Forktail

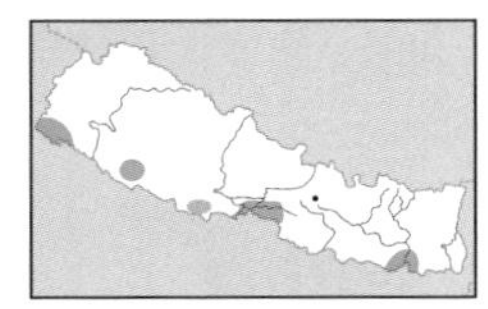

Hodgson's Bushchat *Saxicola insignis* 17 cm

Very local winter visitor and passage migrant, mainly to Sukla Phanta Wildlife Reserve; 75–250 m (–1,800 m). **ID** Larger than Common Stonechat, with bigger-looking head and bill. Male has white throat extending to form almost complete white collar, and more white on wing than Common Stonechat. Female has broad buffish-white wing-bars. **Voice** Metallic *teck-teck* call. **HH** Habits similar to Common Stonechat, but often solitary and rather shy. Feeds mainly on ground by dropping from a perch to take insects. Tall grassland and reeds along rivers. Globally threatened. **AN** White-throated Bushchat.

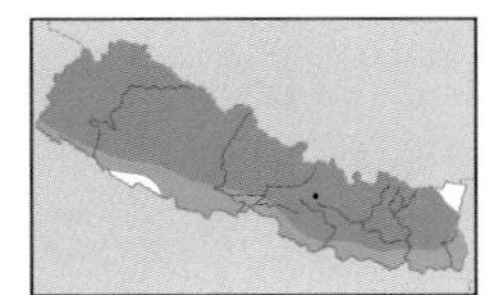

Common Stonechat *Saxicola torquatus* 12.5–13 cm

Common and widespread resident, winter visitor and passage migrant; breeds 365–4,880 m (–5,385 m), winters 75–1,500 m. **ID** Male has black head, white patch on neck, orange breast and whitish rump (features obscured in fresh plumage). Female has streaked upperparts and orange breast and rump. Tail darker than female White-tailed. Three subspecies occur. *S. t. przevalskii* (not illustrated), a winter visitor and breeder in Tibetan plateau region, is largest and has underparts almost entirely deep rufous-orange in both sexes. **Voice** Variable, rather melancholy warbling song; main calls are *hweet* and hard *tsak*, which are often run together. **HH** Often in pairs or family groups. Perches prominently on bush or tall plant, rock or post; alert stance. Frequently flicks wings and jerks tail up and down, while fanning it. Flies or hops to ground, seizes prey and returns to perch; also catches insects in aerial sallies. Breeds in open country with bushes, including high-altitude semi-desert; winters in scrub, reeds and cultivation. **TN** Siberian Stonechat *S. maurus* is often treated as a separate species, but relationships of the various forms are still unclear.

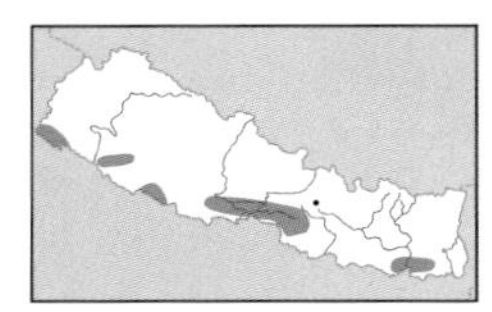

White-tailed Stonechat *Saxicola leucurus* 12.5–13 cm

Local resident; 75–300 m (–915 m). **ID** Male very similar to Common, but inner webs of all but central tail feathers largely white; shows much white in tail in flight. Female has greyer upperparts, with diffuse streaking, and paler grey-brown tail. **Voice** Song is a series of short phrases comprising rapid, squeaky, scratchy, wheezy and creaky notes that may fall and rise; alarm call a *peep-chaaa*. **HH** Habits very similar to Common. Reeds and tall grassland.

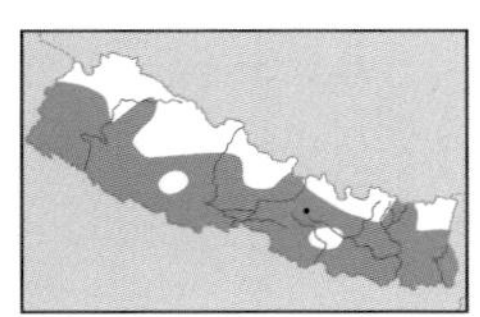

Pied Bushchat *Saxicola caprata* 12.5–13 cm

Widespread resident, common 75–915 m, fairly common to 1,400 m, and occasionally summers up to 2,400 m (–2,850 m). **ID** Male entirely black except white rump and patch on wing; duller due to rufous fringes to body in non-breeding and first-winter plumages. Female has dark brown upperparts and rufous-brown underparts, with rufous-orange rump. **Voice** Brisk, whistling *chip-chepee-chewee chu* song; calls include plaintive *chep chep-hee* or *chek chek trweet*. **HH** Habits very similar to Common Stonechat. Territorial throughout year. Mainly cultivation and open country with scattered bushes or tall grass.

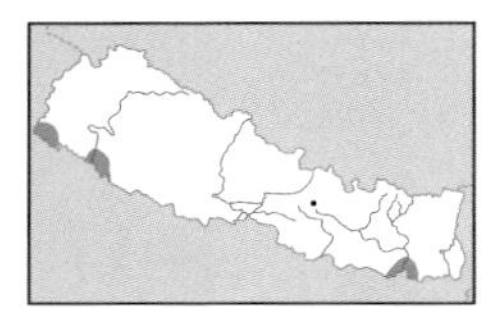

Jerdon's Bushchat *Saxicola jerdoni* 15 cm

Very rare and very local resident; 75–150 m. **ID** Male has blue-black upperparts, including rump and tail, and white underparts. Female and first-winter male similar to female Grey Bushchat, but lack prominent supercilium, and have longer, more graduated tail lacking rufous at sides. **Voice** Song comprises very short series of clear, mellow, mostly downslurred whistles; call is a short plaintive whistle, higher pitched than other chats. **HH** Habits similar to Common Stonechat but less active on wing. Sometimes forages on ground at base of reeds and grass like a babbler. Tall moist/marshy grassland and tall grass and reeds along rivers.

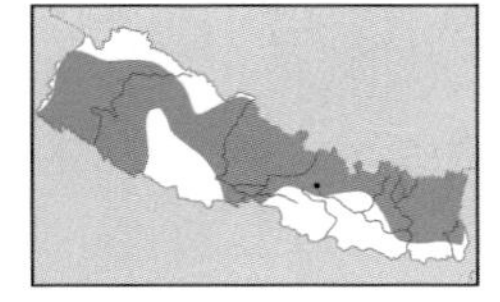

Grey Bushchat *Saxicola ferreus* 15 cm

Common, widespread resident; breeds 1,500–3,355 m, winters 150–2,135 m (–75 m). **ID** Male has white supercilium and dark mask; upperparts grey to almost black, depending on extent of wear; underparts whitish with grey breast and flanks. Female has buff supercilium contrasting with dark brown ear-coverts, and rufous rump and tail-sides. First-winter as fresh-plumaged adult. **Voice** Song brief and repeated, starting with 2–3 emphatic notes and ending with a trill: *tree-toooh tu-treeeh t-t-t-t-tuhr*; calls include *zee-chunk* and sharp *tak-tak*. **HH** Habits similar to Common Stonechat. Territorial on breeding and wintering grounds. Second growth, forest edges and scrub-covered hillsides.

♂ non-br
♀
Hodgson's Bushchat
♂ non-br
♂ br
♀
indicus
Common Stonechat
♂ br
♀
♂ 1st-win
Pied Bushchat
♂ br
♀
White-tailed Stonechat
♂
♀
Grey Bushchat
♂
♀
Jerdon's Bushchat

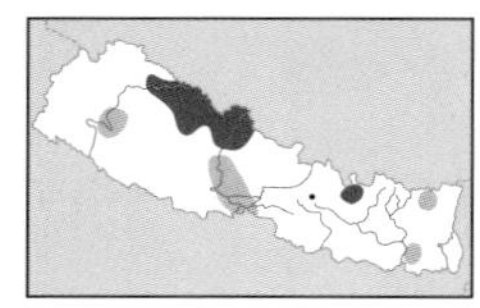

Desert Wheatear *Oenanthe deserti* 14–15 cm

Rare summer visitor to Tibetan plateau and rare passage migrant; 3,200–5,100 m (–150 m on passage). **ID** Comparatively small, well-proportioned wheatear, with largely black tail and contrasting white rump. Male has black throat (partly obscured when fresh) and buff mantle. Female has blackish centres to wing-coverts and tertials in fresh plumage and largely black wings when worn (useful distinction from Isabelline – see Appendix 1). **Voice** Hard *check-check* call; short, variable song, beginning with a mellow, sweet, mournful, downslurred to mostly level whistle and ends with a rapid trill, slow tittering trill or a few broken or quavering whistles. **HH** Like other wheatears, frequently bows whole body and bobs head and forebody, often flicks wings and tail and at times spreads tail and moves it slowly up and down. Forages chiefly by running or hopping on ground, then stopping to seize prey. Has half-upright carriage; fully erect when alert. Breeds in dry semi-desert.

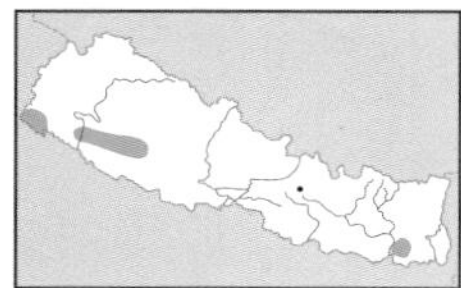

Variable Wheatear *Oenanthe picata* 14.5 cm

Very rare winter visitor and passage migrant; 75–760 m. **ID** Male has black head, breast and upperparts, and white underparts. Females has black replaced by grey. Both sexes show extensive white at sides of tail. See Appendix 1 for comparison with Pied Wheatear. **Voice** Calls include *chek-chek*, agitated indrawn whistles, a single *plow* and double *plew-wit* very similar to Isabelline. **HH** Often perches conspicuously on bushes and walls. Forages by flying to ground from an elevated perch, by hopping on ground or in aerial sallies. Territorial on its wintering grounds. Winters in fallow cultivation, village edges and stony open country.

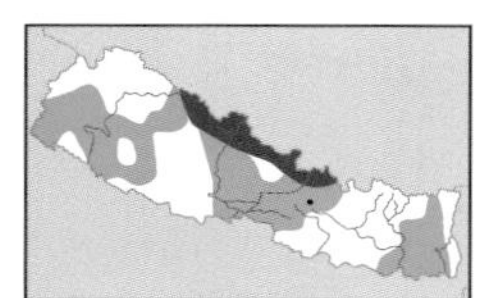

Blue Rock Thrush *Monticola solitarius* 20 cm

Frequent resident and winter visitor; breeds 2,590–4,880 m in Trans-Himalayas, winters 75–1,440 m. **ID** Male indigo-blue, obscured by pale fringes in non-breeding and especially first-winter plumages. Female has bluish cast to slaty-brown upperparts and buff scaling on underparts. **Voice** Short and repetitive song with fluty phrases, often with long pauses; calls include a *chak*, a *veeht veeht* and high *tsee*. **HH** Typically perches very upright on a boulder, frequently wagging tail up and down. Seeks insects mainly on ground among rocks and grass tufts, progressing by long hops, sometimes catches them in air. Breeds in open rocky areas or on steep cliffs; winters along streams and on old buildings.

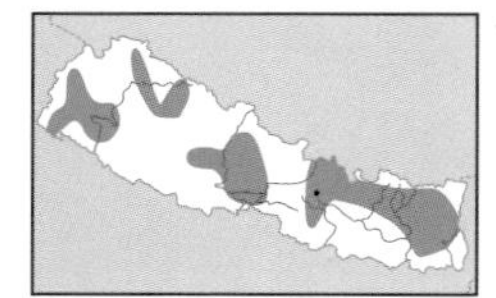

Chestnut-bellied Rock Thrush *Monticola rufiventris* 23 cm

Widespread, locally fairly common resident; breeds 1,800–3,400 m (–4,460 m), winters 915–2,380 m. **ID** Male has chestnut-red underparts and blue upperparts including rump, uppertail-coverts and tail; lacks white in wing. Female has orange-buff lores and neck patch, dark malar stripe, dark barring on slaty olive-brown upperparts, and heavy scaling on underparts. Non-breeding and first-winter male very similar to breeding male, but have fine buff fringes to mantle, scapulars and throat. Juvenile has pale spotting on upperparts; male has blue in wing. **Voice** Undulating and fluty song, more subdued and softer than Blue-capped; calls include rasping jay-like notes. **HH** Often seen perched high in tall forest tree, slowly jerking tail up and down from time to time. Forages mainly on ground; sometimes makes aerial sallies for insects. Open broadleaved and coniferous forest on rocky slopes.

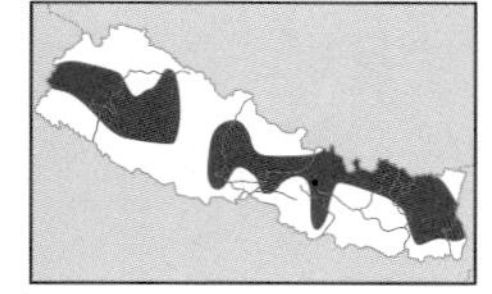

Blue-capped Rock Thrush *Monticola cinclorhynchus* 17 cm

Frequent summer visitor, rare in winter; 1,200–2,135 m (–275 m on passage) **ID** Male has white wing patch and blue-black tail; also blue crown and throat, and orange rump and underparts; pattern and coloration obscured by pale fringes in non-breeding and first-winter plumages. Female has olive-brown upperparts, with barred rump, and whitish underparts boldly scaled and barred with brown; lacks buff neck patch of Chestnut-bellied and blue cast to upperparts of Blue. **Voice** Short, monotonous, undulating, fluty song; calls include a *trigoink*. **HH** Chiefly arboreal, feeding on trunks and branches, occasionally descending to ground. In spring, males perch conspicuously in treetops. Very erect posture and often wags tail up and down. Coniferous forests and open, rocky slopes with stunted, scattered trees.

♂ br
♂ non-br
♂
Variable Wheatear
♀
Desert Wheatear
picata
♀ non-br
♀
♂
1st-win
♂ br
Blue Rock Thrush
♂
♂ br
Chestnut-bellied
Rock Thrush
♀
♂
1st-win
♀
Blue-capped
Rock Thrush

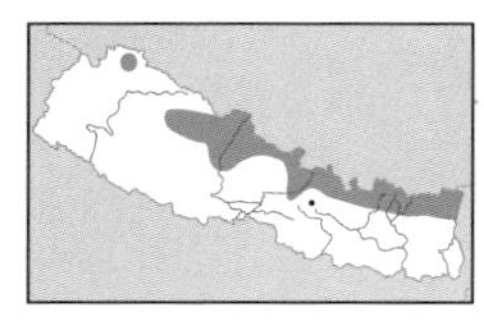

Grandala *Grandala coelicolor* 23 cm

Locally common resident; breeds 3,900–5,500 m, winters 3,000–3,960 m (–1,950 m). **ID** Slim, long-winged, starling-like chat. Adult male almost entirely purple-blue with glistening sheen, with black lores, wings and tail. Adult female and immature male dark brown, streaked white, with blue wash to rump and uppertail-coverts, and white patches on wing. Juvenile similar to female, but darker brown and more boldly streaked; lacks blue on rump and uppertail-coverts. **Voice** Flight call *tew-wee*; repeated *galeb-che-chew-de-dew* song. **HH** Often in large flocks, which circle high overhead. Breeds on rocky slopes and ridges, and stony meadows in alpine zone; winters in similar habitat lower down, may descend below treeline in severe weather.

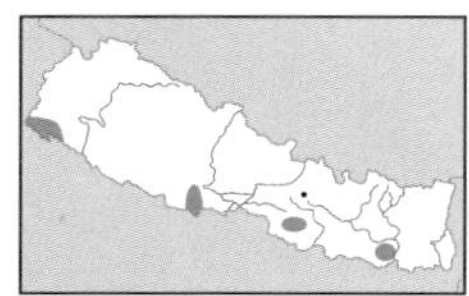

Brown Rock Chat *Cercomela fusca* 17 cm

Rare and very local resident; 75–150 m. **ID** Both sexes brown, with more rufescent underparts and blackish tail. Juvenile darker brown, without rufous. **Voice** Song sweet and thrush-like, can include much mimicry; calls include short, whistling *chee* and harsh *check check*. **HH** Frequently slowly cocks and spreads tail, and bobs body. Old buildings in open country.

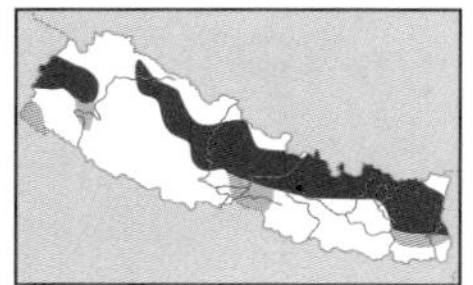

Dark-sided Flycatcher *Muscicapa sibirica* 14 cm

Fairly common and widespread summer visitor and passage migrant; 2,000–3,300 m (–75 m on passage). **ID** From Asian Brown by small dark bill, and longer primary projection (exposed primaries equal to or distinctly longer than tertials). Also darker sooty-brown upperparts, and breast and flanks more heavily marked, with white crescent on neck-sides and narrow white line down centre of belly. Juvenile has finely streaked upperparts, heavy dark mottling on breast and flanks, and orange-buff wing-bar. **Voice** Thin, high-pitched, repetitive phrases, followed by quite melodious trills and whistles. **HH** Perches upright on favoured prominent perch flying out to catch prey, before returning to same perch. Canopy of open broadleaved and coniferous forest, or in clearings.

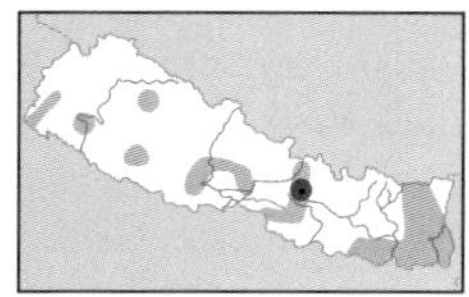

Asian Brown Flycatcher *Muscicapa dauurica* 13 cm

Uncommon summer visitor and passage migrant; rare in winter; breeds 1,000–1,620 m, winters 250–915 m (–75 m on passage). **ID** Grey-brown with short tail, large head, and huge eye with prominent eye-ring. From Dark-sided by larger bill with more extensive orange base to lower mandible, shorter primary projection, and paler underparts (with light grey-brown wash to breast and flanks). Juvenile has prominent buffish spotting on upperparts, whitish underparts with fine dark scaling on breast, and creamy-white wing-bar. **Voice** Song comprises short trills interspersed by two- or three-note whistling phrases louder than Dark-sided; weak trilling *sit-it-it-it* call. **HH** Partly crepuscular. Perches in lower tree branches and makes aerial sallies. Open broadleaved forest, groves and wooded areas.

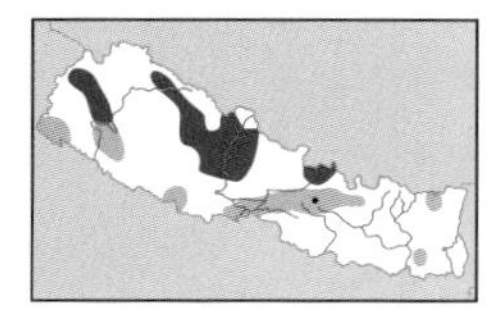

Rusty-tailed Flycatcher *Muscicapa ruficauda* 14 cm

Uncommon partial summer visitor and passage migrant; breeds 2,440–3,655 m; winters 75–915 m. **ID** Rufous uppertail-coverts and tail, resulting in (female) redstart-like appearance. Larger than Asian Brown, with flatter forehead and crown feathers are often slightly raised, giving crested appearance to nape. Further, has rather plain face, with only faint supercilium (back to eye) and indistinct eye-ring, entirely orange lower mandible and cutting edges to upper mandible. Juvenile has buff spotting on upperparts and dark scaling on underparts. **Voice** Song loud and melodious: drawn-out, rising and falling mournful whistle followed by rapid warbling. Calls include bullfinch-like *peu-peu* and short, deep churring. **HH** Unobtrusive. Chat-like habit of bobbing forward and flicking wings at same time. Hunts insects by flitting among foliage and between perches. Forest clearings and edges, open forest.

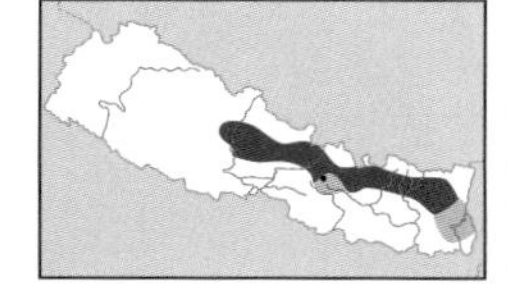

Ferruginous Flycatcher *Muscicapa ferruginea* 13 cm

Very uncommon summer visitor; 2,000–3,300 m (–915 m on passage). **ID** Compact, with large head, large eye and prominent white eye-ring. Adult has blue-grey cast to head (with darker malar stripe), rufous-brown mantle, rufous-orange rump and tail-sides, rufous-orange underparts, and prominent rufous fringes to greater coverts and tertials. Juvenile has orange-buff spotting on upperparts, rufous-orange greater coverts wing-bar and dark scaling on breast. **Voice** Call a quiet accentor-like trill. Probable song comprises very high-pitched notes introduced by sharper, shriller notes, *tsit-tittu-tittu*. **HH** Unobtrusive, usually keeps to middle or lower forest storey. Hawks insects like other brown flycatchers. Humid broadleaved forest.

♂

♀

Grandala

ad

Brown Rock Chat

ad

Dark-sided Flycatcher

ad

Asian Brown Flycatcher

ad

Rusty-tailed Flycatcher

ad

Ferruginous Flycatcher

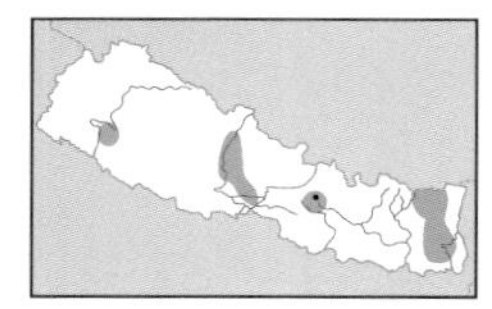

Slaty-backed Flycatcher *Ficedula hodgsonii* 13 cm

Rare and local, mainly winter visitor, possibly also a passage migrant 75–2,000 m. **ID** Small, long-tailed flycatcher with very short bill. Male has deep blue upperparts (blacker on face), bright orange underparts (whiter on belly) and black tail with white patches at base. Lacks any glistening blue in plumage. Female rather nondescript, with olive-brown upperparts, greyish-olive underparts, and poorly defined whitish throat, lores and eye-ring. Juvenile has buff spotting on upperparts and dark scaling on buff underparts, with whitish throat. First-year male resembles female, and breeds in this plumage. **Voice** Particularly sharp rattle, *terrht* call. **HH** Winters in moist, broadleaved and mixed lowland forests.

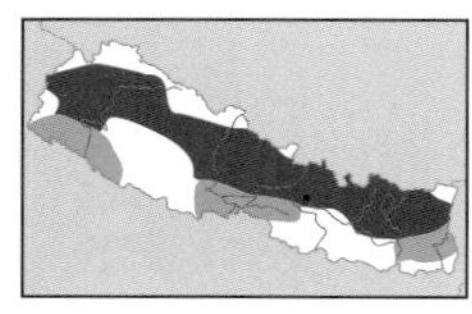

Rufous-gorgeted Flycatcher *Ficedula strophiata* 14 cm

Common, widespread resident; breeds 2,440–3,965 m, mainly winters 915–1,830 m (150–2,135 m). **ID** Male has dark olive-brown upperparts, blackish face and throat, prominent white forehead and eyebrow, small rufous patch in centre of grey breast (can be difficult to see), and large white patches at sides of tail. Female similar, but has less distinct eyebrow, duller and less distinct rufous 'gorget', and paler grey face and throat. Juvenile has tail pattern as adult, with orange-buff spotting on upperparts and dark scaling on underparts. **Voice** Song a thin *zreet-creet-creet-chirt-chirt* calls include a *pee-tweet*, a metallic *pink* and a harsh *trrt*, deeper than similar call of Red-breasted. **HH** Dense and open forests.

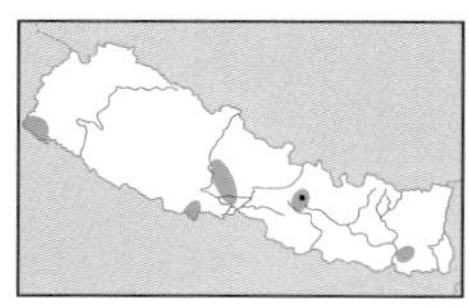

Red-breasted Flycatcher *Ficedula parva* 11.5–12.5 cm

Status uncertain, probably overlooked; 75–1,340 m. **ID** White sides to long blackish tail; bill has distinctly paler base to lower mandible. Male has red throat and upper breast, and creamy-white rest of underparts. Many males lack red throat until second or third year, and resemble females. Female lacks grey cast to crown and face of male, and underparts are creamy-white, suffused buff on breast. First-winter has orange-buff greater coverts wing-bar. **Voice** Calls include a *tic* unlike usual call of Taiga Flycatcher, also a quiet, dry *trrt, trrt*. **HH** Frequently cocks tail. Open wooded areas. **AN** Red-throated Flycatcher.

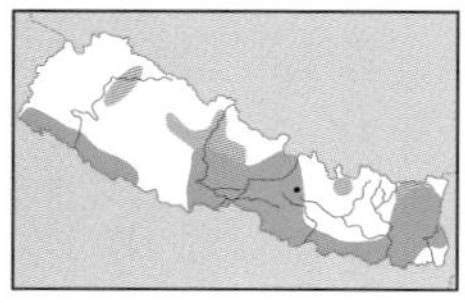

Taiga Flycatcher *Ficedula albicilla* 11–12 cm

Common to fairly common winter visitor, also passage migrant; 75–1,830 m (–2,590 m on passage). **ID** Very similar to Red-breasted. Male has orange restricted to throat bordered below by grey breast-band; female and first-winter have colder grey-brown upperparts than Red-breasted, with pronounced black uppertail-coverts (browner in Red-breasted), and underparts are whiter with grey on breast. Bill mainly dark. **Voice** Calls include a buzzing *drrrrrt*, unlike Red-breasted. **HH** Open forest and scrub at cultivation edges. **TN** Formerly treated as conspecific with Red-breasted Flycatcher *F. parva*, under name Red-throated Flycatcher.

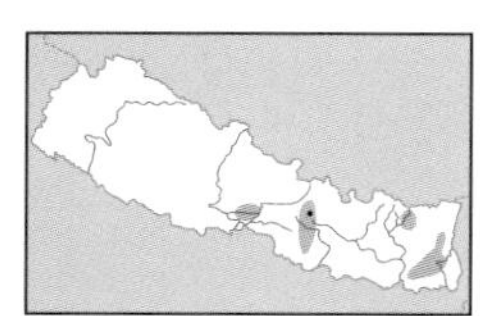

Kashmir Flycatcher *Ficedula subrubra* 13 cm

Very rare visitor; 150–2,135 m. **ID** Very similar to Red-breasted and Taiga Flycatchers, with white-sided black tail that is frequently cocked. Male has deep rufous underparts (extending to lower breast and flanks), diffuse black border to throat and breast, and darker grey-brown upperparts. Female and first-winter male variable, but usually has some mottled rufous-orange on throat and breast, with pronounced grey sides to neck and breast. Can resemble male Red-breasted, but rufous is often more pronounced on breast than on throat, and often continues as a wash onto belly and/or flanks; upperparts also a slightly darker grey-brown. First-winter female lacks orange on underparts. Darker grey-brown upperparts and grey wash to sides of neck and on breast are best distinctions from female Red-breasted; pale base to bill helps separate from Taiga. **Voice** Calls include a staccato rolled twitter, also sharp *chack* and rattling *purr*. **HH** Open broadleaved forest. Globally threatened.

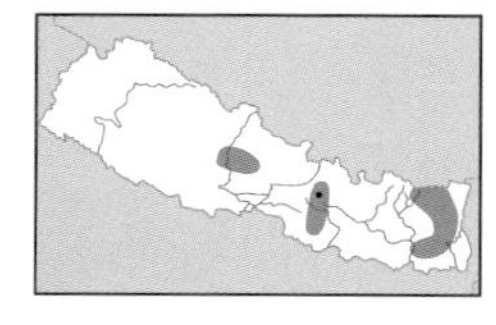

White-gorgeted Flycatcher *Anthipes monileger* 13 cm

Rare resident; 915–3,000 m. **ID** Compact with large domed head and large bill; typically found close to ground. Adult mainly olive-brown, with orange-buff supercilium and large white throat patch enclosed by black gorget. Juvenile lacks black-bordered white throat of adult; has dark brown upperparts streaked warm buff, buff tips to greater coverts, and buffish underparts diffusely streaked dark brown. Pinkish legs and feet in all plumages. **Voice** Thin high-pitched whistling song; calls include metallic *dik*, metallic, scolding rattle and short, plaintive whistle. **HH** Dense undergrowth in moist broadleaved forest. **TN** Formerly placed in genus *Ficedula*.

♂
♀
Slaty-backed Flycatcher
♀
♂
Rufous-gorgeted Flycatcher
♂
♀
Red-breasted Flycatcher
♂
Taiga Flycatcher
♂
Kashmir Flycatcher
♀
ad
White-gorgeted Flycatcher

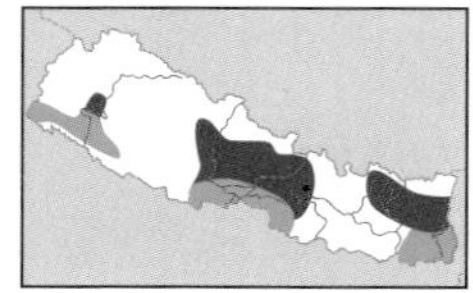

Snowy-browed Flycatcher ***Ficedula hyperythra*** 11 cm

Uncommon resident; breeds 1,500–2,285 m (–3,000 m), winters 275–1,525 m (–75 m). **ID** Small, compact, short-tailed flycatcher with large head, typically found close to ground. Male has short, broad white supercilium, dark slaty-blue upperparts with rufous-brown wings, rufous-orange throat and breast, and white patches at base of tail (can be difficult to see). Female from other small flycatchers by combination of small size and compact shape, dark olive-brown upperparts, orange-buff supercilium and eye-ring, and faint rufous panel on wing. Pinkish legs and feet. Juvenile has orange-buff streaking on upperparts and diffuse dark streaking on underparts. **Voice** Quiet, high-pitched wheezy song thin, repeated *sip* call. **HH** Humid broadleaved forest with dense undergrowth, favours bamboo.

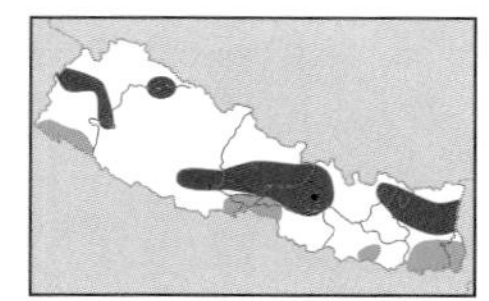

Little Pied Flycatcher ***Ficedula westermanni*** 11 cm

Uncommon resident; breeds 1,200–3,000 m, winters 275–915 m (–75 m). **ID** Small, compact, bull-headed flycatcher, with small dark bill. Male strikingly black-and-white, including long broad white supercilium. Female has brownish-grey upperparts (tinged warm brown on forehead, lores and around eye), and whitish underparts with brownish-grey wash to breast and flanks. Juvenile has buff spotting on upperparts and dark scaling on breast. **Voice** Song is a series of thin, high-pitched notes followed by a rattle; mellow *tweet* call. **HH** Breeds in broadleaved forests; also winters in open wooded country and in tall grassland.

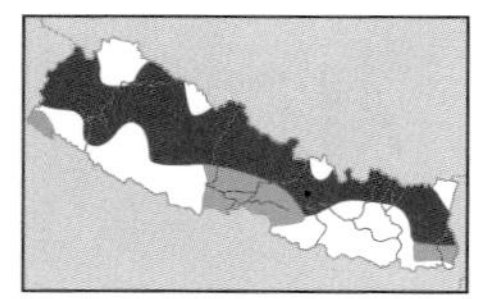

Ultramarine Flycatcher ***Ficedula superciliaris*** 12 cm

Common breeding visitor, 1,800–3,200 m; rare in winter 250–1,500 m (–75 m). **ID** Small, compact, arboreal flycatcher with small bill. Male has deep blue upperparts and sides of neck/breast, and white underparts. Female has greyish-brown upperparts and whitish underparts, with greyish patches on sides of breast; some have blue cast to tail. Coloration of rump and well-defined grey patches on breast-sides separate from female Little Pied. First-year male resembles female, but has blue cast to mantle, wings and tail. Juvenile has buffish-white spotting on upperparts and dark scaling on whitish underparts. Male nominate in west has white supercilium and white patches at base of tail, which are typically lacking in eastern *F. s. aestigma*. **Voice** Feeble, rather disjointed, high-pitched song; calls include rising squeak. **HH** Arboreal. Breeds in broadleaved forest; winters in open woodland and wooded areas.

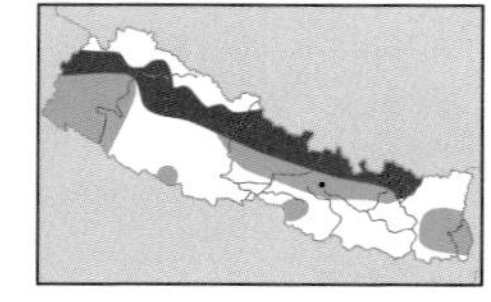

Slaty-blue Flycatcher ***Ficedula tricolor*** 13 cm

Locally common resident; breeds 3,050–3,400 m (–4,000 m), mainly winters 245–1,525 m (160–2,135 m). **ID** Small, slim, long-tailed flycatcher which typically feeds close to (or on) ground, with tail cocked. Male has dark blue upperparts with brighter blue forehead, blue-black sides of head and breast contrasting with white throat, and blue-black tail with white patches at base. Female has well-defined whitish throat contrasting with brownish-buff breast and flanks, warm brown upperparts, and rufous uppertail-coverts and tail. First-year male as female. Juvenile has warm buff streaking on upperparts and dark scaling on buff underparts; tail as female. **Voice** Song comprises 3–4 high-pitched whistles, calls include rapid *tic-tic*. **HH** Breeds in subalpine shrubberies, forest edges and undergrowth; winters in bushes and forest undergrowth.

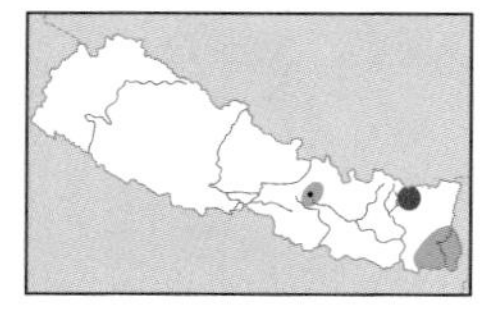

Sapphire Flycatcher ***Ficedula sapphira*** 11 cm

Rare and possibly resident in east; breeds 2,135–2,800 m, to 250 m winter. **ID** Breeding male has bright blue upperparts and sides of breast (with glistening blue crown and rump), orange throat and centre to breast, and white belly and undertail-coverts. Non-breeding and immature male have brown head and mantle, and brownish sides of breast. Female has olive-brown upperparts, orange throat and breast, and rufous rump and tail. Small size, slim appearance and tiny bill separate female from female *Cyornis* flycatchers. Juvenile has orange-buff spotting on brown upperparts and dark-scaled orange-buff breast that contrasts with white belly and flanks. **Voice** Dry rattled call *trrrt*. **HH** Moist evergreen broadleaved forest.

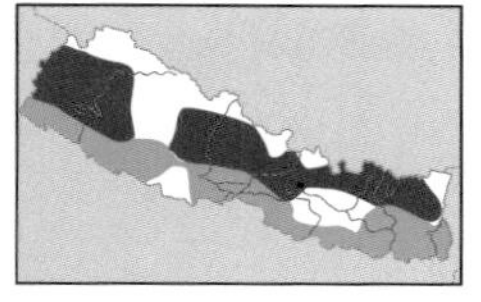

Grey-headed Canary Flycatcher ***Culicicapa ceylonensis*** 13 cm

Common partial migrant; breeds mainly 1,500–2,400 m (1,200–3,100 m); winters 75–1,800 m. **ID** Grey head and breast, greenish mantle, and yellow belly, flanks and vent. Upright stance, crested appearance and flycatcher-like behaviour distinguish it from similarly plumaged *Seicercus* and *Abroscopus* warblers. **Voice** Loud, high-pitched interrogative, repeated *chik...whichee-whichee* song; clear *kitwik...kitwik* and soft *pit...pit...pit* calls. **HH** Forests and wooded areas.

♂
♀
Little Pied Flycatcher
♂
Snowy-browed Flycatcher
♀
superciliaris
♂
aestigma
superciliaris
♂
♀
♂
Ultramarine Flycatcher
Slaty-blue
Flycatcher
♀
superciliaris
1st-win
♀
♂ br
♂ non-br
ad
Sapphire Flycatcher
Grey-headed Canary Flycatcher

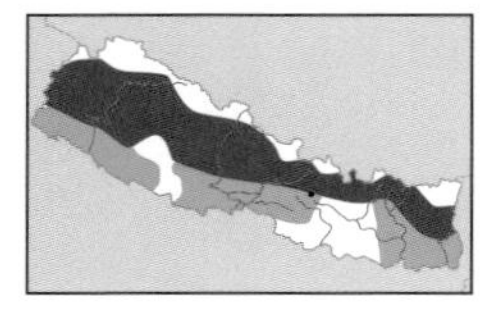

Verditer Flycatcher *Eumyias thalassinus* 16 cm

Partial migrant, common in summer mainly 1,200–2,625 m (1,000–3,200 m); frequent in winter 75–350 m (–915 m). **ID** Male entirely greenish-blue, with brighter forehead and throat, and black lores. Not really confusable with any other flycatcher. Female similar, but duller and greyer, with dusky lores. Female confusable with male (but not female) Pale Blue Flycatcher, but has shorter bill, turquoise-blue upperparts and uniform greyish turquoise-blue underparts (lacking contrast between breast and belly). Juvenile has turquoise cast to grey-brown upperparts, with fine orange-buff spotting and bold orange-buff spotting on brown underparts; wings and tail as adult, with buff tips to coverts. **Voice** Song comprises series of rapid, undulating, strident notes, gradually descending scale. **HH** Conspicuous and confiding. Hunts mainly by sallying forth from an exposed perch. Open forests, clearings, edges and wooded areas.

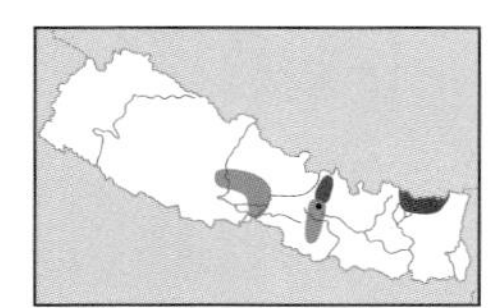

Pale Blue Flycatcher *Cyornis unicolor* 18 cm

Locally frequent resident and summer visitor at higher altitudes; 275–1,525 m. **ID** Male from Verditer by longer bill and pale blue coloration (lacking greenish cast), with distinctly greyer belly. Shining blue forecrown and dusky lores. Female very different from Verditer. Best identified by combination of large size, brownish-grey upperparts, uniform greyish underparts (lacking paler throat, with greyish-white centre of belly and dark buff undertail-coverts), and rufous-brown uppertail-coverts and tail. Juvenile has bold orange-buff spotting on scapulars and heavily scaled underparts; wings and tail as adult, with buff tips to coverts. **Voice** Rich, melodious thrush-like song, unlike other *Cyornis* with descending sequences, *chi, chuchichu-chuchichu-chucchi* usually ending with harsh *chizz*. **HH** Usually frequents middle and upper forest storeys; sometimes near ground. Pursues insects like typical flycatcher, but usually switches perches. Moist, dense broadleaved forest.

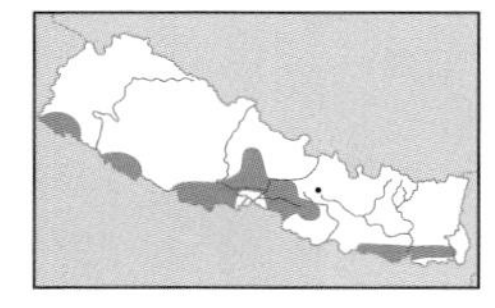

Pale-chinned Flycatcher *Cyornis poliogenys* 14 cm

Locally common resident; 75–455 m. **ID** Similar to females of other *Cyornis*, best identified by greyish crown and ear-coverts, prominent eye-ring, well-defined cream throat, and creamy-orange breast and flanks that grade into belly. **Voice** Song is a high-pitched series of 4–11 notes, slightly rising and falling, sometimes interspersed with harsh *tchut-tchut* notes; repeated *tik* call. **HH** Hawks insects like typical flycatcher. Forages in bushes and undergrowth, sometimes on ground where it resembles a chat. Open broadleaved forest. **AN** Pale-chinned Blue Flycatcher.

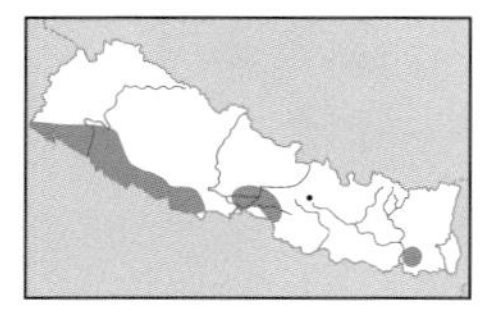

Tickell's Blue Flycatcher *Cyornis tickelliae* 14 cm

Local resident; 150–315 m. **ID** Male from Blue-throated by orange throat. Female has blue-grey cast to upperparts (especially rump and tail), orange breast, and white belly and flanks. **Voice** Song a short, metallic trill of 6–10 notes, the first half descending and second half ascending; *tick tick* call. **HH** Flits actively around bushes, undergrowth and low branches, hawking insects in mid-air. Frequents drier habitats than Blue-throated: open dry broadleaved forests.

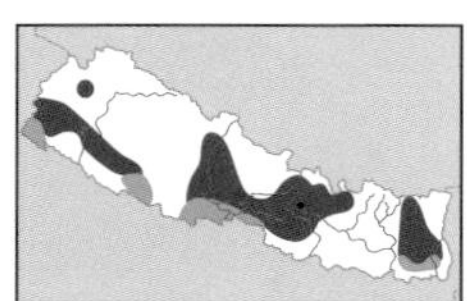

Blue-throated Blue Flycatcher *Cyornis rubeculoides* 14 cm

Partial migrant; uncommon summer visitor, 365–1,500 m (–2,135 m); rare in winter, 75–300 m. **ID** Male has blue throat (some have orange wedge) and well-defined white belly and flanks. Female has narrow and poorly defined creamy-orange throat, and orange breast well demarcated from white belly (compare Pale-chinned). Olive-brown head and upperparts and rufescent tail are best features from female Tickell's. **Voice** Short, sweet song recalls Tickell's but delivery is more rapid and higher-pitched with more trilling notes; calls include *click click* and *chr-r chr-r* alarm. **HH** Frequents middle-storey. Sallies after insects, but does not use regular perch. Open forests and groves.

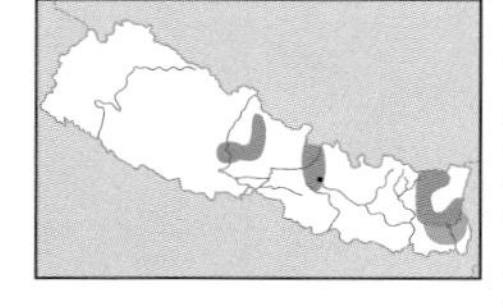

Pygmy Blue Flycatcher *Muscicapella hodgsoni* 10 cm

Rare resident; breeds 2,100–3,500 m, winters 75–3,500 m. **ID** Very small flycatcher with short tail and tiny bill, giving rise to flowerpecker-like appearance. Male has blue upperparts (with bright blue forecrown) and underparts almost entirely orange. Female and first-year male has olive-brown upperparts and orange-buff underparts. **Voice** Song a short, weak *tzzit-che-che-che-heeeee*; call a low *churr*. **HH** Found in canopy and dense bushes in lower forest storey. Very active. Hunts insects by making short sallies, searching leaves, hovering before flowers like a warbler and also dropping briefly to ground. Has curious habit of stretching head forward, dropping and spreading wings, and fanning tail. Dense, moist broadleaved forest.

♀
♂
Verditer Flycatcher
♂
♀
Pale Blue Flycatcher
ad
Pale-chinned Flycatcher
♀
♂
Tickell's Blue Flycatcher
♂
♀
Blue-throated Blue Flycatcher
♂
♀
Pygmy Blue Flycatcher

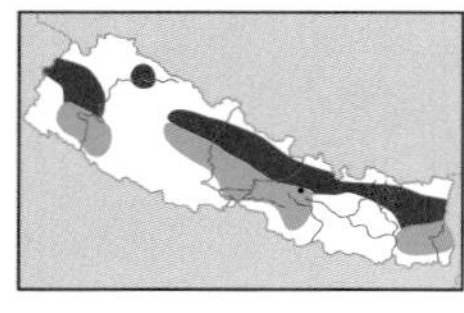

Rufous-bellied Niltava *Niltava sundara* 18 cm

Fairly common resident; breeds 2,135–3,200 m, winters 275–1,830 m (–75 m). **ID** Male has dark blue upperparts and orange underparts, with brilliant blue crown, neck patch, shoulder patch and rump. Female has well-defined oval-shaped throat patch. Also has small blue patch on side of neck (often difficult to see). **Voice** Includes a raspy *z-i-i-i-f-cha-chuk* perhaps song, hard *tic*, thin *see* and low, soft *cha...cha*. **HH** Usually perches quietly in bush, or branch low down in forest, occasionally darting out or dropping to ground to catch insects. Undergrowth in broadleaved forest and second growth.

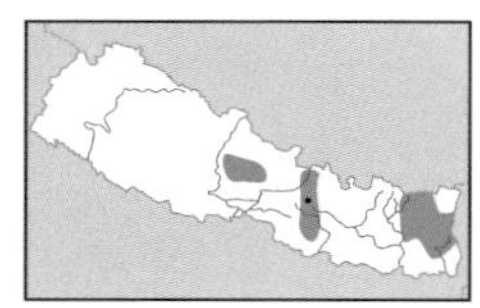

Large Niltava *Niltava grandis* 21 cm

Local, rare resident; 1,525–2,850 m. **ID** Very large, stocky niltava. Male dark blue (often appearing entirely black in poor light), with blackish face and tufted forehead. Brilliant blue crown, neck patch, shoulder patches and rump. Female has blue patch on side of neck (can be obscured), dark olive-brown upperparts with rufescent wings and tail, clearly defined (narrow) buff throat, and rufous-buff forecrown and lores. Lacks white patch on lower throat of female Rufous-bellied. **Voice** Melancholy song of 3–4 ascending whistles, *do-ray-me*, repeated slowly; harsh rattle and unobtrusive nasal *dju-ee*. **HH** Less active than most flycatchers; often uses one perch for long periods, occasionally flicking wings and tail. Dense, moist broadleaved forest, especially near streams.

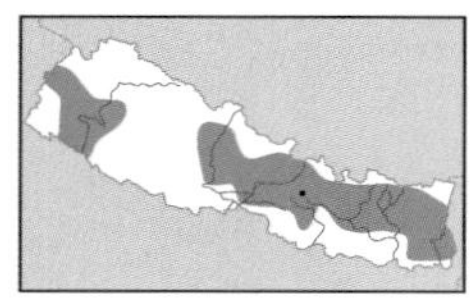

Small Niltava *Niltava macgrigoriae* 13 cm

Locally fairly common resident; breeds 275–2,200 m, up to 1,400 m in winter. **ID** Small size. Male dark blue, with brilliant blue forehead and neck patch. Female dusky-brown with indistinct blue neck patch and rufescent wings and tail; lacks oval throat patch of female Rufous-bellied. **Voice** Very thin, high-pitched song, *twee twee ee twee*, which rises then falls; calls include a high-pitched *see-see* (second note lower) and metallic scolding and churring notes. **HH** Active, but rather shy and elusive, especially when breeding. Partly crepuscular, keeping to shady undergrowth and bushes, flying out occasionally to catch insects. Usually frequents mid-storey of forests. Bushes at track edges, along streams, in forest clearings and broadleaved forests.

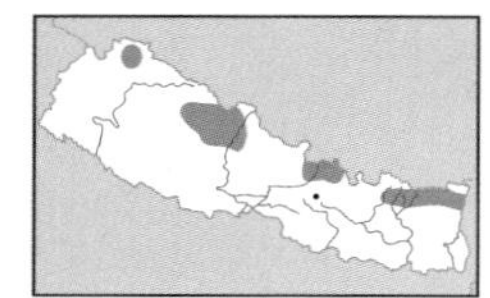

White-throated Dipper *Cinclus cinclus* 20 cm

Locally fairly common resident in north; mainly breeds 3,500–4,500 m (2,590–5,400 m). **ID** From Brown Dipper by white throat and breast contrasting with brown belly; also has brown head and nape merging into blackish-slate mantle, wings and tail. Juvenile from juvenile Brown Dipper by greyer coloration to upperparts, without prominent spotting, and by whiter underparts finely scaled. **Voice** Call an abrupt, rasping *jeet*; song a quiet mixture of grating and twittering notes. **HH** Aquatic; usually seen perched on rock in mid-stream, bobbing up and down. Generally occurs alone, in widely separated pairs or scattered family groups. Highly territorial in winter and summer. Regularly submerges to swim underwater or walk on streambed in search of invertebrates; after popping to surface, often floats downstream on half-spread wings before leaving water. Flies low over water surface on rapidly whirring wings. Broad and shallow (not cascading) streams and small rivers of upper valleys.

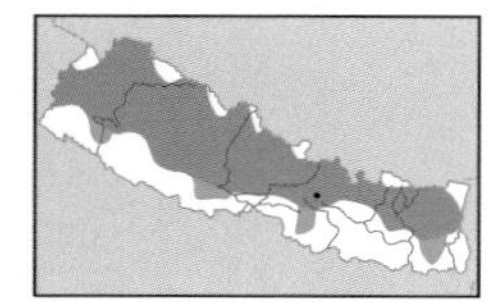

Brown Dipper *Cinclus pallasii* 20 cm

Common, widespread resident; breeds 1,525–4,960 m; winters 455–3,100 m. **ID** Adult entirely brown, lacking white throat and breast of White-throated. Juvenile from juvenile White-throated by browner coloration with conspicuous spotting on upperparts, extensive dark scaling on underparts and more prominent pale fringes to wing feathers. **Voice** Call an abrupt *dzit-dzit*, less harsh than White-throated; song stronger and richer. **HH** Very similar to White-throated. Freshwater streams and rivers ranging from shallow and slow-flowing streams to cascading large mountain rivers.

♂
Rufous-bellied Niltava
♀
♂
♀
Large Niltava
♂
♀
Small Niltava
ad
juv
White-throated Dipper
ad
juv
Brown Dipper

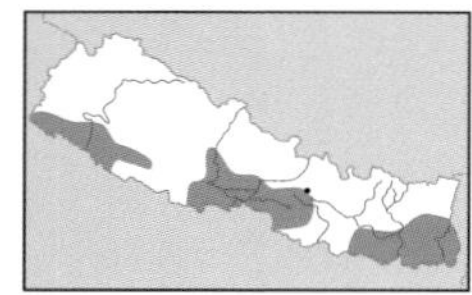

Golden-fronted Leafbird *Chloropsis aurifrons* 19 cm

Resident, fairly common in protected areas, frequent or uncommon elsewhere; chiefly 75–365 m, uncommon to 915 m (–2,285 m). **ID** Adult mainly green with golden-orange forehead (dull in some birds, especially females), purplish-blue throat and broad golden-yellow collar (especially pronounced on breast). Lacks orange on belly of Orange-bellied. Juvenile is all green, with diffuse yellowish patch on forecrown and hint of blue in moustachial region. **Voice** Song is a cheery series of rising and falling liquid chirps, bulbul-like in tone. Wide variety of harsh and whistled notes, also mimics other birds. **HH** Singly, in pairs or family parties according to season. Arboreal, typically inhabiting thick foliage in canopy. Searches leaves for insects, also feeds on berries and nectar. Broadleaved forest and second growth.

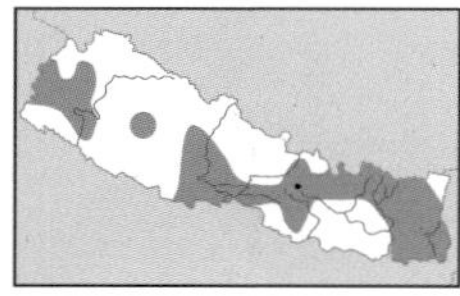

Orange-bellied Leafbird *Chloropsis hardwickii* 20 cm

Local and locally fairly common resident; mainly 1,300–2,135 m (250–2,750 m). **ID** Male striking with black of throat extending onto breast, orange belly and vent, large blue moustachial stripe, and purplish-blue flight feathers and tail. Female largely green, with orange centre of belly and vent, and blue moustachial stripe. Juvenile all green with blue moustachial and usually with hint of orange on underparts. **Voice** Wide variety of harsh and whistled notes, also mimics other birds. **HH** Habits similar to Golden-fronted. Broadleaved forest.

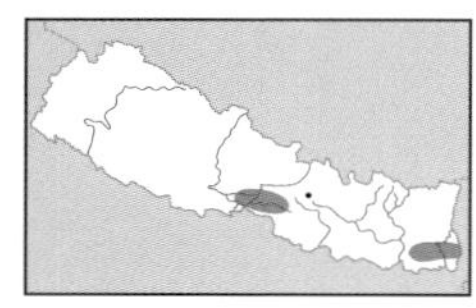

Little Spiderhunter *Arachnothera longirostra* 16 cm

Very rare and local resident; mainly in Chitwan district; 75–305 m. **ID** From Streaked Spiderhunter by smaller size and proportionately longer bill, unstreaked olive-green upperparts, whitish throat and breast merging into pale yellow rest of underparts, whitish crescents above and below eye, and dark moustachial stripe. **Voice** Rapidly repeated *wit-wit-wit-wit*....song; abrasive *itch* call. **HH** Singly or in pairs. Inhabits lower forest storey. Restless and noisy. Especially fond of wild banana flower nectar. Often clings upside-down while probing blossoms. Wild bananas and bamboos in moist broadleaved forest.

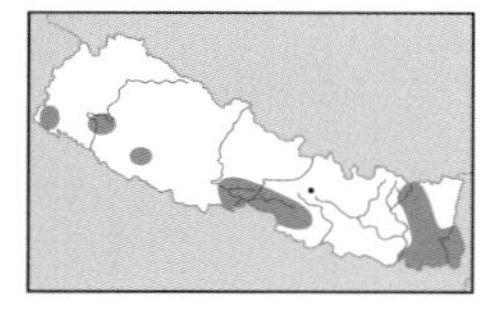

Streaked Spiderhunter *Arachnothera magna* 19 cm

Locally frequent resident in central and eastern Nepal; mainly 250–450 m (120–2,135 m). **ID** From Little by larger size, bold streaking on dark olive-green upperparts, and bold streaking on yellowish-white underparts. Orange legs and feet. **Voice** Strident chattering song; sharp *chirirrik* or *chirik chirik* call. **HH** Usually singly or in pairs, often with itinerant foraging flocks. Forages in upper storey. Fast-moving, flies strongly and swiftly from one tree to next. Very fond of nectar of wild bananas. Moist broadleaved forest with dense undergrowth.

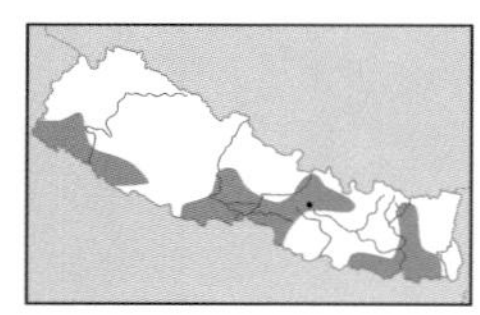

Thick-billed Flowerpecker *Dicaeum agile* 10 cm

Widespread resident, frequent in far west, uncommon to rare further east; 75–800 m all year, summers up to 2,135 m. **ID** From other flowerpeckers by combination of stout bluish-grey bill, indistinct dark malar stripe, lightly streaked breast, comparatively long and broad, fairly dark tail with white tip (can be rather indistinct), and orange-red iris. Juvenile has pinkish bill. **Voice** A *tchup-tchup* call, not as hard as *chick* call of Pale-billed. **HH** Distinctive habit of jerking tail from side to side as it feeds or moves. Arboreal and highly active, feeding mainly on figs of peepul and banyan, also on mistletoe berries. Broadleaved forest and well-wooded country.

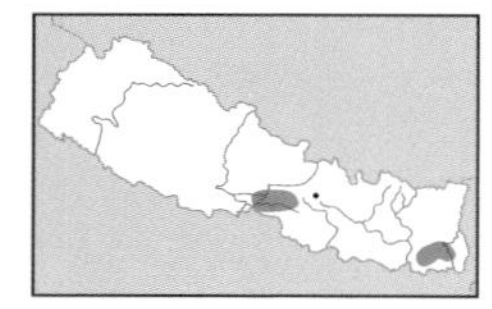

Yellow-vented Flowerpecker *Dicaeum chrysorrheum* 10 cm

Very rare and very local resident; mainly Churia Hills, Chitwan National Park; 245–400 m. **ID** From other flowerpeckers by blackish streaking on white or yellowish-white underparts and orange-yellow vent. Also has blackish malar stripe, curved black bill, whitish supercilium, red eye, bright olive-green upperparts with contrasting blackish primaries, and blackish tail. Juvenile has duller upperparts, and greyish-white underparts with paler yellow vent and less prominent streaking. **Voice** Short distinctive *dzeep*. **HH** Arboreal, especially fond of mistletoe berries. Like other flowerpeckers very active, continually flying about restlessly, twisting and turning in different attitudes when perched, calling frequently. Open forest and forest edges.

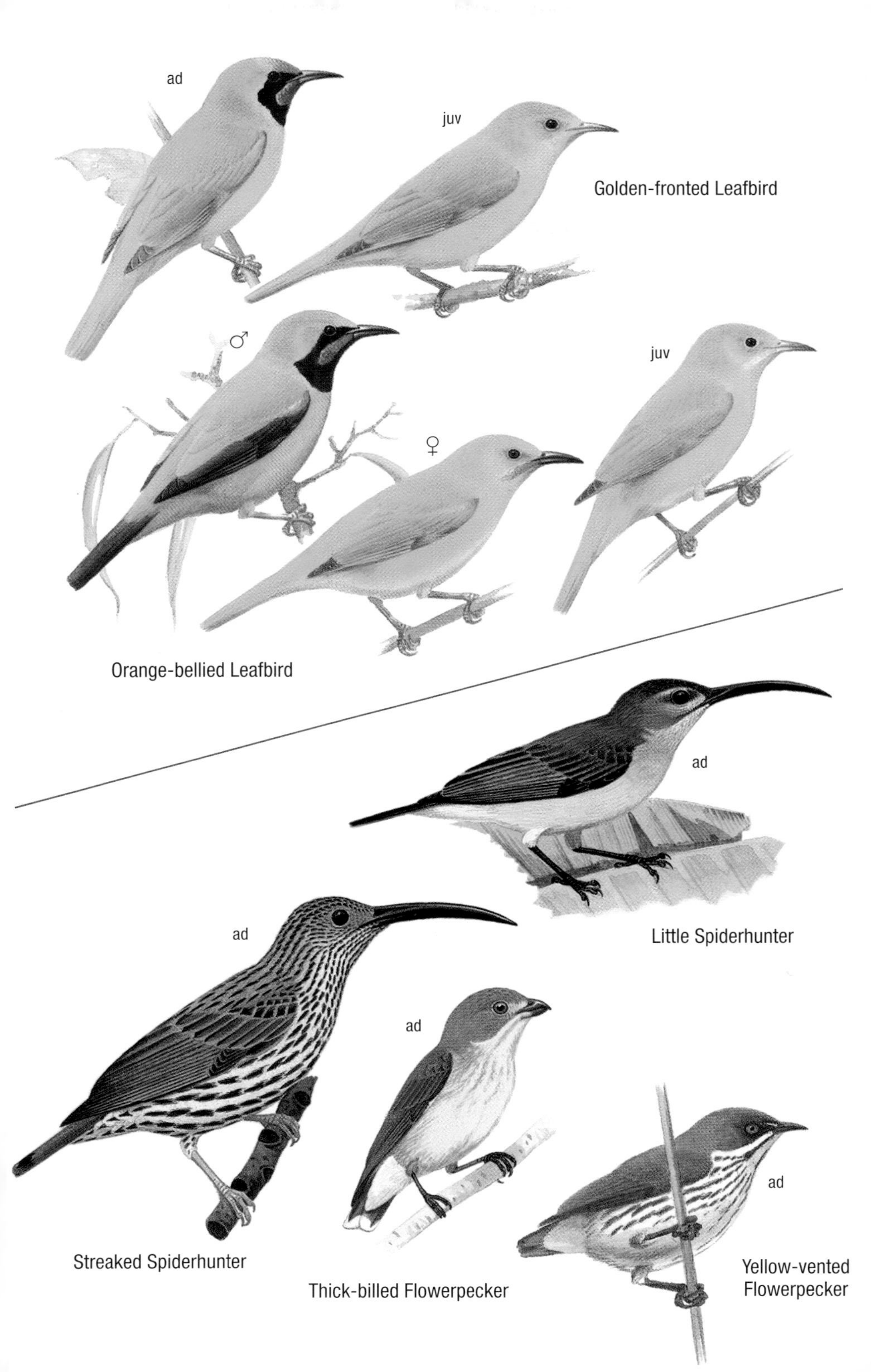
ad
juv
Golden-fronted Leafbird
♂
juv
♀
Orange-bellied Leafbird
ad
Little Spiderhunter
ad
ad
ad
Streaked Spiderhunter
Thick-billed Flowerpecker
Yellow-vented
Flowerpecker

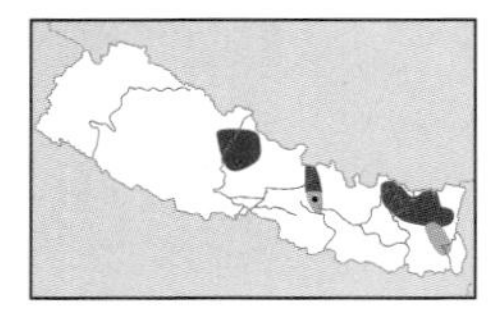

Yellow-bellied Flowerpecker *Dicaeum melanoxanthum* 13 cm

Local, uncommon resident, mainly from west-central Nepal east; breeds 2,350–3,000 m, winters from 1,050 m to at least 1,550 m. **ID** Large, stout-billed flowerpecker. White spots at tip of undertail. Male has bluish-black upperparts and breast-sides, white centre of throat and breast, and yellow rest of underparts. Bright red eye. Female is dull version of male, with olive-brown upperparts, olive-grey sides of breast, and dull olive-yellow belly and vent. Juvenile male similar to female, but has brighter yellow underparts and blue-black cast to mantle and back. **Voice** Agitated *zit-zit-zit-zit* call. **HH** Typical flowerpecker habits, see Fire-breasted, but less active. Broadleaved forest.

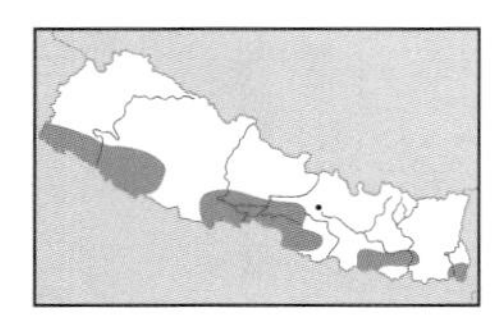

Pale-billed Flowerpecker *Dicaeum erythrorhynchos* 8 cm

Quite widespread resident, fairly common in Chitwan National Park and Parsa Wildlife Reserve, generally uncommon elsewhere; 75–305 m. **ID** From Plain Flowerpecker by pinkish bill. Very plain; greyish-olive upperparts with slight greenish cast, and pale greyish underparts with variable yellowish-buff wash. **Voice** Hurried chittering song is lower pitched than Thick-billed; sharp *chik chik chik* call. **HH** In pairs in breeding season; otherwise in small parties. Constantly on move, all the while calling. Usually in canopy. Strong bounding and dipping flight. Feeds chiefly on mistletoe berries. Open broadleaved forest and well-wooded areas.

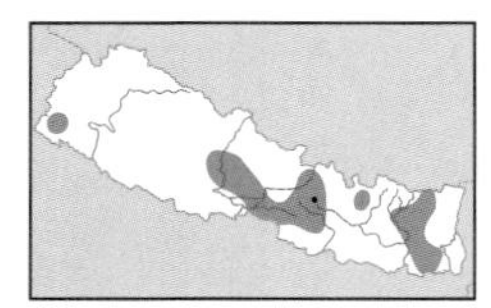

Plain Flowerpecker *Dicaeum concolor* 8.5 cm

Locally frequent resident; mainly 305–1,525 m (150–2,500 m). **ID** From Pale-billed by fine dark bill (with paler base to lower mandible). Olive-green upperparts and edges to flight feathers, and dusky greyish-olive underparts with yellow on throat and belly are further differences from Pale-billed; juvenile has browner upperparts and greyish-white underparts. **Voice** Repeated staccato *tzik*, more piercing than Pale-billed call. Distinctive *tzierr* is repeated as song. **HH** Extremely active and restless, usually in pairs or small parties. Very fond of mistletoe berries, also takes nectar, insects and spiders. Edges and clearings of broadleaved forests, also well-wooded areas. **TN** Often treated as a separate species, *D. minullum*, from Nilgiri Flowerpecker *D. concolor*.

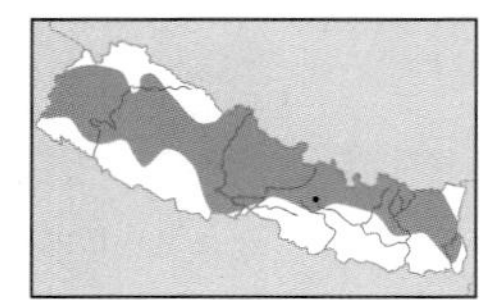

Fire-breasted Flowerpecker *Dicaeum ignipectus* 9 cm

Common and widespread resident; breeds 1,830–2,700 m (915–3,565 m). **ID** Male has dark metallic blue or green upperparts, buff-coloured underparts, scarlet breast patch, and black centre of belly. Female has olive-green upperparts and orange-buff underparts with olive breast-sides and flanks. Juvenile has whiter throat merging into pale greyish-olive of underparts, and duller olive upperparts than female (more like Plain, but lacks supercilium and pale ear-coverts, and is usually at higher elevation). **Voice** Shrill *titty-titty-titty* song; clicking *chip* call. **HH** Typical flowerpecker. Solitary or in pairs in breeding season, occasionally in small parties at other times. Favours mistletoe berries. Arboreal and frequents canopy. Very active, continually flying about restlessly and calling frequently. Broadleaved forest and second growth.

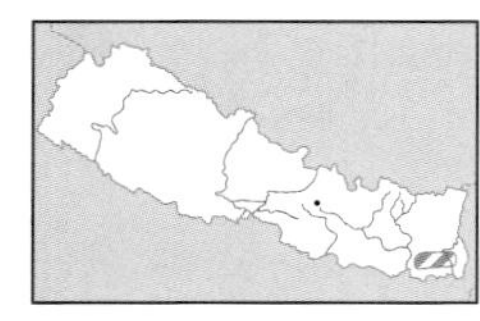

Scarlet-backed Flowerpecker *Dicaeum cruentatum* 9 cm

Probably a former resident, no recent records; two old records from far east; 305 m, 2,135 m. **ID** Male has scarlet upperparts and black 'sides' to whitish underparts. Female has scarlet rump contrasting with blackish tail, olive-brown upperparts and faint buffish wash to whitish underparts. Juvenile as female, but lacks scarlet rump (usually with hint of orange); bright orange-red bill, and fine whitish supercilium to eye. **Voice** Thin, repeated *tissit, tissit*....song; hard, metallic *tip..tip..tip* etc. **HH** Typical flowerpecker habits, see Fire-breasted. Broadleaved forest and second growth.

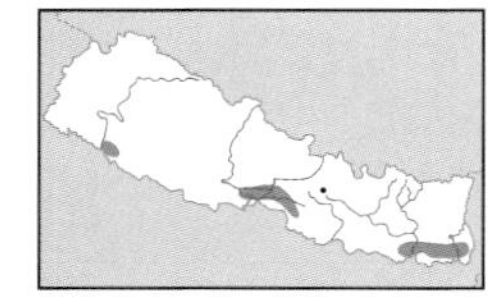

Ruby-cheeked Sunbird *Chalcoparia singalensis* 11 cm

Rare and very local resident; now mainly Chitwan National Park, also far east; 100–455 m. **ID** Shorter, straighter bill than other sunbirds, with rufous-orange throat and yellow underparts. Male has metallic green upperparts and 'ruby' cheeks. Female lacks 'ruby' cheeks and has dull olive-green upperparts. Juvenile entirely yellow below. **Voice** Disyllabic *wee-eeast* with rising inflection. Probable song is a rapid, high-pitched *switi-ti-chi-chu...tusi-tit...swit-swit...switi-ti-chi-chu...switi-ti-chi-chu*. **HH** Small active sunbird, continually flitting about on low branches and bushes. Unlike other sunbirds, sometimes forms small parties in winter. Open forest, forest edge; favours evergreen biotope. **TN** Formerly placed in genus *Anthreptes*.

Yellow-bellied Flowerpecker

Pale-billed Flowerpecker

Plain Flowerpecker

Fire-breasted Flowerpecker

Scarlet-backed Flowerpecker

Ruby-cheeked Sunbird

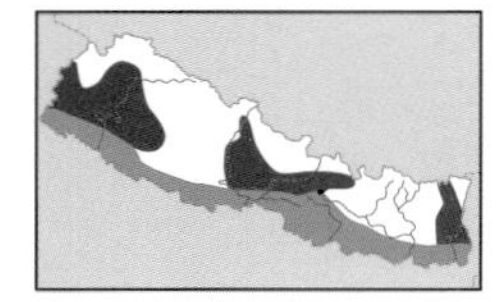

Purple Sunbird *Cinnyris asiaticus* 10 cm

Common, widespread resident to at least 365 m, mainly a summer visitor 900–2,135 m. **ID** Male is metallic blue-green and purple becoming blacker on belly and vent. Female has uniform yellowish underparts, with faint supercilium and darker mask (whiter below in worn plumage). Eclipse male as female, but has broad blackish stripe down centre of throat and breast, metallic blue wing-coverts and glossy black wings and tail. Juvenile is brighter yellow on entire underparts than female. **Voice** Pleasant descending *swee-swee-swee swit zizi-zizi* song; buzzing *zit* and high-pitched upward-inflected and slightly wheezy *swee* or *che-wee*. **HH** Typical sunbird, see Green-tailed, but probably more insectivorous than other species. Sometimes makes short aerial sallies like a flycatcher. Male sings from bare treetop while jerking from side to side and raising its wings to reveal bright orange pectoral tufts. Open deciduous forest and gardens. **TN** Formerly placed in genus *Nectarinia*.

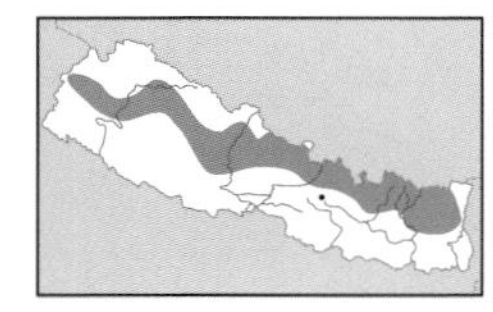

Mrs Gould's Sunbird *Aethopyga gouldiae* 10 cm

Uncommon, widespread resident; breeds chiefly 2,500–3,655 m, winters from 1,830 m to at least 2,700 m. **ID** Male has metallic purplish-blue crown, ear-coverts and throat; crimson sides of neck, mantle and back (reaching yellow rump); yellow belly, and blue tail. Female has pale yellow rump-band, yellow belly, short bill and prominent white on tail. Juvenile male similar to female, but has bright yellow breast and belly. **Voice** Quick-repeated *tzip* call; *tshi-stshi-ti-ti-ti* and a lisping *squeeeeee* that rises in middle in alarm. **HH** Typical sunbird, see Green-tailed, but forages at all levels from bushes up to canopy. Rhododendron and other flowering forest trees and shrubs.

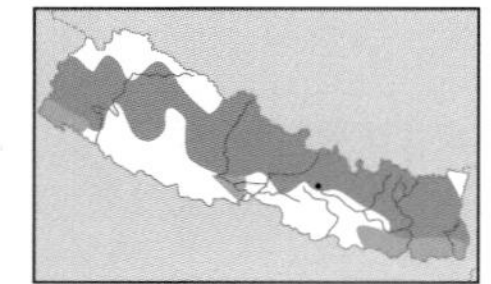

Green-tailed Sunbird *Aethopyga nipalensis* 11 cm

Common, widespread resident; breeds chiefly 1,830–3,000 m (–3,505 m), winters 915–2,745 m (–305 m). **ID** Male from Mrs Gould's by maroon mantle and olive-green back, dark metallic blue-green crown and throat, and blackish sides of head. Blue-green uppertail-coverts and tail (can appear blue, but not purplish-blue as in Mrs Gould's). Female has greyish-olive throat and breast (very grey in some), becoming yellowish-olive on belly and flanks. Lacks well-defined yellow rump-band (although rump and uppertail-coverts are yellowish-green). Two intergrading races occur: *A. n. horsfieldi* (in west) has only narrow maroon band on mantle compared to nominate of centre and east. **Voice** Song *tchiss....tchiss-iss-iss-iss*; loud *chit chit* call. **HH** Typical sunbird, though often associates with foraging flocks in non-breeding season. Arboreal and feeds mainly on blossoms of flowering trees and shrubs. Flits and darts actively from flower to flower, clambering over blossoms, often hovering momentarily, and clinging acrobatically to twigs. Oak–rhododendron and mixed forest and second growth.

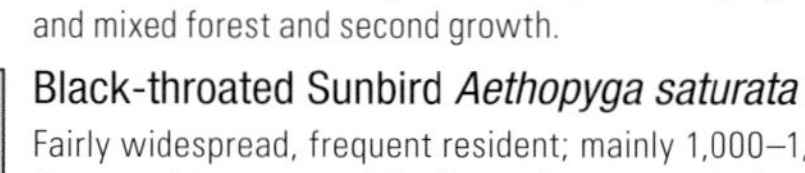

Black-throated Sunbird *Aethopyga saturata* 11 cm

Fairly widespread, frequent resident; mainly 1,000–1,830 m (305–2,200 m). **ID** Male has black throat and breast, greyish-olive underparts and crimson mantle. Female has pale yellow rump-band; from Mrs Gould's by greyish-olive underparts (without yellow); longer, dark and noticeably downcurved bill, and narrow or indistinct pale tips to outer tail feathers. **Voice** Rapid *ti-ti-ti-ti-ti-ti-ti-ti-ti-ti-ti*. **HH** Typical sunbird, see Green-tailed. Bushes in open forest, at dense forest edges and in second growth.

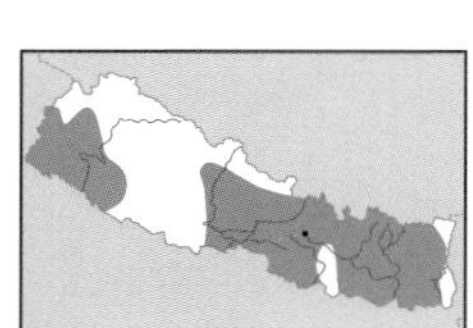

Crimson Sunbird *Aethopyga siparaja* 11 cm

Widespread resident; locally fairly common 75–915 m, frequent up to 1,200 m (–2,100 m). **ID** Male has crimson mantle, scarlet throat and breast, and yellowish-olive belly. Female has yellowish-olive underparts; lacks yellow rump or prominent white on tail. Immature male as female but has red throat and breast. **Voice** Rapid, tripping song of 3–6 sharp, clear notes, *tsip-it-sip-it-sit*. **HH** Typical sunbird, see Green-tailed. Especially fond of red flowers. Forages mainly low down in bushes. Bushes in open broadleaved forest, groves and gardens.

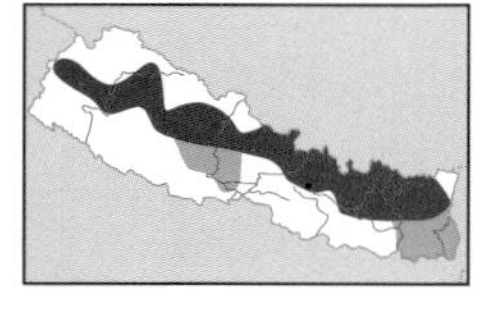

Fire-tailed Sunbird *Aethopyga ignicauda* 12 cm

Fairly common, widespread resident; breeds mainly 3,000–4,000 m, winters chiefly 1,050–2,135 m (610–2,895 m). **ID** Male has scarlet nape and mantle, and very long scarlet tail. Female similar to female Green-tailed, but has straighter bill, squarer tail (lacking white tips) with trace of brownish-orange at sides, and more noticeable olive-yellow on rump (not forming prominent band). Eclipse male similar to female, but has brighter yellow belly, and scarlet uppertail-coverts and tail-sides. **Voice** High-pitched monotonous *dzidzi-dzidzidzidzi* song. **HH** Typical sunbird, see Green-tailed, but especially vivacious. In breeding season, male's long red tail swirls in flight. Breeds in rhododendron shrubberies; winters in broadleaved and mixed forests.

♂ eclipse
♀
juv
♂ br
Purple Sunbird
♂
♀
Mrs Gould's Sunbird
♀
♂
nipalensis
♀
♂
Black-throated Sunbird
Green-tailed Sunbird
♀
♂ non-br
♂ br
♂
♀
Fire-tailed Sunbird
Crimson Sunbird

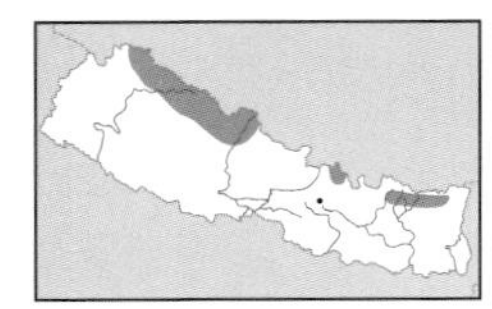

Tibetan Snowfinch ***Montifringilla adamsi*** **17 cm**

Common resident in far north, local winter visitor; breeds 4,200–5,400 m, winters 4,000–4,930 m (–2,530 m). **ID** Adult from other snowfinches in region by largely white greater and median coverts, and white fringes to inner secondaries and outermost tertial (forming broad white panel). Additional features are whitish-tipped black throat with plain grey-brown lores and forehead, and dull grey-brown coloration to head and upperparts. Adult non-breeding has pale (rather than blackish) bill. Juvenile lacks black throat, is more buffish in coloration, with buffish (rather than white) sides to tail. From other juvenile snowfinches by buffish-white wing panel. **Voice** Rapid sparrow-like, staccato chattering song usually delivered from prominent perch. **HH** Rocky, high-altitude semi-desert, often near villages and upland cultivation.

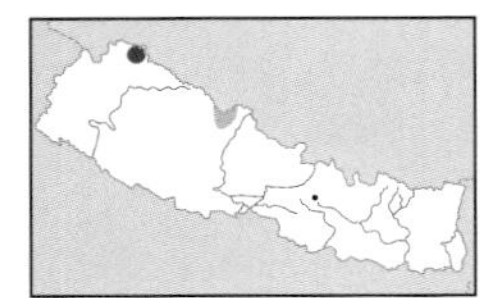

White-rumped Snowfinch ***Onychostruthus taczanowskii*** **17 cm**

Very rare; in far north; 4,815–5,400 m (–2,530 m). **ID** Adult from other snowfinches by white rump (very conspicuous in flight) and pale greyish coloration to upperparts (with prominently streaked mantle and scapulars). Additional features are black lores, white throat, white forehead and supercilium, white sides to tail (lacking dark terminal bar), and diffuse white panel at base of secondaries and inner primaries (with secondaries broadly tipped whitish). Bill pale yellowish-horn with darker tip. Juvenile has warmer brown mantle and wings, buffish breast and flanks, and buff sides to tail. **Voice** Simple weak song of wheezes, whistles and clipped notes delivered in a circling undulating song flight. **HH** Open stony Tibetan steppe. **TN** Formerly placed in genus *Pyrgilauda*.

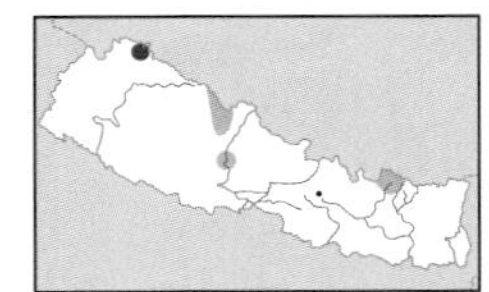

Rufous-necked Snowfinch ***Pyrgilauda ruficollis*** **15 cm**

Very rare; 3,290–4,850 m. **ID** Adult from Blanford's Snowfinch by greyish-white forehead (lacking blackish centre), fine black malar stripe, white throat, and conspicuous streaking on mantle and scapulars. White patch in median coverts (visible in flight) and broad buffish greater coverts wing-bar. Female has duller rufous 'neck' and less white in wing. Juvenile duller, with buffish tinge to breast and flanks, warm cinnamon tinge to ear-coverts and neck-sides, and buffish median and greater coverts wing-bars. From juvenile Blanford's by dark malar stripe and streaked mantle and scapulars. **Voice** Song is an erratic repetition of simple sparrow-like notes, *dishu-tchelu-tischu-delu* etc. **HH** Open stony areas and short grassland in Tibetan steppe country.

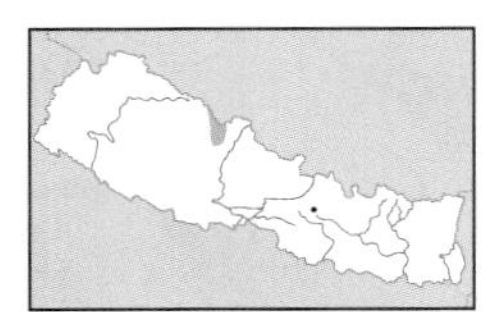

Blanford's Snowfinch ***Pyrgilauda blanfordi*** **15 cm**

Very rare; 4,250–5,350 m. **ID** Adult from Rufous-necked by black centre to white forehead, black 'spur' in front of eye dividing white supercilium, and black chin and centre of throat. Rather uniform wing-coverts (lacking prominent wing-bars), white panel in secondaries, and mantle and scapulars are unstreaked (or, rarely, very faintly streaked). Also has stouter bill. Juvenile from juvenile Rufous-necked by unstreaked mantle and scapulars. **Voice** Rapid twittering call. **HH** Tibetan steppe country. **AN** Plain-backed Snowfinch.

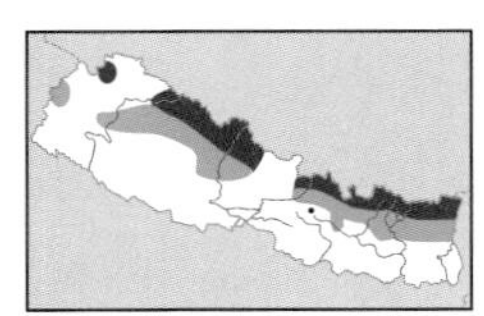

Plain Mountain Finch ***Leucosticte nemoricola*** **15 cm**

Common, widespread resident; breeds 4,200–5,250 m; winters 2,000–3,900 m (–1,300 m). **ID** Mantle boldly streaked with pale 'braces'. Median coverts dark-centred, with bold white fringes, and greater coverts tipped white (forming well-defined wing-bar) with variable dark central panel. Unstreaked grey rump contrasts with mantle/back, and prominent white tips to uppertail-coverts. Tertials and inner secondaries edged white, forming narrow panel on closed wing, but less prominent than on Brandt's Mountain Finch. Juvenile is warmer rufous-buff on head, mantle and underparts than adult, and mantle is less heavily streaked; rufous-buff fringes to tertials and tips to coverts. First-winter retains rufous on head. **Voice** Soft twittering *chi-chi-chi-chi*; sharp twitter *rick-pi-vitt* song. **HH** Breeds in high-altitude steppe. Winters in open forest and upland cultivation.

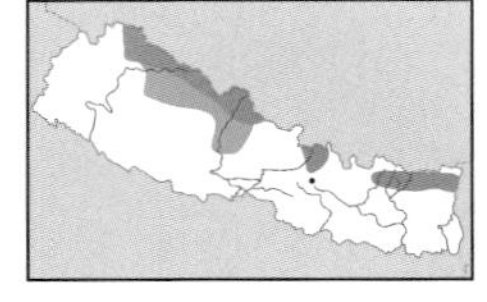

Brandt's Mountain Finch ***Leucosticte brandti*** **16.5–19 cm**

Frequent resident in arid north; mainly 4,200–5,250 m (2,350–6,000 m). **ID** From Plain in all plumages by unstreaked to lightly streaked mantle, and rather pale and comparatively uniform wing-coverts. Broad white edges to primary-coverts and tertials/secondaries, forming more striking white panel on wing than Plain, with more prominent white edges to outer rectrices. Adult breeding has sooty-black head and nape; male has pink on rump (indistinct on female). Less black on head in non-breeding plumage. Juvenile and first-summer have buffish head and mantle, and tertials and wing-coverts have warm buff edgings. **Voice** Loud *twit-twitt, tweet-ti-ti* or *peek-peek* or harsh *churr.* **HH** High-altitude steppe.

ad
Tibetan Snowfinch
juv
ad
White-rumped Snowfinch
ad
Rufous-necked Snowfinch
ad
Blanford's Snowfinch
ad
1st-sum
juv
♂ br
Plain Mountain Finch
Brandt's Mountain Finch

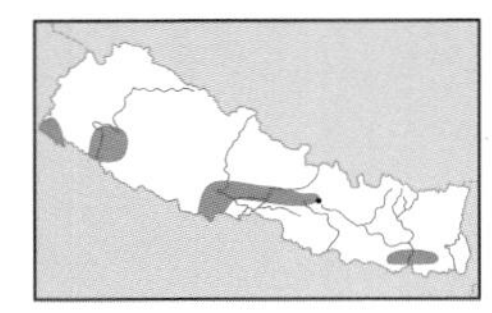

Red Avadavat *Amandava amandava* 10 cm

Local resident; 75–305 m (–1,380 m). **ID** Breeding male mainly red with irregular white spotting. Non-breeding male and female have grey-brown upperparts and buffish-white underparts; best identified by red bill, red rump, and white tips to wing-coverts and tertials. Juvenile lacks red in plumage; buff wing-bars and tertial fringes, pink bill base, and pink legs and feet separate from juvenile munias. **Voice** Weak, high-pitched warbling song; calls include a thin *teei* and variety of high-pitched chirps and squeaks. **HH** Forages for seeds on ground or by clinging to stems and pulling seeds directly from heads. Gregarious outside breeding season and roosts communally. If disturbed, typically fly up together in close-knit flock and move off with fast, whirring wingbeats to nearby cover. Flight of individuals is undulating, but flocks maintain a fairly direct course. Tall grassland, reedbeds and scrub near cultivation.

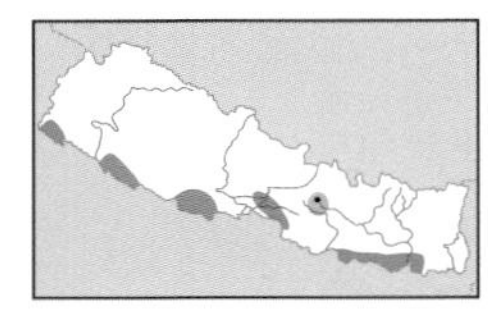

Indian Silverbill *Euodice malabarica* 11–11.5 cm

Local resident; 75–305 m. **ID** Male has fawn-brown upperparts, whitish face and underparts with barred flanks, long and pointed black tail, and white rump and uppertail-coverts. Female duller with plainer face and flanks are less barred. Juvenile lacks barring on flanks, has dark mottling on rump, and tail shorter and more rounded. **Voice** Contact call *tchrip!* or *tchreep!*; repeated *chir-rup!* flight call; song a series of short, abrupt trills. **HH** Habits similar to Red Avadavat. In non-breeding season, roosts communally in old nests, of weavers or their own. Prefers drier habitats than other estrildid finches in Nepal: dry cultivation, grassland and thorn scrub. **TN** Formerly placed in genus *Lonchura*.

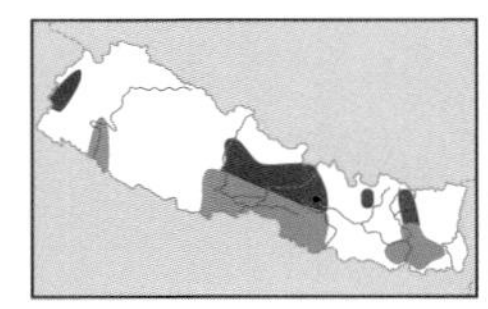

White-rumped Munia *Lonchura striata* 10–11 cm

Local resident; breeds 75–2,135 m; non-breeding season 75–1,220 m. **ID** Dark breast, streaked upperparts and white rump are best features. Furthermore, has rufous-brown on side of head/neck, rufous-brown to whitish fringes on dark brown breast, and faint brownish streaking to greyish-buff belly; juvenile barred brown-buff on throat and breast. **Voice** Song is a rising and falling series of twittering notes; calls include twittering *tr-tr-tr*, *prrit* and *brrt*. **HH** Habits similar to Red Avadavat. Open wooded areas and scrub.

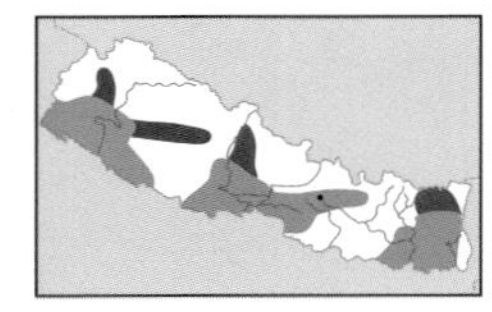

Scaly-breasted Munia *Lonchura punctulata* 10.7–12 cm

Locally common widespread resident; 75–1,525 m (–2,680 m). **ID** Adult has chestnut-brown face, throat and upper breast, whitish underparts boldly scaled black, and olive-yellow to rufous-orange on uppertail-coverts and edges of tail. Juvenile has uniform brown upperparts and buff to rufous-buff underparts, with whitish belly (probably indistinguishable from juvenile Black-headed). **Voice** Typical song a series of *klik-klik-klik* or *tit-tit-tit* notes followed by short series of whistles and churrs, ending with longer *weeee*; contact calls include repeated *tit-ti tit-ti* and loud *kit-teee kit-teee*. **HH** Habits similar to Red Avadavat. Often roosts in nests, either those used previously for breeding or nests built for roosting. Open secondary forest, bushes and cultivation.

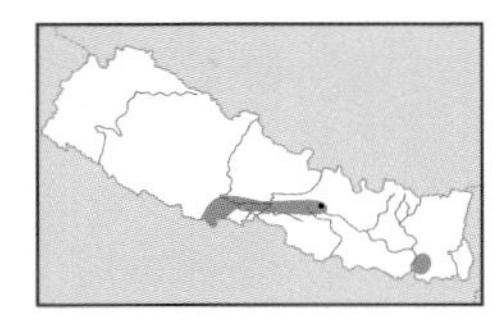

Black-headed Munia *Lonchura malacca* 11.5 cm

Introduced species, local resident; 75–1,220 m (–1,370 m). **ID** Adult has black head and upper breast, rufous-brown upperparts, and black belly centre and undertail-coverts. Lower breast and flanks white. Juvenile has uniform brown upperparts and buff to whitish underparts. **Voice** Song includes thin, evenly rising, high, whining nasal whistle; calls include a nasal, slightly downturned, abrupt squeaky-toy *nyek, nyek...*. **HH** Habits similar to Red Avadavat. Marshes, tall grassland and cultivation. **AN** Tricoloured Munia.

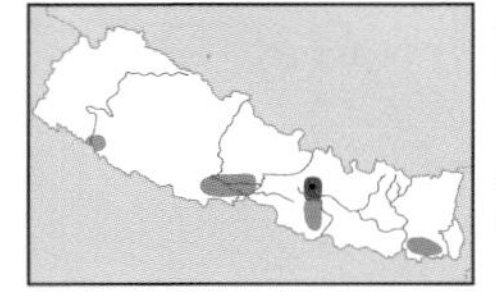

Chestnut Munia *Lonchura* (*malacca*) *atricapilla* 11 cm

Rare and local resident; 75–1,220 m (–1,370 m). **ID** Similar to Black-headed but has chestnut lower breast and flanks. Juvenile warmer buff on underparts than Black-headed. **Voice** Varied, short, squeaky, nasal flight calls. **HH** Habits similar to Red Avadavat. Grassland, marshes and cultivation. **TN** Treated here as conspecific with Black-headed Munia *L. malacca*.

♂ br
juv
♀
Red Avadavat
juv
ad
Indian Silverbill
ad
White-rumped Munia
ad
juv
Scaly-breasted Munia
ad
juv
Black-headed Munia
ad
Chestnut Munia

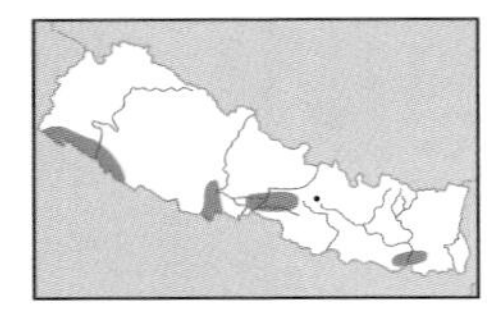

Black-breasted Weaver *Ploceus benghalensis* 14 cm

Local resident; 75–245 m. **ID** Breeding male has yellow crown and black breast. Throat can be black or white, and some variants have white ear-coverts and throat. In non-breeding male, female and juvenile plumages, breast-band blotchy or restricted to small patches at sides, and may show indistinct, diffuse streaking on lower breast and flanks. In these plumages, has yellow supercilium (often white behind eye), distinct yellow patch on side of neck, and yellow submoustachial stripe (with black malar); similar to Streaked Weaver, except crown, nape and ear-coverts more uniform; rump also indistinctly streaked and, like nape, contrasts with heavily streaked mantle. **Voice** Soft, barely audible *tsi tse tsisik tsisik tsik tsik tsik* song; soft *chit-chit* call. **HH** Habits very similar to Baya Weaver. Nest similar to Baya, but top dome is interwoven into a number of standing reed stems and vertical entrance tube is shorter. Tall, moist, seasonally flooded grassland and reedy marshes.

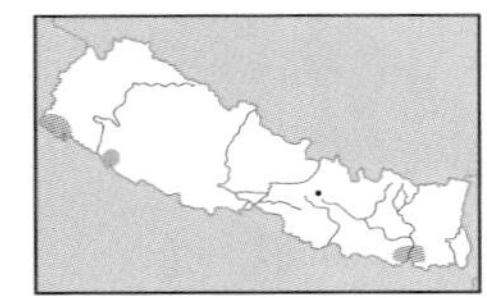

Streaked Weaver *Ploceus manyar* 14 cm

Rare and local, probably a migrant; 75–200 m. **ID** Breeding male has yellow crown, dark brown head-sides and throat, and heavily streaked breast and flanks. Other plumages typically show boldly streaked underparts. However, can be only lightly streaked on underparts, especially juvenile, when best separated from Baya Weaver by combination of yellow supercilium and neck patch, heavily streaked crown, dark or heavily streaked ear-coverts, and pronounced dark malar and moustachial stripes. If streaking is absent on underparts, streaked crown, nape and rump are best features from Black-breasted. **Voice** Soft, continuous trill, *see-see-see-see-see* ending in *o-chee* constitutes song; Loud *chirt chirt* call. **HH** Habits very similar to Baya. In winter chiefly feeds on flowering heads of *Phragmites*, bulrushes and *Saccharum* grasses by clinging to upright stems. Reedy marshes.

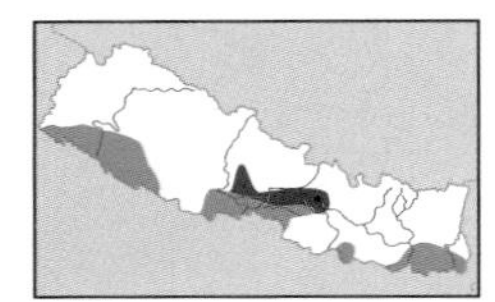

Baya Weaver *Ploceus philippinus* 15 cm

Locally common resident; 75–1,370 m. **ID** Breeding male nominate has yellow crown, dark brown ear-coverts and throat, unstreaked yellow breast, and yellow streaking on mantle and scapulars. Breeding male *burmanicus* (in east) has greyer face, buff or pale grey throat, and buff breast. Non-breeding male, female and juvenile usually have unstreaked buffish underparts; streaking can be as prominent as on poorly marked Streaked, but has less distinct and buffish supercilium, lacks yellow neck patch, and lacks pronounced dark moustachial and malar stripes. Non-breeding male, female and juvenile *burmanicus* more rufous-buff on supercilium and underparts. **Voice** Song a soft *chit chit chit* followed by a drawn-out wheezy whistle, *chec-ee-ee*; *chit-chit-chit* call. **HH** Highly gregarious throughout year. Feeds extensively on ripening cereal crops and insect crop pests. Forages by hopping on ground and picking seeds from tops of upright grass stems. Roosts communally all year. Colonial breeder, building a retort-shaped nest with long vertical entrance tube, suspended from a branch usually of an isolated tree. Open country near water with scattered bushes and tall trees, and cultivation.

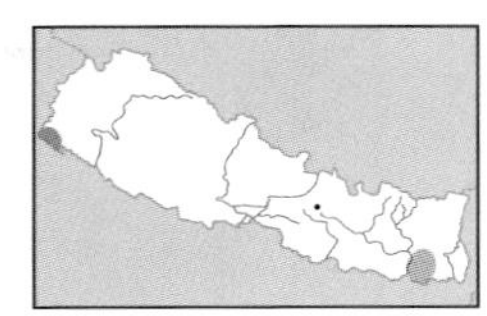

Finn's Weaver *Ploceus megarhynchus* 17 cm

Very local resident and summer visitor in far west, very rare migrant in far east; 75–120 m. **ID** Large weaver with heavy bill and long tail. Male breeding from other weavers by bright yellow head with dark-brown ear-coverts, golden-yellow underparts, and yellow rump and uppertail-coverts. Mantle and back boldly streaked dark brown. Dark patches on breast, can show as complete breast-band. Female breeding and first-year male have pale yellow to yellowish-brown head, and pale yellow to buffish-white underparts; mantle rich brown with dark streaking. Adult non-breeding lacks yellow and is similar to Baya Weaver; upperparts darker grey-brown with head more uniform. **Voice** Song louder and harsher than Baya, *twit-twit-tit-t-t-t-t-trrrrrr*. **HH** Habits similar to Baya. Nests colonially; nest a large untidy ball supported by twigs and reed stems; nests often linked together. Grassland with scattered trees. Globally threatened.

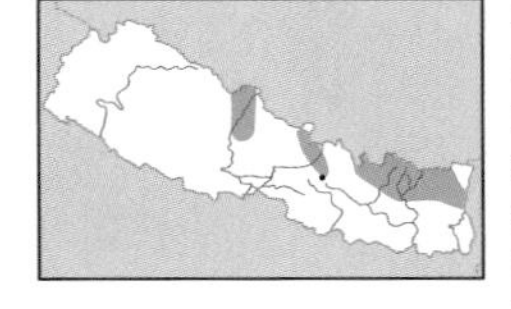

Maroon-backed Accentor *Prunella immaculata* 16 cm

West-central Nepal eastwards, locally frequent winter visitor, 1,830–2,700 m (–4,400 m). **ID** Adult has grey head and breast, white scaling on forehead, yellow iris, maroon-brown mantle and grey panel on wing. Juvenile has similar wing pattern to adult, but has streaked upperparts and underparts. **Voice** Very high and feeble *tzit* call, often doubled. **HH** In small parties in winter. Forages inconspicuously on ground. Moist forest of rhododendrons or mixed conifers/rhododendrons; forest clearings and edges, and edges of terraced fields at forest margins.

♂ br
non-br
non-br
Black-breasted Weaver
♂ br
♂ non-br
♂ br
philippinus
♀
philippinus
♀
Streaked Weaver
♂ br
burmanicus
♂ non-br
Baya Weaver
♂ br
♀ br
Finn's Weaver
non-br
ad
Maroon-backed Accentor

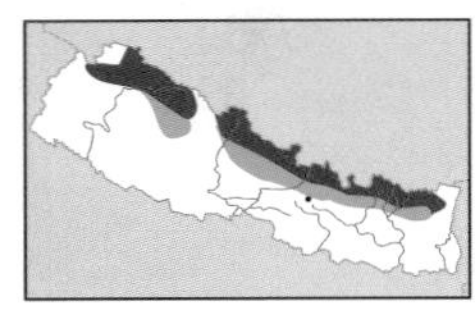

Alpine Accentor *Prunella collaris* 15.5–17 cm

Fairly common resident and winter visitor; breeds 4,200–5,500 m (–7,900 m), winters 2,440–3,795 m. **ID** From smaller Altai by comparatively uniform grey head and breast, diffusely streaked mantle, and (white-tipped) blackish coverts that form dark panel on wing. Chestnut streaking on flanks but, unlike Altai, streaks usually merge so flanks appear wholly chestnut. Both species show white or buffish tips to tail feathers in flight. Juvenile has adult wing pattern, but dark mottling on crown and nape, and dark brown streaking on buffish underparts. **Voice** Varied, melodious song with ringing, whistling and squeaking notes; calls include rolling *churrupp*. **HH** Singly, in pairs or family groups, according to season, tame and confiding. Forages by hopping quietly on ground. Breeds on open stony slopes, rocky pastures; near upland villages in winter.

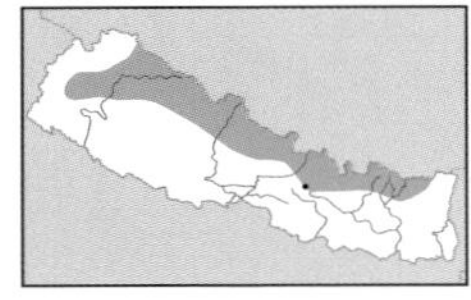

Altai Accentor *Prunella himalayana* 15–15.5 cm

Fairly common winter visitor; 1,300–4,300 m. **ID** From larger Alpine by more extensive white throat and diffuse rufous spotting on breast (extending onto flanks). Typically has diffuse malar stripe of black spots, and black gorget across lower throat, which can be well defined on some. Although variable, often shows brownish ear-coverts and greyish supercilium. Like Alpine, has white tips to wing-coverts, but usually lacks striking black panel on wing. Mantle more heavily streaked and can show prominent pale 'braces'. **Voice** Double *tee-tee* flight call. **HH** Often in tight flocks. Grassy and stony slopes and plateaux.

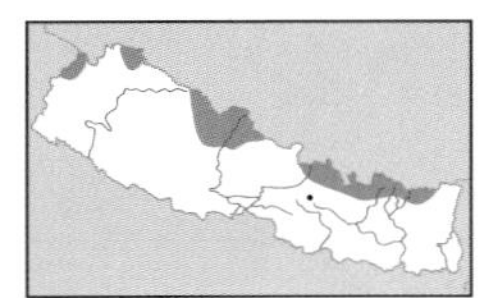

Robin Accentor *Prunella rubeculoides* 16–17 cm

Fairly common resident; breeds 4,200–5,400 m, winters 2,655–3,960 m at least. **ID** Adult has rusty-orange band on breast. From Rufous-breasted by uniform brownish-grey head including throat, diffusely streaked upperparts, and unstreaked white belly (with limited streaking on flanks). Juvenile very similar to juvenile Rufous-breasted, but has more diffusely streaked underparts (streaking hardly apparent on belly and flanks) and poorly defined supercilium. **Voice** High-pitched *tzwe-e-you, tzwe-e-you* song; ringing *pi-pi-pi-pi* alarm call. **HH** Habits similar to Alpine. Breeds in bushes near streams or pools; winters in dry, stony areas and upland villages.

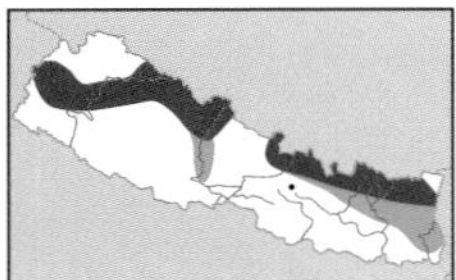

Rufous-breasted Accentor *Prunella strophiata* 15 cm

Common, widespread resident; breeds 3,500–4,930 m, winters 1,600–3,650 m at least. **ID** Adult has rusty-orange band on breast. Best separated from Robin Accentor by striking head pattern: prominent supercilium (whitish in front of eye, broader and rufous behind), blackish ear-coverts and sides to crown, and whitish throat (with variable black streaking, typically forming diffuse malar stripe and band on lower throat). Also has black streaking on grey neck-sides, and dark streaking on belly and flanks. Some variation, as some birds have duller (streaked) and less extensive rufous on breast, and less well-marked head pattern. More heavily streaked underparts and more prominent supercilium (offset by brown ear-coverts and crown-sides) separate juvenile from juvenile Robin Accentor. **Voice** Melodious warbling song, reminiscent of Eurasian Wren, but not as shrill or vehement; penetrating *trr-r-rit trrr-r-it* call. **HH** Habits similar to Alpine. Breeds in dwarf shrubbery near treeline; winters in upland pastures with scattered bushes and bushes around fallow fields.

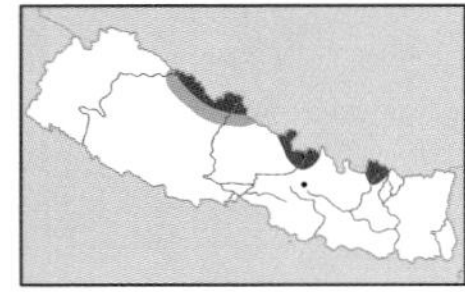

Brown Accentor *Prunella fulvescens* 14.5–15 cm

Resident, fairly common in north-west; uncommon or rare further east to Sagarmatha National Park; winters 2,300–3,800 m at least, breeds up to 5,200 m. **ID** Adult has broad whitish supercilium (contrasting with blackish ear-coverts), grey-brown upperparts with darker streaking, and orange-buff underparts with whiter throat. Juvenile has more heavily streaked upperparts with rufous-buff cast, brown mottling on supercilium and crown, browner ear-coverts (with less striking head pattern) and brown-streaked breast. **Voice** Variable warbling *tuk-tileep-tilee-tileep-tileep* song; weak, ringing trilling call, *si-si-si-si*. **HH** Habits similar to Alpine. Low scrub on dry rocky slopes, also around upland villages and nearby fallow fields in winter.

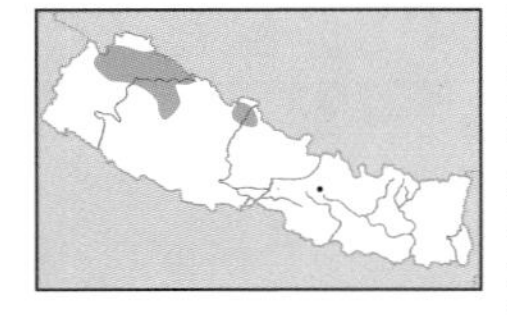

Black-throated Accentor *Prunella atrogularis* 14.5–15 cm

Fairly common winter visitor to north-west; 2,440–3,050 m. **ID** Broad white to orange-buff supercilium and submoustachial stripe, black chin and upper throat, blackish crown-sides and ear-coverts, dark brown streaking on mantle, and orange-buff breast and flanks. First-winter can have black throat partly obscured by pale fringes or, rarely, completely absent. Heavy dark streaking on mantle and flanks, and mottled ear-coverts, separate such birds from Brown. **Voice** Weak, ringing trill *si-si-si-si*. **HH** Habits similar to Alpine. Bushes near cultivation and dry scrub-covered hills.

ad
Alpine Accentor
ad
Altai Accentor
ad
Robin Accentor
ad
1st-win
Rufous-breasted Accentor
ad
Brown Accentor
ad
1st-win
Black-throated Accentor

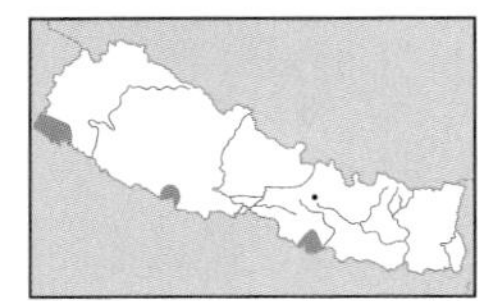

Singing Bush Lark *Mirafra cantillans* 14 cm

Local resident; 100–150 m. **ID** Stocky, stout-billed, broad-winged lark with slight crest and rufous panel in wing. Comparatively uniform brownish-buff ear-coverts, weak and rather restricted spotting on upper breast, and whitish throat with brownish to rufous-buff breast-band. Tail has whitish outer feathers. Bill distinctly shorter and stouter than Oriental Skylark, and has shorter crest. **Voice** Sweet, full song, with much mimicry, delivered from top of bush or in flight. **HH** Habits like Bengal, but compared to other *Mirafra* larks has more varied song and distinctive display flight, towering high on winnowing or flickering wings. Also sings when perched. Dry grasslands.

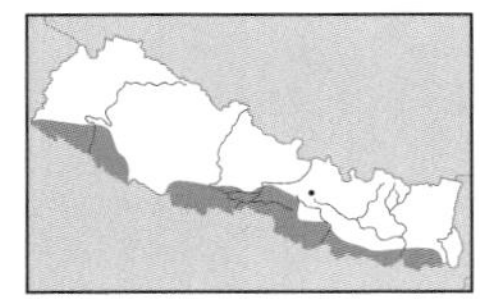

Bengal Bush Lark *Mirafra assamica* 15 cm

Common, widespread resident; 75–275 m. **ID** Diffusely streaked brownish-grey upperparts, buffish supercilium, and dirty rufous underparts (with paler throat and greyish flanks). Prominent rufous wing panel. **Voice** Song a repeated series of thin, high-pitched disyllabic notes, usually delivered in prolonged flight; call a series of variable, thin, high-pitched short notes. **HH** Walks and runs on ground; strong and undulating flight like other larks. Fallow cultivation and short grassland.

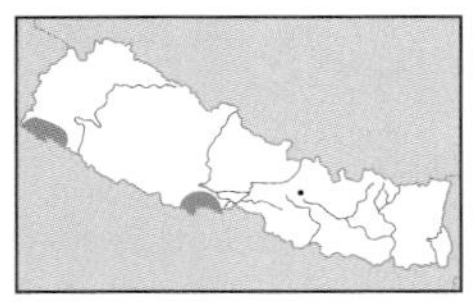

Rufous-tailed Lark *Ammomanes phoenicura* 16 cm

Rare, local resident; 75–100 m. **ID** Dusky grey-brown upperparts, rufous-orange underparts and underwing-coverts, and dark spotting/streaking on throat and breast. Also rufous-orange uppertail-coverts, and rufous-orange tail has broad and well-defined dark terminal bar. **Voice** Aerial flight song comprises sweet *tee-hoo* phrases with low-pitched husky whistles and chirrups. **HH** Runs in rapid zigzags. Cultivation, fallow and ploughed fields, stubbles and open country with scattered bushes and stony outcrops.

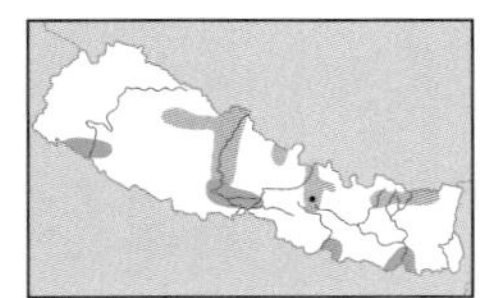

Greater Short-toed Lark *Calandrella brachydactyla* 14 cm

Locally frequent, mainly a passage migrant, also winter visitor; 45–4,575 m (–5,000 m). **ID** Stouter bill than Hume's, with more prominent supercilium and eye-stripe, warmer upperparts with more prominent streaking, prominent dark centres to median coverts, warmer-coloured breast often with well-defined streaking, and different call. Dark breast-side patches often apparent, like Hume's Short-toed Lark. *C. b. dukhunensis* (recorded in Nepal), which is more easily separated from other small larks, has warm sandy-buff upperparts, variable rufous-buff breast-band, and rufous-buff ear-coverts and wash on flanks. *C. b. longipennis* (possibly occurs) has slightly colder and greyer upperparts, and breast washed brownish-buff (thus more similar in coloration to Hume's). **Voice** Dry *tchirrup* or *chichirrup* flight call. **HH** Gregarious in winter, running and flying about restlessly. Open stony and short-grass areas; also fallow cultivation.

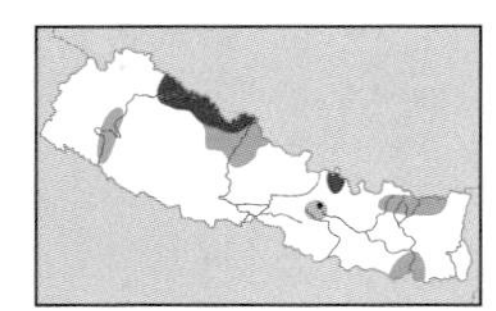

Hume's Short-toed Lark *Calandrella acutirostris* 14 cm

Resident and passage migrant, common in northern Dolpo in summer and rare or uncommon elsewhere; breeds 3,660–4,575 m (–4,800 m); winters down to 75 m. **ID** Greyer and less heavily streaked upperparts than Greater, with pinkish uppertail-coverts. Head pattern usually less pronounced than Greater, with rather uniform ear-coverts, dark lores (pale in Greater), and less pronounced supercilium and eye-stripe. Bill yellowish with pronounced dark culmen and tip. As Greater, dark breast-side patch usually apparent, but has greyish-buff breast-band. **Voice** Full, rolling *tiurr* flight call; song in flight is mellow, variable *tee-leu-ee-lew*. **HH** Habits like Greater Short-toed. Breeds in semi-desert with scattered bushes; winters in fallow cultivation and open wasteland.

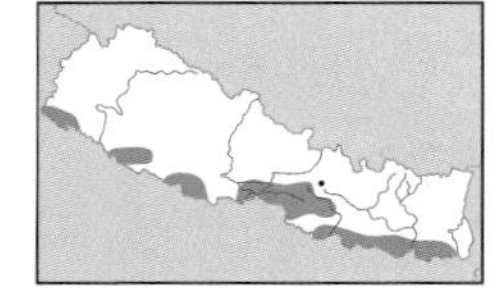

Sand Lark *Calandrella raytal* 12 cm

Locally common resident; 75–305 m. **ID** Small stocky lark with comparatively short tail, rather rounded wings and distinctive, rather jerky and fluttering flight. Additional features from other *Calandrella* larks include finer bill, rather uniform sandy-grey upperparts (streaking most prominent on crown), whitish underparts with fine sparse streaking on breast (and no dark patches on breast-sides), and prominent blackish sides to tail contrasting with white outer tail feathers and very pale rump and uppertail-coverts. Primaries extend beyond tertials on closed wing (primaries equal to length of tertials in other *Calandrella* species in Nepal). **Voice** Rolling, deep and guttural *prr... prr* call; song in flight a series of short, rapidly delivered and repeated undulating, warbling notes. **HH** Runs in zigzagging spurts. Banks of lakes and rivers.

ad
Singing Bush Lark
ad
Bengal Bush Lark
ad
Rufous-tailed Lark
ad
dukhunensis
ad
longipennis
Greater Short-toed Lark
ad
Hume's Short-toed Lark
ad
Sand Lark

PLATE 132: LARKS AND WAGTAILS

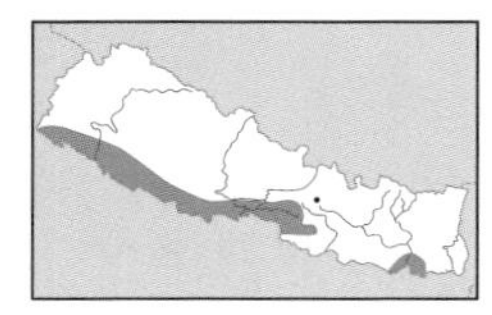

Crested Lark *Galerida cristata* 18 cm

Uncommon to rare resident; 75–275 m. **ID** From Oriental Skylark by larger size, more prominent and erect crest, broader rounded wings, rufous-buff outer tail feathers and underwing, and different vocalisations. Lacks any hint of rufous panel in wing. **Voice** Calls include a fluty *du-ee* and similar *tuee-tuu-teeooo*; prolonged and complex song, with much mimicry, interspersed by variations of call. **HH** Riverbeds and dry cultivation.

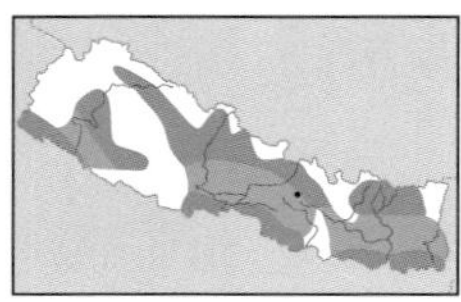

Oriental Skylark *Alauda gulgula* 16 cm

Fairly common, widespread, breeding resident 75–150 m, also breeds 2,500–3,600 m and winter visitor 1,280–1,700 m. **ID** Well-streaked upperparts and breast, with a small crest. Additional features that aid identification from other larks are fine bill (compared to bush larks), buffish-white outer tail feathers and indistinct rufous wing panel. See Appendix 1 for differences from Eurasian Skylark. **Voice** Song comprises bubbling warbles and shorter whistling notes with much variation; grating, throaty *bazz, bazz* call. **HH** Grassland and cultivation.

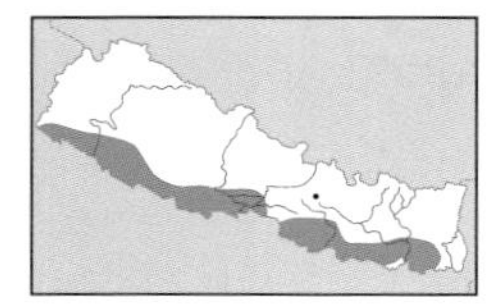

Ashy-crowned Sparrow Lark *Eremopterix griseus* 12 cm

Fairly common, widespread resident; 75–305 m. **ID** Male has grey crown, and black throat, supercilium and underparts. Upperparts fairly uniform sandy-grey. Female from other larks by combination of stout greyish bill, uniform head (lacking dark eye-stripe), rather uniform upperparts (with almost unstreaked mantle and scapulars), indistinct and diffuse breast streaking, and blackish underwing-coverts (latter can be difficult to see in field). **Voice** Display flight song comprises short flute-like *tweedle-deedle-deedle* as bird rises and drawn-out whistle, *wheeh*, in descent. **HH** Open dry areas including cultivation, stony scrub and ploughed fields.

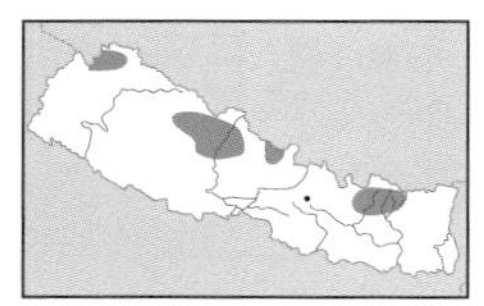

Horned Lark *Eremophila alpestris* 18 cm

Locally fairly common resident in far north; mainly 3,965–5,490 m (2,600–5,900 m). **ID** Male has black-and-white head pattern, with black mask, 'horns' and band on crown. Also black breast-band and sandy upperparts with vinous cast to nape. Female similar but mask is duller, crown and mantle are heavily streaked, and lacks vinous cast to nape. Juvenile has suggestion of dark mask, upperparts spotted yellowish-buff, and underparts have pale yellowish wash. **Voice** Song comprises *tsit-tsit-tsit* notes, followed by short warbling phrases and longer whistles; quiet *tsit-tsit* call. **HH** Desert-like steppe on stony ground with scattered bushes.

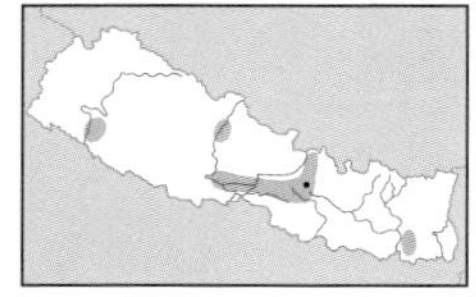

Forest Wagtail *Dendronanthus indicus* 18 cm

Very rare passage migrant; 75–1,750 m. **ID** A forest-dwelling wagtail; from others by combination of broad yellowish-white median and greater coverts wing-bars and white patch on secondaries, double black breast-band (lower band broken in centre of breast), olive upperparts, white supercilium and whitish underparts. Sexes alike. **Voice** Strident metallic *pink* or *dzink-dzzt* call. Repetitive intense Great Tit-like *see-sawing* song. **HH** Characteristic habit of swaying its tail and hind body from side to side with a rather deliberate motion, instead of wagging tail up and down like other wagtails. Glades and paths, chiefly in broadleaved forest.

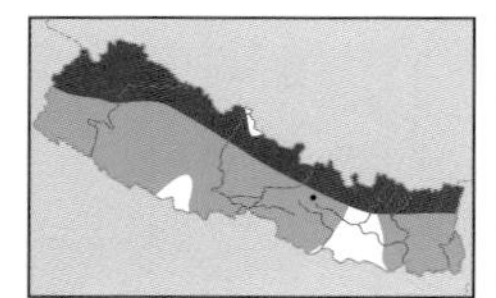

Grey Wagtail *Motacilla cinerea* 19 cm

Common, widespread resident; breeds 1,100–3,550 m (–4,115 m), non-breeding season 75–1,500 m. **ID** Much longer tailed than other wagtails (especially obvious in flight). In all plumages, white supercilium, grey upperparts and yellow vent and undertail-coverts. In flight, narrow white wing-bar and yellow rump. At rest, whitish fringes to tertials but otherwise blackish wings lacking broad fringes to coverts of Yellow and Citrine. Breeding male has black throat, with rest of underparts yellow. Female breeding lacks well-defined black bib, but may show black mottling on chin/throat. Adult non-breeding and first-winter have white throat and pale yellowish to buffish-white underparts, with yellow vent. Juvenile much as non-breeding, but has brownish cast to upperparts, buffish supercilium and dark mottling on breast-sides. **Voice** Sharp *stit* or *zee-fit* call; song comprises series of call-like notes. **HH** Breeds by fast-flowing mountain streams; winters along slower-moving streams.

ad
Crested Lark
ad
Oriental Skylark
♂
♀
Ashy-crowned Sparrow Lark
♀
♂
Horned Lark
ad
Forest Wagtail
♀ / non-br
1st-win
♂ br
Grey Wagtail

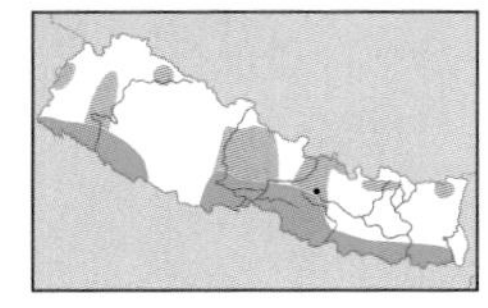

Yellow Wagtail *Motacilla flava* 18 cm

Locally fairly common, mainly winter visitor, 75–915 m, also passage migrant; 75–1,350 m (–3,800 m). **ID** Male breeding has olive-green upperparts and yellow underparts, with considerable subspecific variation. Female in breeding plumage usually shows some features of breeding male. First-winter typically has brownish-olive upperparts, whitish underparts with variable yellowish wash, and buff or whitish median and greater coverts wing-bars and fringes to tertials. Some, however, have greyish upperparts and whitish underparts (lacking any yellow), and can closely resemble first-winter Citrine; identified by: narrower white supercilium that does not extend around ear-coverts, grey forehead concolorous with crown, dark lores resulting in complete dark eye-stripe, pale base to lower mandible, and narrower white wing-bars. Juvenile has dark malar stripe and band on breast. Four subspecies have been recorded, and *lutea* is likely to occur. *M. f. beema* has pale bluish-grey head and white supercilium. *M. f. leucocephala* has whitish head. *M. f. feldegg* (including *melanogrisea*) has black head. *M. f. thunbergi* (including *plexa*) has dark slate-grey crown with darker ear-coverts. *M. f. lutea* has yellowish head. *M. f. 'superciliaris'* is an intergrade, probably between *beema* and *feldegg*, and looks like the latter, but has white supercilium. **Voice** Typical call is a loud disyllabic *tswee-ip*, usually quite distinct from Citrine; some races have harsher *tsreep* call more closely resembling Citrine. **HH** Shallow water, marshlands, damp grasslands. **AN** Western Yellow Wagtail (if Eastern is split).

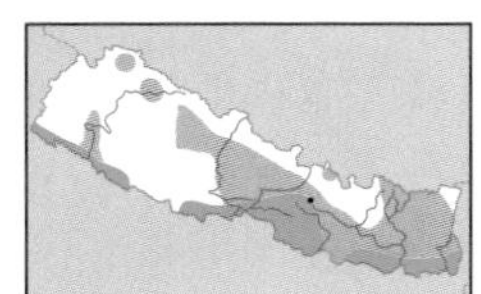

Citrine Wagtail *Motacilla citreola* 19 cm

Locally fairly common winter visitor, 75–915 m; widespread frequent passage migrant; 75–5,200 m. **ID** Adult male breeding from Yellow by yellow head and underparts, black (*M. c. calcarata*) or grey (nominate) mantle, and broad white wing-bars. Female breeding and adult non-breeding best separated from Yellow by broad yellow supercilium that surrounds ear-coverts to join yellow of throat, broad white wing-bars (narrower when worn), and greyish crown and mantle. Juvenile lacks any yellow, and has brownish crown, ear-coverts and mantle, buffish supercilium (with dark upper edge) and surround to ear-coverts, and buffish-white underparts with gorget of black spots across breast. First-winter has grey upperparts and is similar to some first-winter Yellow, but note broader white supercilium, which usually surrounds ear-coverts, pale brown forehead, pale lores, all-dark bill, broader white wing-bars and white undertail-coverts; lacks black breast-band and has different call compared to White Wagtail. By early November, first-winter Citrine has yellowish supercilium, ear-coverts surround and throat. **Voice** Harsh, *brrzzreep* call, typically more buzzing than Yellow. **HH** Streambeds, wet fields, marshes, riverbanks and edges of pools and lakes.

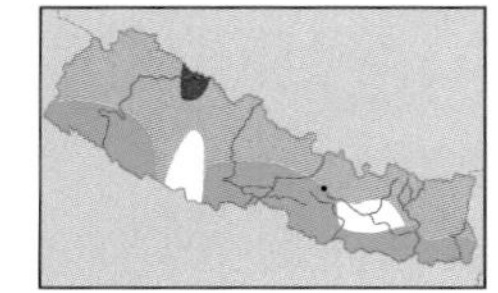

White Wagtail *Motacilla alba* 19 cm

Widespread; common passage migrant and winter visitor; *alboides* breeds 2,400–4,600 m, species winters 75–1,500 m, up to 5,550 m on passage. **ID** Much variation in adults in breeding plumage (sexes similar): *M. a. alboides* has black head, mantle and breast, with white forehead and face patch. Upperparts of female variably mixed with grey. *M. a. personata* similar to *alboides*, but has grey mantle. *M. a. alba* has grey mantle, white forehead and face, and black hindcrown/nape, throat and breast. *M. a. leucopsis* has black mantle and back, with head pattern as *alba* but a white throat. *M. a. ocularis* has grey mantle and is much like *alba*, but has black eye-stripe in all plumages. *M. a. baicalensis* has grey mantle and is much like *alba*, but has white chin and upper throat contrasting with black breast. Much variation in non-breeding and first-winter plumages, although non-breeders of some races retain characteristics of breeding plumage. Juvenile *alboides* has grey head, mantle and breast with whitish supercilium. **Voice** Call a loud *tslee-vit*; song a lively twittering and chattering with call-like notes. **HH** Breeds by running waters in open country in hills and mountains; winters in open country near water.

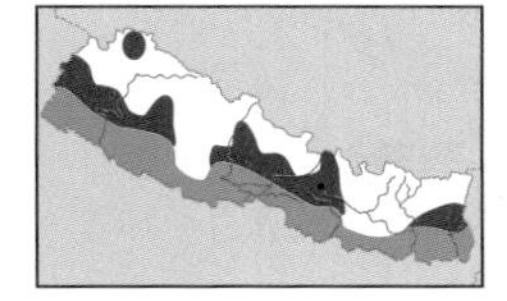

White-browed Wagtail *Motacilla maderaspatensis* 21 cm

Fairly common widespread resident; 75–915 m, uncommon up to 1,700 m, also locally at 3,400–3,750 m. **ID** Very large wagtail. Combination of black mantle and black head with white supercilium separates it from all subspecies of White. Sexes similar, and show no variation in non-breeding plumage. First-winter similar, but has greyer crown and mantle. Juvenile has brownish-grey head, mantle and breast, with white supercilium. **Voice** Distinctive, loud *chiz-zat* call; song comprises clear, high-pitched jumble of loud, pleasant whistling notes. **HH** Riverbanks in broad valleys, pools and lakes.

♂ br
beema
♀
juv
beema
1st-win
1st-win
Yellow Wagtail
hunbergi
1st-win
♂
lutea
lutea
♂
hunbergi
calcarata
♂ br
1st-win
(late)
♂
♂
feldegg
superciliaris
Citrine
Wagtail
♂
calcarata
leucocephala
1st-win
(early)
♀
♂ br
♂ non-br
alboides
calcarata
calcarata
citreola
1st-win
♂ br
personata
calcarata
juv
personata
♂ br
♂
alboides
juv
leucopsis
ad
♂
ocularis
♂
alba
juv
♂
alboides
baicalensis
White Wagtail
White-browed Wagtail

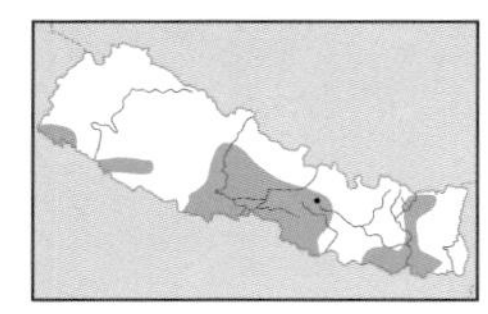

Richard's Pipit *Anthus richardi* 17 cm

Locally fairly common, quite widespread winter visitor; 75–1,500 m. **ID** Large size, upright stance, larger bill, longer legs and hindclaw, and call are best features from otherwise similar Paddyfield Pipit. **Voice** Distinctive loud, explosive *schreep* call; also shorter *chup*. **HH** On ground progresses with swift runs combined with strutting walk on strong legs. When flushed, typically gains height and distance with deep undulations. May hover above ground, fluttering with dangling legs. Moist grassland and cultivation.

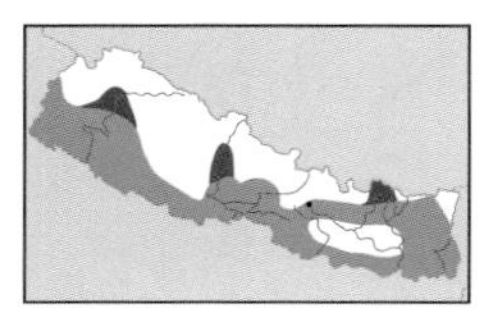

Paddyfield Pipit *Anthus rufulus* 15 cm

Common and widespread resident; 75–1,830 m (–2,440 m). **ID** Smaller than Richard's, with different call; when flushed, has comparatively weak flight. Juvenile/first-winter Tawny can appear very similar, but lores of Paddyfield often pale (can be dark) and shows warm ginger-buff wash on breast and flanks (underparts more uniform cream-white on Tawny); flight call differs. **Voice** Calls include a weak *chup-chup-chup* or *chip-chip-chip*; song given perched or in aerial display a repetitive *chip-chip-chip*. **HH** Forages by running rapidly on ground, wagging tail. Less powerful gait than Richard's. When flushed, usually flies short distance. Edges and banks of terraced fields, short grassland, green fodder crops, ploughed and fallow fields, and stubbles.

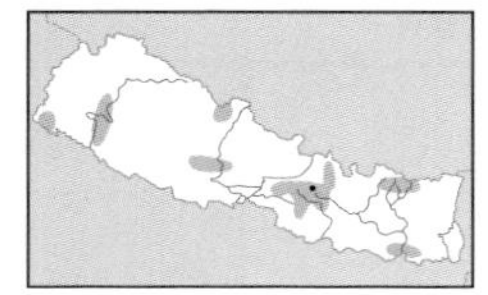

Blyth's Pipit *Anthus godlewskii* 16.5 cm

Uncommon winter visitor and passage migrant; 75–4,710 m (on passage). **ID** Compared to Richard's, very subtle differences are slightly smaller size and more compact appearance, shorter tail, shorter hindclaw, shorter and more pointed bill, shorter legs, and call. Shape of centres to adult buff-fringed median coverts distinctive (square-shaped black centres with broad pale buff tips; centres diffuse and more triangular in Richard's). Best distinctions from Paddyfield are call, larger size and pattern of adult median coverts (Paddyfield as Richard's Pipit). **Voice** Diagnostic powerful, wheezy *spzeeu* call; also a mellow *chup* or *chep* resembling Paddyfield or Tawny.
HH Gait resembles Tawny; lacks Richard's strutting walk, flight similar to Richard's but without fluttering pause before landing. Marshes, grassland and cultivation.

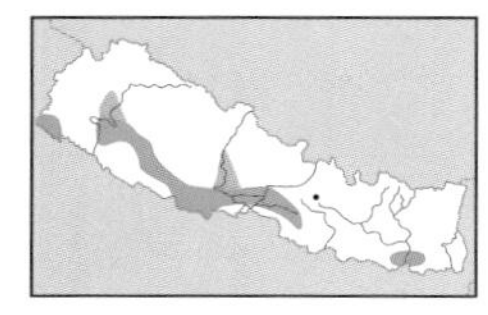

Tawny Pipit *Anthus campestris* 16 cm

Locally uncommon winter visitor 75–305 m, passage migrant 75–1,370 m. **ID** Adult and first-winter have plain or only very faintly streaked upperparts and unstreaked or only very lightly streaked breast. Juvenile plumage can be retained until midwinter, in which upperparts and breast are noticeably streaked. Subtle differences in juvenile from Paddyfield are dark lores and eye-stripe contrasting with supercilium, which tends to be broader and square-ended, and more uniform buffish-white underparts. **Voice** Distinctive loud *tchilip* call; softer *chep* similar to Paddyfield and Blyth's. **HH** Like other pipits, walks and runs swiftly with occasional pauses, frequently wags tail; undulating flight less powerful than Richard's. Stony semi-desert and fallow cultivation.

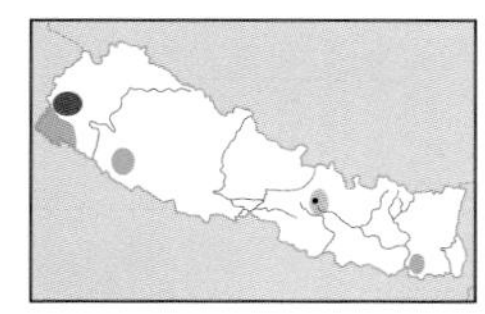

Long-billed Pipit *Anthus similis* 20 cm

Rare, possibly resident, also very local winter visitor; 75–1,700 m. **ID** Considerably larger than Tawny, with larger and darker bill and shorter-looking legs. Like Tawny has dark lores. Lacks distinct dark malar and moustachial, and has darker and greyer upperparts, deeper orange-buff underparts, rufous fringes to tertials and coverts, and rufous-buff outer edge to tail. **Voice** Deep *chup* and loud ringing *che-vlee* calls; slow, measured *chirrit-chirrit-teeweeh-pr-chirrit-chirrit-teeweeh* song. **HH** Non-gregarious. Upright posture when alert, more horizontal when feeding. Flicks tail upwards while fanning it; does not wag tail like most pipits. Breeds on rocky slopes; winters in dry areas: grassland, scrub and cultivation.

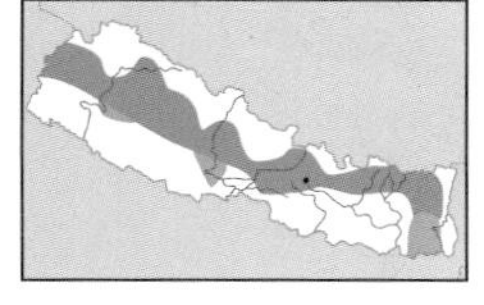

Upland Pipit *Anthus sylvanus* 17 cm

Locally fairly common resident; breeds 1,830–2,900 m, non-breeding season 1,350–2,000 m. **ID** Large, heavily streaked pipit with short broad bill, and rather narrow, pointed tail feathers. Fine black streaking on underparts, whitish supercilium; ground colour of underparts varies from warm buff (fresh plumage) to rather cold and grey (worn). **Voice** Sparrow-like chirp call. Distinctive song: high-pitched penetrating two-note whistle, *whit-tsee, whit-tsee*, repeated monotonously. **HH** Steep rocky and grassy slopes, and abandoned cultivation with scattered trees.

ad
Richard's Pipit
ad
Paddyfield Pipit
ad
Blyth's Pipit
1st-win
juv
ad
Tawny Pipit
ad
Long-billed Pipit
ad worn
ad fresh
Upland Pipit

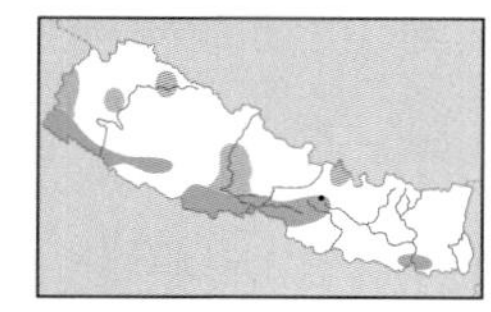

Tree Pipit *Anthus trivialis* 15 cm

Uncommon and fairly widespread; winter visitor, mainly to lowlands, and passage migrant; 75–3,050 m. **ID** Buffish-brown to greyish ground colour to upperparts (lacking greenish-olive cast of Olive-backed Pipit), and buffish edges to wing feathers (greenish-olive in Olive-backed). Head pattern typically less prominent, although can appear similar. *A. t. haringtoni* has not been recorded but probably occurs. **Voice** Call comprises harsher *teez* than Olive-backed, but much overlap; song louder and far-carrying, ending in finch-like trill *chik-chik.....chia-chia-wich-wich-tsee-a-tsee-a tsse-a*. **HH** Groves, grasslands, fallow fields and stubbles.

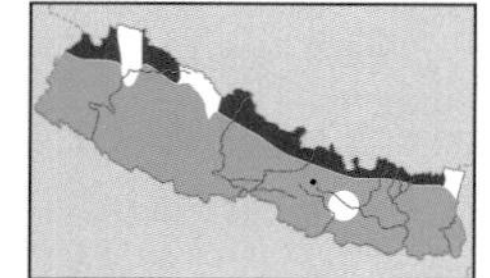

Olive-backed Pipit *Anthus hodgsoni* 15 cm

Common, widespread resident and winter visitor; breeds 2,200–4,000 m (–1,800 m), winters 75–2,560 m. **ID** Greenish-olive cast to upperparts and edges to wing feathers. More striking head pattern than Tree, with more prominent supercilium (buffish in front of eye and white behind), stronger dark eye-stripe and has distinct whitish spot and blackish patch on rear ear-coverts (lacking on Tree). Worn upperparts become more greyish-olive, and breast loses warm buff wash. *A. h. yunnanensis*, a widespread winter visitor, is much less heavily streaked on upperparts than resident nominate. **Voice** Weak *see* flight call, fainter than Tree; song given from treetop or in flight comprises choppy, jangly, pleasant series of chirps of different types and on different pitches. **HH** Breeds in open forest clearings, scrub with scattered trees, and high-altitude shrubberies; winters in shady glades.

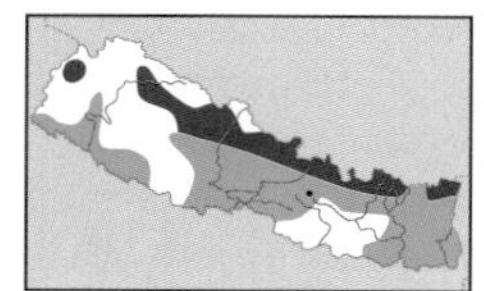

Rosy Pipit *Anthus roseatus* 15 cm

Fairly common altitudinal migrant; breeds 4,000–5,050 m, winters 760–1,500 m (–75 m). **ID** Always has boldly streaked upperparts, olive cast to mantle, and olive to olive-green edges to wing feathers. Adult breeding has mauve-pink wash to underparts with irregular black spotting on breast and flanks; pinkish to buff supercilium is very prominent, with broad dark eye-stripe and moustachial stripe, and whitish eye-ring. Female less pink below with heavier breast streaking than male. In non-breeding plumage, heavily streaked underparts and dark lores separate from Water and Buff-bellied Pipits. **Voice** Weak *seep-seep* call, slightly less strident than Water and similar to Buff-bellied; song comprises *tit-tit-tit-tit-tit-teedle-teedle* as bird rises, *sweet-sweet-sweet* on descent. **HH** Breeds in wet places in alpine meadows; winters in marshes, damp grassland and cultivation.

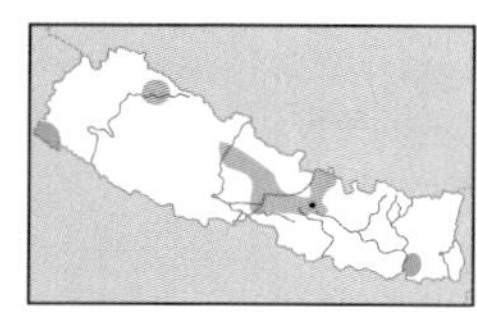

Red-throated Pipit *Anthus cervinus* 15 cm

Winter visitor and passage migrant; frequent at Koshi, rare elsewhere; winters 75–915 m, up to 5,180 m on passage. **ID** Adult has reddish throat and upper breast, usually paler on female and autumn/winter birds. First-winter similar to nominate Tree, but typically has more heavily streaked upperparts, often with pale 'braces', heavily streaked rump, well-defined and broad white wing-bars, strongly contrasting blackish centres and whitish fringes to tertials, more pronounced dark malar patch, and more boldly streaked breast and (especially) flanks. Very different call, browner upperparts and absence of olive in wing separate from non-breeding Rosy. **Voice** Typical call is a long drawn-out *seeeeee*, but can be abbreviated. **HH** Marshes, wet grassland and stubble.

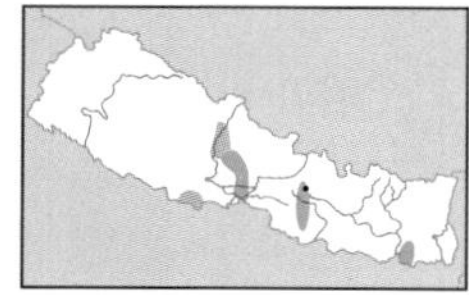

Buff-bellied Pipit *Anthus rubescens* 15 cm

Rare winter visitor and passage migrant; 75–2,715 m. **ID** Lightly streaked upperparts, lack of olive-green on wing and pale lores separate it from Rosy in all plumages. In breeding plumage, has deeper orange-buff wash to underparts than Water; breast and flanks more heavily (but irregularly) spotted black and has more pronounced malar stripe and malar patch, although there is overlap (some Buff-bellied being more lightly streaked). In non-breeding plumage, compared to Water, has darker greyish-brown upperparts that are slightly less prominently streaked; black malar stripe and patch, and bold black spotting/streaking on breast and flanks, all much more pronounced. **Voice** Call similar to Rosy, but perhaps thinner and sharper. **HH** Paddy stubbles, marshes and damp grassy edges of rivers and lakes.

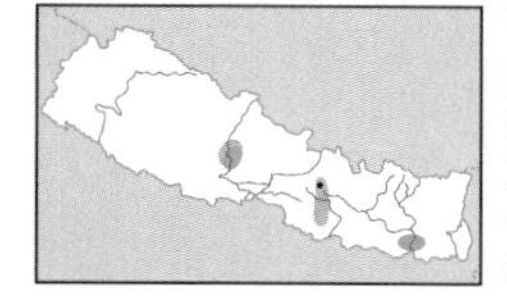

Water Pipit *Anthus spinoletta* 15 cm

Rare winter visitor and passage migrant; 75–1,370 m (–3,400 m on passage). **ID** In all plumages, lightly streaked upperparts, lacks olive-green on wing, has dark legs, and usually has pale lores; underparts less heavily marked than Rosy and Buff-bellied Pipits. Orange-buff wash to supercilium and underparts in breeding plumage. **Voice** Call similar to Rosy, but perhaps thinner and sharper. **HH** Marshes, wet grassland, irrigated cultivation and paddy stubbles.

ad
ad
trivialis
haringtoni
Tree Pipit
ad fresh
hodgsoni
Olive-backed Pipit
ad
ad worn
yunnanensis
♂ br
♂ br
♀ br
non-br
1st-win
Red-throated Pipit
Rosy Pipit
br
br
non-br
non-br
Buff-bellied Pipit
Water Pipit

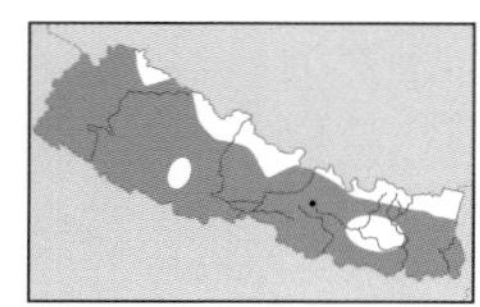

House Sparrow *Passer domesticus* 15 cm

Abundant, widespread resident; 75–2,135 m (–4,350 m). **ID** Breeding male has grey crown with chestnut sides and nape, and black throat and upper breast; duller in non-breeding plumage when head pattern and black throat/breast partly obscured by pale fringes. Female has pale buff supercilium, dark brown streaking on buffish mantle, and unstreaked greyish-white underparts (faintly washed buff on flanks). Juvenile as female, with broader buff-brown fringes to upperparts; juvenile male has greyish chin. **Voice** Monotonous *chirrup* call; *chur-r-r-it-it* alarm; song comprises long series of *chirrup*, *cheep* and *churp* notes. **HH** Markedly gregarious outside breeding season, forming large communal roosts. Forages mainly by hopping on ground, also perches on ripening cereal heads and pecks at seeds. Villages, towns and cities; also cultivation in winter.

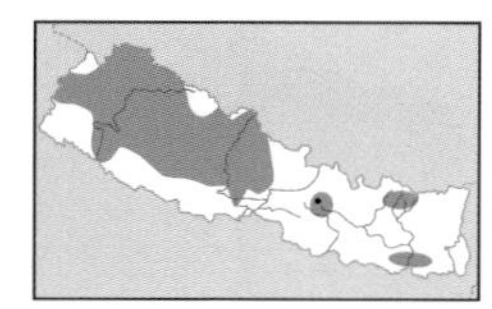

Russet Sparrow *Passer rutilans* 14.5 cm

Resident, widespread and locally fairly common in west, rare further east; mainly 915–4,270 m (75–4,350 m). **ID** Slimmer than Eurasian Tree, with finer bill. Breeding male has bright chestnut mantle and variable yellow wash on ear-coverts and underparts; lacks black cheek patch of Tree. Head and mantle coloration only slightly obscured in non-breeding plumage. Female has more prominent supercilium and eye-stripe than female House; also unstreaked buff or rufous-brown scapulars, rufous-brown lower back and rump, and faint yellowish wash to underparts. **Voice** Call similar to House, but more musical and softer. **HH** Usually in pairs or loose flocks in breeding season, large flocks in winter. Often in upland villages where House Sparrow is absent.

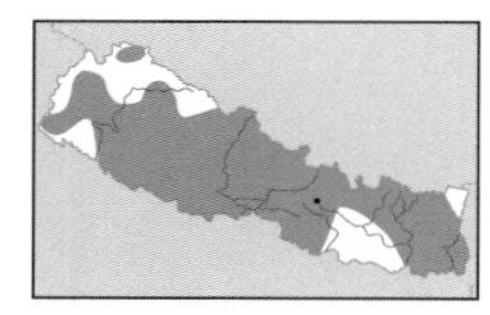

Eurasian Tree Sparrow *Passer montanus* 14 cm

Abundant, widespread resident; breeds 75–4,270 m, winters 75–3,795 m. **ID** Adult has dull chestnut crown, black spot on whitish ear-coverts, small black throat patch not extending to breast, and white collar separating chestnut nape from brown-streaked mantle. Sexes alike. Juvenile similar to adult, but has paler chestnut crown and diffuse black patches on ear-coverts and throat. **Voice** Monotonous *chip chip* call, harder than similar call of House; song is a running-together of this note, interspersed by *tsweep* or similar calls. **HH** Habits very similar to House. Suburbs and fields at edges of towns and villages.

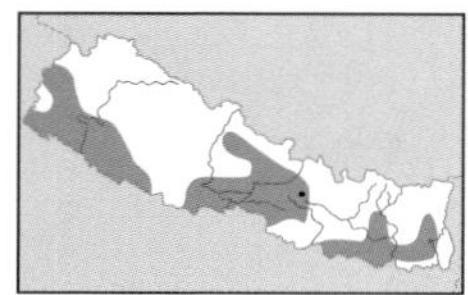

Chestnut-shouldered Petronia *Gymnoris xanthocollis* 13.5 cm

Frequent resident, especially in west; 75–305 m (–1,525 m). **ID** Male from other sparrows by finer bill, uniform (unstreaked) brownish-grey head and upperparts, and yellow patch on lower throat. Chestnut lesser coverts and prominent white double wing-bar. Female similar, but has brown lesser coverts, buff-tinged tips to median coverts, and yellow on throat is faint or absent. Juvenile similar to female, but upperparts more sandy-brown, with pale buffish supercilium, and lacks any yellow on throat. **Voice** Similar call to House Sparrow; song repetitive series of *chip chillup* and *chalp* notes, more liquid than House. **HH** Usually in pairs or small flocks in breeding season when forages more in trees than *Passer* sparrows. Roosts communally. Has undulating flight, more dipping than other sparrows. Open dry forest, thorn scrub and trees at cultivation edges. **TN** Formerly placed in genus *Petronia*. **AN** Yellow-throated Sparrow.

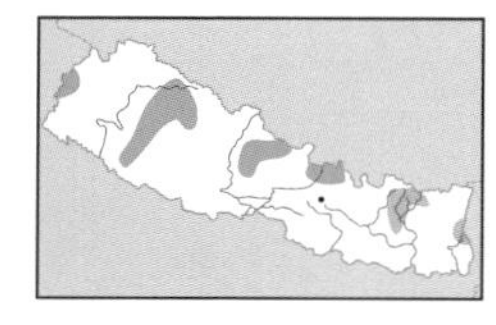

Common Chaffinch *Fringilla coelebs* 16 cm

Winter visitor, locally frequent in west; mainly 2,000–2,750 m (1,555–3,050 m). **ID** Double whitish wing-bars in all plumages. Lacks white rump of Brambling. Male has blue-grey crown and nape, orange-pink face and underparts, and maroon-brown mantle; brighter in breeding plumage. Female duller than female Brambling, with greyish-brown upperparts and dull greyish-buff underparts. **Voice** Metallic *chink* call; quiet *chap* in flight. **HH** Singly or in small flocks. Seeks seeds on ground; hops and walks with distinctive short, quick steps, accompanied by slightly nodding head. Upland fields with nearby bushes and coniferous forest.

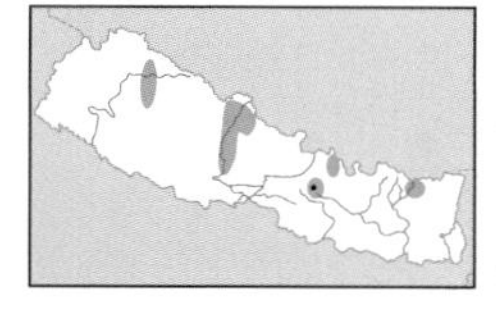

Brambling *Fringilla montifringilla* 16 cm

Rare and erratic winter visitor; 2,135–3,050 m (1,500–4,100 m). **ID** Male non-breeding and female from Common Chaffinch by orange breast and flanks (contrasting with white belly), orange wing-bar and white rump. Head and mantle of male become blacker towards breeding season as feather fringes are lost with wear, and bill becomes black. **Voice** Chaffinch-like *chup* call, usually also a distinctive deep nasal *zweee*. **HH** Habits similar to Chaffinch. Flight has more pronounced undulations than that species. Upland fields bordering bushes and coniferous forest.

♂
♀
House Sparrow
♂
♀
Russet Sparrow
ad
Eurasian Tree Sparrow
♂
♀
Chestnut-shouldered
Petronia
♀
♂ non-br
Common Chaffinch
♂ br
♀
♂ non-br
Brambling

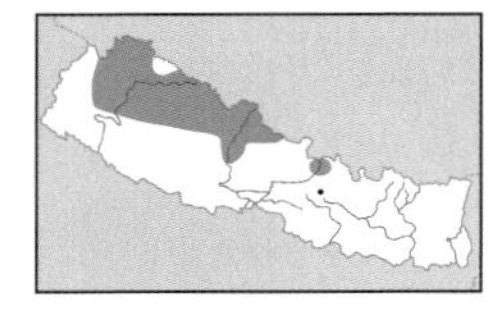

Red-fronted Serin *Serinus pusillus* 12.5 cm

Fairly common resident in north-west; breeds 2,440–4,575 m, down to 2,135 m in winter. **ID** Male has scarlet forehead and largely black head and breast. Mantle and belly/flanks boldly streaked black, and washed olive-yellow. Buffish-orange wing-bars and olive-yellow rump/uppertail-coverts. Female generally duller, with less red on forehead, and black of head and breast is browner. In fresh plumage (autumn/winter), black of head on both sexes partly obscured by buff fringes and forehead duller. Juvenile has cinnamon-brown crown, ear-coverts and throat, with crown lightly streaked; buffish wing-bars, and (as adult) yellowish edges to wing and tail feathers. **Voice** Song comprises melodious rippling trill interspersed with twittering; rapid, ringing *trillit-drillt* and soft *dueet* call. **HH** Gregarious all year. Breeds in open steppe-like landscape with scattered dwarf bushes; winters on open stony and bushy slopes, also upland stubbles.

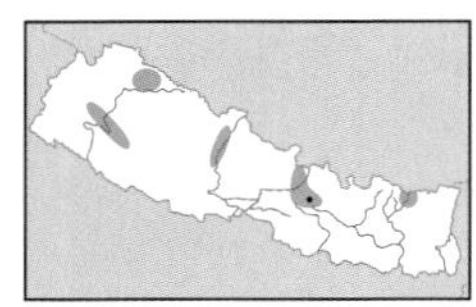

Tibetan Serin *Serinus thibetanus* 12 cm

Rare winter visitor; 1,050–3,500 m. **ID** Lacks yellow panels on wing in all plumages. Adult male has olive-green upperparts, yellow supercilium and border behind ear-coverts, yellowish-green rump, and yellow underparts. Wing and tail feathers broadly edged yellowish-green. Female has blackish streaking on darker greyish-green upperparts, more clearly defined wing-bars, paler yellow throat and breast, and whitish belly (with black flanks and breast streaking). Juvenile duller green, tinged brownish-buff on upperparts, with duller rump, buff fringes to greater coverts, paler (more heavily streaked) underparts. **Voice** Soft chattering interspersed by a wheezy *twang*. **HH** In flocks in winter. Feeds mainly in treetops in Himalayan Alder and mixed pine forests. **TN** Formerly placed in genus *Carduelis*. **AN** Tibetan Siskin.

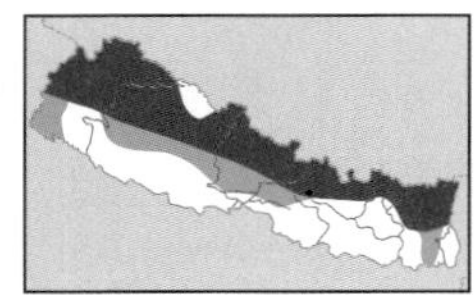

Yellow-breasted Greenfinch *Carduelis spinoides* 14 cm

Common and widespread resident; breeds 2,440–3,700 m (–4,400 m), winters 915–1,850 m (–250 m). **ID** Male has blackish-olive upperparts, yellow supercilium and crescent behind ear-coverts, yellow underparts and rump, and broad yellow panels on wing. Adult female has paler olive upperparts with faint dark streaking, less distinct head pattern, and duller yellow underparts. Juvenile has heavy streaking on buffish-olive upperparts and buffish-yellow underparts; faint yellowish supercilium and submoustachial stripe. **Voice** Twittering call followed by harsh *tsswee*; song comprises extended and more varied version of call, often in display flight. **HH** Gregarious all year. Forest edges with shrubbery, bushy areas near field edges, herbs at field edges or on pebble areas along large rivers, and scattered bushy clumps above treeline.

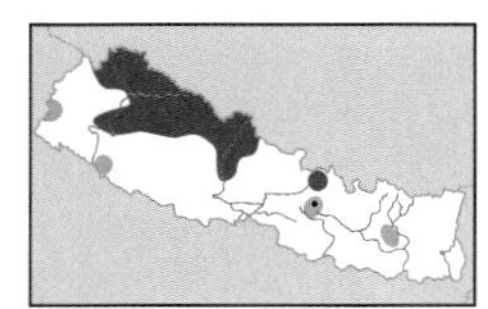

European Goldfinch *Carduelis carduelis* 13–15.5 cm

Resident, uncommon in north-west, rare elsewhere; breeds 2,650–3,650 m (2,450–4,250 m), winters 1,920–2,440 m (–75 m). **ID** Adult largely grey-brown with red face, yellow panel on black wings with white on tertials, and white rump. Juvenile lacks red face of adult; upperparts and breast faintly streaked, and has buffish tips to coverts and tertial markings. **Voice** Liquid twittering flight call; song comprises varied mix of twittering, interspersed by repeated *tew-tew-tew* and *tuwee-it* phrases. **HH** Perches acrobatically on seed heads and hangs upside-down on taller plants. Flight light and bouncing, accompanied by continual twittering. Open mixed forests, often in or near edges and clearings; seed-rich harvested fields.

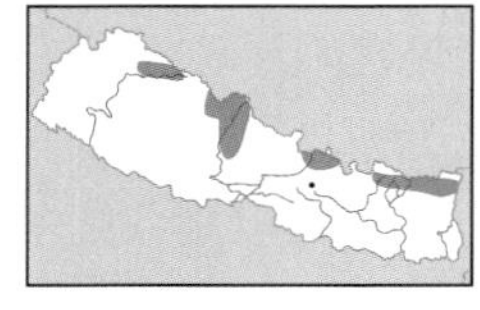

Twite *Carduelis flavirostris* 13–13.5 cm

Rare visitor, possibly resident; 3,965–4,575 m (–2,715 m). **ID** Heavily streaked, with buff wing-bars and white edges to wings and tail. Throat and breast rich buff. Small bill yellowish (black in breeding male). Male has pinkish rump (obscured in non-breeding plumage). **Voice** Weak, nasal, twittering call interspersed by distinctive series of wheezy *tweee* notes. **HH** Feeds chiefly on small seeds picked from ground. Boulder-strewn alpine meadows and stony hills; also grassy or stony flats and at village edges.

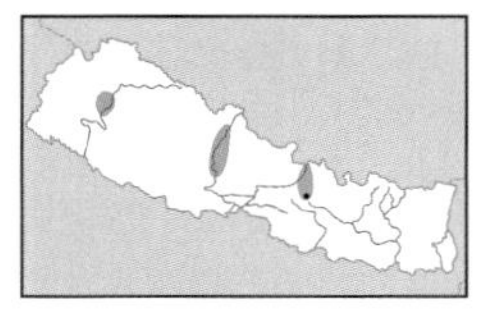

Spectacled Finch *Callacanthis burtoni* 17–18 cm

Very uncommon, erratic and local winter visitor; 2,135–3,355 m. **ID** Wings black, with bold whitish tips to greater coverts and flight feathers, and white tip to black tail. Male has pinkish-red 'spectacles', maroon-brown mantle and pinkish-red wash to underparts. 'Spots' on wing-coverts pinkish-white. Female has paler head with orange-yellow 'spectacles', olive-brown mantle and buffish-brown underparts with yellowish wash to breast. Juvenile has buff eye-patch. **Voice** Loud, clear *pwee* and a softer *tew-tew*. **HH** Feeds unobtrusively, mainly on ground, also in low bushes. Oak and hemlock forests in winter; rhododendron and fir forests in spring.

♀
juv
♂
Red-fronted Serin
♂
♀
Tibetan Serin
♂
♀
juv
Yellow-breasted Greenfinch
♂
juv
caniceps
European Goldfinch
♂
♀
Twite
♂
♀
Spectacled Finch

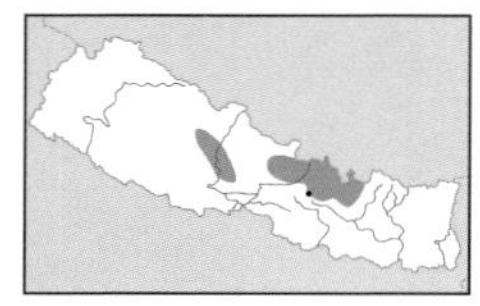

Blanford's Rosefinch *Carpodacus rubescens* 15 cm

Rare and local, possibly resident; breeds 2,745–3,050 m; winters 2,135–3,050 m. **ID** Slimmer bill than Common Rosefinch; male duller red on head and underparts (latter with greyish cast). First-summer male browner, especially on mantle, wings and underparts. Female from female Dark-breasted by stouter bill, more uniform wings (lacking prominent pale wing-bars and tips to tertials), more uniform upperparts, reddish or bright olive cast to rump, and paler underparts. **Voice** Short, thin, high-pitched *sip* call; also a series of abrupt, rising and falling notes, *pitch-ew, pitch-it, chit-it, chit-ew*.... **HH** In small flocks in non-breeding season; singly or in pairs when breeding. Glades in coniferous and mixed conifer–birch forest.

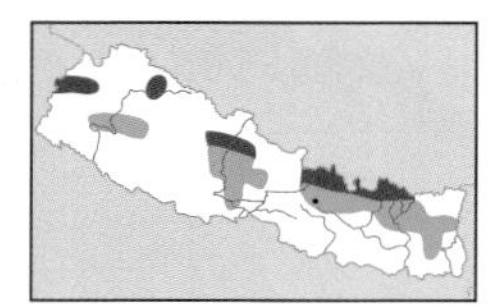

Dark-breasted Rosefinch *Carpodacus nipalensis* 15–16 cm

Fairly common resident; breeds 3,050–3,900 m (–4,270 m), winters 1,830–2,745 m (–1,370 m). **ID** Slim, with slender bill. Male has maroon-brown breast-band, dark eye-stripe and maroon-brown upperparts. Forehead, supercilium, throat and belly bright pink. Female lacks supercilium, has unstreaked underparts, diffusely streaked mantle, buffish wing-bars and tips to tertials, and olive-brown rump and uppertail-coverts. First-summer male similar to female, but has maroon-brown upperparts. **Voice** Calls include a plaintive wailing double whistle and a *cha-a-rr* alarm; song a monotonous chipping. **HH** In pairs or small flocks according to season. Forages on ground or in bushes. Breeds in open oak–rhododendron and fir–rhododendron forest, shrubberies above treeline and steep grassy slopes with scattered bushes or boulders; winters in forest clearings, and cultivation with nearby bushes.

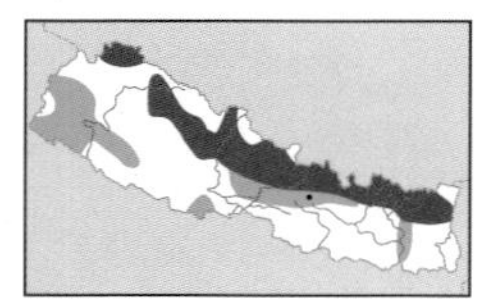

Common Rosefinch *Carpodacus erythrinus* 14.5–15 cm

Common and widespread resident; breeds 2,650–4,300 m; winters 100–2,000 m. **ID** Compact, with short, stout bill. Male has red head, breast and rump. Female and first-year male have streaked upperparts and underparts, rather plain face with beady black eye, and double wing-bar. Migrant nominate subspecies has less red in male, and female is less heavily streaked, compared to resident subspecies. **Voice** Distinctive, clear, rising *ooeet* call; alarm a *charp chay-eeee*; song a monotonous clear whistling *weeeja-wu-weeeja*. **HH** Gregarious in non-breeding season, in pairs when breeding. Forages by hopping on ground and clambering among foliage of bushes and trees. Sometimes perches inactively for long periods. Undulating flight like other rosefinches. Breeds in high-altitude shrubbery and open forest; winters in cultivation with bushes and open wooded country.

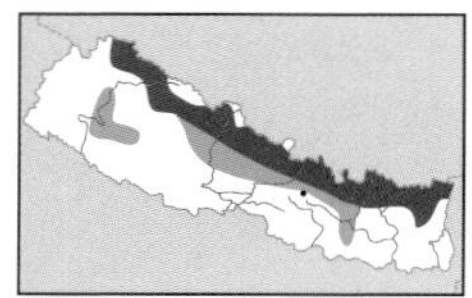

Beautiful Rosefinch *Carpodacus pulcherrimus* 15 cm

Common, widespread resident; breeds 3,600–4,650 m, winters 2,100–3,300 m. **ID** Small, compact rosefinch. Male has lilac-pink supercilium, rump and underparts, cold pinkish-grey coloration to heavily streaked crown and mantle, and pronounced shaft-streaking on underparts (especially flanks and undertail-coverts). **Voice** Flight call is harsh *chaannn*, other calls include a soft *trip* and soft twitter. **HH** Habits similar to Common. Breeds in high-altitude shrubberies; winters on bush-covered slopes and upland terraced cultivation with bushes.

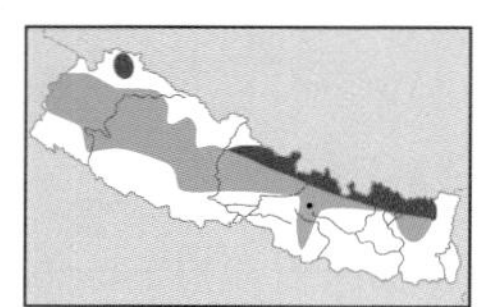

Pink-browed Rosefinch *Carpodacus rodochroa* 14–15 cm

Fairly common, widespread resident; breeds 3,050–3,965 m, winters 915–3,000 m. **ID** Small, compact rosefinch. Male has pink supercilium, rump and underparts, maroon-pink crown and ear-coverts, and pinkish-brown mantle. Female and first-year male have prominent buff supercilium contrasting with dark ear-coverts, brownish-buff mantle, tawny rump, and strong tawny wash from breast to undertail-coverts. **Voice** Loud *per-lee* call; song is sweet and lilting. **HH** Feeds unobtrusively on ground or in bushes. Breeds in dense bushes near forest edges that offer good cover, and rhododendron and juniper shrubberies; winters in oak forest and on bush-covered slopes.

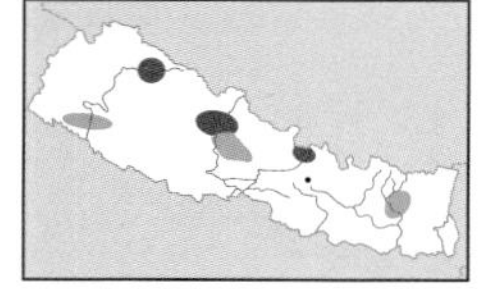

Vinaceous Rosefinch *Carpodacus vinaceus* 13–16 cm

Rare, possibly resident; breeds 3,050–3,200 m; winters 1,065–3,050 m. **ID** Male mainly dark crimson, with diffuse dark streaking on mantle. Bright pink supercilium and pinkish-white tips to tertials. Female from other female rosefinches by absence of supercilium (plain-faced appearance), whitish tips to tertials (can be lacking when worn), and warm brownish-buff coloration to streaked underparts (almost concolorous with upperparts). **Voice** Calls include a hard *pwit* or *zieh* with a whiplash quality; song a simple *pee-dee, be do-do*. **HH** Forages on ground or low down in bushes. Perches quietly in bushes for long periods. Dense forest undergrowth and glades with much bamboo.

♂
♀
Blanford's Rosefinch
♂
♀
Dark-breasted Rosefinch
♂
♂
roseatus
erythrinus
♀
♀
Common Rosefinch
♂
♀
Beautiful Rosefinch
♂
♀
Pink-browed Rosefinch
♂
♀
Vinaceous Rosefinch

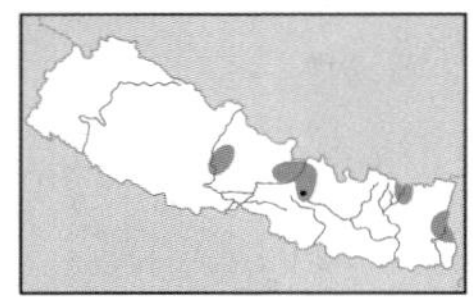

Dark-rumped Rosefinch *Carpodacus edwardsii* 16–17 cm

Rare, possibly resident; 2,440–3,635 m. **ID** Male confusable with Spot-winged, but pink tips to coverts and tertials much less prominent, lacks prominent pink on rump, and has dark breast and flanks. Female from Pink-browed by larger size, stockier appearance, stouter bill, darker brown upperparts and deeper brownish underparts (with breast usually a shade darker than belly), while pale fringes to tertials have distinctly paler tip to outer edge (more even on Pink-browed). Pale tips to coverts and tertials less prominent than on Spot-winged, with darker (browner) underparts, less distinct supercilium, paler ear-coverts, and more finely and sparsely streaked throat and breast. First-summer male has pink wash to supercilium, ear-coverts and breast. **Voice** Abrupt, short, rather shrill, high-pitched, metallic *tswii* call. **HH** Breeds in high-altitude shrubberies and fir forest; winters in open forest, and bamboo and scrub thickets.

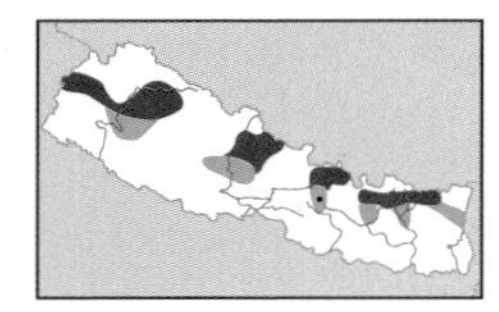

Spot-winged Rosefinch *Carpodacus rodopeplus* 15 cm

Frequent resident; breeds 3,050–4,000 m, winters 2,000–3,050 m. **ID** Male has pink supercilium and underparts, maroon upperparts (splashed pink) and pinkish tips to wing-coverts and tertials. Female has prominent buff supercilium, buff tips to tertials, and fulvous underparts with bold streaking on throat and breast. Head pattern more striking than female Dark-rumped; also paler and more heavily streaked throat, and more prominent wing-bars and tips to tertials. **Voice** Nasal *churr-weeee* call, the second part higher pitched. **HH** Breeds in shrubberies above treeline and alpine meadows; winters in bushes and bamboo thickets in forest and damp forest ravines.

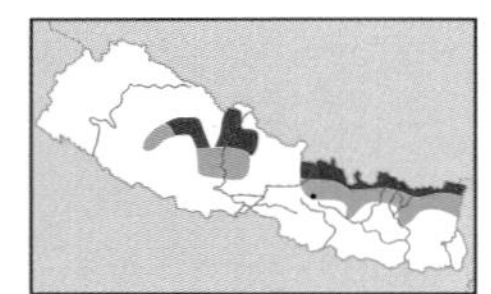

White-browed Rosefinch *Carpodacus thura* 17 cm

Resident, uncommon in west, frequent or fairly common in central and eastern Nepal; breeds 3,800–4,200 m, winters 2,440–3,660 m (–1,830 m). **ID** Large size and long-tailed appearance. Male has pink-and-white supercilium, pink rump and underparts, prominent pinkish median coverts wing-bar, and heavily streaked brown upperparts. Female and first-year male have prominent supercilium, ginger-brown throat and breast, heavily streaked underparts, and (heavily streaked) olive-yellow (female) or reddish-brown (first-year male) rump. **Voice** Calls include sharp, buzzing *deep-deep, deep-de-de-de-de* and loud, harsh whistling *pwit pwit*; weak twittering song. **HH** Breeds in high-altitude shrubberies and open forest; winters on open hillsides with bushes.

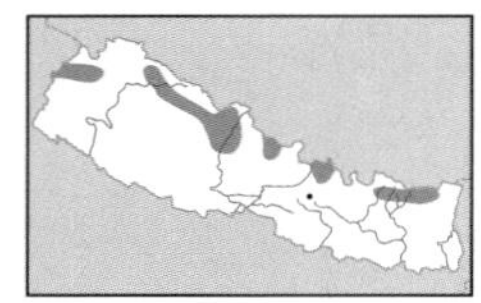

Streaked Rosefinch *Carpodacus rubicilloides* 19 cm

Locally fairly common resident in far north; breeds up to 3,500–4,700 m, winters 2,700–3,660 m (–2,400 m). **ID** Large with long tail. Male from Great by deeper crimson-pink head and underparts (with smaller white spots), prominently streaked upperparts and dark centres to wing feathers. Female nondescript, lacking supercilium. From Great by darker grey upperparts with heavier streaking, more heavily streaked ear-coverts and underparts, and darker centres to wing-coverts and tertials. **Voice** Calls include loud *twink* and melancholy *dooid dooid*; song comprises slowly descending *tsee-tsee-soo-soo-soo*. **HH** Breeds in dry high-mountain steppe with scattered bushes north of main Himalayas; winters in nearby valleys.

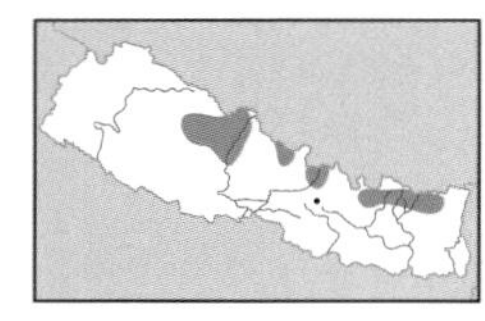

Great Rosefinch *Carpodacus rubicilla* 19–20 cm

Uncommon resident in far north; 3,660–5,550 m (–2,650 m). **ID** Large with long tail. Male from Streaked by paler rose-pink head and underparts, and paler sandy-grey upperparts only faintly streaked. Female and first-year male similar to female Streaked, but have paler, sandy-brown upperparts (only faintly streaked), less heavily streaked ear-coverts and underparts, paler, more uniform wing-coverts and tertials, and unstreaked rump. In flight, both sexes show white edge to outer tail feathers (barely apparent on Streaked). **Voice** Loud *tooey tooey* contact call; rapid twittering flight calls; brief *twee-twee-twe-cho-chush-u* song. **HH** Higher alpine zone on sparsely vegetated and rocky ground; also *Caragana* scrub in winter. **AN** Spotted Great Rosefinch, if local form *severtzovi* is split.

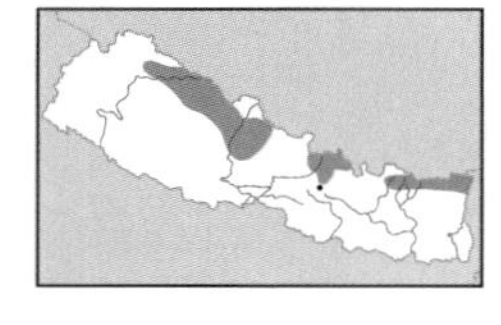

Red-fronted Rosefinch *Carpodacus puniceus* 20 cm

Frequent resident; breeds 4,265–5,490 m, winters 2,745–4,575 m. **ID** Large size, conical bill and short tail. On male, red of plumage contrasts with brown crown, eye-stripe and upperparts. Female and first-year male lack supercilium, and have dark and heavily streaked upperparts; breast variably washed pale yellow, and rump and uppertail-coverts more olive than back, or are yellow. **Voice** Calls include bulbul-like cheery whistle *are-you-quite-ready*; short *twiddle-le-de* song. **HH** Steep rocky mountainsides with dwarf scrub and grassy, stony slopes above treeline.

♂
♀
Dark-rumped Rosefinch
♂
Spot-winged Rosefinch
♀
♂
♀
White-browed Rosefinch
♂
♀
Streaked Rosefinch
♂
♀
severtzovi
Great Rosefinch
♂
♀
Red-fronted Rosefinch

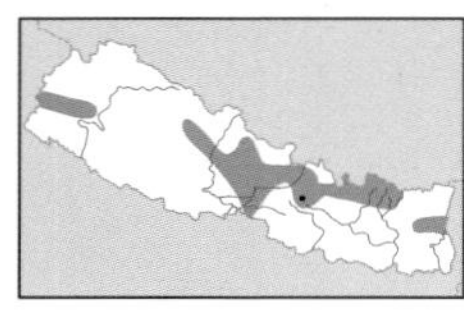

Brown Bullfinch *Pyrrhula nipalensis* 16–17 cm

Uncommon resident; mainly 1,830–3,050 m (250–3,200 m). **ID** Adult has grey-brown mantle, grey underparts, narrow white rump and long tail. Additional features include narrow black surround to bill, black lores and scaling on forehead and crown, white patch below eye. Sexes similar, but male has crimson-pink outer edge to inner tertial (yellow in female). Juvenile has brownish-buff upperparts, warm buff underparts and lacks adult head pattern. **Voice** Mellow *per-lee* and soft whistling twitter; repeated mellow *her-dee-a-duuee* song. **HH** Moist broadleaved forests, especially of oaks.

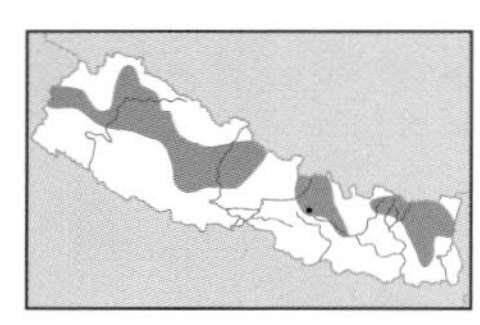

Red-headed Bullfinch *Pyrrhula erythrocephala* 17 cm

Fairly common, widespread resident; breeds 3,050–4,000 m, winters 1,830–3,865 m. **ID** Male has orange crown, nape and breast, and grey mantle. Female has yellow crown and nape. Wing-bars of both sexes greyish-white. First-year male similar to female, but has olive-yellow breast and upper flanks. Juvenile similar to female, but browner, and head and upperparts warm brown. **Voice** Soft, mellow *pew-pew* call; low, mellow *terp-terp-tee* song. **HH** Breeds in conifer and rhododendron forests and edges, also shrubberies; winters in rhododendron, and oak/conifer/rhododendron forests.

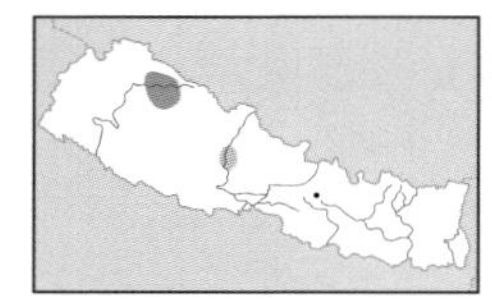

Black-and-yellow Grosbeak *Mycerobas icterioides* 22 cm

Very rare and local resident, mainly recorded in Rara National Park, 3,050 m. **ID** Male very similar to male Collared. Black of plumage duller (less glossy) and has black thighs. Nape, upperparts and underparts generally a purer, paler lemon-yellow. Female very different from female Collared: mantle and breast pale grey and concolorous with head, and belly, flanks and rump pale peachy-orange. Immature male similar to adult female, but has yellow rump and blackish wings and tail, with patches of black on scapulars and throat. **Voice** Throaty whistle *pi-riu pi-riu pi-riu*; song a rich, clear *prr-trweeet-a-troweeet* or *tookiyu tookiyu*. **HH** Habits like Collared. Coniferous forest.

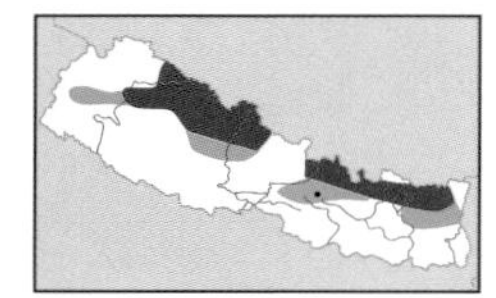

Collared Grosbeak *Mycerobas affinis* 22 cm

Widespread resident; locally fairly common; breeds mainly 3,000–3,900 m, winters down to 2,440 m (–1,050 m). **ID** Male very similar to male Black-and-yellow, although black is strongly glossed, has yellow thighs, yellow is more golden-yellow, and nape has strong orange cast. Female very different from female Black-and-yellow: grey head well demarcated from olive-yellow underparts, and has greyish-olive mantle, largely yellowish-olive wings and olive-yellow rump contrasting with black tail. Immature male resembles adult male, but duller and mantle mottled black. **Voice** Mellow, rapid *pip-pip-pip-pip-pip-pip-ugh* call; rising whistling *ti-di-li-ti-di-li-um* song. **HH** In pairs or small parties according to season, frequently calling noisily to each other. Often perches on tops of tall trees, but also feeds in low bushes and on ground. Flight swift and undulating. Coniferous and coniferous/broadleaved forest.

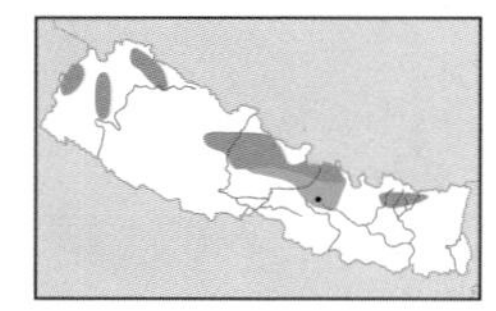

Spot-winged Grosbeak *Mycerobas melanozanthos* 22 cm

Rare, local resident; breeds 3,050–4,000 m, winters 1,400–2,135 m. **ID** Stocky, short-tailed and stout-billed grosbeak. Both sexes have broad white tips to greater coverts, tertials and secondaries. Male has black head and upperparts (including rump) and lemon-yellow underparts. Female has bold blackish streaking on yellow head and body, and striking head pattern with yellow supercilium, broad black stripe through ear-coverts and black malar stripe. Juvenile as female, but underparts whiter. Immature male similar to female but head mainly black. **Voice** Rattling *krrrr* or *charrarauk* call; loud, melodious whistling *tew-tew-teeeu* song. **HH** Habits like Collared. Breeds in coniferous/broadleaved forest; winters in broadleaved forest.

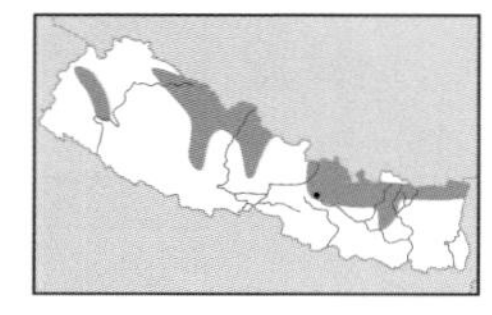

White-winged Grosbeak *Mycerobas carnipes* 22 cm

Widespread resident, locally fairly common; breeds mainly 3,050–4,720 m (–4,600 m), winters down to 2,745 m. **ID** Long-tailed grosbeak. Male has black head, mantle and breast, and olive-yellow rump and rest of underparts; yellow tips to greater coverts and tertials, and large white patch at base of primaries (very prominent in flight). Female resembles male, with distinctive white patch at base of primaries, but black of plumage replaced by sooty-grey, grey underparts become dull yellowish-olive on lower belly and vent, and has dull olive-yellow rump. Juvenile similar to adult female, but browner, with paler fringes to upperparts and olive-yellow tips to median (and greater) coverts. First-year as female; first-summer male has yellower rump, vent and belly, and more olive mantle. **Voice** Soft, nasal *schwenk* or squawking *wit-wet-et et* call; *wet-et-et-un-di-di-di-dit* song. **HH** Habits like Collared. Juniper shrubberies near treeline and forest with junipers.

ad
♂
1st-sum
♂
juv
♀
Brown Bullfinch
Red-headed Bullfinch
♂
♂
Black-and-yellow Grosbeak
♀
Collared Grosbeak
♀
♂
♂
♀
♀
White-winged Grosbeak
Spot-winged Grosbeak

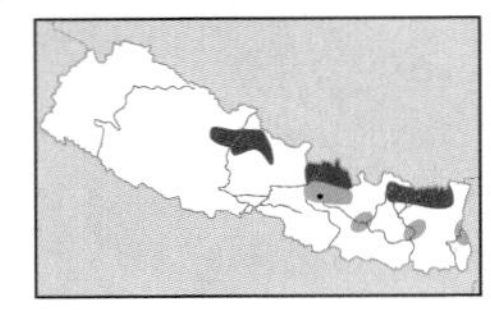

Crimson-browed Finch *Propyrrhula subhimachala* 19–20 cm

Locally frequent resident; breeds 3,150–4,000 m, winters 2,590–3,050 m. **ID** Short, stubby bill. Male has red forehead, throat and upper breast, greenish upperparts and greyish underparts. Female has olive-yellow forehead and supercilium, greyish belly and greenish-olive upperparts. **Voice** Melodic sparrow-like *chirp*; bright and varied warbling song, also a *ter-ter-tee*. **HH** Quiet, rather sluggish and shy. Feeds in bushes and small trees chiefly on seeds and berries, especially barberries. Breeds in shrubberies; also in conifer–oak forest in non-breeding season; favours junipers.

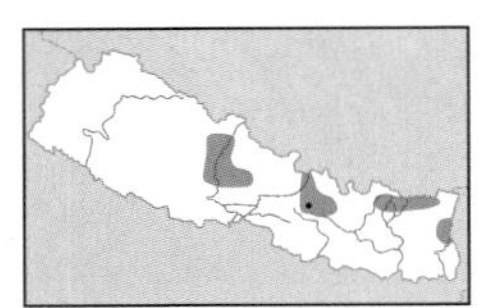

Scarlet Finch *Haematospiza sipahi* 18 cm

Uncommon, local resident; mainly 1,220–2,560 m (520–3,100 m). **ID** A stocky, short-tailed finch, with stout pale bill. Male mainly bright scarlet, with darker wings and tail. Female has bright yellow rump; upperparts olive-green with diffuse dark feather centres, and may show white mottling on crown and mantle; underparts similar, but greyer, and paler towards vent. Some female-plumaged birds have orange rump and orange-brown tinge to crown, nape and mantle. **Voice** Loud, pleasant *too-eee* and a *kwee-i-ur* calls; clear, liquid *par-ree-reeeeeee* song. **HH** Singly, in pairs or flocks, depending on season. Forages on ground, in bushes and treetops. Flight swift, powerful and undulating. Edges, clearings and ravines in broadleaved forest, especially near streams.

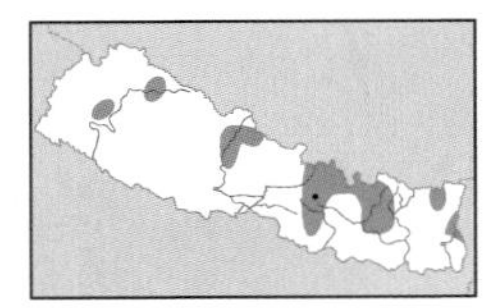

Red Crossbill *Loxia curvirostra* 16–17 cm

Locally frequent, probably resident; 2,590–3,660 m (–2,100 m). **ID** Dark bill with crossed mandibles and deeply forked tail. Male rusty-red, with darker wings and tail. First-winter/summer male duller orangey-red, with much variation (some being greenish-yellow). Female olive-green, with brighter greenish-yellow rump, and dark wings and tail. Juvenile buffish and boldly streaked on upperparts and underparts; mandibles are not initially crossed. **Voice** Loud, hard *chip chip* call; song comprises series of call notes. **HH** Gregarious when not breeding, forming restless, noisy flocks. Feeds in tops of tall conifers; most easily detected by call. Very agile when foraging, often using its bill as an aid. Coniferous forest, especially hemlock. **AN** Common Crossbill.

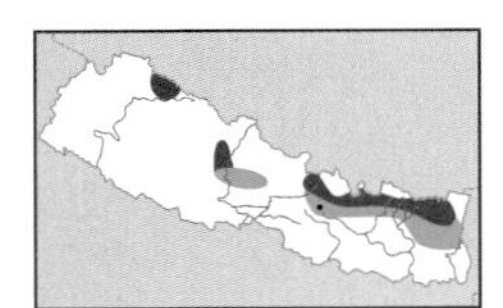

Golden-naped Finch *Pyrrhoplectes epauletta* 15 cm

Very uncommon resident; breeds 3,260–3,355 m, winters 1,525–3,000 m. **ID** Small, stocky finch with fine bill. White 'stripe' on tertials in both sexes. Male black, with orange crown and nape, and orange flash at sides of breast. Female has olive-green head, grey mantle, and rufous-brown wing-coverts and underparts. Juvenile as female, but duller. First-winter male can show scattered orange feathers on nape and black feathers on underparts. **Voice** Thin, high-pitched *teeu*, *purl-lee* and squeaky *plee-e-e* calls; song comprises rapid *pi-pi-pi-pi* and soft, bullfinch-like piping. **HH** Very quiet and unobtrusive. Forages on ground or in bushes, often low down in foliage. Breeds in subalpine rhododendron shrubberies; winters in dense undergrowth in oak–rhododendron forests.

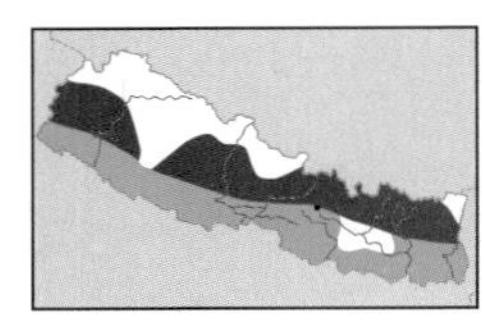

Crested Bunting *Melophus lathami* 17 cm

Fairly common, widespread resident; breeds 1,220–2,440 m and possibly lower, winters 75–1,460 m. **ID** Always has crest and chestnut on wing and tail; tail lacks white. Male has bluish-black head and body (paler fringes when fresh). Female and first-winter male streaked on upperparts and breast; first-winter male darker and more heavily streaked than female, with olive-grey ground colour to underparts. **Voice** Call *tip* or *pink*; *tsri-tsri-tsi-tsu-tsu-tsu* song, last three notes descending. **HH** Singly, in pairs or small, loose parties according to season. Forages on ground; perches readily on tops of bushes or boulders in upright posture. Undulating flight. Dry rocky and grassy hillsides, and terraced cultivation.

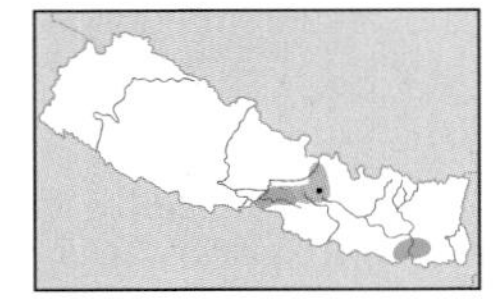

Black-headed Bunting *Emberiza melanocephala* 16–18 cm

Rare winter visitor and passage migrant, frequent in Koshi area, rare elsewhere; 75–105 m (–2,050 m). **ID** Larger than Red-headed Bunting (see Appendix 1) with longer bill. Male has black head and chestnut mantle. Female when worn may show ghost pattern of male; fresh female almost identical to Red-headed, but indicative features (not always apparent) include rufous fringes to mantle and/or back, slight contrast between throat and greyish ear-coverts, and more uniform yellowish underparts. Immature has buff underparts and yellow undertail-coverts. **Voice** Call *pyiup* or *tyilp*; *plut* flight call. **HH** Typical bunting, gregarious in winter, feeding and roosting in flocks. Forages by hopping or creeping on ground, on seeds, especially grass seeds, also insects in summer. Undulating flight. Crop fields, grasslands and sugarcane plantations.

♂
Crimson-browed Finch
♀
♀ variant
♂
Scarlet Finch
♀
♀
♂
♂
Red Crossbill
juv
♀
Golden-naped Finch
♂ br
♂ non-br
♀ worn
imm
Black-headed Bunting
♂
♂
1st-win
♀
Crested Bunting

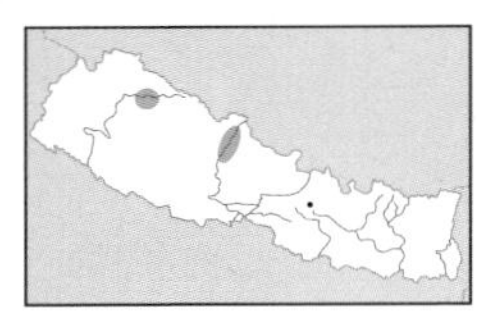

Yellowhammer *Emberiza citrinella* 16.5 cm

Very rare, local and erratic winter visitor; 1,100–2,745 m. **ID** Most show yellow on supercilium, throat and belly, which help separate from Pine Bunting. Compared to Yellow-breasted Bunting, head pattern less striking, has less prominent wing-bars and is longer tailed. Some first-winter females lack yellow, and are very difficult to separate from female Pine, but belly never white, and has yellowish (rather than whitish) edges to primaries and base of tail. Hybrids with Pine can show varied intermediate characters. **Voice** As Pine Bunting. **HH** Typical bunting, see Black-headed. Upland cultivation.

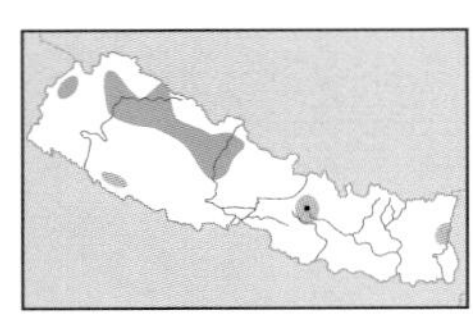

Pine Bunting *Emberiza leucocephalos* 17 cm

Irregular winter visitor to west, locally frequent; mainly 2,440–3,050 m (–250 m). **ID** Male has chestnut supercilium and throat, whitish crown and ear-coverts spot, whitish gorget below chestnut throat, and chestnut streaking on breast and flanks, although pattern obscured in winter. Female has greyish supercilium and nape/neck sides, dark border to ear-coverts, usually some chestnut streaking on breast/flanks, and white belly. Chestnut rump and streaking on breast/flanks, and different call, are best distinctions from female Commmon Reed Bunting (see Appendix 1). **Voice** Calls include short *dzik*, rasping *dzuh* and rolling *prul-lullu*. **HH** Typical bunting, see Black-headed. Fallow fields in dry hills.

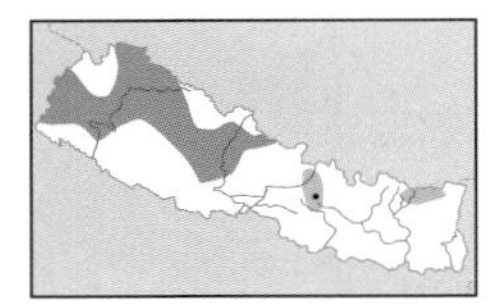

Rock Bunting *Emberiza cia* 16 cm

Widespread resident in west, locally common, mainly 2,440–4,000 m, occasionally winters to 1,800 m. **ID** Male has grey head and breast, with black head markings. Rump and rest of underparts deep rufous. Female is dull version of male, with less pronounced head pattern. Juvenile warm buff, heavily streaked brown, with dark border to ear-coverts; rufous tinge to rump and belly. **Voice** Sharp *tsee* and soft *yip* calls; fast, ringing *zit ziterit zit zit ziterit zit* song. **HH** Typical bunting. Breeds on open, dry grassy and rocky slopes, edges of open conifer forests and terraced cultivation; winters in stubble and fallow fields.

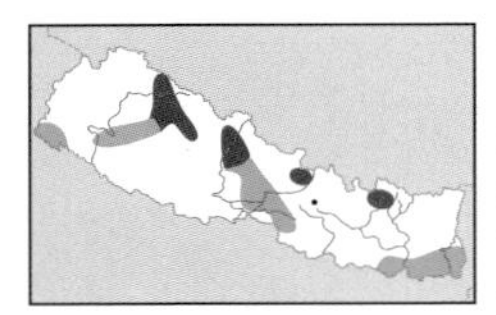

Chestnut-eared Bunting *Emberiza fucata* 16 cm

Rare, status uncertain, probably mainly winter visitor and passage migrant; summers 2,135–2,300 m (–5,000 m), winters 75–915 m. **ID** Adult has chestnut ear-coverts, black breast streaking and usually some chestnut on breast-sides. Some first-winters rather nondescript, but plain head with warm brown ear-coverts and prominent eye-ring distinctive. **Voice** Rapid, twittering *zwee zwizwezwizizi trup-trup* song; explosive *pzick*, higher-pitched *zii* and lower-pitched *chutt* calls. **HH** Typical bunting. Dry rocky, grassy or bushy slopes; also winters in grassland with bushes.

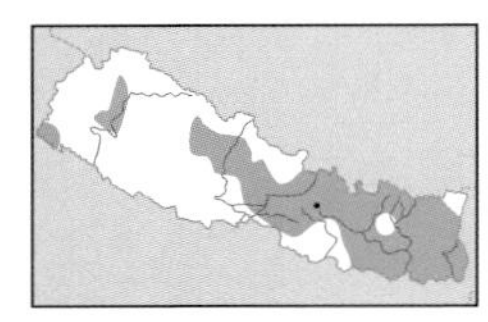

Little Bunting *Emberiza pusilla* 13 cm

Widespread, locally uncommon winter visitor; 75–2,000 m (–3,560 m). **ID** Small size. From Common Reed (see Appendix 1) by chestnut ear-coverts (and often supercilium and crown-stripe), and absence of dark moustachial stripe. Has more pointed bill than Reed, and uniform grey-brown mantle, lightly streaked dark brown (lacking pale 'braces'), more finely streaked breast and flanks, and more prominent pale median and greater coverts wing-bars. **Voice** Sharp *tzic* call. **HH** Typical bunting. Stubble, ploughed or grass fields.

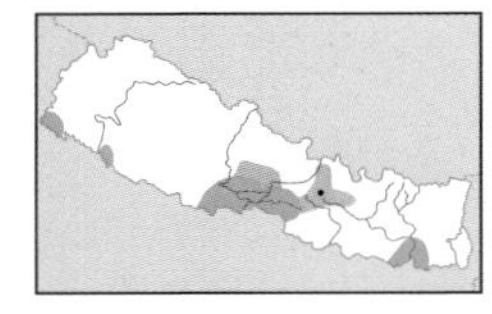

Yellow-breasted Bunting *Emberiza aureola* 15 cm

Very local and uncommon, mainly passage migrant, small numbers overwintering; 75–915 m (–3,400 m). **ID** A stocky, comparatively short-tailed bunting. More direct, less undulating flight compared to other buntings, appearing more weaver- or sparrow-like. Male has black face, chestnut breast-band (obscured when fresh) and white inner wing-coverts. Female has striking head pattern (broad yellowish supercilium and pale crown-stripe), boldly streaked mantle (pale 'braces' often apparent) and prominent white median coverts bar. Juvenile as female, but underparts paler yellowish-buff, with fine, dense streaking on breast and flanks. **Voice** Calls include soft *chup* and metallic *tick* like Little. **HH** Typical bunting. Cultivation, stubbles and grassland. Globally threatened.

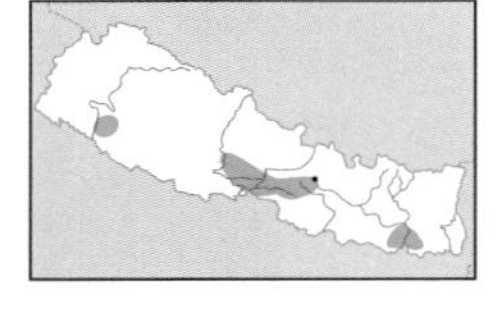

Black-faced Bunting *Emberiza spodocephala* 15 cm

Rare and local winter visitor; 75–250 m (–1,280 m). **ID** Male has greenish-grey head with blackish lores and chin, and yellow underparts. Non-breeding male has yellow submoustachial stripe and throat. Female has yellowish supercilium, yellow throat, olive rump and white on tail (see Chestnut Bunting in Appendix 1). **Voice** Soft *tsip* or sharper *tzit* calls. **HH** Typical bunting. Paddyfields and marsh edges.

♂ non-br
♂ br
Yellowhammer
♀
♂ br
♂ non-br
♀
Pine Bunting
♂
♀
Rock Bunting
♀
♂
1st-win
Chestnut-eared Bunting
1st-win
Little Bunting
♀
♂ non-br
♂ br
♂ non-br
♂ br
juv
Black-faced Bunting
♀
Yellow-breasted Bunting

APPENDIX 1:
VAGRANTS, EXTIRPATED AND RECENTLY RECORDED SPECIES

Extirpated species are those not recorded since the 19th century. A few recently recorded species whose status is uncertain are also included.

Rain Quail *Coturnix coromandelica* — 18 cm

Vagrant. **ID** Male similar to male Common Quail, but has more strongly marked head pattern, variable black breast patch and streaking on flanks, and cinnamon sides to neck and breast. Female smaller than female Common, with unbarred primaries. **Voice** A loud, metallic and high-pitched *whit-whit* repeated 3–5 times. **HH** Habits similar to Common Quail. During breeding season in monsoon can be heard frequently throughout day. Crops, grassland, scrub.

Jungle Bush Quail *Perdicula asiatica* — 17 cm

Extirpated, only recorded in 19th century. **ID** Male has barred underparts, rufous-orange throat and supercilium (with supercilium edged white above and below), white moustachial stripe, brown ear-coverts and orange-buff vent. Female has vinaceous-buff underparts, with head pattern like male. **Voice** A harsh grating *chee-chee-chuck, chee-chee-chuck*. **HH** In coveys of up to 20 outside breeding season. Birds in a covey bunch together and squat low before suddenly bursting into flight in all directions with loud whirring of wings. Dry grass and scrub and deciduous forest; chiefly on dry and stony ground.

Fulvous Whistling-duck *Dendrocygna bicolor* — 51 cm

Vagrant. **ID** Larger than Lesser Whistling-duck, with bigger, squarer head and larger bill. Adult from adult Lesser by warmer rufous-orange head and neck, dark blackish line down hindneck, dark striations on neck, more prominent streaking on flanks, indistinct chestnut-brown patch on forewing, and white band on uppertail-coverts. Often associates with Lesser. **Voice** Very noisy in flight and at rest; a repeated whistle *k-weeoo*. **HH** Freshwater marshes.

Bean Goose *Anser fabalis* — 66–84 cm

Vagrant. Black bill with orange band, and orange legs. Slimmer neck and smaller, more angular head than Greylag Goose; head, neck and upperparts darker and browner. In flight lacks pale grey forewing of Greylag. **Voice** Honking and cackling flight call, similar to other geese, but contains distinctive wittering *hank-hank* phrase. **HH** Habits similar to Greylag. Large rivers.

Greater White-fronted Goose *Anser albifrons* — 66–86 cm

Vagrant. **ID** Adult from Greylag by broad white band at front of head, browner coloration, black barring on belly, and orange legs and feet. More uniform upperwing in flight, darker back and rump, and darker base to tail than Greylag. Juvenile lacks white frontal band and barring on belly; more similar to Greylag, but smaller and less stocky, browner, with darker feathering at base of bill, dark tip (nail) to bill, and orange legs and feet. **Voice** Cackling and honking flight call is higher pitched than Bean Goose and contains distinctive musical *lyo-lyok* phrase. **HH** Habits similar to Greylag. Large rivers.

Tundra Swan *Cygnus columbianus* — 115–140 cm

Vagrant. **ID** Adult white with black-and-yellow bill; yellow of bill typically as oval-shaped patch (compare Whooper Swan). Juvenile smoky-grey with pinkish bill. Smaller, with shorter neck and more rounded head than Whooper. **Voice** Similar to Whooper, with much honking and yelping, but calls typically higher-pitched. **HH** Large rivers.

Whooper Swan *Cygnus cygnus* — 140–165 cm

Only recorded in 19th century. **ID** Adult white with black-and-yellow bill; yellow of bill extends as wedge towards tip. Juvenile smoky-grey with pinkish bill. Longer neck and more angular head shape than Tundra. **Voice** Much honking and trumpeting, deeper and stronger than Tundra. **HH** Lakes and large rivers.

Mandarin Duck *Aix galericulata* — 41–49 cm

Vagrant. **ID** Male spectacular. Most striking features are reddish bill, orange 'mane' and 'sails', white stripe behind eye, and black and white stripes on side of breast. Female and eclipse male mainly greyish with white 'spectacles' and white spotting on breast and flanks. In flight, dark upperwing and underwing, with white trailing edge and white belly. **Voice** Silent, except in display. **HH** Large rivers.

Eastern Spot-billed Duck *Anas zonorhyncha* — 53 cm

Vagrant. **ID** From Indian Spot-billed Duck by lack of red loral spot, diffusely marked breast, more uniform sooty-black upperparts and flanks, blue (rather than green) speculum, dark grey tertials (with whitish fringes), and dusky bar on cheeks. **Voice** As Mallard. **HH** Similar habitat to Indian Spot-billed and habits similar to Mallard.

Baikal Teal *Anas formosa* 39–43 cm

Vagrant. **ID** Male has striking head pattern, black-spotted pinkish breast, white vertical stripe on sides of breast, black undertail-coverts and chestnut-edged scapulars. Female has complex (albeit variable) head pattern: typically, a dark-bordered white loral spot, buff supercilium broken above eye by dark crown, and white throat that curves up to form half-moon-shaped cheek-stripe. Both sexes have grey forewing and broad white trailing edge to wing in flight (like Northern Pintail). Eclipse male similar to female, but has darker and more rufous fringes to mantle, rufous breast and flanks, and less well-defined loral spot. **Voice** Chuckling *wot-wot-wot* by male. **HH** Large rivers.

Pink-headed Duck *Rhodonessa caryophyllacea* 60 cm

May be extinct. Extirpated from Nepal, only recorded in 19th century. **ID** Long neck and body, and triangular head. Male has pink head and hindneck, and dark brown foreneck and body; bill pink. Female similar, but paler, dull brown body, greyish-pink head, and brownish crown and hindneck. In flight, pale fawn secondaries, contrasting dark forewing, and pale pink underwing with dark body. **Voice** Male has low, weak whistle and female a low quack. **HH** Shy and secretive. Feeds by dabbling on surface, but can also dive; occasionally perches in trees. Secluded pools and marshes. Globally threatened.

Greater Scaup *Aythya marila* 40–51 cm

Vagrant. **ID** Larger and stockier than Tufted Duck, with more rounded head and no crest. Bill larger and wider, with smaller black nail at tip than Tufted. Male has grey upperparts contrasting with black rear end and green gloss to blackish head. Female has broad white face patch, which is less extensive on juvenile/immature. Female usually has greyish- white vermiculations ('frosting') on upperparts and flanks. Eclipse/immature male has brownish-black head, neck and breast, and variable patch of grey on upperparts. **Voice** Silent, except in display. **HH** Feeds mainly by diving, loafs on open water when not feeding. Large lakes and rivers.

Long-tailed Duck *Clangula hyemalis* 36–47 cm

Vagrant. **ID** Small, stocky duck with stubby bill and pointed tail. Swims low in water and partially opens wings before diving. Both sexes have dark upperwing and underwing in flight. Winter male mainly white, with dark cheek patch and breast, and long tail. Female and immature male variable; usually have dark crown and pale face with dark cheek patch.**Voice** Silent, except in display. **HH** Feeds by diving, often remaining underwater for many seconds. Lakes and large rivers. Globally threatened.

Smew *Mergellus albellus* 38–44 cm

Vagrant. **ID** Small, stocky 'sawbill' with square-shaped head. In flight, both sexes have dark upperwing with white wing-covert patch. Male mainly white, with black face, crest-stripe, breast-stripes and back. Flanks grey. Female, and first-winter and eclipse male, have chestnut cap and white cheeks, and mainly dark grey body. **Voice** Generally silent. **HH** Feeds diurnally, mainly by diving, flock members typically submerging in unison or quick succession. Large rivers and lakes.

Red-breasted Merganser *Mergus serrator* 52–58 cm

Vagrant. **ID** Male has spiky crest, white collar, ginger breast and grey flanks. Female, and eclipse/immature male, more closely resemble respective plumages of Goosander, but have slimmer appearance, with slimmer bill, and narrower head with weaker and more ragged crest. Chestnut of head and upper neck duller and contrasts less with grey lower neck and breast, throat only slightly paler, and has browner body. In flight, white wing patch broken by black bar, unlike on Goosander. **Voice** Silent, except in display. **HH** Habits similar to Goosander. Large rivers and lakes.

Red-throated Loon *Gavia stellata* 53–69 cm

Vagrant. **ID** Upturned-looking bill and rounded head. In non-breeding plumage, paler grey and whiter than similar Black-throated (not recorded in Nepal). Grey of crown and hindneck paler and less extensive compared to Black-throated and does not contrast so strongly with white of ear-coverts and foreneck. Red throat and uniform grey-brown upperparts in breeding plumage. **Voice** Silent except in breeding season. **HH** Feeds exclusively by diving. Large rivers and lakes. **AN** Red-throated Diver.

Greater Flamingo *Phoenicopterus roseus* 125–145 cm

Vagrant. **ID** Larger than Lesser Flamingo (not recorded in Nepal), with longer, thinner neck and longer legs. Bill larger and less prominently kinked. Adult from adult Lesser by paler pink bill with prominent dark tip, and pink facial skin. Pinkish-white head, neck and body, although Lesser can be similar. Comparatively uniform crimson-pink upperwing- coverts contrast in flight with paler, whitish body. Immature has greyish-white head, neck and body; brown-streaked coverts, bill grey tipped black, and legs grey (pinker with age).

Juvenile has brownish head, neck and body, with heavy brown streaking to upperparts. **Voice** Flocks utter a goose-like *ka-ka*. **HH** Immerses head in shallow water with bill inverted and filters food. Shallow brackish lakes, mudflats and saltpans.

Glossy Ibis *Plegadis falcinellus* 55–65 cm

Vagrant. **ID** Small, dark ibis with rather fine downcurved bill. Graceful in flight, with extended slender neck, somewhat bulbous head, and legs and feet projecting well beyond tail. Adult breeding deep chestnut, glossed purple and green; narrow white surround to bare lores. Adult non-breeding duller, with white streaking on dark brown head and neck. Juvenile similar to adult non-breeding, but dark brown with white mottling on head, and only faint greenish gloss to upperparts. **Voice** Usually silent away from breeding colonies. **HH** Walks in shallow water or wades belly-deep while probing rapidly into water and mud. Freshwater marshes and large lakes, flooded grassland and paddyfields.

White-bellied Heron *Ardea insignis* 127 cm

Extirpated, only recorded in 19th century. **ID** Large size, very long neck, huge dark bill, and large, dark legs and feet. Grey head with white throat, and white-striped grey foreneck and breast contrasting with white belly. In flight, uniform dark grey upperwing, and white underwing-coverts contrasting with dark grey flight feathers. In breeding plumage, greyish-white nape plumes, grey back plumes and white-striped breast plumes; lores and orbital skin yellowish-green. Juvenile has browner upperparts and streaked appearance to upperparts. **Voice** Loud, very donkey-like croaking bray, *ock, ock, ock, ock, urrrrrr*. **HH** Shy. Singly or in pairs. Foraging tactics similar to those of other diurnal herons. Rivers, marshes and lakes in tropical and subtropical forest. Globally threatened.

Great White Pelican *Pelecanus onocrotalus* 140–175 cm

Vagrant. **ID** Adult and immature have black underside to primaries and secondaries, contrasting strongly with white (or largely white) underwing-coverts. Pouch yellow or orange-yellow. Adult generally cleaner and whiter than Spot-billed Pelican. Adult breeding has white body and wing-coverts tinged pink, bright orange-yellow pouch, pinkish skin around eye and short drooping crest. Adult non-breeding has duller bare parts, and lacks pink tinge and white crest. Immature has variable brown on wing-coverts and scapulars. Juvenile has largely brown head, neck and upperparts, including upperwing-coverts, and brown flight feathers; upperwing appears more uniform brown and underwing shows pale central panel contrasting with dark inner coverts and flight feathers; greyish pouch becomes yellower with age. **Voice** Usually silent away from breeding colonies. **HH** Singly, in small flocks or huge concentrations on larger lakes and lagoons. Often fishes cooperatively by swimming forward in semicircular formation, driving fish into shallows; each bird then scoops up fish from water into its pouch. Roosts in flocks, usually on open sandbanks. Large lakes and rivers.

Indian Cormorant *Phalacrocorax fuscicollis* 63 cm

Vagrant. **ID** Smaller and slimmer than Great Cormorant, with thinner neck, slimmer oval-shaped head, finer-looking bill and proportionately longer tail. In flight, appears lighter, with thinner neck and quicker wingbeats. Larger than Little, with longer neck, oval-shaped head and longer bill. Adult breeding glossy black, with blue eyes, dark facial and gular skin, tuft of white behind eye, scattering of white filoplumes on neck. Non-breeding lacks white plumes; whitish throat, yellowish gular pouch and browner-looking head, neck and underparts. Immature has brown upperparts and whitish underparts. Flies and swims like typical cormorant, see Little. **Voice** Usually silent except near nest. **HH** Frequently fishes with Little. Large lakes and rivers.

Barbary Falcon *Falco* (*peregrinus*) *pelegrinoides* 33–44 cm

Vagrant. **ID** Adult has pale blue-grey upperparts, buffish underparts with only sparse streaking and barring, rufous on crown and nape, fine pale supercilium and narrow dark moustachial stripe. In flight, underwing appears pale with dark wingtips and crescent-shaped carpal patch. Possibly confusable with Red-necked Falcon, but larger and stockier, lacks barring on underparts, has more evenly barred tail, and at rest wingtips reach tail tip (fall short in Red-necked). Juvenile has darker brown upperparts with narrow rufous-buff fringes, heavily streaked underparts and underwing-coverts, and only hint of rufous on forehead and supercilium. **Voice** Generally silent away from nest. **HH** Fast and powerful falcon. Semi-desert; favours open, rocky country.

Indian Vulture *Gyps indicus* 89–103 cm

Vagrant. **ID** Key features of adult are sandy-brown body and upperwing-coverts (Griffon is more rufescent), blackish head and neck with sparse white down on hindneck (Griffon has more extensive white down), white downy ruff, yellowish bill, and lacks pale streaking on underparts; in flight, lacks broad whitish band on median underwing-coverts of Griffon, and has whiter rump and back. See Slender-billed for differences from that species. Much smaller and less heavily built than Himalayan, with darker head and neck, white ruff, and dark legs and feet. Juvenile has feathery buff neck-ruff, dark bill and cere with pale culmen, and head and neck have whitish down; from juvenile Griffon by pale culmen, darker brown upperparts with more pronounced pale streaking, and paler and less rufescent coloration to streaked underparts. Best features of juvenile from juvenile White-rumped are paler and less clearly streaked underparts, paler upper- and underwing-coverts, and whitish rump and back. **Voice** Hissing and cackling sounds. **HH** Habits like White-rumped. Cities, towns and villages. Globally threatened. **AN** Long-billed Vulture.

Slaty-breasted Rail *Gallirallus striatus* 27 cm

Vagrant. **ID** Longish bill with red base (stouter and straighter than Water Rail). Legs olive-grey. Adult has chestnut crown and nape, slate-grey foreneck and breast, white barring and spotting on upperparts, and white barring on ark grey belly, flanks and undertail-coverts. Juvenile duller, with crown and nape dark-streaked olive-brown (some with rufous tinge), upperparts paler olive-brown and more sparsely marked with white, and underparts browner with less pronounced barring on flanks. **Voice** A sharp *cerrk*. **HH** Reedy marshes and paddyfields.

Water Rail *Rallus aquaticus* 23–28 cm

Vagrant. **ID** Longish and slightly downcurved bill with red at base. Legs pinkish. Adult has dark-streaked olive-brown upperparts, grey underparts, black-and-white barring on flanks, and white undertail-coverts. Juvenile has upperparts similar to adult, but underparts buff with extensive dark brown mottling, flank barring brown and buff, and undertail-coverts rufous-buff. See Brown-breasted Rail for differences from that species. **Voice** Call resembles a squealing pig, rises in pitch then dies away. **HH** Reedy marshes. **AN** European Water Rail.

Spotted Crake *Porzana porzana* 22–24 cm

Vagrant. **ID** Profuse white spotting on head, neck and breast. Stout bill, irregularly barred flanks and unmarked buff undertail-coverts. Adult has yellowish bill with red base, and grey head and breast. Sexes similar, but female has less grey on head, neck and breast, and is more profusely spotted with white. Juvenile like female but has buffish-brown head and breast, and bill is brown. **Voice** Song can be heard on migration: a swishing *h-wet... hwet*, resembling a whiplash. **HH** Reedy marshes.

Eurasian Oystercatcher *Haematopus ostralegus* 40–46 cm

Vagrant. **ID** Black and white, with broad white wing-bar. Bill and eye reddish, and legs pinkish. White collar in non-breeding plumage. Broad white wing-bar in flight. **Voice** A piping *pi... peep... peep... peep* and *pi-peep*. **HH** Larger rivers.

White-tailed Lapwing *Vanellus leucurus* 26–29 cm

Vagrant. **ID** Blackish bill, large dark eyes and very long yellow legs. Plain head. Striking black-and-white pattern to wings. Tail all white, lacking black band of other *Vanellus*. Juvenile has dark subterminal marks and pale fringes to feathers of upperparts, and paler neck and breast than adult; crown mottled dark brown. **Voice** Calls include a *pet-oo-wit* and *pee-wick* recalling Northern Lapwing. **HH** Usually feeds in shallow water, by pecking at surface and by foot-dabbling; also probes on land. Freshwater marshes.

Grey Plover *Pluvialis squatarola* 27–30 cm

Vagrant. **ID** White underwing and black axillaries. Stockier, with stouter bill and shorter legs than Pacific Golden. Whitish rump and prominent white wing-bar. Extensive white spangling to upperparts in breeding plumage; upperparts mainly grey in non-breeding (in all plumages lacks golden spangling of Pacific Golden). **Voice** A mournful *chee-woo- ee*. **HH** Habits similar to Pacific Golden, but usually less gregarious; often in pairs or small groups, with other waders. Large lakes and rivers.

Greater Sand Plover *Charadrius leschenaultii* 22–25 cm

Vagrant. **ID** Larger and lankier than Lesser Sand Plover, with longer and larger bill, usually with pronounced gonys and more pointed tip (bill longer than distance between bill base and rear of eye). Longer legs paler, with distinct yellowish or greenish tinge. In flight, feet project noticeably beyond tail, has more pronounced dark subterminal band to tail, and broader white wing-bar on primaries. **Voice** In flight a trilling *prrrirt* or *kyrrrr... trrr*, softer and longer than Lesser. **HH** Gait and feeding behaviour typical of other plovers. Large rivers.

Whimbrel *Numenius phaeopus* 40–46 cm

Vagrant. **ID** Smaller than Eurasian Curlew, with shorter bill, often with more marked downward kink. Prominent whitish supercilium and crown-stripe, contrasting with blackish eye-stripe and sides to crown, resulting in more striking head pattern. Juvenile as adult. **Voice** Flight call distinctive, *he-he-he-he-he-he-he*, flat-toned and laughter-like. **HH** Feeds chiefly by picking from surface of open mud, also by probing. Large rivers.

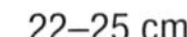

Terek Sandpiper *Xenus cinereus* 22–25 cm

Vagrant. **ID** Longish, upturned bill and short yellowish legs. In flight, prominent white trailing edge to secondaries and grey rump and tail. Adult breeding has blackish scapular lines. Juvenile similar to adult breeding, but has buff fringes and dark subterminal marks to feathers of upperparts. **Voice** Flight call a soft pleasant whistle *hu-hu-hu* and a sharper *twit-wit-wit-wit* recalling Common. **HH** A very active feeder, running erratically to chase prey; also probes deeply, and sometimes feeds in shallow water. Large rivers.

Ruddy Turnstone *Arenaria interpres* 23 cm

Vagrant. **ID** Short bill and orange legs. In flight, white stripes on wings and back, and black tail-band. In breeding plumage, complex black-and-white neck and breast pattern and much chestnut-red on upperparts; duller and less strikingly patterned in non-breeding plumage. Juvenile similar to adult non-breeding but has buff fringes to upperparts and blackish breast. **Voice** A rolled *trik-tuk-tuk-tuk* or *tuk-er-tuk*; a sharp *chick-ik* or *kuu* when flushed. **HH** Runs actively on shore, turning over pebbles to catch small invertebrate prey sheltering below; also probes soft sand. Flight strong and direct. Large rivers.

Red Knot *Calidris canutus* 23–25 cm

Vagrant. **ID** Stocky, with short, straight bill. Adult breeding has brick-red underparts. Adult non-breeding whitish on underparts and uniform grey on upperparts. Juvenile has buff fringes and dark subterminal crescents to upperparts, and buff wash on breast and flanks. **Voice** A low short *knutt... knutt*; often silent. **HH** Feeds chiefly by probing soft mud and picking from surface. Large rivers.

Sanderling *Calidris alba* 20 cm

Vagrant. **ID** Stocky with short bill. Very broad white wing-bar. Adult breeding variable; initially mottled grey and black, head and breast become more rufous with wear. Rufous birds possibly confusable with Little and Red-necked Stints, but Sanderling considerably larger, with broader wing-bar, has rufous centres to scapulars and coverts, patterned tertials, and lacks hind toe. Sides of head distinctly streaked compared to Red-necked. Non-breeding pale grey above and very white below. Blackish lesser wing-coverts, especially noticeable in flight but also show at rest as black patch at bend of wing (unless concealed by breast feathers). Juvenile checkered black-and-white above with buff wash to streaked sides of breast. **Voice** Call a liquid *plit*. **HH** Extremely active; runs swiftly, stopping suddenly to catch tiny prey or probe in sand. Large rivers.

Long-toed Stint *Calidris subminuta* 13–15 cm

Vagrant. **ID** Long yellowish legs, longish neck and upright stance recall miniature Wood Sandpiper. In all plumages, prominent supercilium and heavily streaked foreneck and breast. Adult breeding and juvenile have prominent rufous fringes to upperparts and rufous crown; juvenile has very striking mantle V. In winter, upperparts more heavily marked than Little. **Voice** Call a soft *prit* or *chirrup*, similar to but less purring than Curlew Sandpiper. **HH** Often feeds with other stints; runs energetically to pick up tiny invertebrates. Large rivers and lakes.

Curlew Sandpiper *Calidris ferruginea* 18–23 cm

Vagrant. **ID** White rump. More elegant than Dunlin, with longer, more downcurved bill, and longer legs. Adult breeding has chestnut-red head and underparts. Adult non-breeding paler grey than Dunlin, with more distinct supercilium. Juvenile has strong supercilium, buff wash to breast and buff fringes to upperparts. **Voice** Flight call a low, purring *prrriit*. **HH** Feeds in wet sand in similar manner to Dunlin, also by wading in deeper water. Large rivers.

Red-necked Phalarope *Phalaropus lobatus* 18–19 cm

Vagrant. **ID** Typically seen swimming. More delicately built than Grey Phalarope (not recorded in Nepal), with finer bill. Adult breeding has white throat and red stripe on side of grey neck. Adult non-breeding has prominent black mask and cap, with dark line down hindneck; darker grey upperparts than Grey, with white edges to mantle and scapular feathers, forming fairly distinct lines. Juvenile has dark grey upperparts with orange-buff mantle and scapular lines. **Voice** Utters a single *twick*. **HH** Swims buoyantly, spins around, darts erratically here and there. Large lakes.

Mew Gull *Larus canus* 43 cm

Vagrant. **ID** Smaller and daintier than Caspian, with shorter and finer bill; in flight, wings proportionately longer and slimmer. Adult has darker grey mantle than Caspian with more black on wingtips; bill yellowish-green, with dark subterminal band in non-breeding plumage. Head and hindneck heavily marked in non-breeding (unlike adult non-breeding Caspian). First-winter/first-summer have grey mantle; unbarred greyish greater coverts forming mid-wing panel, narrow black subterminal tail-band and well-defined dark tip to greyish/pinkish bill are best distinctions (in addition to structural differences) from second-year Caspian (which has grey mantle). Second-winter has black on primary-coverts but otherwise similar to adult. **Voice** A nasal *keow* and drawn-out shrill *glieeoo*. **HH** Larger rivers. **AN** Common Gull.

Slender-billed Gull *Chroicocephalus genei* 43 cm

Vagrant. **ID** Gently sloping forehead, longish neck and longer bill compared to Black-headed. In flight, both neck and tail appear longer than Black-headed. Adult has white head throughout year (may show grey ear-coverts spot in winter), deep red bill (often looking blackish, paler in winter), pale iris (dark in Black-headed), and variable pink flush on underparts. First-winter/first-summer from Black-headed by paler and less distinct dark eye-crescent and ear-spot (sometimes completely lacking), pale iris, paler orange bill (with dark tip smaller or absent), paler legs, less prominent dark trailing edge to inner primaries, and more extensive white on outer primaries resulting in more prominent white 'flash' on wing. Juvenile has grey-brown mantle and scapulars, with pale fringes (generally paler and lacking ginger-brown coloration of juvenile Black-headed). **Voice** Slightly deeper than Black-headed. **HH** Larger rivers. **TN** Formerly placed in genus *Larus*.

Mottled Wood Owl *Strix ocellata* 48 cm

Rare and very local, probably resident; 120 m. **ID** A distinctive owl with dark eyes, black concentric barring on pale facial discs, white and rufous mottling on upperparts, and whitish underparts barred dark brown and mixed rufous. Prominent white 'half-collar' on upper breast. Back and wings barred and mottled dark brown, and greyish-white with some rufous (mainly on coverts and inner edges of remiges). **Voice** Calls include spooky, quavering *whaa-aa-aa-aa-ah*. **HH** Open wooded areas, groves around villages and cultivation.

Long-eared Owl *Asio otus* 35–37 cm

Vagrant. **ID** From Short-eared at rest by erect ear-tufts, orange-brown coloration to facial discs, orange (rather than yellow) eyes, greyish (rather than buff) ground coloration to upperparts, and more heavily streaked belly and flanks. Further differences in flight are orange-buff base coloration to primaries and tail feathers, with more pronounced dark barring, lack of prominent dark carpal patch, more rounded wings and shorter tail. **Voice** Territorial call of male a long drawn, subdued *oo* or *hu*. **HH** On passage and in winter frequents wooded areas.

Sykes's Nightjar *Caprimulgus mahrattensis* 23 cm

Vagrant. **ID** Small, grey nightjar. Finely streaked crown, black marks on scapulars, large white patches on sides of throat and irregular buff spotting on nape forming indistinct collar. Compared to Indian Nightjar, which is similarly proportioned, crown much less heavily marked, lacks well-defined rufous-buff nuchal collar, scapulars relatively unmarked, and central tail feathers more strongly barred. **Voice** Continuous churring song; low, soft *chuck-chuck* in flight. **HH** Variety of habitats in winter including grassland.

Rufous-necked Hornbill *Aceros nipalensis* 90–100 cm

Extirpated, only recorded in 19th century. **ID** Male has bright rufous head, neck and underparts, black-and-white tail, scarlet gular pouch and blue orbital skin. Female mainly black, with red throat and white terminal band to tail. In flight, both sexes show white wingtips. Immature similar to adult male, but neck browner and bill smaller and unmarked. **Voice** A short, repeated, monosyllabic bark, higher pitched but otherwise similar to Great Hornbill. **HH** Broadleaved evergreen forest. Globally threatened.

Silver-breasted Broadbill *Serilophus lunatus* 18 cm

Extirpated, only recorded in 19th century. **ID** Large head, crested appearance, large eye, stout bluish bill, upright stance and sluggish movements give rise to distinctive jizz. Blackish supercilium, yellow eye-ring, pale chestnut tertials and rump, complex white and blue pattern to black wings, and black tail. White patch at base of primaries in flight. Female has broken white necklace. Juvenile similar to adult but has dark bill. **Voice** A *ki-uu*, like a rusty hinge. **HH** Broadleaved evergreen and semi-evergreen forest.

Ashy Minivet *Pericrocotus divaricatus* 20 cm

Vagrant or very rare winter visitor. **ID** Grey and white, lacking any yellow or red in plumage. In flight, whitish wing-bar, especially prominent on underwing. Male has black cap with white forehead. Female has grey 'cap', with black lores and narrow white forehead and supercilium. Immature has little or no white on head, brownish-grey upperparts, white tips to tertials and greater coverts, and faint brownish scaling on neck-sides and breast. **Voice** Flight call more rasping than other minivets, an unmelodious *tchue-de... tchue- dee-dee... tchue-dee-dee*. **HH** Light forest.

Swinhoe's Minivet *Pericrocotus cantonensis* 20cm

Vagrant or very rare winter visitor. **ID** Male like Ashy but hindcrown dark grey, white behind eye, upperparts tinged brown, breast and belly washed vinous-brownish, rump pale drab brownish, white shafts to tail feathers; wing patch (if present) pale yellowish-buff. Compared to male, female paler above, rump less sharply contrasting and wing patch (if present) yellower. Female from Ashy by paler rump, browner upperparts, no dark forehead band, underparts less clean. **Voice** Calls very similar to Ashy Minivet. **HH** Light forest.

Black-naped Oriole *Oriolus chinensis* 27 cm

Rare visitor. **ID** Larger, stouter bill than Slender-billed. Black mask typically broader on nape than Slender-billed (similar width in some; poorly defined in immature). Male has yellow mantle and wing-coverts concolorous with underparts (brighter than Slender-billed). Mantle and wing-coverts olive in female (as Slender-billed). Female and immature inseparable from Slender-billed by plumage. Heavier bill and diffuse nape-band best features from immature Indian Golden. **Voice** A cat- or jay-like squeal; song like Indian Golden. **HH** Broadleaved forest.

Tibetan Lark *Melanocorypha maxima* 21 cm

Resident? Breeds in Limi valley, upper Humla. 4870–5000 m. **ID** Has black patch on side of breast with rather small-headed, long-necked appearance, and has much longer, thinner bill. Has rather uniform head, lacking prominent supercilium or eye-stripe (crown and nape are diffusely streaked), and crown, ear-coverts and rump can be distinctly rufous. Breast is diffusely spotted and/or washed with grey. Has white trailing edge to secondaries, and white sides and broad white tip to outer tail feathers. Juvenile has dark blackish-brown upperparts with whitish fringes, and yellowish underparts with diffuse dark spotting on breast. **Voice** Song is a disjointed, staccato, hiccupping stutter, also includes some pleasant warbled phrases and is rich in mimicry. Call is a coarse, rippled *tchu-lip.* **HH** Male sings from the top of a grassy hump, twitching open his wings excitedly. Alpine steppe and alpine marshy grassland.

Eurasian Skylark *Alauda arvensis* 18 cm

Status uncertain. Winter visitor or vagrant. **ID** Larger than Oriental Skylark, although very similar to some in plumage, with slight crest. Told from that species by longer tail, noticeable extension of primaries beyond tertials on closed wing (tertials almost reach primary tips on Oriental), white outer tail feathers (more buffish on Oriental), pronounced whitish trailing edge to secondaries, and proportionately shorter and stouter bill. Upperparts are paler sandy-brown (than on most subspecies of Oriental), and underparts whiter. Lacks rufous cast to ear-coverts, and the (indistinct) rufous panel on wing of Oriental. **Voice** Calls include a grating *chirriup* and liquid *trruwee.* **HH** Short grassland and cultivation.

Mountain Tailorbird *Phyllergates cuculatus* 13 cm

Resident? Far east. **ID** From Common Tailorbird by brighter orange-rufous forecrown, yellowish supercilium contrasting with dark grey eye-stripe, grey ear-coverts, grey nape and sides of breast, and bright yellow belly and undertail-coverts. Confusable with Broad-billed Warbler, but has long (slender) bill, less extensive rufous on crown, grey nape, yellowish supercilium, and white throat. Juvenile has olive-green crown and nape (concolorous with mantle) and diffuse eye-stripe. **Voice** Song is a thin, high-pitched and melodious whistle of 4–6 notes. **HH** Moist temperate broadleaved evergreen forest and secondary growth.

Bohemian Waxwing *Bombycilla garrulus* 18 cm

Vagrant. **ID** Mainly fawn-brown, with prominent crest. Black mask and throat, grey rump and uppertail-coverts contrasting with blackish tail that has broad yellow tip, and waxy red and yellow markings on coverts. Starling-like appearance in flight. **Voice** Often utters distinctive soft ringing trill in flight. **HH** Open country with fruiting trees and bushes.

Eastern Orphean Warbler *Sylvia crassirostris* 15 cm

Vagrant. **ID** Larger and bigger-billed than Lesser Whitethroat; more ponderous, with heavier appearance in flight. Adult has blackish crown, pale grey mantle, blackish tail and pale iris, white in male (always dark in Lesser). First-year has crown concolorous with mantle, darker grey ear-coverts and dark iris, and can appear similar to Lesser. Orphean often shows darker-looking uppertail, eye-ring is absent or indistinct, and has greyish centres and pale fringes to undertail-coverts; these features are variable and difficult to see in the field. **Voice** Strong, varied thrush-like warbling song. **HH** Scrub and groves. **TN** Usually considered race of Orphean Warbler *S. hortensis.*

Hume's Whitethroat *Sylvia althaea* 13.5 cm

Vagrant. **ID** Slightly larger, with larger and stouter bill than Lesser Whitethroat, with darker purer grey mantle, darker grey crown and blackish ear-coverts, on some whole crown can be dark grey (resulting in dark-hooded appearance) and can show strong greyish wash to sides of breast. **Voice** Calls include a subdued hard clicking *tek tek* also a churr. **HH** Scrub and open forest. **TN** Formerly treated as conspecific with Lesser Whitethroat.

Asian Stubtail *Urosphena squamiceps* 11 cm

Vagrant. **ID** Very short tail, otherwise similar to Pale-footed Bush Warbler but has more rufescent upperparts, longer and more prominent buffish supercilium, and brownish-black eye-stripe that almost reaches hindcrown. White underparts, long pale pinkish legs and large feet. **Voice** Call is a sharp *stit* in alarm. **HH** Broadleaved forest.

Baikal Bush Warbler *Bradypterus davidi* 13 cm

Vagrant. **ID** From Spotted Bush Warbler by paler brown upperparts, brownish wash on sides of neck and on breast (lacks grey breast-band of Spotted), typically weaker and browner spotting on throat and breast, whitish supercilium, and pale base to lower mandible (bill all dark in Spotted). Weakly spotted birds similar to some Chinese Bush Warblers, but have shorter tail and more boldly patterned undertail-coverts. **Voice** Calls include a hard, raspy, irregularly spaced *tshuk*. **HH** Wet marshes and reedbeds.

Brown Bush Warbler *Bradypterus luteoventris* 13.5 cm

Only recorded in 19th century. **ID** Very similar to Chinese, but has warmer brown upperparts (with slight rufescent cast), warm rufous-buff ear-coverts, sides of neck and breast, and flanks, contrasting with silky-white rest of underparts. Also has indistinct rufous-buff supercilium to eye (concolorous with lores) and lacks prominent pale tips to undertail-coverts. Lacks spotting on throat and breast. Juvenile has yellowish wash to underparts. **Voice** Song a quiet, rapid *tututututututu...* etc. **HH** Grassy hills and undergrowth in pine forest.

Lanceolated Warbler *Locustella lanceolata* 12 cm

Vagrant. **ID** From Grasshopper Warbler by bold, fine streaking (almost spotting) on throat, breast and flanks, and stronger and better-defined streaking on upperparts. In addition, smaller, with shorter tail and stouter bill, and in fresh plumage has warmer brown upperparts and warmer buff flanks, typically lacking yellow on underparts. Some can be very similar to Grasshopper: streaking on undertail-coverts less extensive but blacker and clearer cut, and tertials darker with clear-cut pale edges. **Voice** Calls include a sharp metallic *pit*, faint *tack* and shrill *cheek-cheek-cheek-cheek*. **HH** Tall grassland.

Grasshopper Warbler *Locustella naevia* 13 cm

Vagrant. **ID** From Rusty-rumped by olive-brown upperparts including rump and tail, and less prominent supercilium. From Lanceolated by (usually) unmarked or only lightly streaked throat and breast, and less heavily streaked upperparts (especially rump and uppertail-coverts). Some Grasshopper Warblers are as heavily streaked as a lightly streaked Lanceolated (which see for subtle differences). First-winter can have yellowish wash on underparts. **Voice** Calls include a hard *sit*. **HH** Tall grassland. **AN** Common Grasshopper Warbler.

Rusty-rumped Warbler *Locustella certhiola* 13.5 cm

Vagrant. **ID** Larger and more robust than Grasshopper; more prominent supercilium contrasting with greyer crown, more heavily streaked mantle with rufous tinge, rufous rump and uppertail-coverts, cleaner and narrower fringes to tertials (often whiter tips), rufous olive-brown breast-sides and flanks (contrasting with white of rest of underparts), and unstreaked undertail-coverts. In flight, dark, white-tipped tail contrasts with rufous uppertail-coverts. Juvenile has yellowish wash to underparts, light breast spotting, less distinct supercilium, and more olive-brown and less heavily streaked crown and mantle. **Voice** A sharp metallic *pit* and hard, drawn-out, descending rattle *trrrrrrrrr*. **HH** Reedbeds. **AN** Pallas's Grasshopper Warbler.

Oriental Reed Warbler *Acrocephalus orientalis* 18 cm

Vagrant. **ID** Smaller, with shorter and squarer tail, than Clamorous. Often has streaking on sides of neck and breast, and well-defined whitish tips to outer rectrices. Streaking on underparts can, however, be lacking or invisible in field, while a few Clamorous can show faint streaking. **Voice** Calls similar to Clamorous. **HH** Reedbeds.

Black-browed Reed Warbler *Acrocephalus bistrigiceps* 13 cm

Vagrant. **ID** From Paddyfield by broader and more clear-cut supercilium and more pronounced blackish lateral crown-stripes; also shorter tail and longer primary projection beyond tertials, with dark grey (rather than pale brown) legs and feet. Rufescent above, with warm buff sides of breast and flanks, in fresh plumage (upperparts olive-brown in worn plumage). **Voice** Call a soft repeated *chuk*. **HH** Tall grassland.

Moustached Warbler *Acrocephalus melanopogon* 12.5 cm

Vagrant. **ID** From other *Acrocephalus* regular in subcontinent by combination of broad white square-ended supercilium (broadens behind eye, and contrasts with blackish crown-sides and eye-stripe) and boldly streaked rufous-brown mantle. Distinctive habit of cocking tail above back in chat-like fashion. **Voice** Calls include a *trr-trr* and hard *tcht*; song varied and scratchy, although contains clear notes and trills including a subdued *lu-lu-lu-lu*. **HH** Tall reeds and grassland.

Radde's Warbler *Phylloscopus schwarzi* 12 cm

Vagrant. **ID** Similar to Dusky, with long buffish-white supercilium contrasting with dark eye-stripe. Stout bill, and orange-tinged legs and feet; call different from Dusky, a nervous *prit prit*. In fresh plumage, can show greenish-olive cast to upperparts and buffish-yellow cast to supercilium and underparts, which distinguish from Dusky. **Voice** Nervous, twittering *twit-twit* or *prit-prit*, or sharp *chuck chuck* or *tschak*. **HH** Undergrowth and bushes.

Red-faced Liocichla *Liocichla phoenicea* 23 cm

Extirpated, only recorded in 19th century. **ID** Striking, mainly olive-brown liocichla with crimson ear-coverts and sides to neck, black supercilium, crimson panel in wings, crimson undertail-coverts, and rufous-orange tip to black tail (undertail appears entirely rufous-orange). **Voice** Loud, plaintive, cheerful song of 3–8 notes: *chewi-ter-twi-twitoo*; *chi-cho-choee-wi-chu-chooee* etc.; also a mewing *jji-uuuu*. **HH** Undergrowth in moist forest and thickets.

Black-breasted Parrotbill *Paradoxornis flavirostris* 19 cm

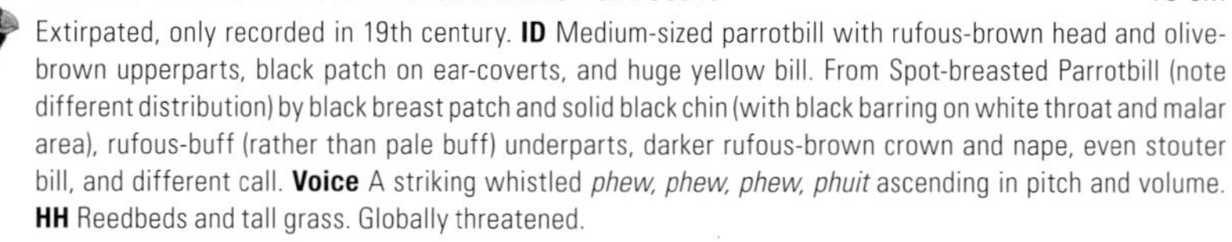

Extirpated, only recorded in 19th century. **ID** Medium-sized parrotbill with rufous-brown head and olive-brown upperparts, black patch on ear-coverts, and huge yellow bill. From Spot-breasted Parrotbill (note different distribution) by black breast patch and solid black chin (with black barring on white throat and malar area), rufous-buff (rather than pale buff) underparts, darker rufous-brown crown and nape, even stouter bill, and different call. **Voice** A striking whistled *phew, phew, phew, phuit* ascending in pitch and volume. **HH** Reedbeds and tall grass. Globally threatened.

Asian Glossy Starling *Aplonis panayensis* 20 cm

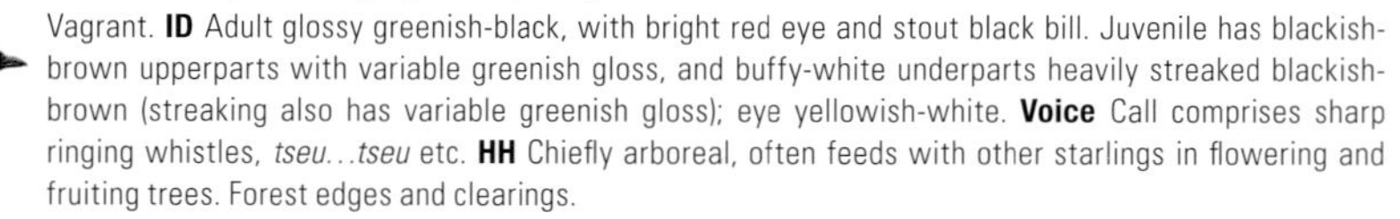

Vagrant. **ID** Adult glossy greenish-black, with bright red eye and stout black bill. Juvenile has blackish-brown upperparts with variable greenish gloss, and buffy-white underparts heavily streaked blackish-brown (streaking also has variable greenish gloss); eye yellowish-white. **Voice** Call comprises sharp ringing whistles, *tseu...tseu* etc. **HH** Chiefly arboreal, often feeds with other starlings in flowering and fruiting trees. Forest edges and clearings.

Daurian Starling *Agropsar sturninus* 19 cm

Vagrant. **ID** A small stocky starling with short tail and stout bill. Adult male has pale grey head, nape and underparts, purplish-black hindcrown patch and mantle, white tips to median coverts and rear scapulars forming prominent white V from behind, and glossy greenish-black wings with greyish-white tips to inner greater coverts and tertials. Female and juvenile duller; wing-bars and tips to scapulars less prominent in juvenile. **Voice** Call when flushed a slow, soft drawn-out *chirrup* resembling similar call of Common Starling. **HH** Open wooded areas. **TN** Formerly placed in genus *Sturnus*. **AN** Purple-backed Starling.

White-shouldered Starling *Sturnia sinensis* 20 cm

Vagrant. **ID** Adult male has silky-grey head and body, white forehead and throat, white scapulars and wing-coverts forming large white patch and contrasting with black of rest of wing. Body may have rusty-orange wash. Female and juvenile browner, with less or no white on wing. Pale grey uppertail-coverts and black tail with greyish-white sides and tip help separate birds lacking white in wing from Chestnut-tailed. **Voice** Unknown in region. **HH** Riverine forest. **TN** Formerly placed in genus *Sturnus*.

Grey-sided Thrush *Turdus feae* 24 cm

Vagrant. **ID** Superficially resembles Eyebrowed, with white supercilium, dark lores, and white crescent below eye. Adult male has rufescent-olive upperparts, including crown and ear-coverts, and grey underparts, becoming paler on belly and vent. Female is similar to male, but has white throat, and centre to breast and belly are whiter; has brown-streaked malar stripe, grey of breast and flanks is variably washed with orange-buff, and coloration of underparts can approach that of the dullest Eyebrowed. Best told by rufescent coloration to crown, ear-coverts and sides of neck (these areas have distinct greyish cast on Eyebrowed). First-winter similar to female, but with pale tips to greater coverts. **Voice** Unrecorded in region. **HH** Forest. Globally threatened.

Green Cochoa *Cochoa viridis* 28 cm

Vagrant. **ID** Adult mainly green, with blue crown and nape (and blackish mask), faint black scaling on mantle, pale blue panels on wing, and pale blue tail with black tip. Sexes similar, but female has green at base of secondaries. Juvenile wing and tail patterns as adult, but has white crown with black scaling, orange-buff spotting and dark scaling on upperparts and underparts, and buff tips to wing-coverts. **Voice** A pure, drawn-out monotone whistle, thinner and weaker than Purple. **HH** Dense moist, broadleaved evergreen forest.

Siberian Blue Robin *Luscinia cyane* 15 cm

Vagrant. Male has dark blue upperparts, black sides to throat and breast, and white underparts. Female has olive-brown upperparts, pale buff throat and breast, the latter faintly scaled dark brown, and usually has blue uppertail-coverts and tail. Female similar to female Indian Blue Robin but lacks orange-buff on breast and flanks of that species. First-winter male similar to female, but has blue on mantle. **Voice** Quiet, hard *tuk* and a loud *se-ic*. **HH** Shy and skulks in dense bushes.

Daurian Redstart *Phoenicurus auroreus* 15 cm

Vagrant. **ID** Adult male (worn) has prominent white wing patch, blackish mantle, and black of throat does not extend to breast. Adult male (fresh) and first-winter male have black of mantle and coverts partly obscured by brown fringes (mantle appears brown, diffusely streaked black), and grey of crown and black of throat are duller owing to dark grey fringes. Adult female similar to female Black Redstart, but has prominent white wing patch and darker centre to rufous tail. **Voice** Calls include a short, penetrating, monosyllabic indrawn whistle, like an unoiled bicycle, and a hard, deep *gak*, the two often combined. **HH** Posture and manner typical of genus. Bushes.

Isabelline Wheatear *Oenanthe isabellina* 16.5 cm

Vagrant. **ID** Rather plain sandy-brown and buff. Head and bill look rather large, and legs long. Tail shorter than that of Desert Wheatear with more white at base and sides. Wings sandy-brown with contrastingly dark alula (lacking black centres to coverts and tertials/secondaries of Northern Wheatear). Sexes similar. **Voice** Hard *chak*. **HH** Open stony ground.

Northern Wheatear *Oenanthe oenanthe* 15 cm

Vagrant. **ID** Breeding male has blue-grey upperparts, black mask and pale orange breast. Breeding female greyish to olive-brown above; never shows dark grey/black on throat. Shows more white at sides of tail than Isabelline. **Voice** Calls include a short whistle and a very short *stik*. **HH** Open stony ground and cultivation.

Red-tailed Wheatear *Oenanthe chrysopygia* 14.5 cm

Vagrant. **ID** In all plumages from other wheatears in region by rufous-orange lower back and rump, and rufous sides to tail. Male has greyer crown and mantle and black lores compared to female. **Voice** Low grating call. **HH** Open, stony ground. **TN** Formerly treated as conspecific with Kurdish Wheatear *O. xanthoprymna*.

Pied Wheatear *Oenanthe pleschanka* 14.5 cm

Vagrant. **ID** Male from nominate Variable Wheatear (the race in Nepal) by greyish-white crown and nape. Always shows black edge to outer tail feathers (lacking in Variable) and often has only a narrow and broken terminal black band (broad and even on Variable). Non-breeding and first-winter have pale fringes to upperparts and wings (with white crown, and black of face and throat, of male partly obscured). Distinct from Variable which lacks prominent pale fringes to body and wing feathers in fresh plumage. Breeding female similar to female Variable, except tail pattern. **Voice** Harsh *zack-zack*. **HH** Open stony ground.

Brown-breasted Flycatcher *Muscicapa muttui* 14 cm

Vagrant. **ID** Compared to Asian Brown has larger bill with entirely pale lower mandible, pale legs and feet, rufous-buff edges to greater coverts and tertials, rufescent tone to rump and tail, more pronounced brown or grey-brown breast-band, and warmer brownish-buff flanks. Juvenile streaked rufous-buff on upperparts and breast. **Voice** A pleasant, feeble song; call a thin *sit*. **HH** Dense thickets in forest.

Spanish Sparrow *Passer hispaniolensis* 15.5 cm

Vagrant. **ID** Breeding male from male House by chestnut crown, fine white supercilium, extensive black on breast becoming black streaking on flanks and belly, and largely black mantle with distinct buffish-white 'braces'. Non-breeding plumage duller, with head and body pattern partially obscured by pale fringes (although features distinct from House, especially dark flank markings, still apparent). Female very similar to female House, but generally has longer whitish supercilium, fine grey streaking on breast and flanks, and buffish-white 'braces' on mantle. **Voice** Call similar to House, but has more metallic or squeaky notes. **HH** Cultivation, especially cereal crops, marshes and reedbeds.

Eurasian Siskin *Carduelis spinus* 11–12 cm

Vagrant. **ID** Most likely to be confused with female Tibetan Serin. Male distinguished by black crown and chin, and black-and-yellow wings. Female differs from female Tibetan in wing pattern (yellowish patches at base of secondaries and primaries) and brighter yellow rump; bill marginally longer and slimmer. **Voice** A *tet-tet* and distinctive wheezy *toolee*. **HH** Conifers.

Eurasian Linnet *Carduelis cannabina* 13–14 cm

Vagrant. **ID** From Twite by greyish crown, ear-coverts and nape, browner mantle, whitish rump and larger greyish bill. Male has chestnut-brown mantle and wing-coverts, with variable dark streaking on mantle. In breeding plumage, crown, nape and ear-coverts purer grey and has crimson on forehead and breast. Female and first-winter have duller brown mantle, with dark brown streaking, and dark streaking on breast and flanks. **Voice** Rapid *chi-chi-chi-chi* call. **HH** Open stony slopes and upland meadows. **AN** Common Linnet.

Mongolian Finch *Eremopsaltria mongolica* 14–15 cm

Vagrant. **ID** Usually shows pronounced whitish or buffish panels on wing and pronounced pale edges to tail. Breeding male has pinkish-red on head and breast, pink edges to greater coverts and primaries, prominent white panels on wing, and pink rump. Non-breeding male and female have less pink in plumage and pale wing panels less distinct in female. Juvenile lacks pink in plumage and has buff fringes to tertials and greater coverts. **Voice** Often silent, gives quiet four-noted rattle in flight. **HH** Dry stony slopes. **TN** Formerly placed in genus *Bucanetes*.

Grey-headed Bullfinch *Pyrrhula erythaca* 17 cm

Vagrant. **ID** Male from Red-headed by grey crown and nape, deeper orange-red underparts, black band on upper rump, and longer tail. Female from female Red-headed by grey crown and nape, fawn-brown mantle, pinkish-brown underparts, and blackish band on upper rump. Juvenile very similar to female, but crown and nape more greyish-olive, mantle dull grey-brown, underparts dull brown, and has broad buffish-brown tips to greater coverts. **Voice** A soft *soo-ee* or triple whistle. **HH** Mixed forest.

White-capped Bunting *Emberiza stewarti* 15 cm

Vagrant. **ID** Male has grey head, black supercilium and throat, and chestnut breast-band; pattern obscured in winter. Female has rather plain head with pale supercilium; crown and mantle diffusely streaked, underparts finely streaked and washed buff, and has chestnut rump. Juvenile similar to female, but head rather pale with indistinct streaking. **Voice** Distinctive twittering *chus-chua-chua* call. **HH** Dry grassy and rocky slopes and fallow fields.

Grey-necked Bunting *Emberiza buchanani* 15 cm

Vagrant. **ID** In all plumages, pinkish-orange bill and rather plain head with whitish eye-ring. Male has blue-grey head with buffish submoustachial stripe and throat, deep rusty-pink breast and belly, and diffusely streaked sandy-brown mantle with pronounced rufous scapulars. Rump sandy-grey. Female very similar to male, but generally paler, with buffish cast to grey head and nape (often with some streaking). First-winter/juvenile often have only slight greyish cast to head, underparts are warm buff (with variable rufous cast), and crown and underparts faintly streaked. **Voice** Call a soft click. **HH** Dry rocky and bushy hills.

Rustic Bunting *Emberiza rustica* 14–15 cm

Vagrant. **ID** Striking head pattern (broad supercilium, dark sides to crown and border to ear-coverts), rufous streaking on breast and flanks, white belly, rufous on nape, and prominent white median coverts bar. Crown feathers frequently raised. **Voice** Sharp *tzic* call, similar to Little Bunting. **HH** Damp grassland.

Chestnut Bunting *Emberiza rutila* 14 cm

Vagrant. **ID** Small size and small, fine bill. Chestnut rump and little or no white on tail. Male has chestnut head and breast, with coloration obscured in fresh plumage (especially first-winter). Female has buff throat and yellow underparts; head pattern less striking than Yellow-breasted. **Voice** Call like Little Bunting. **HH** Rice stubbles, forest clearings and open forest and scrub.

Red-headed Bunting *Emberiza bruniceps* 16 cm

Vagrant. **ID** Smaller than Black-headed, with shorter, more conical bill. Male has rufous head and yellowish-green mantle. Female when worn may show rufous on head and breast, and yellowish crown and mantle, and are distinguishable from female Black-headed. Fresh female almost identical to Black-headed, but indicative features include paler throat than breast, with hint of buffish breast-band, and forehead and crown often virtually unstreaked. Immature often inseparable from Black-headed but may exhibit some features mentioned above. **Voice** Call a *tliip* or rather harsh *prrit*. **HH** Cultivation, especially of cereals.

Common Reed Bunting *Emberiza schoeniclus* 14–15 cm

Vagrant. **ID** Male has black head and white submoustachial stripe, obscured by fringes when fresh. Female has buff supercilium, brown ear-coverts, and dark moustachial and malar stripes reach bill (compare Little). **Voice** Call a distinctive *tseeu* and harsh *chirp*. **HH** Reedbeds and irrigated crops.

APPENDIX 2:
SPECIES COLLECTED BY HODGSON NOT DEFINITELY RECORDED IN NEPAL

The following species were collected by Brian Hodgson in the 19th century and formerly listed as coming from Nepal, but may well have originated in India close to the Nepalese border, see Introduction.

Oriental Dwarf Kingfisher *Ceyx erithaca*

Tiny kingfisher. Orange head with violet iridescence and black upperparts with variable blue streaking. Juvenile duller, with whitish underparts and orange-yellow bill.

Dark-rumped Swift *Apus acuticauda*

Similar in shape and appearance to Fork-tailed Swift, but has blackish rump, and lacks distinct pale throat.

Oriental Bay Owl *Phodilus badius*

Oblong-shaped, vinaceous-pinkish facial discs. Underparts vinaceous-pink, spotted black; upperparts chestnut and buff, spotted and barred black. Call a series of eerie, upward-inflected whistles.

Indian Bustard *Ardeotis nigriceps*

Very large bustard. In all plumages, has greyish or white neck, black crown and crest, uniform brown upperparts, and white-spotted black wing-coverts. Upperwing lacks extensive area of white.

Chestnut-bellied Sandgrouse *Pterocles exustus*

Pin-tailed, with dark underwing and blackish-chestnut belly. Female has buff banding on upperwing-coverts and no black gorget on throat, which are useful distinctions at rest from Black-bellied.

Black-bellied Sandgrouse *Pterocles orientalis*

Large, stocky and short-tailed. Has black belly, and white underwing-coverts contrast with black flight feathers.

Painted Sandgrouse *Pterocles indicus*

Small, stocky and heavily barred. Underwing dark grey. Male has chestnut, buff and black bands on breast, and unbarred orange-buff neck and inner wing-coverts. Female heavily barred all over, with yellowish face and throat.

Collared Treepie *Dendrocitta frontalis*

Black face and throat, grey nape and underparts, rufous lower belly and vent, rufous rump, and black tail.

Blue-fronted Robin *Cinclidium frontale*

Long, graduated tail, lacking any white or rufous. Male deep blue, with glistening blue forehead. Female has dark brown tail and uniform pale brown underparts.

Rufescent Prinia *Prinia rufescens*

Large bill with paler lower mandible. Grey cast to crown, nape and ear-coverts in summer. In non-breeding plumage has more rufescent mantle and edgings to tertials, and has stronger buffish wash to throat and breast, compared to Grey-breasted. Buzzing call.

White-spectacled Warbler *Seicercus affinis*

White eye-ring, grey crown and supercilium, well-defined blackish lateral crown-stripes, greenish lower ear-coverts, and yellow lores and chin.

Greater Rufous-headed Parrotbill *Paradoxornis ruficeps*

Rufous head, white underparts. Larger and longer billed than Lesser. Lores and ear-coverts deep rufous-orange (lores whitish in Lesser).

Yellow-throated Fulvetta *Alcippe cinerea*

Yellow supercilium, black lateral crown-stripes, greyish crown, and yellow throat and breast.

APPENDIX 3:
SPECIES DELETED FROM THE NEPAL LIST

Large Blue Flycatcher *Cyornis magnirostris* 14cm

In the last edition this species was included (as Hill Blue Flycatcher *C. banyumas*, but is now split as a separate species) for Nepal, based on a specimen collected in the Marsyangdi Valley in August 1950. This has been re-identified as a juvenile Rufous-bellied Niltava. To date, no confirmed sight records or photographs are known for Nepal and the species is therefore removed from the Nepal list.

INDEX